Unforgettable In So Many Ways.

At first glance, you'll see the difference - the smooth lines, the classic Sea Ray style, the unforgettable presence on the water. It's hard to take your eyes off it. . .isn't it? This is the way boating was meant to be. And it doesn't take long to understand what all Sea Ray Sport Cruiser owners share. . . a great appreciation for boating at its best and a constant quest for the perfect weekend of sun and fun.

Start today by visiting your local Sea Ray Dealer, call 1-800-SRBOATS or visit www.searay.com

Sea Ray®

The Standard of Excellence™

"Raymarine to serious boaters, come in."

*The **NEW** Ray101 Handheld VHF*

*The **NEW** Ray48 Fixed-mount VHF*

Raymarine. When communication counts.

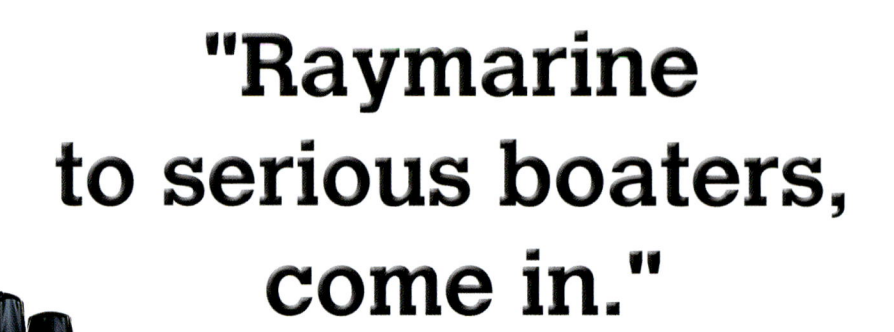

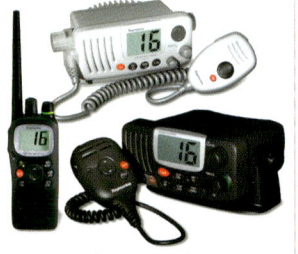

Raymarine Radios.
New features. New options. Maximum performance.

Raymarine radios combine feature-rich functionality and bold new styling. Now available in black or white, Ray fixed-mount radios offer features like Class D Digital Selective Dialing. The Ray101 handheld VHF features the best battery technology in its class, with a 3-hour charge time. The Raymarine bold, backlit LCD displays let you read data loud and clear. Find the Raymarine radio that speaks to you. Go to raymarine.com today.

Raymarine®

www.raymarine.com

RADAR   NAVAIDS   FISHFINDERS   AUTOPILOTS   COMMUNICATIONS   SOFTWARE   INSTRUMENTS   INTEGRATION

POWERED BY
VOLVO PENTA

2004 VISTA SERIES > 248 268 **288** 298 328 348 378

CRUISING'S BETTER WITH THE RIGHT COMPANY.

FOUR WINNS CRUISERS — BUILT FOR THE WAY YOU BOAT.

FOUR WINNS
www.fourwinns.com

marine industry
CSI³ award 2003
The Only Recreational Boat Company To Receive Three CSI Awards

Enjoy Carefree Boating with **Genmar FirstMate**
www.genmarfirstmate.com

Not Everyone Sees what a Tiara Owner Sees

Tiara
YACHTS

© 2002 Tiara Yachts

BUILT IN HOLLAND, MICHIGAN AND SWANSBORO, NORTH CAROLINA IN SEVERAL LENGTHS AND STYLES, FROM 29' TO 52'. CONTACT US AT 1-800-843-3172 OR TIARAYACHTS.COM

Lake Huron
including Georgian Bay and the North Channel
PORTS O' CALL

Lakeland
BOATING

O'MEARA-BROWN PUBLICATIONS, INC.
CHICAGO, ILLINOIS

Lake Huron
including Georgian Bay and the North Channel
PORTS O' CALL

PUBLISHER
Walter "Bing" O'Meara

EDITORIAL CONSULTANT
Robert Pollock

EDITORIAL STAFF
Matthew M. Wright, editor
Cyndee Miller, managing editor
Dave Mull, editorial director
Jenny Llakmani, special projects editor
Brian Jerauld, associate editor

CONTRIBUTING WRITERS
Jim Moodie, Oliver Moore III, Robert Pollock, Dave Wallace, John Wright

DESIGN STAFF
Edward C. Crosby, production director
Tara O'Hare, art director
Brook Rhodes, associate art director

BUSINESS STAFF
Mark Conway, associate publisher
Linda O'Meara, vice president/marketing director
Patti McCleery, regional sales manager
Sharon P. O'Meara, circulation director
Mary Currier, circulation assistant

Published by
O'Meara-Brown Publications, Inc.

10 9 8 7 6 5 4

Copyright © 2004 O'Meara-Brown Publications, Inc. All rights reserved. The publisher takes no responsibility for the use of any of the materials or methods described in this book, nor for the products thereof. The name *Lakeland Boating* and the *Lakeland Boating* logo are trademarks of O'Meara-Brown Publishing.

Questions regarding the content or the ordering of this book should be addressed to:
Lakeland Boating
727 S Dearborn, Suite 812
Chicago, IL 60605
312-276-0610
lb@omeara-brown.com
www.lakelandboating.com

Lakeland BOATING

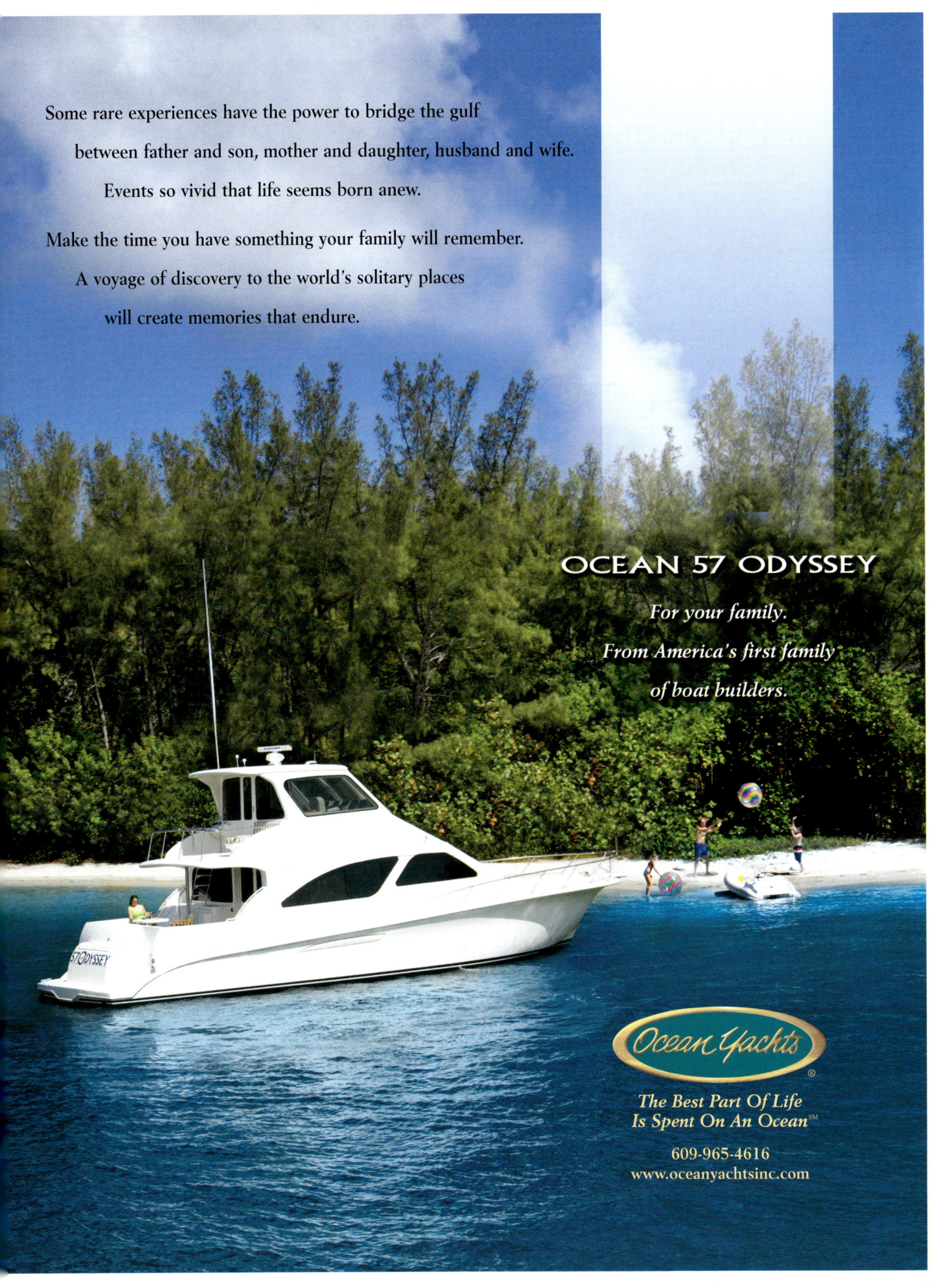

Some rare experiences have the power to bridge the gulf

between father and son, mother and daughter, husband and wife.

Events so vivid that life seems born anew.

Make the time you have something your family will remember.

A voyage of discovery to the world's solitary places

will create memories that endure.

OCEAN 57 ODYSSEY

*For your family.
From America's first family
of boat builders.*

Ocean Yachts

*The Best Part Of Life
Is Spent On An Ocean*℠

609-965-4616
www.oceanyachtsinc.com

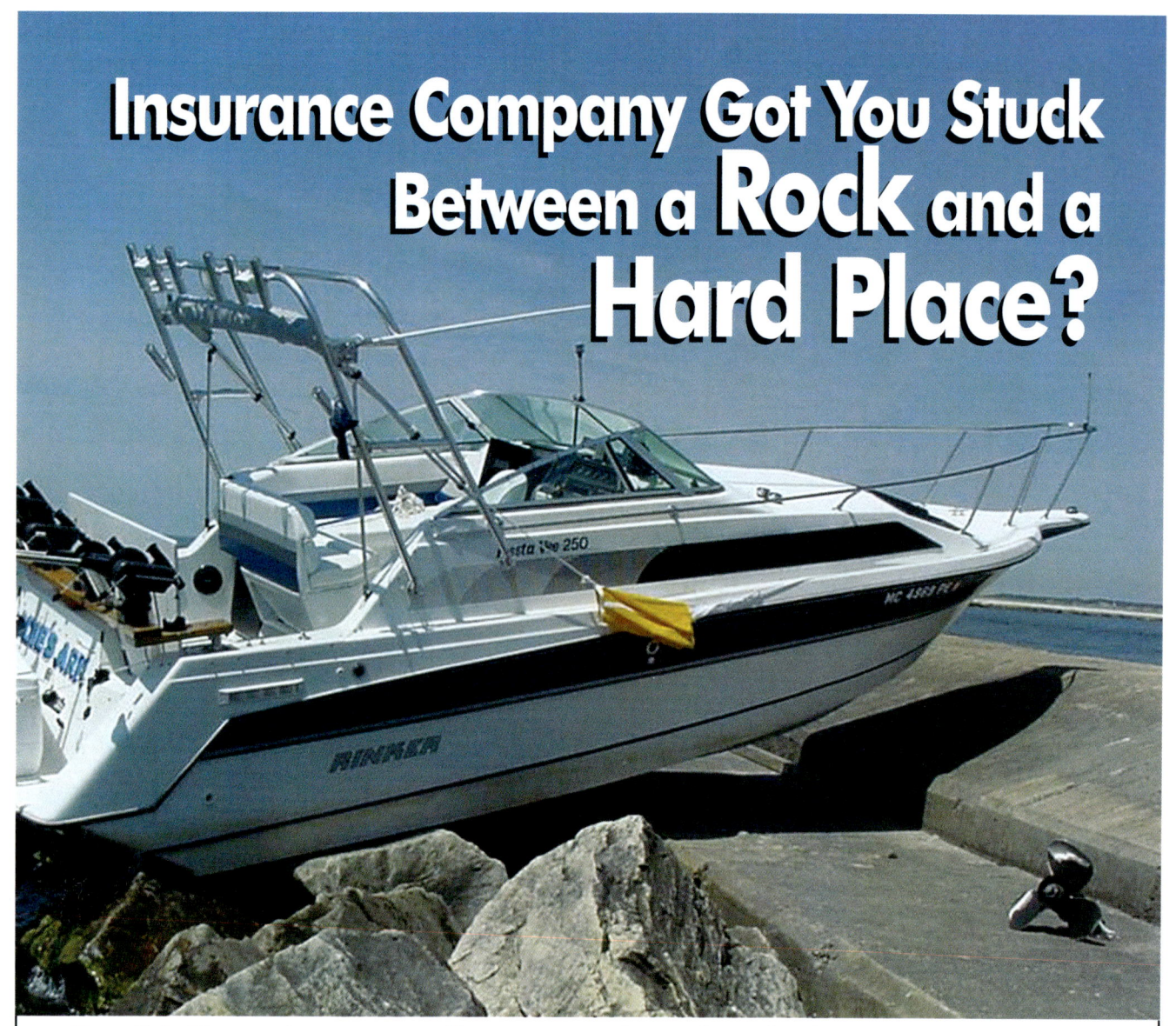

Insurance Company Got You Stuck Between a Rock and a Hard Place?

BoatU.S. Marine Insurance will get you and your boat back in the water FAST, with our 24-hour emergency dispatch and rapid claims service.

Low Cost—
Plus special discounts for USPS, USCG Auxiliary, and other approved boating safety courses, up to 20% savings.

Broad Coverage—
Five policy types—Yacht, Boat Saver, Bass, PWC or Liability Only. All BoatU.S. Policies come with Full Salvage Coverage.

Damage Avoidance Program—
Seaworthy magazine, sent to you free four times a year, provides you with insight on how to avoid situations that can jeopardize the safety of your boat and crew.

For a free, no obligation quote call
800-283-2883
Priority Code 4851
or apply online at **BoatUS.com**

BoatU.S.
MARINE INSURANCE

Contents

Fraser Hill 3
How to Use This Book 16
Anchoring Basics 19
Forecast for Lake Huron 20
I Do Declare!: Customs 22
Boaters Resource Guide 26
Distance Chart 27

Section One
Lake Huron

Map of Lake Huron Ports 28
Mackinac Island 30
Mackinaw City 36
Cheboygan 40
Hammond Bay Harbor 46
Rogers City 48
Presque Isle 52
Alpena 56
   Shipwrecks of Thunder Bay 59
Harrisville 60
Oscoda & Au Sable 64
East Tawas & Tawas City 68
Au Gres 72
   Surviving Lightning 75
Bay City 76
Sebewaing 80
   Helicopter Rescue 83
Caseville 84
Port Austin 88
Harbor Beach 92
Port Sanilac 94
Lexington 98
   Considering Yacht Clubs 101
   Water Pollution Regulations 101
Port Huron & Sarnia 102
Port Franks 106

Lakeland Boating · Lake Huron Cruising Guide **9**

NORDHAVNS ARE FAMOUS FOR SURVIVING ROUGH CONDITIONS, INCLUDING A SOFT STOCK MARKET AND WEAK ECONOMY

The Nordhavn 40, on its way around-the-world.

Experienced boaters are familiar with the record-setting accomplishments of Nordhavn trawlers. For more than 15 years, these rugged, luxurious yachts have been crossing oceans and circumnavigating the globe, handling the roughest conditions.

But Nordhavn owners have the added security of knowing their yacht can also protect them from stormy economic conditions. Thanks to their timeless design and uncompromising, robust construction, Nordhavns hold their value like no other yacht.

Indeed, many Nordhavns are sold years later for nearly what the owners originally paid for them. And while we don't promote buying and selling Nordhavns for a profit, some owners have done just that.

How is this possible? Take one step aboard a Nordhavn, and feel the difference. Spend time in the engine room and wheelhouse,

Joint Decision, *a Nordhavn 57, survived a grounding while exploring the wilds of Alaska, successfully finding deeper water and suffering no damage.*

and see the superb engineering that goes into each vessel. Take a look at its standard equipment list, and recognize the world's finest brands. Nothing but the best goes on a Nordhavn.

So whether you're looking for a new yacht or a previously owned one, be sure to consider a Nordhavn. It's a safe bet, in any condition.

For more information, visit our website, www.nordhavn.com or call and speak directly to those who are responsible for designing and building these timeless, ocean-going yachts.

NORDHAVN
POWER·THAT·IS·OCEANS·APART ®

Pacific Asian Enterprises • Box 874 • Dana Point, CA, U.S.A. 92629-0937 • Tel: (949) 496-4848 • Fax: (949) 240-2398 • www.nordhavn.com
Nordhavn Yachts and Brokerage Sales West • 24703 Dana Drive, Dana Point, CA, 92629 • Tel: (949) 496-4933 • Fax: (949) 496-1905
Nordhavn Yachts and Brokerage Sales NE • One Washington Street • Newport, RI 02840 • Tel: (401) 846-6490 • Fax: (401) 846-6408
Nordhavn Yachts and Brokerage Sales SE • 450 SW Salerno Road Ste 201 • Stuart, FL 34997 • Tel: (772) 223-6331 • Fax: (772) 223-3631

CHOOSE FROM 35 TO 76 FEET
Now with 12 models in our fleet, you will surely find the right Nordhavn for you.
35CP ~ N40 ~ N43 ~ N46 ~ N47 ~ N50 ~ N55 ~ N57 ~ N62 ~ N64 ~ N72 ~ N76

Contents

Crossing the Lake .109
Grand Bend 110
   Preparing for Heavy Weather 113
   Hypothermia: The Facts 113
Bayfield 114
Goderich 118
Kincardine 122
   Abandoning Ship 125
Port Elgin 126
Southampton 130
West Coast of the Bruce Peninsula 134
South Baymouth .138
Providence Bay 140
Les Cheneaux Islands 142
Cedarville 143
Hessel 146
St. Ignace 149

Section Two
THE NORTH CHANNEL

Map of the North Channel Ports 152
DeTour Village 154
Drummond Island 158
Cockburn Island 163
Manitoulin Island 166
Meldrum Bay 167
Campbell Bay 170
Gore Bay 172
   Great Lakes Cruising Club 175
Kagawong 176
West Bay 178
Little Current 180
Sheguiandah 184
Manitowaning 186
   Lake Huron-Area Airports 189
Wikwemikong 190

LAKELAND BOATING · LAKE HURON CRUISING GUIDE   **11**

Contents

Powwows 193
Killarney 194
Lansdowne Channel 200
Baie Fine and the Pool 201
Heywood Island 204
Whitefish Falls 205
Benjamin Islands Group 206
Spanish 212
Whalesback Channel 215
Spragge 218
Blind River 220
Thessalon 224
   Fish Recipes 227
Bruce Mines 228
   Angler's 10 Commandments 231
Hilton Beach 232
   Lake Huron-Area Maritime Museums ...235
Desbarats 236
Richards Landing 238
Sault Ste. Marie 241

Section Three
GEORGIAN BAY

Map of Georgian Bay Ports 246
Tobermory 248
   Legend of Flowerpot Island 250
   Chi-Cheemaun Ferry 252
   Fathom Five National Marine Park 252
Wingfield Basin & Dyer's Bay 253
Lions Head & MacGregor Bay 256
   Lake Huron Fishing Hotspots 260
Wiarton 262
   Trailerboating the North Country 265
Owen Sound 266
Meaford 270
Thornbury 274

12   LAKELAND BOATING · LAKE HURON CRUISING GUIDE

IF WE COULD, WE WOULD.

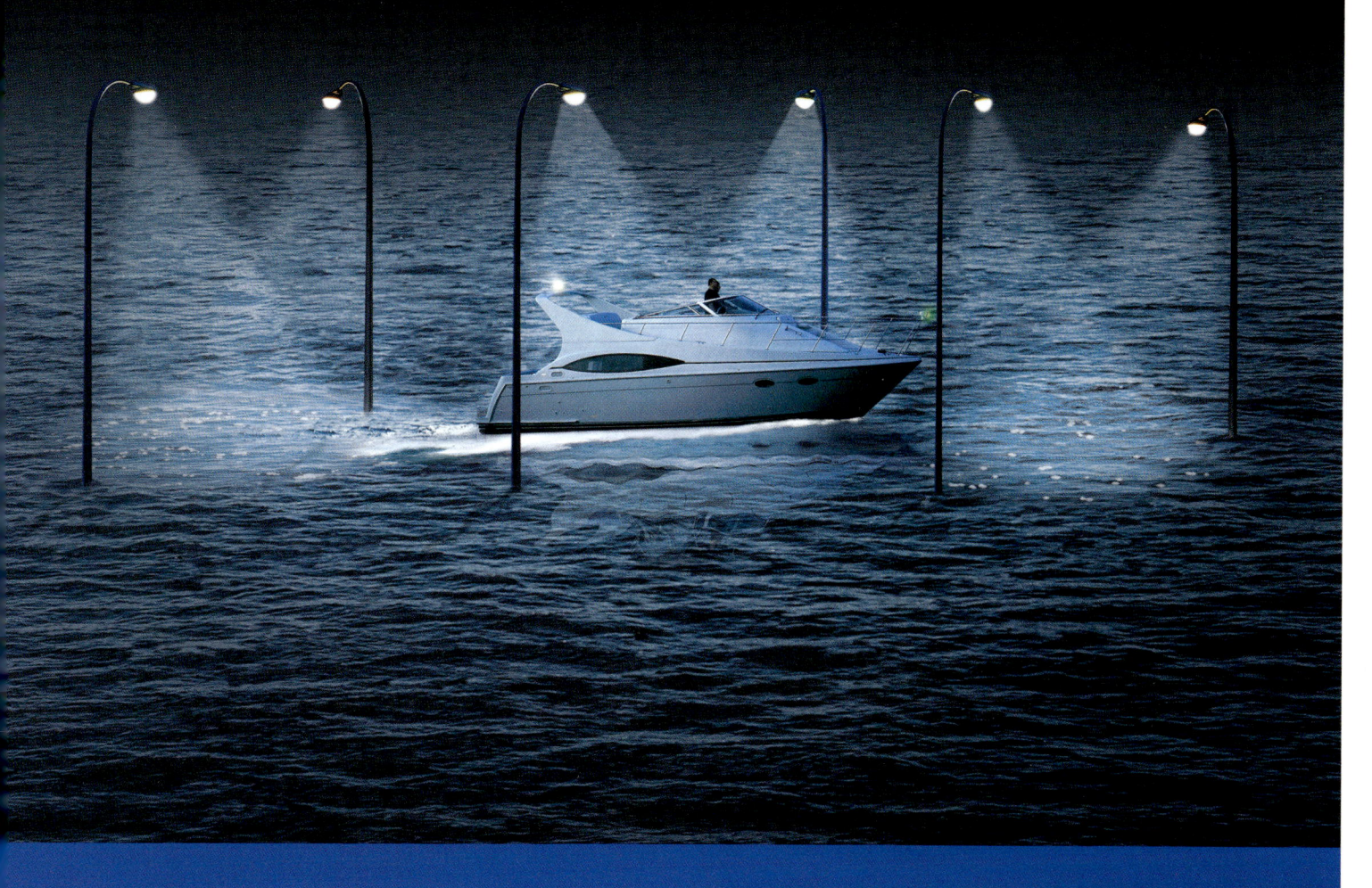

A SUPERIOR PROPULSION SYSTEM DESERVES A SUPERIOR CUSTOMER CARE PROGRAM.
LEARN ABOUT THE MOST COMPREHENSIVE PROGRAM OF ITS KIND AT WWW.VOLVOPENTA.COM/US

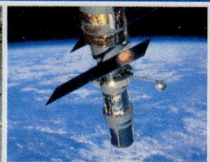

FIRST LAUNCH
A highly informative christening exercise for new Volvo Penta diesel owners that assures surprise-free boating from the start.

WARRANTY
2 years of the most complete coverage available for marine propulsion systems, plus 3 more years on base engine components for diesels 6 liters and up.

VOLVO ACTION SERVICE
The gold standard of marine customer support, available 24/7/365. It works just like a concierge service for boaters.

72 HOUR PARTS
If the servicing dealer doesn't have the part you need in stock, we guarantee delivery to him within 72 hours or the part is free.*

RAPID RESPONSE
This is the most prompt, attentive and professional service program available. It benefits owners of Volvo Penta diesel engines.

SEAKEY
The marine industry's only satellite-based system dedicated to boat security, boater safety and boating enjoyment.

VOLVO PENTA

*See Volvo Penta of the Americas, Inc.'s express limited warranty for North America for details (Publication No. 796733-9).
©2003 Volvo Penta of the Americas, Inc. Volvo Penta and Volvo Action Service are registered trademarks of AB Volvo. Rapid Response is a service mark of Volvo Penta of the Americas, Inc. SeaKey and First Launch are registered trademarks of Volvo Penta of the Americas, Inc.

Contents

VHF Frequencies	277
Collingwood	278
Wasaga Beach	282
Penetanguishene	284
Midland	288
Victoria Harbour	293
Waubaushene	295
Port Severn	296
30,000 Islands	299
Trent-Severn Waterway	298
Honey Harbour	300
South Bay Cove	301
Sans Souci	306
Parry Sound	308
Killbear	312
Snug Harbour	315
Dillon Cove	317
Pointe au Baril Station	319
Bayfield Inlet	323
Byng Inlet	326
Northeastern Shore	330
Bustard Islands	331
Bad River Channel	333
Beaverstone Bay	335
Collins Inlet	338
Club Island	342

Indexes/Reference

Reference List	344
NOAA & CHS Chart Locator	345
Port Index	346
Editorial Index	347
Advertiser Index	347

DO YOU DREAM IN COLOR OR BLACK AND WHITE?

FIESTA VEE 312

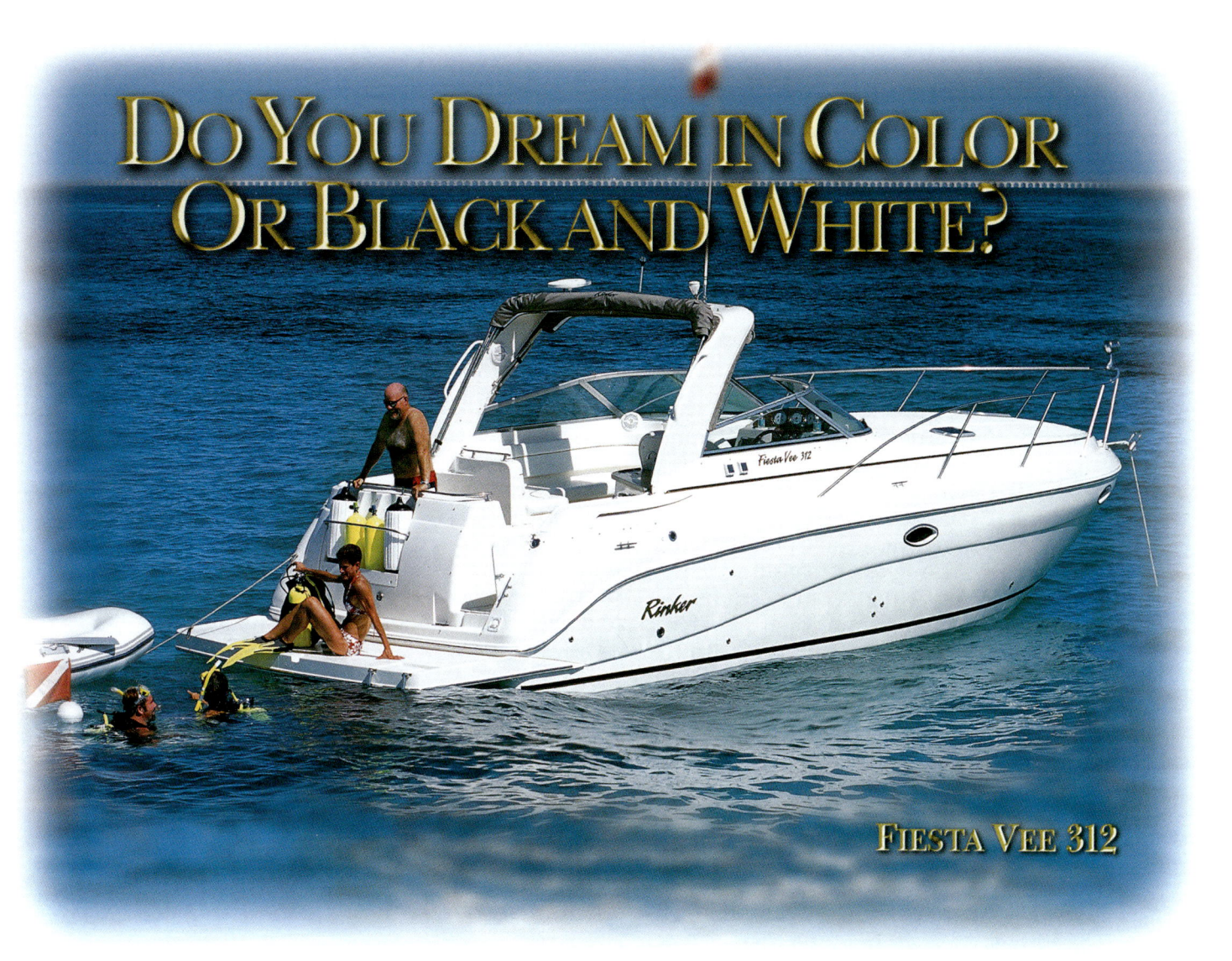

When you dream about buying a new cruiser, do you see everything in technicolor from diving to desserts, or are you more concerned about the hard facts on the standard features? Relax, with the Fiesta Vee 312, your dream has come true.

The Fiesta Fee 312 makes overnighting easy. From the roomy cockpit complete with its own entertainment center and blender to the spacious main salon accommodations, you'll discover the new cruiser of your dreams.

Many of the standard features on the Fiesta Vee 312 are either optional or simply not available on our competitors' boats. Our factory-engineered Dinghy Package is a perfect example. It's sad to say, a lot of express cruisers don't even have a dinghy package available.

EXPERIENCE THE VALUE FOR YOURSELF.

Go to rinkerboats.com and download the Fiesta Vee 312 Comparison Chart, then schedule a sea trial with your local Rinker dealer. You won't have to pinch yourself to see if you are still dreaming.

MERCURY MerCruiser®   www.rinkerboats.com   TIME WELL SPENT™   **Rinker** — America's First Family of Boating

Getting Started

How To Use This Book
Our cruising guide is here to steer you in the right direction.

Lake Huron: Ports o' Call begins with Mackinac Island at the north end and heads counter-clockwise around the big lake before tackling the North Channel and Georgian Bay, each in its own section. We know that you're not likely to read this guide cover to cover—you'll probably jump around to ports you're interested in visiting. Check out the front cover flap with its convenient locator map, which lets you find and bookmark whatever port grabs your interest. This flap, along with the large pull-out chart, complete with waypoints, will help you plan your trip and devise an ideal itinerary.

Once you begin your voyage, we suggest you keep this book right alongside your charts, so that as you approach each port, you can use our aerial perspective and our harbor detail chart to make the final approach an easy one. This feature could be particularly useful should unforeseen circumstances cause you to seek out an unfamiliar port not originally on your itinerary.

The charts that are reproduced for each port are the most recent NOAA or CHS charts available at the time of publication. Some source charts are more current than others. Along with each reproduction we provide the source chart number and the date it was most recently updated. We strongly encourage you to supplement your cruising database with local knowledge and the information provided in the U.S. Coast Guard's Local Notice to Mariners. In some ports, the aerial photograph is more up-to-date than the chart.

In some ports, such as Drummond Island, Beaverstone Bay and Sault Ste. Marie, certain marinas, anchorages or points of interest are too far from the main port to show up well in a single aerial photo. In these instances, we offer alternate shots to make your in-port navigation and choice of marinas as easy as possible. We've attempted to give marine facilities information for every marina that accepts transients in each port. With our restaurant suggestions we've been a tad more selective, generally excluding fast food and most places far from the water. We relied heavily on personal favorites and the suggestions of local marina operators.

As for the usual disclaimer about accuracy, marine facilities are constantly changing, and restaurants and other attractions sometimes close. You may find new places not included in this book. We tried our hardest to bring you the best that Lake Huron, Georgian Bay and the North Channel have to offer—but one of the greatest things about cruising is the new discoveries that await in every port. Did we leave something out? Drop us a line and let us know about your favorite port of call's best-kept secret.

Lakeland Boating
727 S. Dearborn
Suite 812
Chicago, IL 60605
phone: 312-276-0610
fax: 312-276-0619
lb@omeara-brown.com
www.lakelandboating.com
www.glangler.com

Thanks again for bringing *Lakeland Boating* with you on your cruise!

What Every Captain Knows...

As every captain knows, there's no such thing as a foolproof navigational aid.

While the *Lakeland Boating* staff has taken great care to ensure that this guide contains the most up-to-date and accurate information available, conditions change and errors sometimes occur.

The photographs, charts, latitudes and longitudes, and navigational recommendations presented here are for your convenience only. It is not the editors' intention that this guide replace other navigational aids, but instead that it be used in conjunction with them. The prudent mariner does not rely solely on any single navigational aid, but rather on the many that are available. Read *Lakeland Boating's Lake Huron: Ports o' Call*—and chart your own course.

Grand Banks Heritage Series

Grand Banks Eastbay Series

Grand Banks Aleutian Series

A class-defining yacht that has welcomed generations of sailors to yachting, Grand Banks Heritage Series is a fusion of decades-rich tradition and modern boat building. With a performance package that can reach 20 knots, these classic yachts can deliver you to destinations with unparalleled ease. Discover the meaning of grace under power. Discover the Grand Banks Heritage Series.

With distinctive styling, obsessive quality and pure pedigree don't you think its time to discover a yacht that moves you?

GRAND BANKS®
www.grandbanks.com

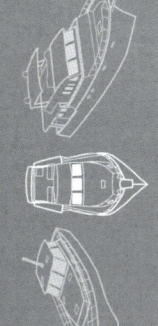

Speeds may vary depending on model and engine package selected at time of purchase. For more information visit us on the web at www.grandbanks.com or call toll free at (800) 809-0909

PAYING TOO MUCH FOR YOUR BOAT INSURANCE?

We are one of the nation's largest Marine Insurance Agencies representing over 27 companies nationwide. We can show you how to **SAVE MONEY** as well as provide the most comprehensive coverage available!

808 W. Lake Lansing Rd.
East Lansing, MI 48823

WORLDWIDE MARINE UNDERWRITERS INC.

PROGRESSIVE BOAT INSURANCE

1-800-339-1235
517-333-4100
Fax 517-333-4200

Hosting Smoke on the Water Poker Run

www.worldwidemarineins.co   www.smokeonthewaterpokerrun.com

YOUR GUIDE TO THE INTERCOASTAL WATERWAY!

From the author of Lakeland Boating's Captain's Logs

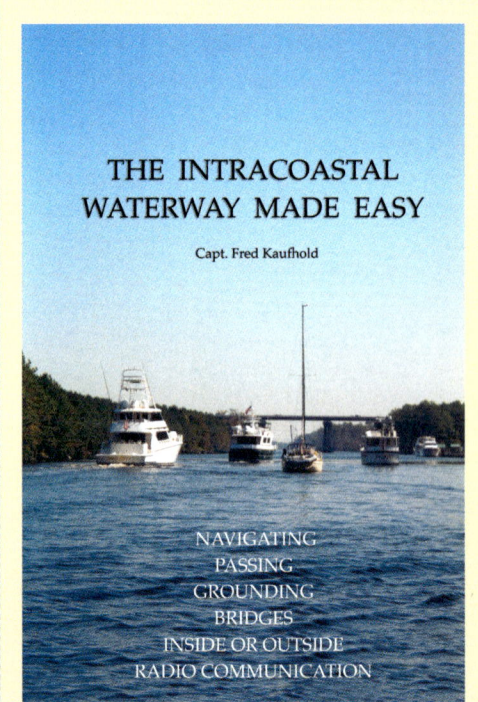

The Intracoastal Waterway Made Easy
By Capt. Fred Kaufhold

Loaded with tips, charts and full-color photos

$12.95

To Order Call: 800-589-9491

Mail in a request:
O'Meara-Brown Publications, Inc.
c/o Retail Services
Box 704
Mount Morris, IL 61054

Price includes shipping and handling
(Illinois residents please add sales tax of 87¢ for a total of $13.82)
We accept MasterCard, Visa, check or money order in US funds

THE INTRACOASTAL WATERWAY MADE EASY
Capt. Fred Kaufhold

NAVIGATING
PASSING
GROUNDING
BRIDGES
INSIDE OR OUTSIDE
RADIO COMMUNICATION

Anchoring Basics

Getting Hooked

No room at the marina?
Discover the simple pleasures of anchoring in the boonies.

Nothing extends the range of a cruising boat more than a good set of anchors. This is especially true when exploring areas such as the North Channel and Georgian Bay, where marinas can be scarce but quiet coves and hidey-holes abound.

We recommend, if you have the storage space, that you carry a few anchors of different designs and weights. Certain types of anchors perform better on particular types of bottom.

Today's best cruising anchors fall into five broad categories: the stock-in-head, the plow style (also called Bruce and CQR), the traditional stock anchors (often called yachtsman or Herreshoff-pattern anchors), grapnels and mushrooms.

Weight gets an anchor to the bottom quickly and helps it dig in. However, once a modern anchor is properly dug into the bottom, almost all of the holding power is determined by engineering.

Weight is also important for an anchor's ability to reset with shifting wind. Anchors with legendary resetting ability—CQR and Bruce, for example—are heavy iron or steel castings. Lightweight anchors may pull out of the bottom and drag before they reset.

The length of rope and chain between the boat and the anchor is called the rode. Nylon rope is best for the rode because it stretches under strain, reducing the force of the violent yank that can occur when the boat fetches up under the press of wind, current or waves. The elastic quality of a nylon rode also keeps the anchor buried.

A length of chain between the nylon rope and the anchor greatly improves any anchor's holding ability. The chain should be at least 15 feet long, although one foot of chain for every foot of the boat's waterline length is recommended. Short lengths of plastic-coated chain aren't cost-effective.

In a protected situation, the rode should be five to seven times the depth of the water. Should the weather deteriorate, the length of the rode (also called "scope") should be 10 to 15 times the depth.

Here are some special techniques to use when anchoring:

Using a second anchor off the stern can reduce the swinging room needed in a crowded anchorage. Let out rode on the bow anchor until the boat is over the spot where you want the stern anchor. Lower the stern anchor and pull the boat forward by hauling in the bow anchor rode. Secure both rodes when the boat is properly positioned.

Taking lines ashore is another effective way of controlling the swinging of an anchored boat. A convenient tree can make a natural bollard, and an anchor with its flukes driven into the shoreline ground creates an instant deadman.

Two anchors can be set off the bow for additional holding power. They should be set 45 degrees apart. If the boat swings beyond 180 degrees, however, the two rodes will become tangled. This can make hoisting the anchors difficult.

When using a dinghy to set a second anchor or carry a line ashore, secure the bitter end to the big boat. Coil the rope in the dinghy so it pays out as you row away from the boat.

For more information on anchoring, pick up a copy of the Hearst Marine Books Complete Guide to Anchoring and Line Handling, *written by Dave G. Brown, published by William Morrow in 1996.*

Anchor Selection Table

VG = Very Good; G = Good; F = Fair; P = Poor

Anchor Type	Sand	Mud	Weeds	Gravel	Rock
Stock-In-Head (Danforth)	VG	G	P	P	P
Stock-In-Head (Mud Version)	G	VG	P	P	P
Plow (CQR)	VG	VG	P	G	F
Plow (Bruce)	VG	VG	P	G	F
Traditional	F	F	F	F	G
Grapnel	P	P	P	P	F

Lake Huron Forecast

'Weather Permitting...'

Strategies for avoiding defeat—
or a less-than-ideal vacation—on the
Great Lakes' climatological battlefield.

Like Lake Michigan, Lake Huron is basically a north-south body of water with the potential for significant weather and temperature differences along its length. The lower portion of the lake around Port Huron is a long way, climatologically speaking, from the Straits of Mackinac and the North Channel.

Generally cooler water and the relative complexity of the land masses and current activity at the northern end create more fog than is typically found on other parts of the lake. This is particularly noticeable in the Straits of Mackinac, as well as around the tip of the Bruce Peninsula. Cruisers are advised to enjoy their leisure under these conditions, because much of this fog will burn off by noon in fair weather.

Whether it be waves, rain or fog, nobody should take the weather on this lake for granted. Weather in the Great Lakes basin is affected by three factors: the air masses that move in from other regions, the unique situation of having large bodies of water surrounded by huge land masses, and the moderating influences of the lakes themselves. The prevailing air mass movement is from the west, which alternates between humid air from the Gulf of Mexico and cold, dry air from the Arctic. It all comes together here in the center of the North American weather system.

The prevailing feature of this system is the passage of alternating cyclonic low-pressure cells and anticyclonic highs, carried along on the westerly winds. This is why the precipitation rate is so high and why the changes come so frequently. Meteorologists have referred to the Great Lakes as a "climatological battlefield." Add to this uncertainty the seemingly random influences of the Jet Stream and the pesky El Niño Pacific current that drag up new violence and feed it into the system—and you can see why every cruising log needs to begin with the words, "Weather permitting..."

Actually, with the advent of GPS and radar, it is possible to move almost any time you wish, regardless of poor weather. The disadvantage of this "go anyway" policy in the North Channel and Georgian Bay is that you miss the scenery, which is one of the key reasons for being here. The safest and most satisfying strategy for these waters is to allow enough time for fair weather travel days.

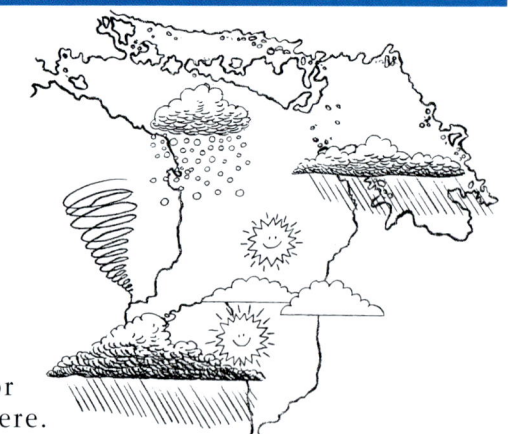

About the only weather factor left to control your cruising destiny is wave height. Much of the best cruising water on Lake Huron is more protected than on some of the other Great Lakes, however, so the chances are good that you will find a higher percentage of travel days available to you in the North Channel and Georgian Bay than on larger bodies of open water, such as Lake Michigan, lower Lake Huron or the east-west reach of Lake Erie.

Plan your moves according to the best weather projections you can get, and don't push your luck. "Better safe than sorry" is a great credo for enjoyable cruising.

There are some general bits of advice that hold up under most conditions. If the wind is offshore, you can be sure of easy going, even after it shifts. Practice "blind" navigation in good weather and learn to trust your instruments. With confidence and caution, you can use those quiet, foggy mornings for travel rather than postponement, particularly along the more predictable shoreline of Lake Huron proper.

Two of the most valuable improvements to arise in modern weather forecasting are long-distance radar and expanded TV coverage available to boaters all along the coast. Cruisers who used to suffer through hours of meteorology training just to understand the secretly coded MAYFOR weather forecasts can now watch the local TV news and see current weather patterns as the weatherman shows the radar picture in full color on the small screen. If you are in an area that offers cable, the Weather Channel is available 24 hours a day.

The challenge of changing weather adds spice to Great Lakes cruising and makes us better seamen in spite of ourselves. It may not always be pleasant, but it is seldom dull—and if you just have the patience to wait long enough, it will change. That's a forecast you can count on.

No reservations.
Plush accommodations.
Your own private island.
Life is good.

The all new **3370 OFFSHORE** delivers high performance with next-generation technology and innovation. 24'– 33' WALKAROUND models available. For more information on our entire Pursuit line, call or visit our website today.

YAMAHA When you want the best

1.800.947.8778
www.pursuitboats.com

CSI award 2003
marine industry
Excellence in Customer Satisfaction

PURSUIT
Ultimately, There Is Pursuit

ACROSS THE

Everything you ever wanted to know about customs but were afraid to ask.

Getting through customs between Canada and the United States isn't as daunting as you might imagine. Just plan well in advance. And if you're going to be crossing the border quite a bit, consider getting a Canpass or Form I-68—they'll make your trips much easier.

Coming into Canada

All recreational boaters arriving in Canada from the United States must call customs to get permission to enter the country. Having Canpasses for everyone onboard makes the process more convenient.

What the heck is a Canpass?

A Canpass–Private Boats permit allows boaters to call customs before they arrive and receive a clearance number to enter Canada. Citizens and legal residents of Canada and the United States can use the pass.

How can I get one?

You can obtain an application from a Canada Customs office or go online to www.ccra.gc.ca, the Canada Customs and Revenue Agency (CCRA) website. Complete the form and send it to the provincial Canpass Processing Centre, along with a non-refundable annual fee of $25 Canadian, and photocopied proof of citizenship or proof of legal residence.

Ontario Canpass Processing Centre
P.O. Box 126
Niagara Falls, ON L2E 6T1
800-842-7647

Okay, I've got one. What now?

If everyone onboard has a Canpass, call the CCRA at 888-CANPASS up to four hours before you arrive.

The vessel's operator must provide the following information to the customs officer:
• estimated time of arrival and docking destination in Canada
• full names, birthdates and citizenship of people on board
• Canpass permit number(s)

---

22 LAKELAND BOATING · LAKE HURON CRUISING GUIDE

BORDER

- purpose and length of stay in Canada (except for returning residents)
- whether there are any goods to declare

You then will be issued a customs clearance number, which you must keep.

What else should I know?

When you anchor, dock or beach your boat anywhere in Canada, you must report to customs.

Customs conducts random on-site verifications and examinations to ensure compliance.

Imported goods must be declared. You can pay for the duty and tax with a credit card, or if you want to use cash, you have to report to the nearest customs office.

If you want to bring a firearm into Canada, ask Canada Customs for the brochure "Importing a Firearm or Weapon into Canada" or visit www.ccra.gc.ca.

If some passengers on a boat have Canpass permits and others don't, the vessel master must follow the procedures for those without the pass.

Private boaters aren't required to pass through Canadian customs to leave Canada.

What if I don't have a Canpass?

A Canpass will make going through customs much more efficient. If you don't have one, you must go to a designated reporting station. (Call 888-CANPASS for a list of approved marinas and docks.) You must then report your arrival to the CCRA by calling the Canpass number. Stay with the vessel, as a customs officer may have to inspect it. If no inspection is required, you will be issued a clearance number, which you must keep a record of. Now you can proceed with your Canadian trip.

Coming into the U.S.

Recreational boaters arriving in the United States from Canada must obtain permission from both Immigration and customs to enter the country. Boaters have two choices: Get a Form I-68 seasonal permit, or head to a port of entry every time you enter the United States.

What are these Form I-68s, and how do I get one?

This permit lets you enter the United States from Canada for recreational purposes for the entire boating season, without reporting for further inspection. Everyone on board must have an I-68. They cost $16 for an individual and up to $32 for a family, and are valid for one year.

You can download an application at uscis.gov/graphics/formsfee/index.htm, pick one up at any U.S. port of entry or call 800-870-3676. Then, you must appear in person at a port of entry for an inspection, an interview and a name query against the Interagency Border Inspection System. You must present valid IDs (see box at end of story), as well as three photos. Note: The agency is particular—the shots must be 1-by-1-inch color pictures of you in a two-thirds left profile, with your left eye and right ear showing. You can't wear earrings or glasses, and no hair can cover your ear. Officials also will take your fingerprints when the application is submitted.

The I-68 program cannot be used by personal recreational vessels with cargo-carrying capacities of 5 net tons or greater, or any commercial watercraft (including charterfishing boats for hire). These must proceed to a port of entry for inspection.

Who can get a Form I-68?

- U.S. or Canadian citizens
- Legally admitted permanent residents of the United States
- Canadian landed immigrants of British Commonwealth nationality
- Canadian landed immigrants who are

nationals of: Andorra, Australia, Austria, Belgium, Brunei, Denmark, Finland, France, Germany, Iceland, Ireland, Italy, Japan, Liechtenstein, Luxembourg, Monaco, the Netherlands, New Zealand, Norway, Portugal, San Marino, Singapore, Slovenia, Spain, Sweden, Switzerland, the United Kingdom and Uruguay

What if I don't have an I-68?

Boats containing even one individual without an I-68 must report at an Outlying Area Reporting Station (OARS) videophone or port of entry. Boaters are required to provide their own ground transportation to reach ports of entry located inland. As always, make sure you have your ID cards (see box).

What about those videophones?

Installed at public marinas along the Canadian border, OARS videophones provide an automated inspection service enabling two-way visual and audio communication between the inspector and the applicant. You'll need a passport or evidence of citizenship combined with one official government photo ID (such as a valid driver's license).

Where are these videophones?

MICHIGAN
Mackinac Island: DNR pumpout station
MINNESOTA
Northwest Angle: Grumpy's Resort, Young's Bay and Jim's Corner
Rainey Lake: Bohman's Landing
NEW YORK
Waddington: Public Town Dock
Ogdensburg: City Marina
Morristown: Public Town Dock
Alexandria Bay: Main Public Dock
Clayton: Front Public Dock
Sackets Harbor: Navy Point Marina
Oswego: Oswego Marina
Olcott: Public Dock Building
Wilson: Tuscarora State Park
Youngstown: Youngstown Yacht Harbor
Lewiston: near the Riverside Inn
North Tonawanda: Pinochle Park
Buffalo: Erie Basin Marina
OHIO
Ashtabula: Ashtabula City Dock
Fairport Harbor: Grand River Marina
Mentor: Mentor Lagoons Marina
East Lake: Chagrin Lagoon Yacht Club
Cleveland: East 55th Street Marina
Conneaut: Conneaut Port Authority
Sandusky: Cedar Point Marina
South Bass Island: Put-in-Bay Docks
PENNSYLVANIA
Presque Isle: State Port
Erie: Lampy Marina and Dobbins Landing
North East Township: North East Marina

What are the Great Lakes ports of entry?

MICHIGAN
Detroit: 313-226-3134
Port Huron: 810-985-9541
Sault Ste. Marie: 906-632-7221
Grand Rapids: 616-456-2515
MINNESOTA
Boaters can pass through customs at an OARS videophone or by calling:
International Falls: 218-283-2541
Warroad: 218-386-1676
NEW YORK
Lake Ontario, Lake Champlain and the St. Lawrence River: 800-825-2851
Lake Erie: 800-927-5015
Toll-free numbers do not work from Canada. Call 716-881-5335.
OHIO
Cleveland: 216-267-3600
Toledo: 419-259-6424
Sandusky: 419-625-0022
Port of Erie: 814-833-1355
PENNSYLVANIA
814-833-1355

For inquiries about ports of entry in Michigan, call 586-307-2160.

For Minnesota and Wisconsin, call 701-775-6259.

For New York, Pennsylvania and Ohio call 716-447-3942.

And for eastern New York, call 802-868-3361.

What should I expect?

Don't even think about showing up without your IDs (see box). You also must provide your vessel's name and length, in addition to the nationality, names, dates of birth and citizenship of all passengers. Pleasureboats 30 feet and over in length must have a customs user fee decal affixed to the boat. To find information and applications, go to www.cbp.gov and do a search for user fee decal. You also can email decals@customs.treas.gov, or phone 317-298-1245 or any port of entry.

What else do I have to do?

Boaters must phone the local customs office each time they cross the border into the United States. Customs agents will either give clearance or ask you to wait at a designated site for inspection. Only the master or designee may go ashore to report arrival. No passenger can leave the boat, and no baggage or merchandise can be removed until a report of arrival is made and a customs inspector grants release.

Sure, that's a lot of information to digest, but visiting your neighboring country is a lot of fun and worth the effort. Don't miss out.

Source: Ontario Marine Operators Association: 705-549-1667; www.omoa.com

What ID is required to cross the border?
Each passenger must present a passport or evidence of citizenship, along with an official government photo ID (such as a valid driver's license).

The only number you need to guide you through the vast waters of boater information...

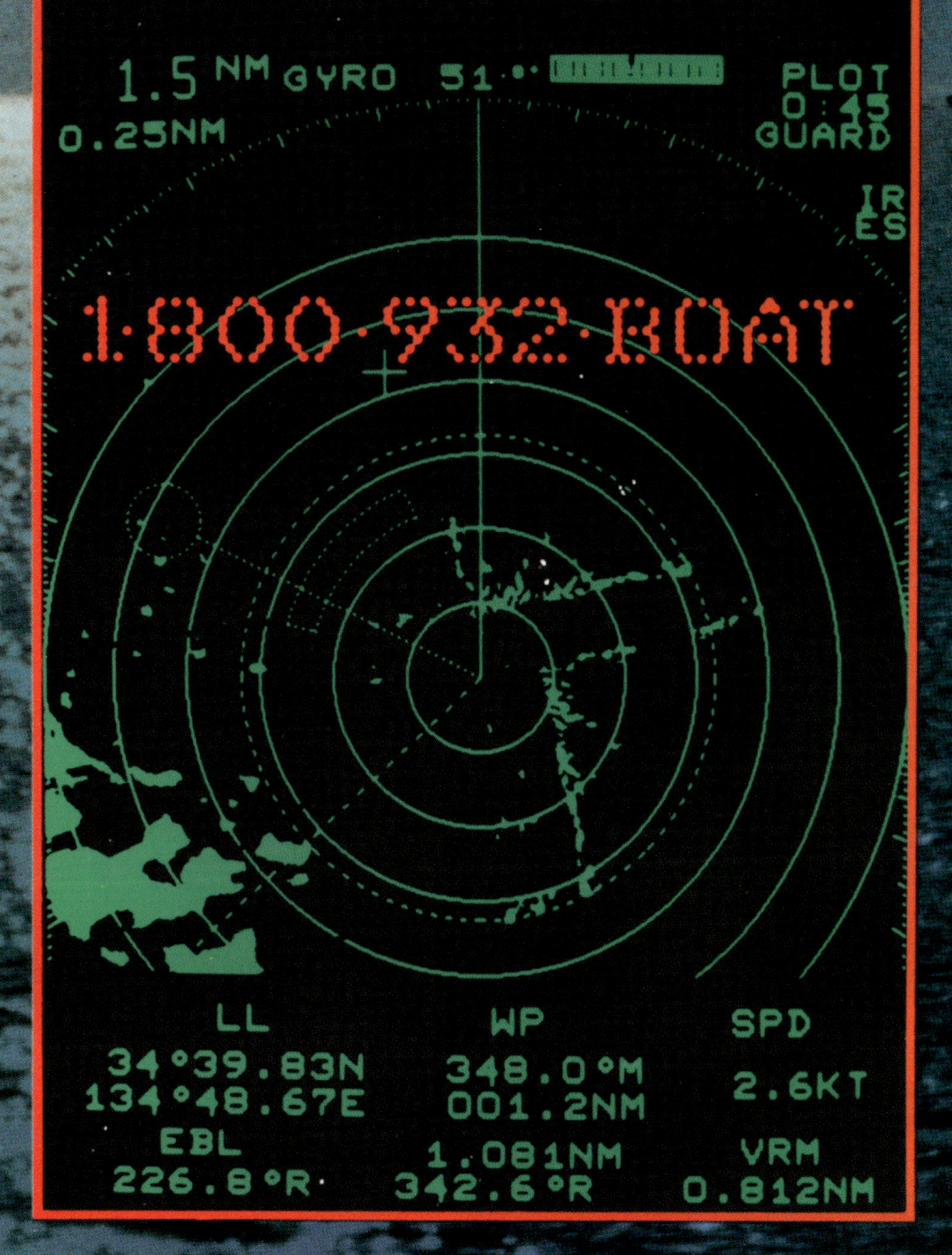

1·800·932·BOAT

We're the Michigan Boating Industries Association...M.B.I.A. We'll answer your questions about where to find a certain boat, boating product or service. We'll also send to you our free recreational boating information kit. Call us at **1-800-932-2628.**

M.B.I.A....making boating better in Michigan

Michigan Boating Industries Association

32398 Five Mile Road, Livonia, MI 48154-6109
734.261.0123 or 800.932.BOAT (2628) in Midwest
Fax: 734.261.0880
email address: boatmichigan@mbia.org
web address: www.mbia.org/

Boaters Resource Guide

Boat Registration
Register your boat where it is used most often.

Illinois 217-782-6302
dnr.state.il.us/admin/systems/boats.htm
Indiana. . . . . 317)-233-6000
www.state.in.us/bmv/watercraft/
Michigan 517-322-1460
www.michigan.gov/sos
Minnesota. 888-MINN-DNR
www.dnr.state.mn.us/licenses/watercraft
Ohio. 614-265-6480
www.ohiodnr.com/watercraft/
New York 518-473-5595
www.nydmv.state.ny.us/dmvfaqs.htm#boats
Pennsylvania 866-BoatReg
sites.state.pa.us/PA_Exec/Fish_Boat/
Wisconsin 608-267-7799
dnr.state.wi.us/org/caer/cs/registrations/boatingregis.htm
Ontario 416-973-8027
www.ccra-adrc.gc.ca

General Info

American Power Boat Association 586-773-9700
www.apba-racing.com
Powerboat racing.

American Red Cross 202-303-4498
www.redcross.org
Safety and disaster relief.

BoatU.S. 703-823-9550
www.boatus.com
Boater safety classes, insurance, towing and loans, plus info on goverment affairs.

U.S. Coast Guard Office of Boating Safety
Customer Info Line. 800-368-5647
Covers queries on boat defects, unsafe boats and boating safety as well as USCG safety courses.
www.uscgboating.org

Federal Communications
Commission 202-418-0680
www.wireless.fcc.gov
Radio licenses and info on PLBs and other devices.

Great Lakes Cruising Club   312-431-0904
www.glcclub.com
Tons of cruising tips from seasoned vets.

National Oceanic and Atmospheric
Administration 301-713-0622
www.nws.noaa.gov
Current weather, including advisories.

U.S. Army Corps of Engineers,
Detroit District 888-694-8313
www.usace.army.mil
Get the lowdown on what projects they're working on.

U.S. Power Squadrons 919-821-0281
www.usps.org
Great seamanship classes and vessel safety checks.

U.S. Coast Guard
Navigation Center 703-313-5800
www.navcen.uscg.gov
It'll guide you whether you use GPS, loran or paper charts.

Damage & Accidents
You must report incidents resulting in:

• Loss of life or personal injury that requires medical treatment beyond first aid.

• Damage to vessel and other property exceeding $200 or a complete loss of vessel.

• People overboard, if injured or dead.

Ontario 705-329-6111
Ontario Provincial Police in Orillia

9th District U.S. Coast Guard . . . 216-902-6000
Report casualties at sea

Distance Chart

Points in this table are arranged in geographical sequence proceeding from the St. Marys River southward along the west shore, and returning northward up the east shore around Georgian Bay and westward through the North Channel.

a From the foot of Grand River Avenue
b From the sailing course point north of Old Mackinac Point
c From abreast of the east end of the U.S. center pier, and (except those marked &) via Middle Neebish and DeTour; distances downbound through West Neebish are 1 mile less
d Distances to Georgian Bay ports (except those marked *, +, &, %) are via the bay entrance from Lake Huron and the St. Marys River points and via Little Current from the North Channel points
* Via False Detour and North Channels
+ Via Mississagi Strait and the North Channel
& Via Lake Nicolet, St. Joseph and North Channels
% Via Potagannissing Bay and the North Channel

This information was taken from *United States Coast Pilot No. 6, 24th Edition*, published by the National Ocean Service, Charting and Geodetic Service, and the National Oceanic and Atmospheric Administration.

	Old Mackinac Point	Sault Ste. Marie	DeTour Village	Port Dolomite	St. Ignace	Mackinac Island	Cheboygan	Rogers City	Stoneport	Rockport	Alpena	Au Sable	East Tawas	Bay City	Saginaw	Harbor Beach	Port Sanilac	Goderich	Kincardine	Southampton	Wiarton	Owen Sound	Meaford	Collingwood	Penetanguishene	Midland	Victoria Harbour	Depot Harbour	Parry Sound	Byng Inlet	Key Harbour	French River	Killarney	Little Current	Gore Bay	Algoma Mills	Thessalon
Port Huron, MI[a]	*238	*238	+250	225	247	243	233	194	182	166	157	117	119	162	175	63	33	65	94	121	228	238	241	258	265	266	267	243	247	229	232	224	213	225	*238	*238	*238
Old Mackinac Point[b]	69	90	45	27	6	7	18	54	78	83	115	142	163	210	223	186	215	211	192	189	217	227	230	247	254	255	256	231	235	211	*205	*196	*167	*143	+116	*101
Sault Ste. Marie[c]	&48		45	68	90	84	84	84	96	107	137	165	185	232	246	208	238	234	213	207	229	239	242	259	266	267	268	+238	&242	&201	&193	&184	&155	&131	&106	&88
DeTour, MI	%24			23	44	39	40	51	61	92	120	140	187	201	164	193	189	168	162	185	194	197	214	221	222	224	198	201	%171	%163	%154	%125	%101	%76	%58
Port Dolomite, MI	*92				28	39	47	70	76	113	143	163	206	219	190	218	203	183	176	207	219	220	238	241	254	243	215	218	199	197	189	*135	+108	*90
St. Ignace, MI	%68					6	31	55	83	115	143	163	210	224	187	216	214	194	191	217	227	229	246	252	2499	256	231	234	*209	*204	*195	*166	*142	*115	*100
Mackinac Island, MI	%63						20	50	77	111	140	159	207	220	184	212	209	189	186	212	221	224	241	248	251	242	226	229	*204	*199	*190	*161	*137	*110	*95
Cheboygan, MI	%63							17	66	101	129	149	196	210	176	202	196	178	175	204	213	216	233	240	241	217	221	*197	*195	*186	*157	*133	*106	*92
Rogers City, MI	*63								40	61	90	110	157	171	134	163	161	142	139	169	177	182	199	206	206	208	183	187	166	163	157	*138	*114	*87	*74
Stoneport, MI	+75									27	48	76	141	154	130	158	146	120	115	151	164	163	181	186	189	190	163	166	147	143	145	137	*118	+92	+73
Rockport, MI	*79										9	62	129	142	106	134	132	113	112	148	158	161	178	185	186	187	162	166	143	136	123	*119	+92	+79
Alpena, MI	*107											31	49	116	129	97	125	124	107	109	156	165	168	192	201	195	193	174	171	152	145	131	*143	*120	*107
Au Sable, MI	*135												21	69	83	58	86	93	85	98	165	174	177	201	189	185	180	183	166	163	160	146	*148	*135
East Tawas, MI	*155													54	68	60	88	95	94	110	191	190	194	210	217	218	220	196	199	182	176	175	*168	*155
Bay City, MI	*202														13	102	130	137	136	153	228	237	240	257	264	265	267	243	245	228	223	222	*215	*202
Saginaw, MI	*216															115	144	151	149	167	241	250	254	271	278	278	280	256	258	242	235	237	228	*216
Harbor Beach, MI	*178																32	47	56	79	175	185	188	205	212	213	214	190	194	176	171	159	*178
Port Sanilac, MI	+207																	47	70	97	200	210	213	230	237	238	239	215	219	201	196	185	171	*178
Goderich, ON	+198																		36	64	174	183	187	207	210	211	213	189	192	174	172	159	138	*191
Kincardine, ON	+173																			30	144	153	156	176	180	181	182	158	162	144	142	128	120	*207
Southampton, ON	+166																				125	134	137	157	161	162	163	140	143	125	123	110	121	140	+224
Wiarton, ON[d]	+188																					29	36	53	73	74	76	71	74	81	88	94	110	122	+225
Owen Sound, ON[d]	+197																						29	46	69	70	71	75	79	89	97	103	119	142	+227
Meaford, ON[d]	+200																							24	52	53	55	68	71	86	96	105	121	140	193
Collingwood, ON[d]	+217																								54	55	57	76	80	97	107	108	120	138	107	155
Penetanguishene, ON[d]	+224																									10	12	69	72	94	104	106	120	139	87	197
Midland, ON[d]	+225																										7	70	73	95	105	107	121	140	111	159
Port McNicoll, ON[d]	+227																											71	75	97	106	108	122	142	174	190
Depot Harbour, ON[d]	193																												6	62	71	73	87	107	140	155
Parry Sound, ON[d]	197																													65	75	77	91	111	143	159
Byng Inlet, ON[d]	115																														27	31	50	70	102	118
Key Harbour, ON[d]	148																															22	42	62	944	110
French River, ON[d]	139																																33	53	86	101
Killarney, ON[d]	110																																	24	56	72
Little Current, ON	86																																		33	48
Gore Bay, ON	61																																			27
Algoma Mills, ON	43																																			

LAKELAND BOATING · LAKE HURON CRUISING GUIDE

Section One

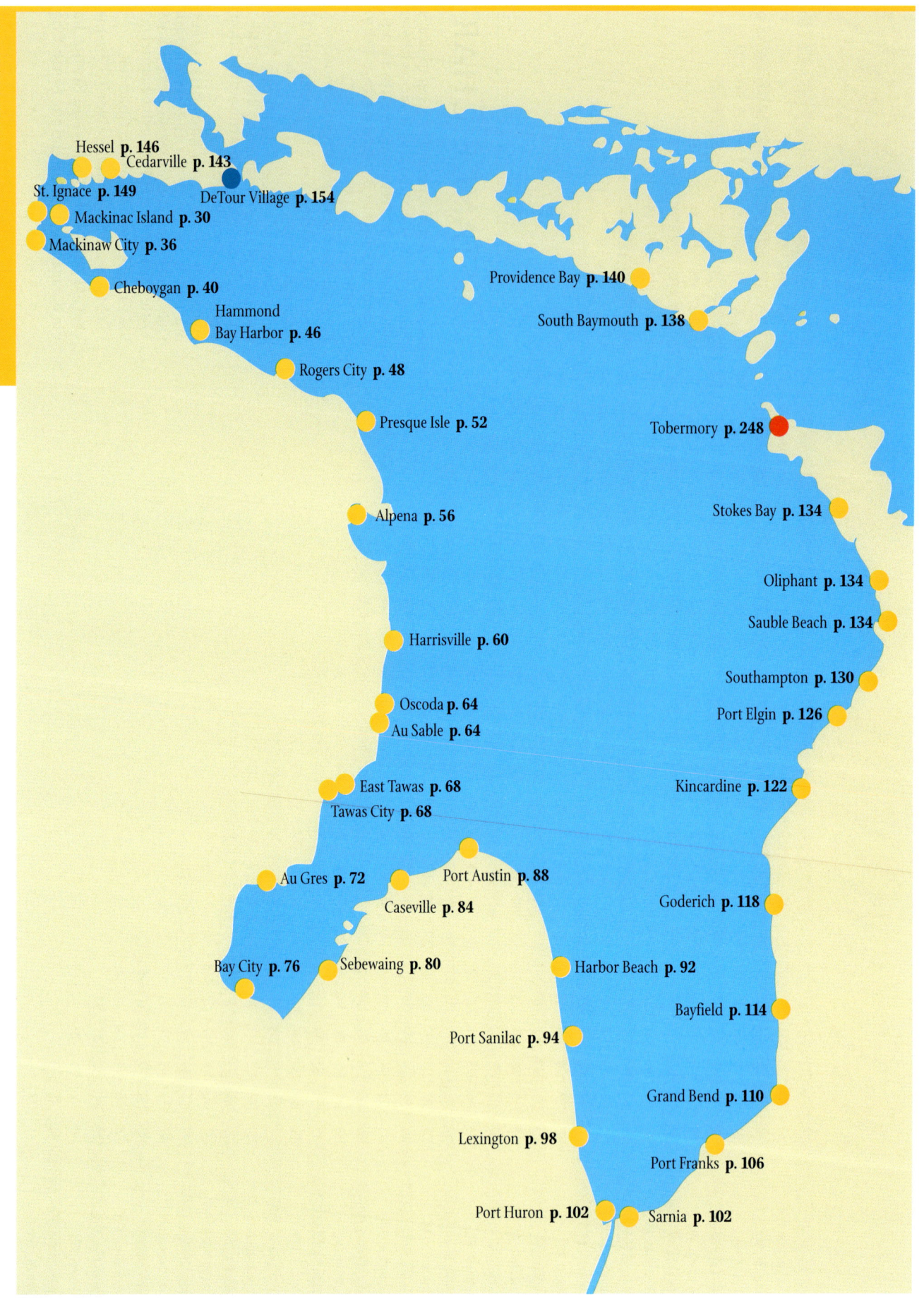

- Hessel **p. 146**
- Cedarville **p. 143**
- DeTour Village **p. 154**
- St. Ignace **p. 149**
- Mackinac Island **p. 30**
- Mackinaw City **p. 36**
- Cheboygan **p. 40**
- Hammond Bay Harbor **p. 46**
- Rogers City **p. 48**
- Presque Isle **p. 52**
- Providence Bay **p. 140**
- South Baymouth **p. 138**
- Tobermory **p. 248**
- Alpena **p. 56**
- Stokes Bay **p. 134**
- Oliphant **p. 134**
- Sauble Beach **p. 134**
- Harrisville **p. 60**
- Southampton **p. 130**
- Oscoda **p. 64**
- Au Sable **p. 64**
- Port Elgin **p. 126**
- East Tawas **p. 68**
- Tawas City **p. 68**
- Kincardine **p. 122**
- Au Gres **p. 72**
- Port Austin **p. 88**
- Caseville **p. 84**
- Goderich **p. 118**
- Bay City **p. 76**
- Sebewaing **p. 80**
- Harbor Beach **p. 92**
- Bayfield **p. 114**
- Port Sanilac **p. 94**
- Grand Bend **p. 110**
- Lexington **p. 98**
- Port Franks **p. 106**
- Port Huron **p. 102**
- Sarnia **p. 102**

28   LAKELAND BOATING · LAKE HURON CRUISING GUIDE

Lake Huron

Bay City, Michigan

LAKELAND BOATING · LAKE HURON CRUISING GUIDE

Mackinac Island, Michigan

Lat 45°50.65
Lon 84°36.87

Make your approach almost due north between the two breakwalls and head for Mackinac Island's only transient marina, Mackinac Island State Dock **A**. The heart of town **B** is just a few steps away. Guarding the high ground is the world-renowned Grand Hotel **C**, as well as historic Fort Mackinac **D**. The island itself is only eight miles in circumference and is bordered by an extension of Huron Street, which makes for a delightful half-day bicycle tour. The airstrip **E** provides easy access to those who wish to connect with a cruise by plane. The lat/lon is for the Round Island Passage Light, just outside the picture beyond the breakwall to the left.

LAKELAND BOATING · LAKE HURON CRUISING GUIDE

Mackinac Island, Michigan
Stop and smell the flowers—and the fudge.

The clip-clop of horses' hooves and the heavenly scent of fudge greet visitors to Mackinac Island as they step ashore. The historic island's ban on automobiles, its landmark **Grand Hotel** and its Victorian-era ambience conspire to take people on a trip back through time to a very beautiful and unusual place.

American Indians considered the island sacred, calling it "the Great Turtle" for the way it rises gently out of the water. The French and British saw it as a strategic position from which to guard the Straits of Mackinac and battled for control of the fur trade and fort here.

Now the only fighting that goes on is among transient boaters jockeying for position on the waiting list for slips. This is a tough port to find space in, especially during the racing season. The marina is small, the tourist attractions are many, and the amenities are minimal. For the faint of heart or the impatient, there is always the option of docking in Mackinaw City or St. Ignace and riding the ferry across. If you decide to do that, you can easily bring your bike with you.

Dockage and pumpout are available at the **Mackinac Island State Dock**. Be warned, space here is at a premium, especially during the height of the summer tourism season. The exclusive **Mackinac Island Yacht Club** is located across the street.

As for the island itself, it has everything a curious cruiser, an eager shopper, a hungry epicurean or a student of history could wish for. Amenities are concentrated in one area, and there are plenty of quiet open spaces for those who want to get away from it all and breathe a little fresh air.

Getting around is easier than the car-addicted mainlander might think. The charming town itself is so compact that you can walk anywhere, and with a bike, you become master of the entire island. If you rent a horse and surrey—a four-wheeled, open-air pleasure carriage—you get transportation with a guided tour.

Certain landmarks are high on every visitor's list. The

Where to Eat

Astor Street Café 906-847-6031
Pasta, shrimp and whitefish are available.

Carriage House 906-847-3321
Fine-dining at the Iroquois Hotel. Veranda seating available.

French Outpost 906-847-3772
Outdoor patio with nightly entertainment. Be sure to order the nachos, which can feed a horse.

Governor's Dining Room 906-847-3420
At Island House, overlooking the docks. Prime rib and seafood specialties. Daily breakfast buffet and nightly entertainment.

Grand Hotel. 906-847-3331
Elegant, with serious dress codes. Great brunch.

Harborview Dining Room. 906-847-6494
At the Chippewa Hotel. Superb dining. View of the water. Drinks in the **Pink Pony Lounge**.

Horn's Gaslight Bar 906-847-6154
Full lunch and dinner menu. Mexican and American fare, plus entertainment every night.

Jockey Club 906-847-3331
An informal restaurant at the first tee of the Grand Hotel's Jewel Golf Course.

Kilwin's 906- 847-6500
Fresh fudge, plus limeade to quench your thirst.

**Martha's Ice Cream
& Sweet Shop.** 906-847-3790
Homemade pastries and frozen yogurt.

Pancake House 906-847-3829
Breakfast, lunch and dinner. Hearty pancakes and eggs any style.

Patrick Sinclair's Irish Pub. 906-847-9916
Great drinks, live music and excellent mussels.

Pilot House Restaurant & Bar 906-847-0270
Lighter fare at the Lakeview Hotel. Serving lunch and dinner.

Point Dining Room. 906-847-3312
At the Mission Point Resort. Casual gourmet dining, including breakfast buffets.

Pub Oyster Bar. 906-847-9901
1890s saloon setting. Breakfast, lunch and dinner.

Round Island Bar 906-847-3312
Also at the Mission Point Resort. Informal, with a great view.

Sarducci Brothers Pizza Co. 906-847-3880
Limited Italian menu. Pan pizzas, whole or slices.

Village Inn Restaurant 906-847-3542
Famed for planked whitefish, steakburgers and more. Eat inside or out.

Waterfront Patio Café 906-847-3869
Family-oriented outdoor café featuring charbroiled burgers, chicken and brats.

Woods 906-847-3331
A Grand Hotel casual restaurant located at Stonecliff. Shuttle service available.

Local events

May

Michilimackinac Pageant
A reenactment of the 1763 American Indian attack on the British-held fort.

Memorial Day
At Fort Mackinac Post Cemetery. Soldiers march and perform a short ceremony.

June

54th Annual Lilac Festival

Taste of Mackinac

Mackinac Celebration
Cottage tour followed by dinner at Grand Hotel.

American Heritage Music Festival

June-August

Governor's Residence Tours
A look inside the 1902 cottage overlooking the harbor.

July

A Star-Spangled 4th
A 38-gun salute at the fort.

An American Picnic
An all-you-can-eat affair on the fort's parade ground.

Chicago-Mackinac Island Yacht Race

Bay View to Mackinac Yacht Race

Campeau Company Encampment
18th-century reenactment.

August

Campaigne Franche de la Marine de Michilimackinac Encampment
18th-century reenactment.

Blacksmith Convention

Snag and Brag Fishing Contest

Wine-Tasting Week

Grand Hotel, famed for its porch—the longest in the world—is one such attraction, though it is also known for having something of an attitude. Non-guests must pay a fee to visit, and everyone must adhere to a dress code in the evening, which includes a tie and jacket for men. For boaters who don't carry such impractical gear on board, other restaurants and hotels in town are more accommodating.

The other landmark that's impossible to overlook is **Fort Mackinac**. Not only does it dominate the hill over the city, but it's staffed by soldiers in circa-1805 uniforms. Courts-martial are re-enacted, cannons boom and cameras click.

The fantastical atmosphere of the island is the heart of its appeal. It's also the reason you should think of Mackinac Island as a recreation stopover and not as a major port. Stop here for the fun, the fudge and a great meal or two.

More Info

Emergency	911
Mackinac Island Medical Center	906-847-3582
Mackinac Island Tourism Bureau	906-847-3783
	www.mackinac.com
Mackinac State Historic Parks	906-847-3328
Mackinac Island Taxi Service	906-847-3323
Arnold Transit Line	906-847-3351
Star Line Ferry	906-643-7635
Shepler's Ferry	906-643-9440
Alford's Drug Store	906-847-3881

Distances from Mackinac Island (statute miles)
St. Ignace: 6 W
Mackinaw City: 7 SW
Hessel: 17 NE
DeTour Village: 39 ENE
Beaver Island: 48 W

MACKINAC ISLAND
MICHIGAN
Scale 1:10,000

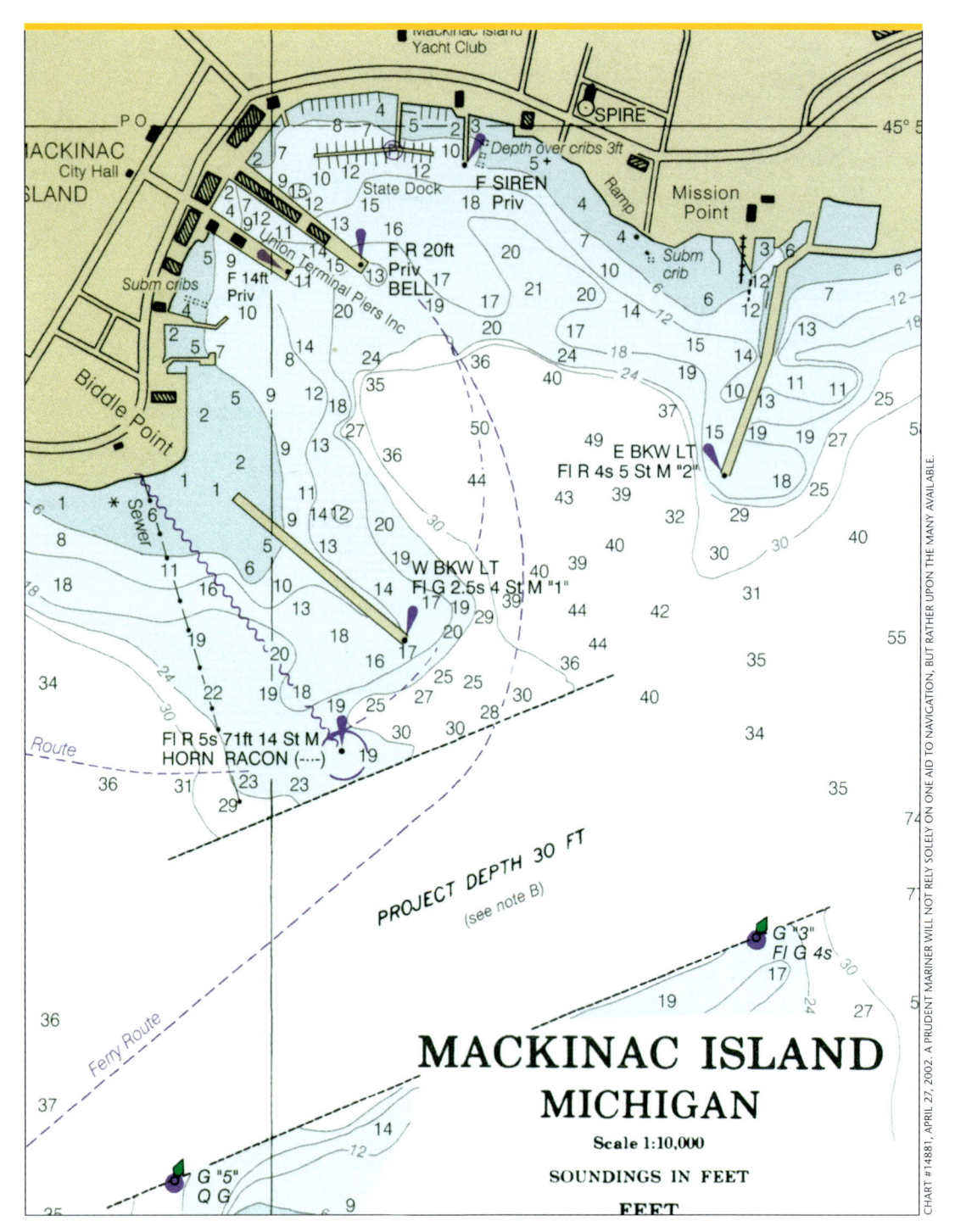

SOUNDINGS IN FEET

Facilities information subject to change. We suggest you call ahead.

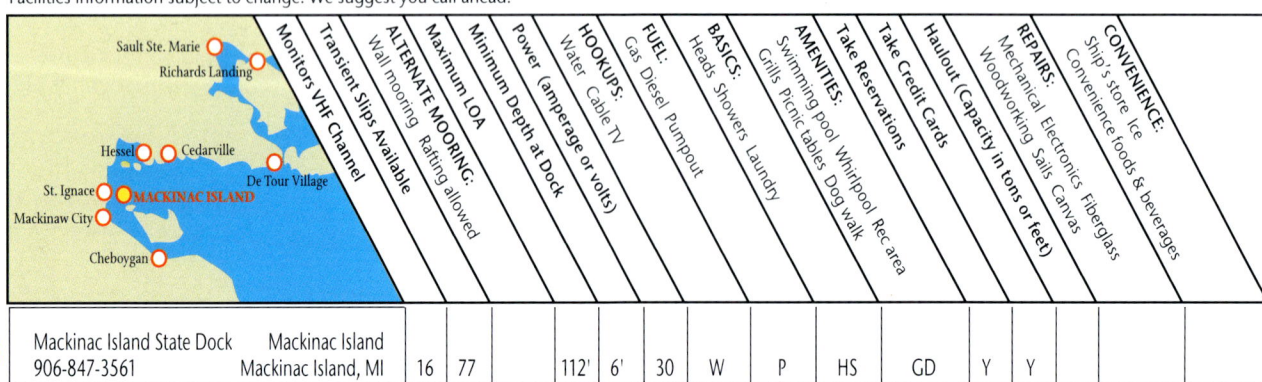

	Location	Monitors VHF Channel	Transient Slips Available	ALTERNATE MOORING: Wall mooring / Rafting allowed	Maximum LOA	Minimum Depth at Dock	Power (amperage or volts)	HOOKUPS: Water / Cable TV	FUEL: Gas / Diesel / Pumpout	BASICS: Heads / Showers / Laundry	AMENITIES: Swimming pool / Whirlpool / Rec area / Grills / Picnic tables / Dog walk	Take Reservations	Take Credit Cards	Haulout (Capacity in tons or feet)	REPAIRS: Mechanical / Electronics / Fiberglass / Woodworking / Sails / Canvas	CONVENIENCE: Ship's store / Ice / Convenience foods & beverages
A	Mackinac Island State Dock 906-847-3561 — Mackinac Island, Mackinac Island, MI	16	77		112'	6'	30	W		P	HS			GD	Y	Y

34   LAKELAND BOATING · LAKE HURON CRUISING GUIDE

LAKELAND BOATING · LAKE HURON CRUISING GUIDE

Mackinaw City, Michigan

Lat 45°48.92
Lon 84°43.71

The stately bridge that crosses the Straits of Mackinac serves as a dramatic dividing line between the great cruising waters on either side. To the west **A** lies northern Lake Michigan and the lure of Beaver Island, Harbor Springs and Door County. On Lake Huron to the east **B**, the exploring cruiser will find the challenges of the North Channel and Georgian Bay. You could easily spend an entire vacation cruising either one of these rugged and unspoiled waterways and leave eager to come back for more. One port of call that serves this area is Mackinaw City **C** with its fine municipal marina **D** at the foot of Central Avenue, which is the downtown shopping area. Just across the straits lies St. Ignace **E**, an alternative stopover.

Distances from Mackinaw City (statute miles)
Cheboygan: 15 SE
Rogers City: 49 SE
Presque Isle Harbor: 67 SE
Beaver Island: 44 WSW
Charlevoix: 55 SSW

LAKE HURON

Mackinaw City, Michigan
Hop a ferry—or stick around
and explore this bustling, historic port.

Mackinaw City's strategic position has made it a crossroads for everyone from voyageurs to modern-day cruisers. The earliest French explorers were quick to understand the military potential of the location, and since the beginning of the 20th century, Mackinaw City has been the jumping-off point for scenic Mackinac Island. Today it offers a well-maintained public marina with ample docking and the hustle and bustle of a well-groomed tourist town.

Mackinaw City is easy to locate from any point of sail. Just home in on the Mackinac Bridge, the largest suspension bridge in the world, which is visible from 20 miles away on either lake. Then plot a course to Old Mackinac Point. Coming up the South Channel, be sure to mark the numerous wrecks off Point Nipigon.

The **Mackinaw City Municipal Marina** is the northernmost of two basins. The eastern breakwall is marked with a flashing green light atop a 25-foot standard. Make an immediate turn to starboard, keeping watch for the heavy ferry traffic, and hail the dockmaster for a berth assignment. The facility can handle anything up to megayachts. There is no provision for anchoring out. The marina underwent a $400,000 renovation in 1999 and boasts modern restrooms and showers, a map room and an office with a fax and Internet hookups. In July and August, it stays open 24 hours.

The town is laid out in an L shape. The street leading off to the left is "motel row," accommodating visitors to Mackinac Island.

If you haven't been to Mac City in the last few years, you might not recognize it. After a $25 million renovation in 2000, the town now boasts more than 100 shops, a five-screen cinema and an 850-seat Branson, Missouri-style theater. Strolling musicians roam the streets, a bandshell presents live music four or five nights a week and there's a laser light show every evening at **Mackinaw Crossings,** the newly developed shopping center and entertainment complex. A trolley service will get you around, but most of the sights are within walking distance of the marina.

One of the most interesting diversions in the area is **Colonial Michilimackinac State Park,** which has been painstakingly restored on the original site; an archeological dig is ongoing. Stroll through the stockade gates and you are transported to a British and French colonial town circa 1705. The settlement was the capital of the fur trade until the French surrendered it and all of New France to the British in 1761. The British abandoned the site and built a more defensible fort on Mackinac Island. The fur trade went with it, setting off a decline that lasted until tourism blossomed in the late 19th century.

Fort Michilimackinac was the scene of Chief Pontiac's great ruse and the subsequent massacre of the British in 1763. The battle, a modern version of the classical Greek

LAKELAND BOATING · LAKE HURON CRUISING GUIDE   37

Where to Eat

Audie's Restaurant. 231-436-5744
Leisurely, relaxed, white-tablecloth dining. Whitefish, steaks, salads, cocktails and seafood served by a friendly staff. Even the ride there can be glamorous—a limo will pick you up.

Embers Restaurant & Lounge 231-436-5773
This pleasant restaurant near the waterfront is open for breakfast through dinner, with buffets as well as à la carte selections. The specialities here include whitefish, homemade soups and bread pudding.

Mama Mia's Pizza 231-436-5534
Downstairs from the bridge museum you'll find Mexican and Italian entrées. The popular taco pizza merges these two cuisines in a unique and delicious fashion.

Pancake Chef 231-436-5578
This is a popular breakfast spot with a scrumptious buffet. And fret not—if you've got a hankering for a tall stack of blueberry pancakes for lunch or dinner, you're happily accommodated.

Scalawags. 231-436-7777
This cheery downtown restaurant is worth seeking out. Serving delicious fresh fish and chips, lovingly prepared using the best ingredients.

Squealy Downing's Family Restaurant 231-436-7330
Featuring a buffet with fantastic home cooking. A full menu, with charbroiled steaks, seafood, pizza, specialty sandwiches and cocktails. Plus, the place is decorated with images of pigs. Free delivery.

---

tale of the Trojan horse, is reenacted every Memorial Day with a cast of hundreds. The cunning chief induced the British to allow his braves to enter the fort under the pretext of challenging the British to a game of bagataway, or lacrosse, which was invented by the Indians. Feigning a mis-aimed pass, one of the Native Americans sent the ball flying over the walls. When the soldiers opened the gates to retrieve it, Chief Pontiac's warriors swarmed in. Only the French were spared.

Other goings-on in the park include the Voyageurs and Fur Trade Rendezvous in the middle of July and the French and Indian War Encampment in early August.

Head three and half miles southeast of the city to reach **Historic Mill Creek**, a 625-acre nature park with a water-powered sawmill. While here, enjoy educational talks and nature tours. Staff members, dressed as 1820s sawyers and millwrights, run the mill and use the lumber to reconstruct the **Millwright's House**.

The **Mackinac Bridge Museum**, a storehouse of fascinating facts about the construction of this breathtaking span, is located upstairs from **Mama Mia's Pizza**. The bridge statistics are mind-boggling. For instance, the five-mile-long bridge weighs as much as 4,444 Statues of Liberty or 12 Washington Monuments—that's more than 1 million tons! Five million rivets were used to hold it together, and the location of each and every one is mapped because the rivets are subject to inspection. The supporting cables are hung from towers 522 feet high. Spinning the cables required 42,000 miles of wire. If you are feeling energetic, take the annual Labor Day walk across the 19,243-foot-long bridge.

Local events

May
Memorial Weekend Pageant
More than 400 cast members re-enact events that took place between the French, British and Indian tribes on June 2, 1763.

Memorial Day Observance
At Fort Mackinac Post Cemetery. Soldiers march, perform a short ceremony and salute.

June
Mackinaw Kite Festival

July
4th of July

August
Mackinaw City Antique Show

Corvette Crossroads Auto Show

September
Labor Day Bridge Walk
The only day of the year that walking the "Mighty Mac" is allowed.

More Info

Emergency. 911
Visitors Bureau. 800-666-0160
Mackinac State Historic Parks 231-436-5563
www.mackinac.com/historicparks
Mackinaw Crossings 231-436-5030
Star Line Ferry 906-643-7635
Shepler's Ferry. 231-436-5023

MACKINAW CITY MICHIGAN

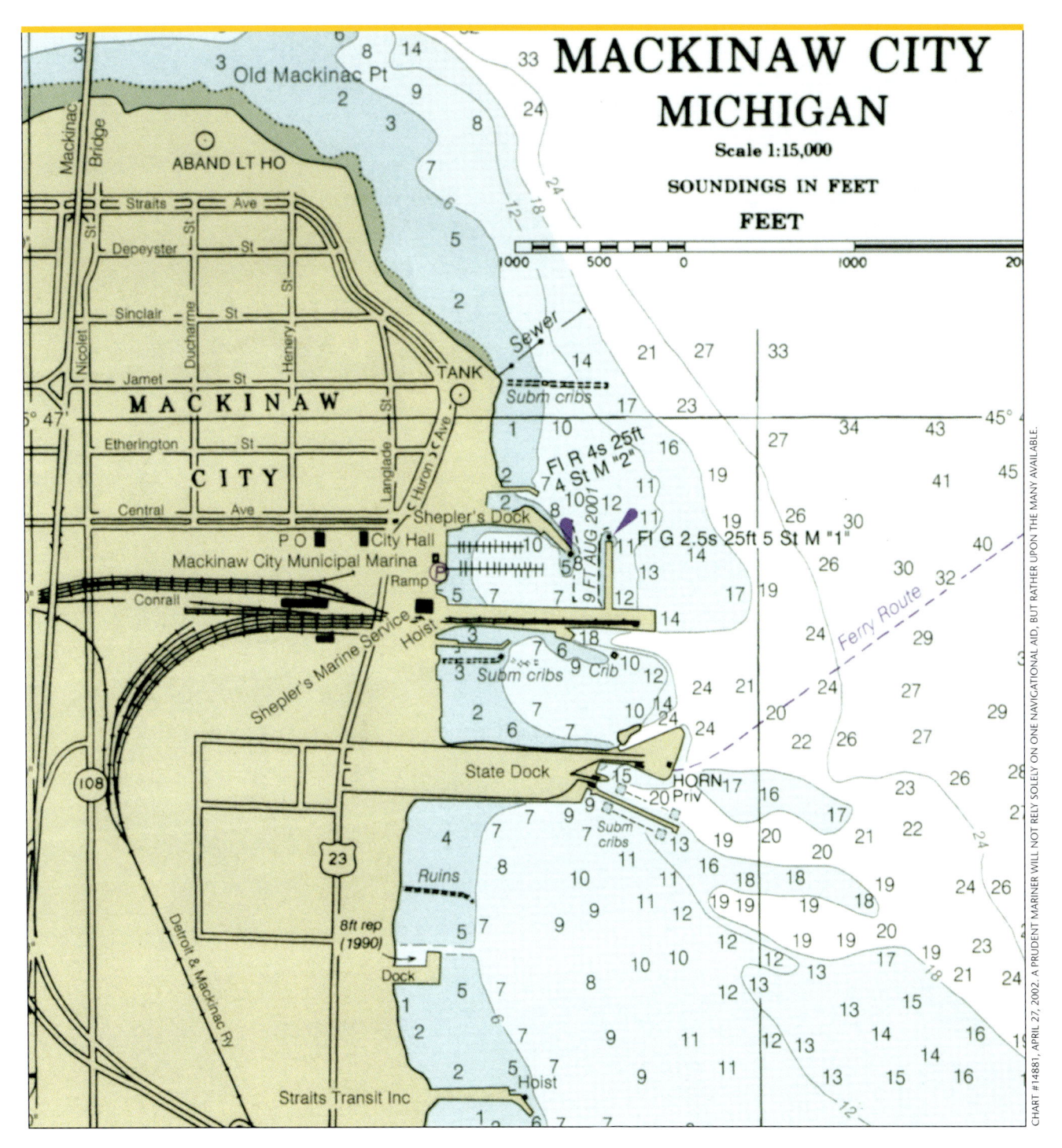

Scale 1:15,000

SOUNDINGS IN FEET

FEET

Facilities information subject to change. We suggest you call ahead.

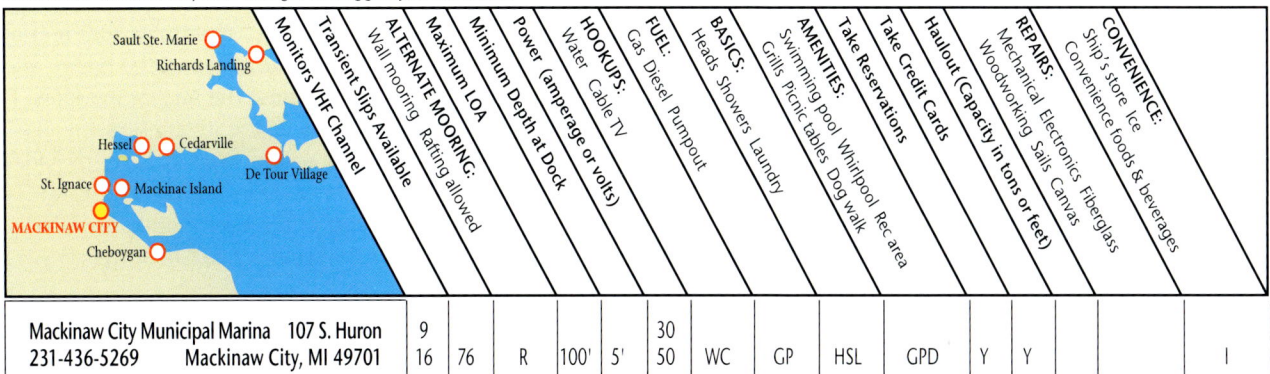

	Monitors VHF Channel	Transient Slips Available	ALTERNATE MOORING: Wall mooring / Rating allowed	Maximum LOA	Minimum Depth at Dock	Power (amperage or volts)	HOOKUPS: Water / Cable TV	FUEL: Gas / Diesel / Pumpout	BASICS: Heads / Showers / Laundry	AMENITIES: Swimming pool / Whirlpool / Rec area / Grills / Picnic tables / Dog walk	Take Reservations	Take Credit Cards	Haulout (Capacity in tons or feet)	REPAIRS: Mechanical / Electronics / Fiberglass / Woodworking / Sails / Canvas	CONVENIENCE: Ship's store / Ice / Convenience foods & beverages
D Mackinaw City Municipal Marina 107 S. Huron 231-436-5269 Mackinaw City, MI 49701	9 16	76	R	100'	5'	30 50	WC	GP	HSL	GPD	Y	Y			I

LAKELAND BOATING · LAKE HURON CRUISING GUIDE   39

Cheboygan, Michigan

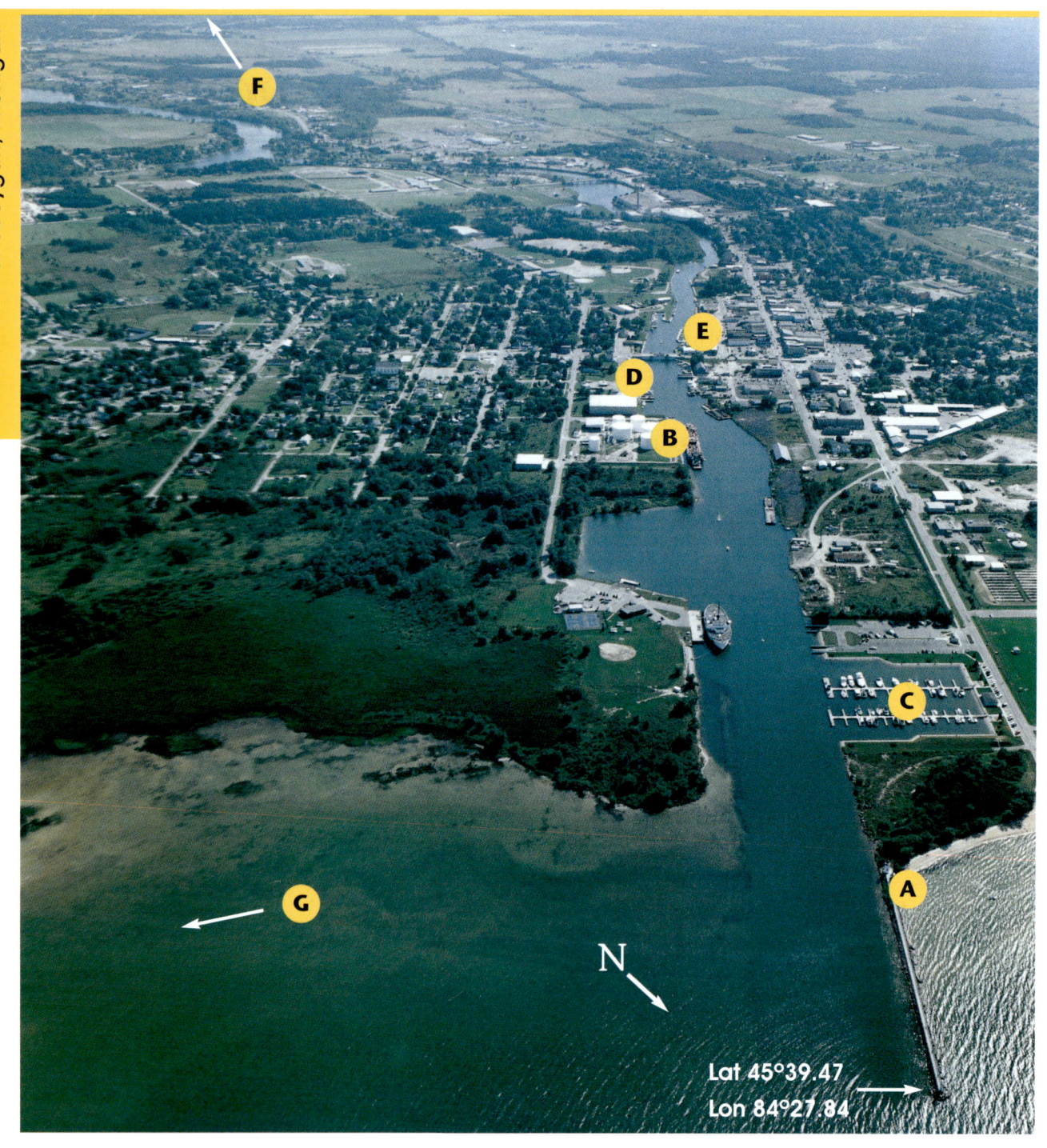

When approaching Cheboygan, look for the flashing buoy No. 2 and proceed on a course of 212.5°T up the channel and into the Cheboygan River. The 100-year-old Crib Light lighthouse **A**, located at the base of the breakwall that extends into the lake, is a distinctive marker. Keep to the right until you reach the Coast Guard icebreaker *Mackinaw* **B** to avoid shallow depths. Just beyond the light and to starboard, the Cheboygan County Marina **C** offers the most transient slips in town, full hook-ups and fuel. Walstrom Marine, Inc. **D** is located to port upstream just before the highway bridge. Cheboygan City Marina **E** maintains 600 feet of broadside dockage to starboard just beyond the bridge and offers easy access to city attractions. The Anchor In Marina **F** (not pictured), located upriver at the junction of the Black and Cheboygan rivers, just before Mullett Lake, provides transient slips, fuel and service. Duncan Bay Boat Club **G**, approximately one mile east of the river mouth, is on heading 125°M from red 2 on the approach to Cheboygan.

LAKE HURON

Cheboygan, Michigan
Hospitality and history on the Inland Waterway.

At the head of the Inland Waterway and close to the Straits of Mackinac, Cheboygan offers the only harbor on the northern coast that is completely protected from winds in all directions. Its charming Victorian downtown and wide range of ship's stores and repair facilities make it a perfect place to pull in for a night or longer.

The entrance to the Cheboygan River is well-marked, as is the ship channel, on a course of 212.5°T. The Fourteen Foot Shoal Light is a distinctive landmark. The historic 100-year-old tower of the **Crib Light lighthouse** was relocated to the shore end of the north breakwater when the light was dismantled in 1984. The marina on the starboard bank just inside the river is the **Cheboygan County Marina**. It offers easy access to a large park with a sandy beach. Upstream, just before the highway bridge on the port bank, is **Walstrom Marine, Inc.**, which offers all services and has a 70-ton Travelift. On the upstream side of the bridge (which monitors channel 16 from 6 a.m. to 6 p.m.) and to starboard are the two docks of the **Cheboygan City Marina**, one in Washington Park, the other a half-block away on Water Street. They offer quick access to the bright lights of the big city. The marina has vehicle and trailer parking available by fee or permit.

As you continue upriver and through the locks, **Anchor In Marina** is the last transient facility before the Cheboygan River meets Mullett Lake.

Most historians agree that Cheboygan comes from an old Chippewa word meaning "a place for going through." This was a reference to what is now known as the Inland Waterway. Long ago, this was an important thoroughfare for American Indians, French voyageurs and trappers wanting to reach Lake Michigan while avoiding the dangerous waters of the straits. Today, boats less than 60 feet can navigate most of its 40 miles. Boats less than 26 feet can make it to the end at Conway and hitch a trailer lift for the 10-mile overland ride to Harbor Springs or Petoskey.

Cheboygan's glory days began in the 1800s as the lumbermen and their sawmills kept moving north in search of untouched timber stands. The town hit its peak in 1890, when eight big mills cutting day and night shipped 127 million feet of lumber, which was loaded aboard the schooners and the steam barges called "rabbits" to build the cities of the east. Its famed mountain

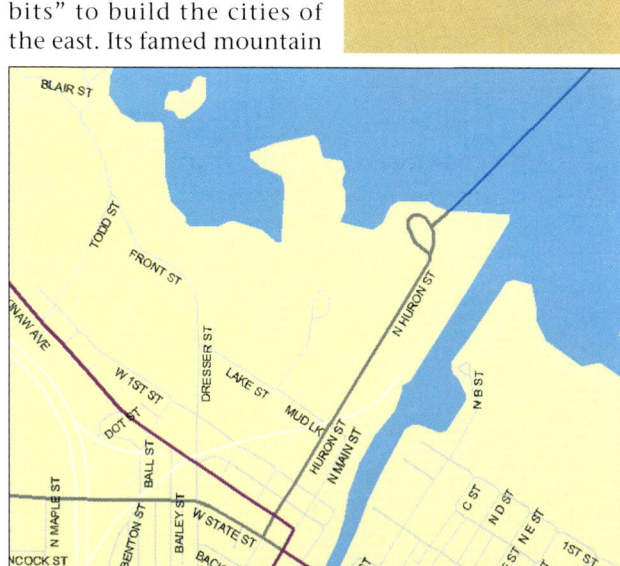

Distances from Cheboygan (statute miles)
Mackinac Island: 16 W
Hammond Bay: 30 S
Rogers City: 40 S
DeTour Village: 36 NE
Bois Blanc Island: 6 N

Where to Eat

The Boathouse.................. 231-627-4316
Situated behind the city docks past the highway bridge, this is a favorite for convenience, as well as good food.

Carnation Restaurant 231-627-5324
Beef lovers head to this local institution for steaks, ribs and porkchops. Dinghy to the town dock and walk a block to the main street.

Hack-Ma-Tack Inn 231-625-2919
The menu features whitefish and prime rib, served up in an attractive early American setting. It is located in a wooded area, overlooking the Cheboygan River, with a 400-foot dock.

Pappas' Restaurant 231-627-2459
This downtown eatery attracts a faithful following because of the ambience, as well as the bill of fare. The specialty of the house is broiled whitefish.

LAKELAND BOATING · LAKE HURON CRUISING GUIDE 41

Local events

June
Classic Car Show
Summer Craft Show

July
Free Summer Concerts
July 4th Parade
Arts Festival

August
County Fair
Sidewalk Sales
Antique Daze
Northern Michigan's Largest Garage Sale

of sawdust covered 12 acres and was taller than the topmasts of the now-vanished schooners. In those days, the red-light district occupied what would be a third of today's town.

Cheboygan still has a far more socially acceptable remnant of yesteryear's nightlife in the form of its **Opera House**. Built in 1877 and lovingly restored in 1984, the Opera House is just two blocks from the city docks. Its perfect acoustics have allowed it to attract top entertainers and symphony companies.

Golfers will enjoy the 18-hole course at the **Cheboygan Golf & Country Club**. And if the Coast Guard icebreaker *Mackinaw* is at her dock just inside the river mouth, be sure to go aboard for the tour. Six—count 'em, six—2,000-hp engines supply the kick that keeps the shipping lanes free of ice in late spring and early fall. The *Mackinaw* is scheduled to become a floating museum; she is too costly to operate and too large to leave the Great Lakes for duty elsewhere.

The Inland Waterway

The Inland Waterway, a 40-mile stretch of water that begins in Cheboygan, makes for a delightful day trip or weekend excursion. It's particularly lovely in the fall, when it's less crowded and you can enjoy the spectacular foliage. A 5-foot depth and a 30-foot width are maintained throughout the waterway.

Pass through the lock and follow the markers as the bottle-green Cheboygan River meanders into Mullett Lake. Head south down the Indian River to the village of the same name, en route to Burt Lake. From here, head northwest, maneuvering the hairpin turns of the Crooked River. Watch for deer, loons and bald eagles as you inch along the dark river. It widens into a mile-long marshy area known as Hay Lake. Just below Alanson, there's a swinging bridge, the only one of its kind in Michigan. Another small lock lies before the entrance to Crooked Lake. Most boaters turn around here, a few miles from Petoskey and Lake Michigan.

Arm yourself with a recent chart, a good depth-sounder and a sense of adventure.

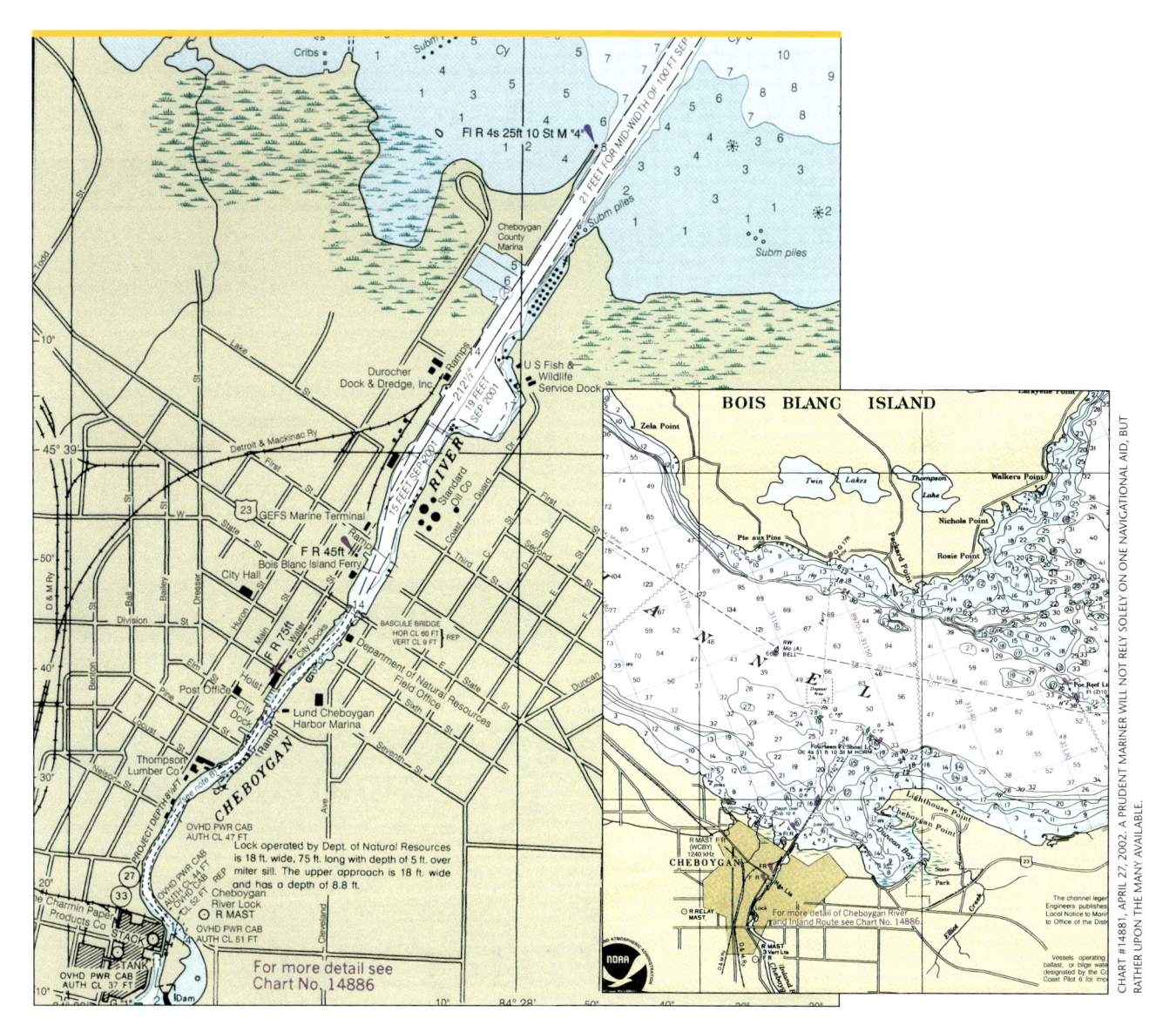

Facilities information subject to change. We suggest you call ahead.

			Monitors VHF Channel	Transient Slips Available	ALTERNATE MOORING: Wall mooring Rafting allowed	Maximum LOA	Minimum Depth at Dock	HOOKUPS: Power (amperage or volts) Water Cable TV	FUEL: Gas Diesel Pumpout	BASICS: Heads Showers Laundry	AMENITIES: Swimming pool Whirlpool Rec area Grills Picnic tables Dog walk	Take Reservations	Take Credit Cards	Haulout (Capacity in tons or feet)	REPAIRS: Mechanical Electronics Fiberglass Woodworking Sails Canvas	CONVENIENCE: Ship's store Ice Convenience foods & beverages
F	Anchor In Marina 231-627-4620	724 S. Main St. Cheboygan, MI 49721	9 16	8	W	44'	7'	30 W	GP	HSL		GPD	Y	Y	30	MEFW	SI
E	Cheboygan City Marina 231-627-4944	P.O. Box 245 Cheboygan, MI 49721	9 16	15	W	90'	4'	30 W		HS		GPD	Y	Y			
C	Cheboygan County Marina 231-627-4944	1080 N. Huron St. Cheboygan, MI 49721	9 16	39		90'	6'	30 50 WC	GDP	HSL		RGPD	Y	Y			I
G	Duncan Bay Boat Club 231-627-2129	902 Boat Club Dr. Cheboygan, MI 49721	16	15		69'	8'	30 50 WC	P	HSL		SGPD	Y	Y			IC
D	Walstrom Marine, Inc. 231-627-7105	113 E. State St. Cheboygan, MI 49721	16	10	WR	70'	16'	50 W	P	HSL		PD	Y	Y	70	MFWS	S

LAKELAND BOATING · LAKE HURON CRUISING GUIDE   43

For a different kind of outing, take a cruise to the outlying Bois Blanc Island (pronounced "Bob-Lo Island"), situated four miles offshore. This 23,000-acre wilderness, dotted by the occasional cottage and summer resort, contains a half-dozen scenic lakes. The island's harbor is located on the southern side.

Lighthouse junkies should cruise by Lighthouse Point; the light was built in 1857. If you are looking for a more refined alternative to Cheboygan's downtown docking, head for the **Duncan Bay Boat Club** (www.duncanbay.org), an immaculate facility that can accommodate large yachts. The entrance to the basin is directly opposite Cheboygan Point in Duncan Bay. A long breakwater marks the western side of the channel. Be sure to follow the private markers in, because there are shoals on either side of the dredged channel. Duncan Bay has a luxurious clubhouse with a swimming pool that overlooks the water.

The Cheboygan County Fair is held in late July and early August. This is the real McCoy, complete with livestock judging, baking contests, a carnival and all the hoopla of a typical rural American agricultural celebration.

More Info

Emergency	911
Community Memorial Hospital	231-627-5601
Chamber of Commerce	231-627-7183
	www.cheboygan.com
Rivertown Taxi	231-627-3333

Anchor In Marina

ONE OF NORTHERN MICHIGAN'S LARGEST MARINA STORAGE FACILITIES

Specializing in Wood & Fiberglass Restoration

- Complete Repairs
- 60,000 lb Marine Lift
- Marine Accessories
- Mechanical Repairs
- Fiberglass Repair
- Paint and Varnish Work
- In/Outside Winter Storage
- Summer Slip Rentals
- In & Out Summer Storage
- Overnite Dockage

Complete Mechanical Repair

CAN HANDLE BOATS UP TO 60,000 LBS.

VISA • MasterCard • DISCOVER

GROWING TO SERVE NORTHERN MICHIGAN'S BOATING NEEDS

Open Year Round

231-627-4620

NOAA and Canadian Chart Dealer
US 27 HWY SOUTH
Located on the Inland Waterway Where the Cheboygan & Black River Meet
CHEBOYGAN, MI

WALSTROM
MARINE ■ YACHTING
SINCE 1946

Walstrom Marine is only 1/4 mile from the river entrance, conveniently located ahead of Cheboygan's locks and bridges. We are adjacent to the home port of Coast Guard cutter Mackinaw, at the base of northern Michigan's 40-mile inland waterway. Restaurants, parks, shopping and entertainment venues are all within walking distance.

Walstrom Marine employs certified gas and diesel technicians, and is equipped with a 70-ton Marine Travelift. We provide dockage, as well as heated, unheated and outside storage.

Our marina is an ideal take-off point for excursions:
- North Channel
- Mackinac Island
- Le Cheneaux Islands
- Cedarville & Hessel
- Chain of Lakes inland waterway
- Sault Ste. Marie
- Drummond & Beaver Islands
- Many other beautiful destinations...

Tiara • Hatteras • Mainship • Pursuit • Avon

113 East State Street, Cheboygan, MI 49721
Phone: 231-627-6681 • Fax: 231-627-8091
Email: cheboygan@walstrom.com

DON'T MISS OUR NEXT 50 YEARS

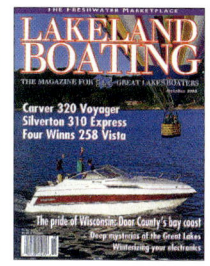

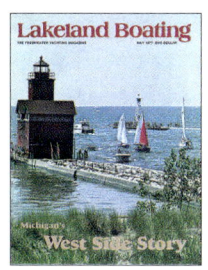

**SUBSCRIBE TODAY
CALL
800-827-0289**

XL3 Charts...
Clearly the best value.

$199.00 MSRP PER CARTRIDGE

Huge coverage.
All the Great Lakes on two chips!
Same detail. Same price.

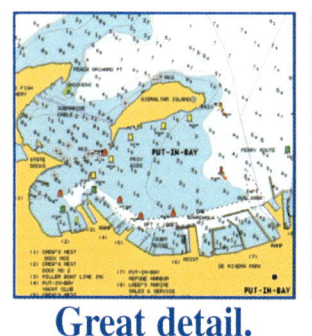

Great detail.
Bathys, soundings, navaids,
port plans, services and more.

Easy to use.
CF & MMC cartridges come
ready to use, just plug and play.

Popular plotters.
Check our site for a growing
list of compatible plotters.

NAVIONICS®

Clearer Charts for Serious Fishing and Cruising

www.navionics.com • 800-848-5896

EAGLE   HUMMINBIRD   L° LOWRANCE   NORTHSTAR   Raymarine

Hammond Bay Harbor, Michigan

Lat 45°35.62
Lon 84°09.67

Hammond Bay Harbor is midway from Cheboygan and Rogers City. To avoid hazards, hold a half-mile offshore. Enter the channel from the northwest between the two stone breakwalls **A**, which are clearly marked with green and red lights, as well as daymarks. Stick to the middle to avoid the shoaling around the tip of the east breakwater. Hammond Bay Refuge Harbor **B**, run by the state of Michigan, maintains a ramp **C** and 13 slips for transients. Gas and pumpout are available. Depths inside the harbor are about 9 feet. The nearby beach **D** and forest **E** are an ideal setting for some outdoor rest and relaxation.

Facilities information subject to change. We suggest you call ahead.

	Monitors VHF Channel	Transient Slips Available	ALTERNATE MOORING: Wall mooring / Rafting allowed	Maximum LOA	Minimum Depth at Dock	Power (amperage or volts)	HOOKUPS: Water / Cable TV	FUEL: Gas / Diesel / Pumpout	BASICS: Heads / Showers / Laundry	AMENITIES: Swimming pool / Whirlpool / Rec area / Grills / Picnic tables / Dog walk	Take Reservations	Take Credit Cards	Haulout (Capacity in tons or feet)	REPAIRS: Mechanical / Electronics / Fiberglass / Woodworking / Sails / Canvas	CONVENIENCE: Ship's store / Ice / Convenience foods & beverages
B Hammond Bay Refuge Harbor U.S. 23 North 989-938-6291 Hammond Bay, MI 49759	9	13	WR	60'	6'	20	W	GP	HS	GPD	N	Y			IC

LAKE HURON

Hammond Bay Harbor, Michigan

Go for a hike or relax on the beach in this quiet refuge.

There's not much to do in Hammond Bay Harbor except dock your boat. Yet while there isn't a town or shopping area nearby, this quiet, beautiful spot is unmarred by industry and development. At **Hammond Bay Refuge Harbor,** which is not actually in Hammond Bay but is located about two miles southeast of Nine Mile Point, you can grab a slip or get gas.

Hammond Bay Harbor is located midway between the popular destinations of Cheboygan and Rogers City. The approach from the north or east is straightforward. From the south, be careful to avoid shoals, which extend some distance from the shore. The marina is entered from the north between the east and west breakwaters. A flashing green light, visible for several miles, is located at the north end of the east breakwater, along with a green square daymark atop a 25-foot silver support. Inside the east breakwall, there's a 3-foot shoal, so be careful entering. Once inside the breakwall, either take a slip or anchor out in the protected basin between the breakwall and the docks.

To the south of the marina a sandy beach begins and extends for miles. The forest behind the beach is essentially undeveloped, offering opportunities for hiking and spotting wildlife.

Distances from Hammond Bay Harbor (statute miles)
Cheboygan: 20 N
Rogers City: 21 SE

More Info

Emergency 911
Presque Isle County Tourism 888-854-9700
Zimmer's Shuttle Service 989-734-2255

Where to Eat

Rosa's Squeeze Inn . . . 989-734-2991
Italian and American cuisine. Try the signature chicken à la Rosa, which is sauteed in garlic sauce and sherried bread crumbs and baked with a secret three-cheese blend.

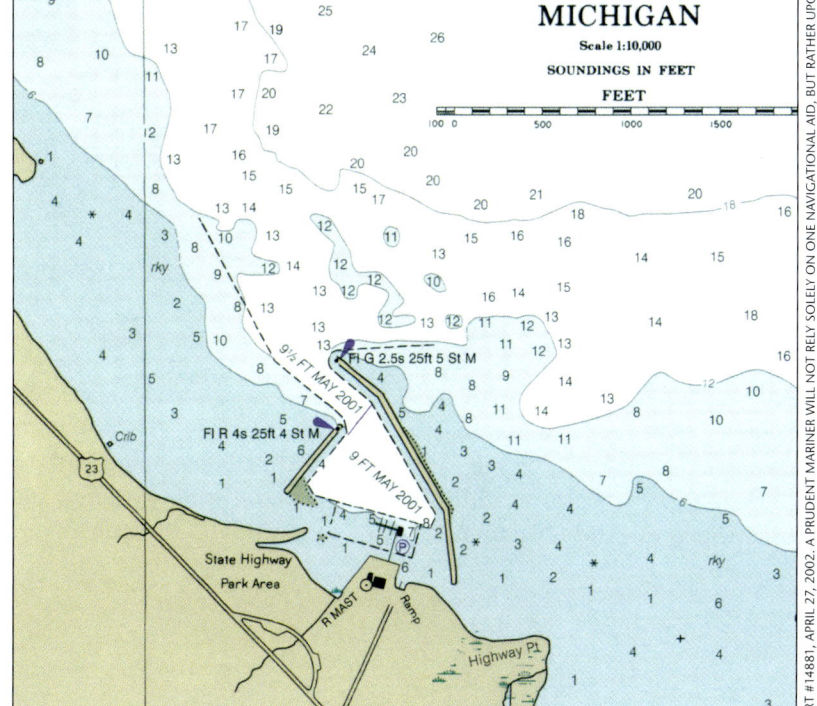

LAKELAND BOATING · LAKE HURON CRUISING GUIDE   47

Rogers City, Michigan

Lat 45°25.33
Lon 83°48.58

N

Rogers City has added a new breakwall to its harbor. The downward-facing light on the east breakwater points to the entrance, which is at the southeast corner of the breakwalls **A**. The revamped Rogers City Marina **B** offers ample transient dockage and full services, including fuel **C** and a launch ramp **D**. Lakeside Park **E** extends south along the shore, with a lovely beach and tennis courts.

48   LAKELAND BOATING · LAKE HURON CRUISING GUIDE

LAKE HURON

Rogers City, Michigan
A renovated harbor ready to welcome boaters.

With a new maritime museum and bicycle paths, plus plenty of shops within walking distance of the revamped harbor, Rogers City is awash with activities for visiting mariners. In the summer, anglers congregate on the new fishing pier and music lovers gather at the nearby bandshell for free concerts on Thursdays and Fridays. "The harbor is one of the most protected on Lake Huron, and other marinas have been patterned after ours," says Ken Rasche, harbormaster at **Rogers City Marina**. The facility, which was renovated in 1995, now boasts new showers, 123 slips, 600 feet of wall space, a larger parking lot and a new on-land fuel dock operational each year through December.

Lakeside Park, adjacent to the marina, is a pleasant spot with a good swimming beach and swim raft. The park also has a playground, a basketball court, an enclosed pavilion and a grill that serves ice cream and Greek food. The marina is connected to more than seven miles of trails and **P.H. Hoeft State Park**.

From the marina and park, the town of Rogers City is a short walk up a boulevard decorated with nautical flags. For a taste of history, walk a few blocks to the new **Great Lakes Lore Museum**, which documents the city's shipping past. Also be sure to visit the city's nearby **Presque Isle Historical Museum**, once the home of the Bradley family, which had deep roots in the limestone industry and helped found the city.

Though home to the world's largest limestone quarry, Rogers City is not a gritty industrial town. Instead, the town, which calls itself the Nautical City (as well as the Gateway to the North Channel and the Salmon

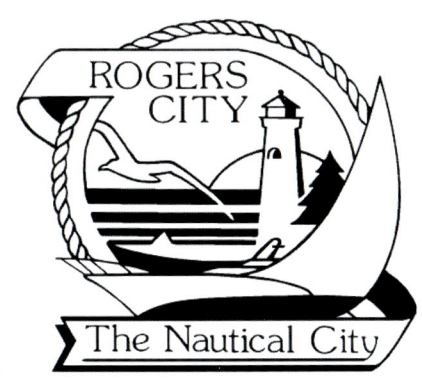

WELCOME BOATERS!
This Summer visit the ROGERS CITY HARBOR

45 25'48"N
83 48'46"W

**ROGERS CITY MICHIGAN "THE NAUTICAL CITY"
ON THE BEAUTIFUL SHORES OF LAKE HURON
FISHING & PLEASURE BOATERS - FAMILY FRIENDLY
U.S. CUSTOM'S PORT OF ENTRY**

AT THE ROGERS CITY HARBOR

Our Guests Enjoy:
- New Floating Docks
- New and Enlarged Comfort Station
- New Fuel Docks for Gas and Diesel
- New Electrical Service
  (120V.30 AMP., 120/240V .50 AMP)
- Six New Launch Ramps
- New Breakwalls

PLUS
- Beautiful new bandshell with Free concerts on Thursdays.
- Free summer music programs on Fridays
- Beautifully landscaped grounds

- Direct access from the harbor to our 6.2 mile biking trail.
- Adjacent city park and playground featuring a new wooden playscape.
- Boardwalk in the park

AND OF COURSE
- The same professional, accommodating and courteous service that our staff has provided for years!

WE LOOK FORWARD TO SEEING YOU!
Rogers City Chamber of Commerce 1-800-622-4148
ROGERS CITY MARINA 989-734-3808

www.rogerscity.com

Where to Eat

Black Bear Café. 989-734-2007
Family dining—breakfast, lunch and dinner—in the Mariner's Mall.

Chee Peng 989-734-2775
Delicious Chinese-Thai food.

Gaslight Restaurant & Lounge. 989-734-4531
A comfortable atmosphere and a full range of lake specialities.

Lighthouse Restaurant & Lounge. . . 989-734-4858
Features homemade soup and a salad bar. Open until midnight.

Nowicki's Sausage Shoppe 989-734-4100
Pizza, hot dogs, soups and subs. Best known for their homemade sausage.

Pavilion Grill. 989-734-4897
A pleasant setting overlooking the marina, providing snacks such as hamburgers, hot dogs and ice cream.

More Info

Emergency. 911
Chamber of commerce 800-622-4148
www.rogerscity.com
Presque Isle County Tourism Council. 888-854-9700
Medicine Shoppe Pharmacy 989-734-4701
Grulke Hardware. 989-734-2909

Capital), goes all out to welcome boaters.

The huge mountains of white limestone awaiting loading are a unique landmark that is clearly visible from any direction. The downward-facing white light on the east breakwater points to the entrance to the manmade harbor, which is at the south corner of the breakwalls. Thanks to dredging, at least 10 feet of water is available in the entry channel and 8 feet is available at the docks.

Rogers City is a U.S. Customs Port of Entry. The marina participates in the Michigan DNR's harbor reservation program, and its staff is trained to use a defibrillator.

Rogers City, which was surveyed under the command of William Rogers in 1868, developed relatively late compared with the other towns on this coast because of its inaccessibility. Although the city is named after Rogers, it was his partner Albert Molitor who opened a timber mill here and founded the village in 1872.

Originally, the city occupied two blocks on First Street, with a row of houses on either side of the road. A saloon was the first building east of Michigan Avenue and a courthouse was conveniently constructed next door. The railroad finally came to town in 1911, transforming the area into a popular summer resort as well as a year-round exporter of forest products and fish to city markets. But its heart and soul are built around the limestone quarries and the lakers that carry the stone to distant ports. Even **St. Ignatius Catholic Church** is built in the profile of a ship.

As experienced seafarers, the townspeople know firsthand the fury of Lake Huron storms. The loss of the *Carl Bradley* in 1958 and the *Cedarville* in 1965 are the most recent in a long line of ship disasters involving local crews.

Rogers City is not a bad place to be weathered in. There's a movie theater and a bowling alley a short walk from the harbor, as well as a library, which provides Internet connections. As for sightseeing, the vast open-pit quarry where limestone is blasted and loaded by conveyor onto waiting ships is worth a visit. A landmark of another kind is **Nowicki's Sausage Shoppe**. Now run by the fourth generation of the family, this store is renowned for its sausages—more than 50 varieties in all. In 1979, one of their wursts made the *Guinness Book of Records* as the longest wienie in the world. Laid out on tables at Lakeside Park, it stretched 8,773 feet and weighed about 3,500 pounds. It was sold, in manageable-sized hunks, in a single day. Another favorite shopping place is **Plath's Meat Market**, which was established in 1913 and has an excellent smoked pork. For the freshest fish, try **Gauthier & Spaulding Fisheries**.

For last-minute provisioning, **Glen's Market** is open 24/7 during the summer; you'll also find other grocery and convenience stores in town. Banks, a post office, a paperback bookstore and numerous other specialty stores are close by. The **Painted Lady** provides hours of browsing limited-edition prints, pottery, antiques and unusual gifts.

Rogers City is also home to several spectacular lighthouses, including **40-Mile Point** and two other nearby lights.

Plenty of summer activities highlight the season. The Nautical City Festival starts at the end of July and runs for an entire week. It is a cornucopia of festivities, ranging from bingo and slow-pitch tournaments to parades, airplane rides and lots of dancing under the big tent. It's followed by the three-day Salmon Tournament, with more than 300 boats participating. September's Posen Potato Festival concludes the summer's festivities.

Local events

June
Merchandise Fair

July
Nautical Festival
Concert Band
Salmon Tournament

Distances from Rogers City (statute miles)
Presque Isle Harbor: 18 SE
Alpena: 40 SE
DeTour Village: 40 N
Meldrum Bay, Ontario: 53 N

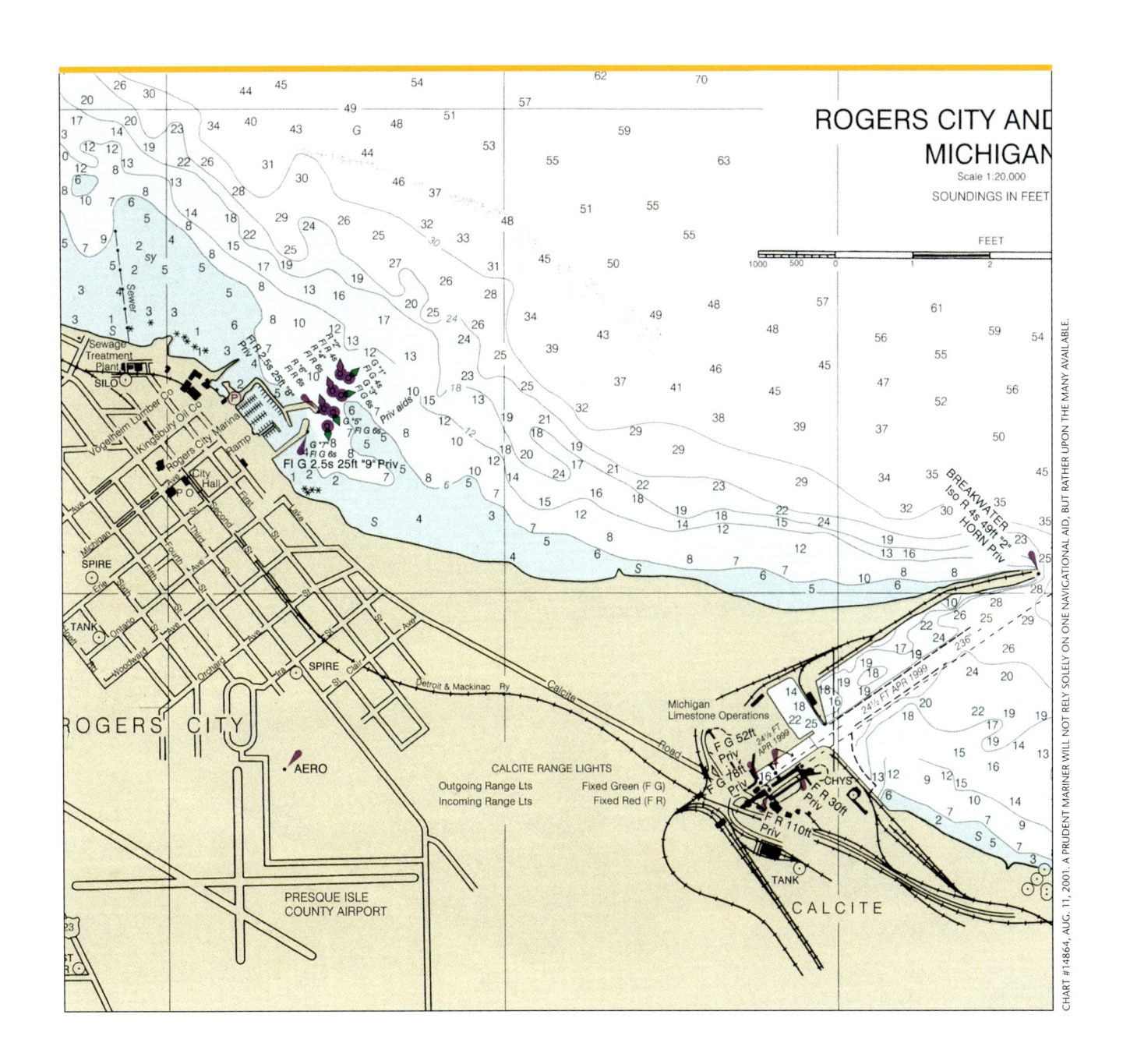

Facilities information subject to change. We suggest you call ahead.

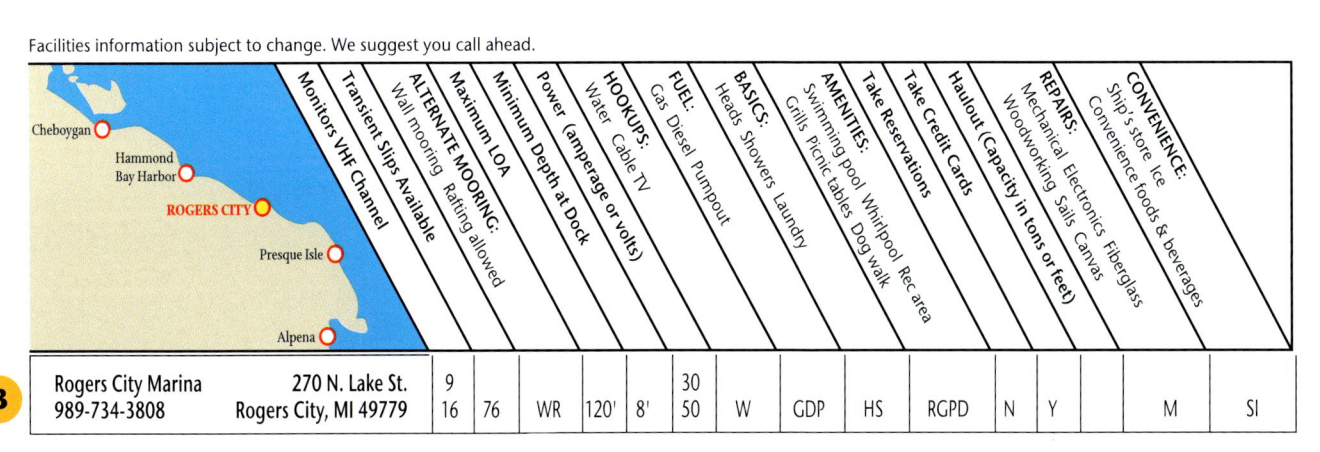

	Rogers City Marina 989-734-3808	270 N. Lake St. Rogers City, MI 49779	Monitors VHF Channel	Transient Slips Available	ALTERNATE MOORING: Wall mooring / Rafting allowed	Maximum LOA	Minimum Depth at Dock	Power (amperage or volts)	HOOKUPS: Water / Cable TV	FUEL: Gas / Diesel / Pumpout	BASICS: Heads / Showers / Laundry	AMENITIES: Swimming pool / Whirlpool / Rec area / Grills / Picnic tables / Dog walk	Take Reservations	Take Credit Cards	Haulout (Capacity in tons or feet)	REPAIRS: Mechanical / Electronics / Fiberglass / Woodworking / Sails / Canvas	CONVENIENCE: Ship's store / Ice / Convenience foods & beverages
B	Rogers City Marina 989-734-3808	270 N. Lake St. Rogers City, MI 49779	9 16	76	WR	120'	8'	30 50	W		GDP	HS	RGPD	N	Y		M	SI

Lakeland Boating · Lake Huron Cruising Guide 51

Presque Isle, Michigan

Lat 45°21.40
Lon 83°29.51

N

On approach from the north, look for the Presque Isle Light **A** on the north side of the peninsula. Round the buoy marking the 5-foot depth and set a course for 274°T to the harbor entrance. Approaching from the south, turn to port just south of the buoy and follow the same course to the opening in the breakwalls. Look for the old Presque Isle Light **B**, which is now a museum. The Presque Isle Marina **C** offers around 80 transient slips, repairs, a fuel dock with both gasoline and diesel **D**, and a launch ramp **E**. There's also a beach **F** for frolicking.

Distances from Presque Isle (statute miles)
Mackinac Island: 72 NW
Alpena: 42 S
Tobermory, Ontario: 84 E

LAKE HURON

Presque Isle, Michigan
A wild wonderland that doubles the fun for lighthouse fans.

The sole natural harbor on the Michigan shore of Lake Huron, **Presque Isle Marina** sits at the doorstep of a unique, unspoiled wilderness area. The full-service facility has plenty of room for transient boaters, and while shoreside services are minimal, the well-appointed **Portage Store & Deli** across the street stocks the staples: meat and seasonal vegetables, plus a well-stuffed deli counter and enough tackle to land a giant salmon.

For cruisers coming from the north, the **Presque Isle Lighthouse** at the north end of the peninsula is a highly visible landmark. From the south, the **Old Presque Isle Lighthouse**, now a museum, is a prime aid to navigation. The Presque Isle area abounds with shipwrecks—and for good reason. Be sure to stand well offshore to avoid outlying rocky shoals. When entering the harbor, note the charted 5-foot spot next to N "2."

The marina is located on the south side of the narrow neck of land that gave the area its name. In French, Presque Isle means "almost an island." Well-marked trails that twist through the woods and dunes make for a memorable hike, and naturalists will enjoy the Presque Isle area because of its unique plant life. The dwarf lake iris is a miniature flower that grows only along the shores of northern lakes Huron and Michigan. Its shoreline habitat has been diminished by development, and it is now a threatened species protected by federal and state laws. The iris' habitat is clearly marked behind the breakwall on the other side of the road. Pitcher's thistle, another rare species found in this area, has its habitat along the back edge of the beach. Wild fauna also abounds, including deer, bears, coyotes, foxes, bald eagles, loons and raccoons.

The Old Presque Isle Lighthouse, which became operational in 1840, is said to have been designed by Jefferson Davis, who later became president of the Confederacy. Financed by a $5,000 government appropriation and built in just two years, the structure features 4-foot-thick walls and a circular stairway carved from blocks of stone. The New Presque Isle Lighthouse, which dates to 1870, was constructed a mile north of the original structure so that it would be visible to ships from the south as well as the north. At 109 feet, it is the tallest light on the Great Lakes, and its original third-order Fresnel lens from Paris is still in place. The old lightkeeper's house, now a museum, features exhibits on 19th-century seafaring and lighthouse-keeping.

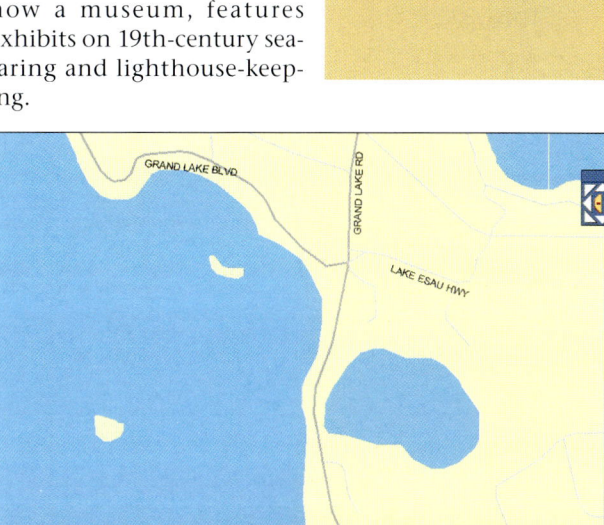

Local events

June
Wooden Boat Show

August
Lighthouse Arts and Crafts Show

More Info

Emergency	911
Alpena General Hospital	989-356-7390
Presque Isle Tourism	888-854-9700
	www.presqueislemi.com
New Presque Isle Lighthouse	989-595-9917
Old Presque Isle Lighthouse	989-595-6979

Where to Eat

The Fireside Inn **989-595-6369**
Built in 1909, when the area was emerging as a summer resort, the inn continues to greet visitors and serve delicious meals.

The Portage Restaurant **989-595-6559**
The fresh fish is mouth-watering. On the other side of the building is a pizza and ice cream shop. On Sundays the restaurant serves breakfast with homemade bread and thick slices of bacon.

LAKELAND BOATING · LAKE HURON CRUISING GUIDE

Another site worth visiting is the **Besser Natural Area**, a preserve dedicated to a primordial stand of white pine that somehow escaped the lumber barons' greedy gaze. A mile-long trail winds through the trees and past a lagoon that was once connected to Lake Huron. Look hard to catch the outline of a wrecked ship that once served the town of Bell.

Other reminders of Presque Isle's past include a cemetery with headstones dating back to 1850 and the **Kauffman Homestead Cabin**, which houses a craft shop. Just south of the boat harbor, a small stone marks the grave of Adeline Sims, a feisty midwife reportedly known for chewing tobacco and cursing like a lumberjack.

The **Presque Isle Harbor Wooden Boat Show** takes place in the middle of June. In recent years, the fascinating flotilla has included *Miss America IX*, the boat that made Gar Wood famous by setting a world record of 77.233 mph in 1930. The festival also includes a regatta of radio-controlled model boats and a street dance.

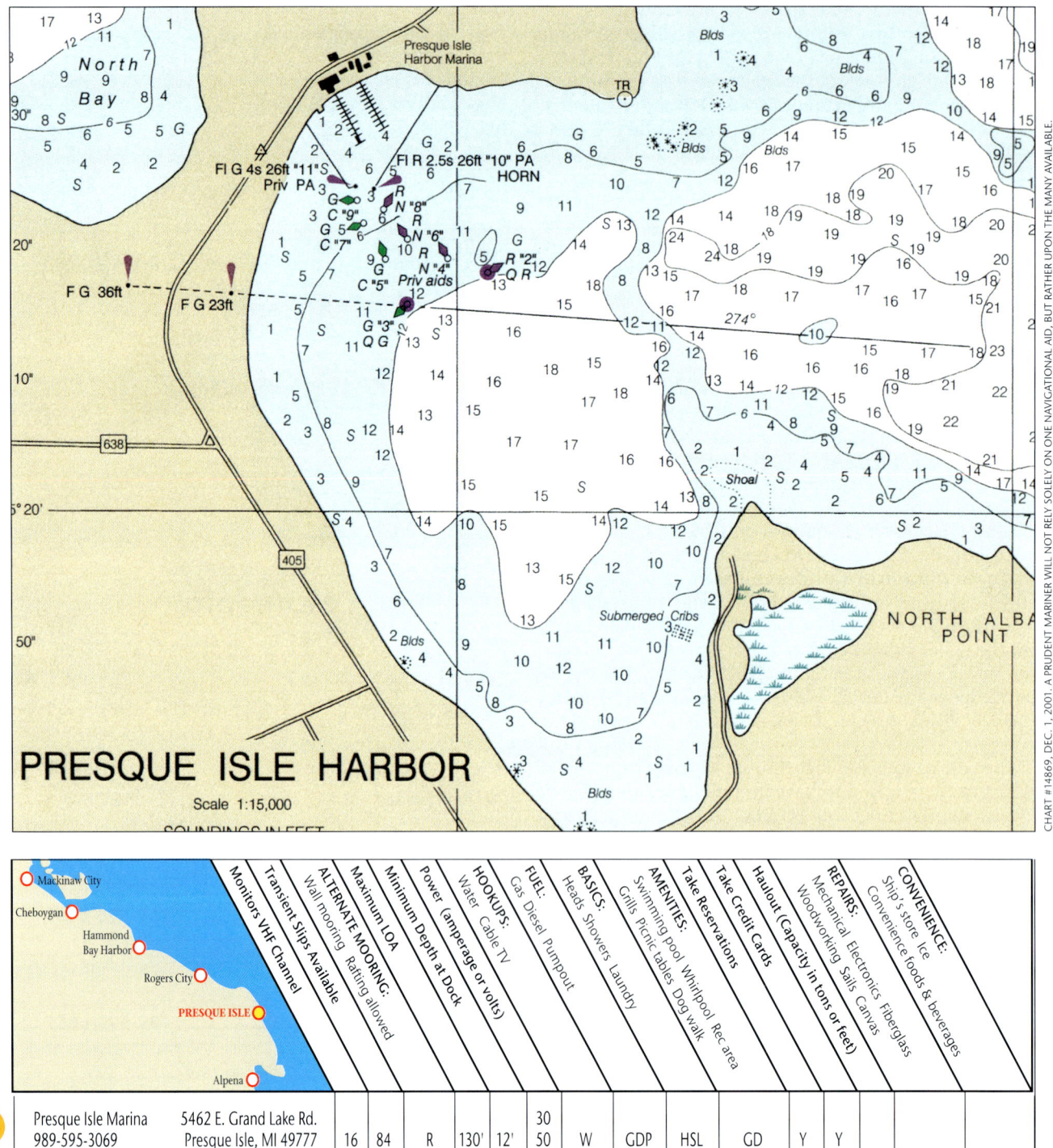

	Presque Isle Marina 989-595-3069	5462 E. Grand Lake Rd. Presque Isle, MI 49777	Monitors VHF Channel	Transient Slips Available	ALTERNATE MOORING: Wall mooring / Rafting allowed	Maximum LOA	Minimum Depth at Dock	Power (amperage or volts)	HOOKUPS: Water / Cable TV	FUEL: Gas / Diesel / Pumpout	BASICS: Heads / Showers / Laundry	AMENITIES: Swimming pool / Whirlpool / Rec area / Grills / picnic tables / Dog walk	Take Reservations	Take Credit Cards	Haulout (Capacity in tons or feet)	REPAIRS: Mechanical / Electronics / Fiberglass / Woodworking / Sails / Canvas	CONVENIENCE: Ship's store / Ice / Convenience foods & beverages
C	Presque Isle Marina 989-595-3069	5462 E. Grand Lake Rd. Presque Isle, MI 49777	16	84	R	130'	12'	30 50	W		GDP	HSL		GD	Y	Y

54 LAKELAND BOATING · LAKE HURON CRUISING GUIDE

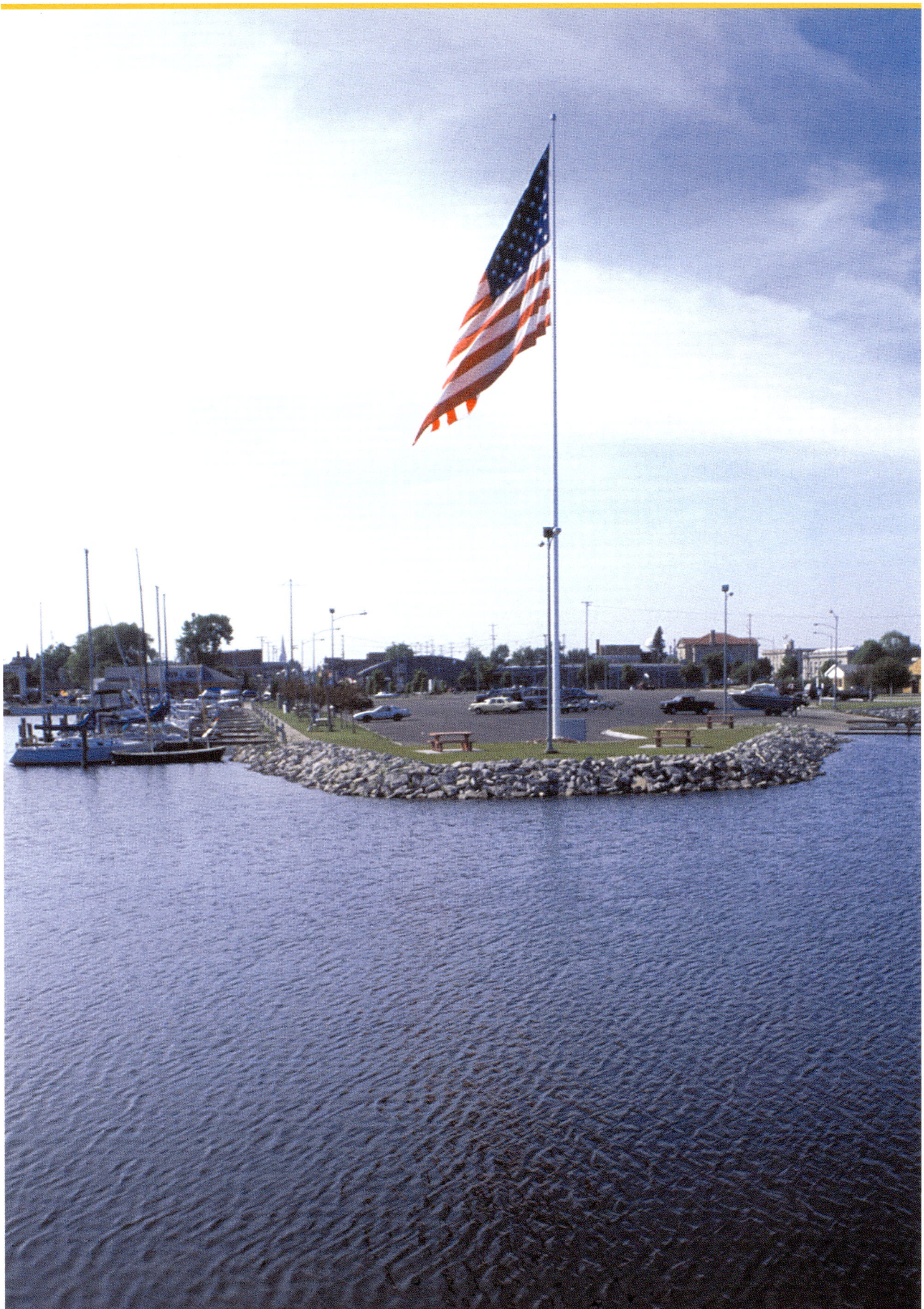

VITO PALISANO

Alpena, Michigan

Lat 45°03.62
Lon 83°25.39

If approaching from the north, head for the Thunder Bay Island Light, then proceed to the Thunder Bay traffic buoy due south of North Point. If approaching from the south, avoid underwater obstructions by keeping offshore and heading for the traffic buoy. Set a course for 304°T and travel nine miles to the entrance to the channel that leads to the river mouth. Alpena Light **A**, to starboard, marks the harbor entrance. Thunder Bay Shores Marine **B** occupies all of an L-shaped basin immediately to port. The facility offers seasonal and transient dockage and operates a fuel dock.

Distances from Alpena (statute miles)
Port Huron: 153 S
Harbor Beach: 68 S
Presque Isle Harbor: 42 N
Cheboygan: 85 NW
Harrisville: 33 S

Facilities information subject to change. We suggest you call ahead.

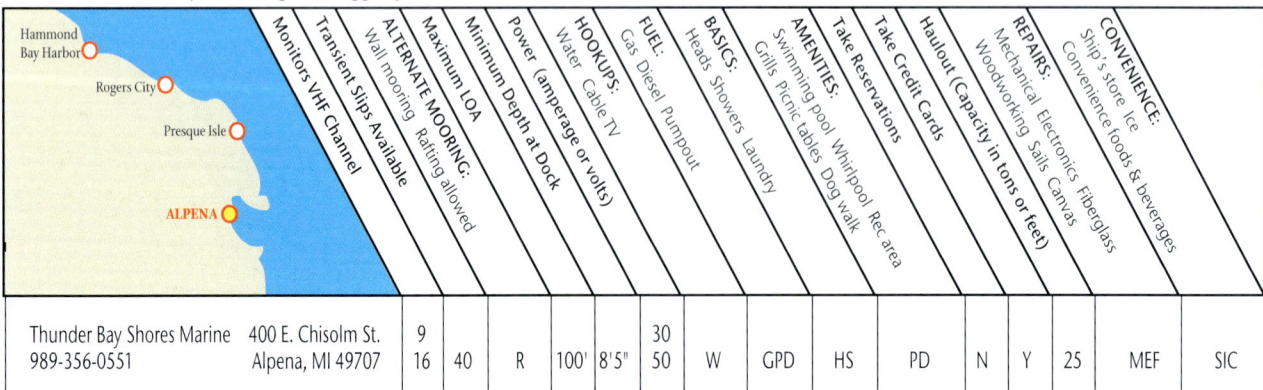

		Monitors VHF Channel	Transient Slips Available	ALTERNATE MOORING: Wall mooring / Rafting allowed	Maximum LOA	Minimum Depth at Dock	HOOKUPS: Power (amperage or volts) / Water / Cable TV	FUEL: Gas / Diesel / Pumpout	BASICS: Heads / Showers / Laundry	AMENITIES: Swimming pool / Whirlpool / Rec area / Grills / Picnic tables / Dog walk	Take Reservations	Take Credit Cards	Haulout (Capacity in tons or feet)	REPAIRS: Mechanical / Electronics / Fiberglass / Woodworking / Sails / Canvas	CONVENIENCE: Ship's store / Ice / Convenience foods & beverages
B	Thunder Bay Shores Marine 400 E. Chisolm St. 989-356-0551 Alpena, MI 49707	9 16	40	R	100'	8'5"	30 50 W	GPD	HS	PD	N	Y	25	MEF	SIC

56 LAKELAND BOATING · LAKE HURON CRUISING GUIDE

LAKE HURON

Alpena, Michigan
Shipwrecks abound near this historic port.

Not merely a stopover for supplies and marine services, Alpena boasts an attractive harbor, a charming downtown with a restored historic district and plenty of diversions. And while it's the largest city in northeastern Michigan, where the grand old homes of the lumber barons still gaze out over the lake, it has nevertheless retained a small-town feel.

Before the arrival of the first Europeans, this part of Michigan was a heavily timbered wilderness inhabited by members of the Ottawa and Chippewa tribes. In 1826, the region was ceded to the United States. Originally called Animickee for the Chippewa chief who signed the secession treaty, the town was later named Alpena, which means "excellent partridge land" in Algonquin.

The first lumber mill was built here in the 1840s, and by the end of the century Alpena was a bustling hub of schooners carrying white pine, as well as steamers working the Thunder Bay fishery. By the first decades of the 20th century, both the forests and the fishery had been depleted, and Alpena might have faded into insignificance if not for an enterprising grocer named John Monaghan, who discovered the richest lode of limestone on the continent under the vanished forest. Today, Alpena is a major producer of cement, and the towering stacks make an easy landmark for cruisers.

When making your approach, be sure to study the chart carefully: the area is studded with reefs and shoals. From the north, set a course to **Thunder Bay Island Light**. From there, it's just under six miles to the Thunder Bay traffic buoy, which is due south of North Point. Coming from the south, it's also safest to make for the traffic buoy before heading into the harbor. The entrance is located nine miles from the buoy on a course of 304°T.

The buoyed and lighted entrance markers lead to the commercial wharves in the Thunder Bay River. **Thunder Bay Shores Marine** lies south of the river entrance, marked by a red tower with a flashing light. While older charts show the entrance as an opening in the south breakwall, more recent construction has relocated the entrance to just inside the river mouth, immediately to port. The marina is also home to a city program that lets visitors check out bikes for free. The **Alpena Yacht Club**, which is also headquartered at the marina, offers privileges to members of affiliated clubs.

Lake Huron's unpredictable weather, its murky fog banks and sudden gales, coupled with rocky shoals, have earned Thunder Bay the unfortunate appellation of Shipwreck Alley. Scores of vessels have ended their careers on the lake floor off Alpena. Thunder Bay's dense population of wrecks—from wooden schooners to sidewheel steamers to modern freighters—prompted its designation as a **National Marine Sanctuary and Underwater Preserve** in 2000. For shipwreck buffs, **R.J. Dives Great Lakes Charters**, located at the harbor, offers glass-bottom boat tours. Visitors can see two shipwrecks on an average trip. In addition, the company also offers scuba diving charters and island tours. Cruisers should not attempt to make the dangerously shallow passage between the three small islands to the north: Thunder Bay, Sugar and Gull.

Local charter companies offer tours of the **Middle Island Light Station,** which include a nature walk and an opportunity to see the restoration in progress. Visitors can even spend the night at the lightkeeper's lodge.

Despite its industrial nature, Alpena is a thoroughly inviting stopover. The spacious harbor is adjacent to **Bay View Park**, which features basketball and tennis courts, a playground, a picnic area and free concerts throughout the sum-

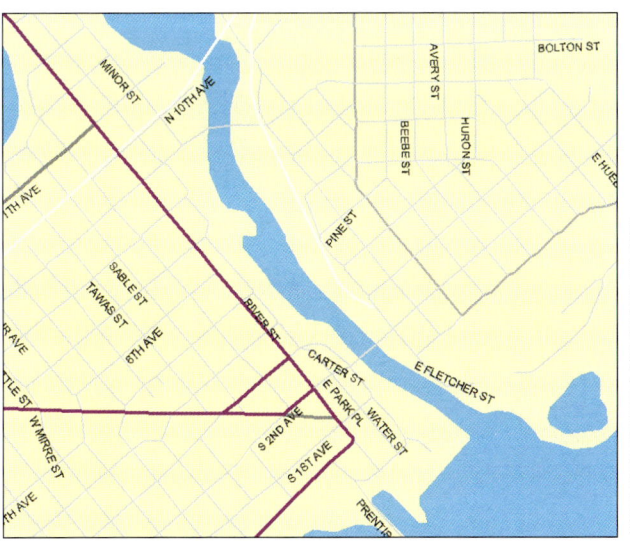

Where to Eat

Jeppettos 989-354-8190
Located on the water at the site of the old Thunderbird Inn. A great view and tasty food.

John A. Lau Saloon 989-354-6898
This historic saloon echoes Alpena's lumbering era of the 1880s. Featuring a microbrewery, great steaks, and an outdoor beer garden.

LAKELAND BOATING · LAKE HURON CRUISING GUIDE 57

Local events

June
Alpena Riverfest

Thunder Bay
Polka Explosion

Jesse Besser Museum
Log Cabin Day

July
Michigan Brown Trout Festival

July-August
Friday Night! Downtown!

August
Alpena County Fair

Antique Tractor and Steam Engine Show

mer. Bay View is one of a dozen city parks and beaches located on a portion of the city's waterfowl sanctuary.

Alpena's historic shopping district boasts the former Sepull's Pharmacy, founded in 1920. Now the **Country Cupboard**, the store features the original tin ceiling and ceiling-high shelves along with antiques and collectables. If you're looking for a little culture, the **Thunder Bay Theater** hosts musicals, comedies and dramas. Also be sure to check out the **John A. Lau Saloon**; **Sweet Tooth's**, a gourmet sweet shop; and **Rudolph's**, a year-round Christmas shop. Also worth a visit is the **Jesse Besser Museum**, which features exhibits on Native Americans, early local industry and 19th-century historic buildings and hosts Sunday planetary shows.

New to downtown is **Art in the Loft**. Located on the top floor of the historic Center Building, the gallery features paintings, photography and pottery. The town also has added Friday Night! Downtown! events. Held in July and August, these family-oriented festivities offer music, entertainment and food.

More Info

Emergency 911
Alpena General Hospital. 989-356-7390
Alpena Area Convention
and Visitors Bureau 800-4-ALPENA
www.alpena.mi.us
Thunder Bay Dive Center 989-356-9336
J&S Cab Company 989-354-4601
Gambles Hardware Store 989-365-6356
Kmart 989-354-5569
R.J. Dives Great Lakes Charters 989-657-DIVE

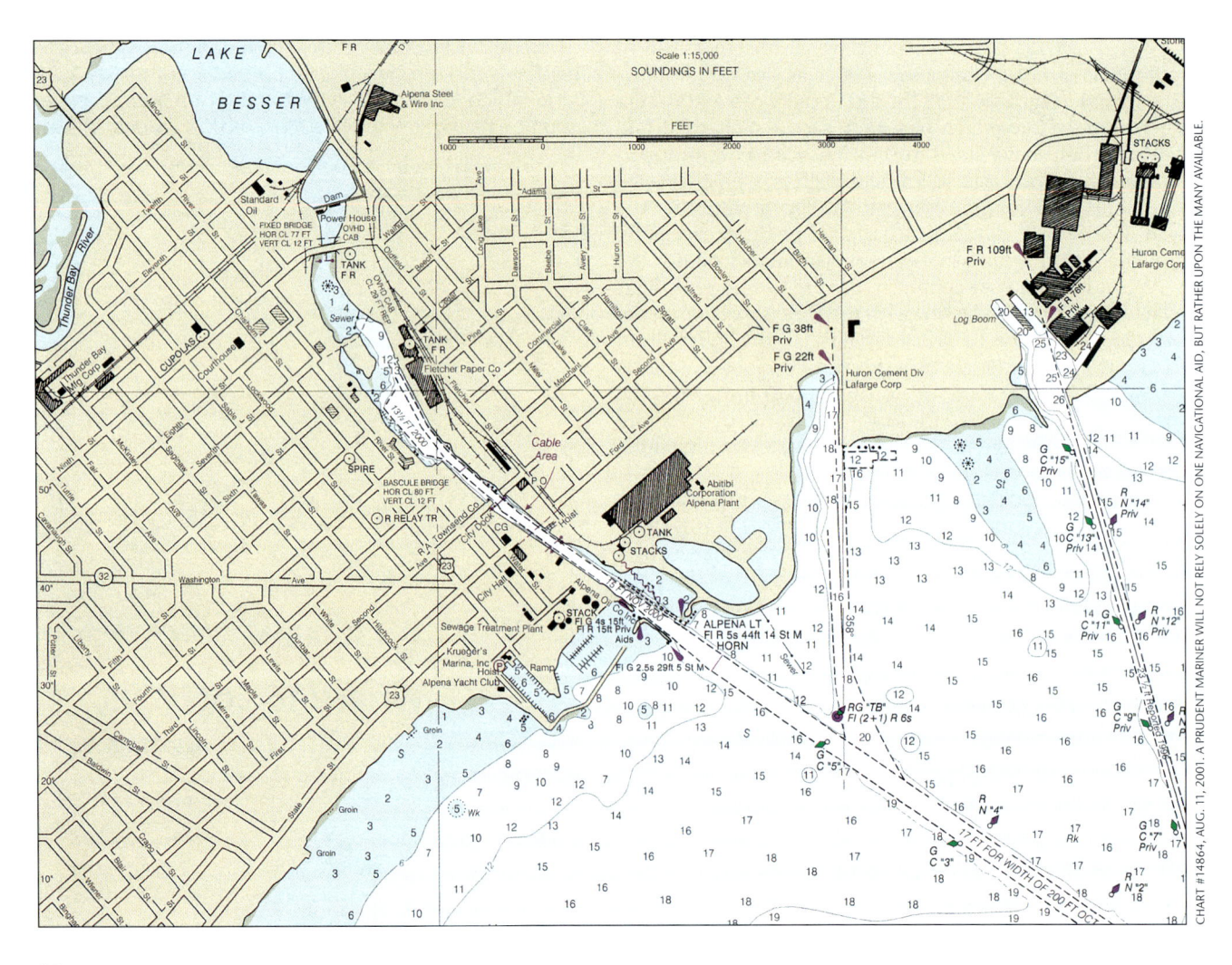

58 LAKELAND BOATING · LAKE HURON CRUISING GUIDE

DIVE RIGHT IN

Sunken Treasures
Take the plunge and visit 10 Thunder Bay shipwrecks.

Underwater historians, divers and ghoulish gawkers can learn all about Thunder Bay's maritime disasters in Frederick Stonehouse's *A Short Guide to the Shipwrecks of Thunder Bay.*

1 *Nordmeer*
On November 19, 1966, the 470-foot freighter *Nordmeer,* proceeding upbound with $1 million worth of cargo, abruptly turned inside a buoy marking Thunder Bay Shoal. With her steel hull torn, she quickly filled and settled on the shallow bottom seven miles from the Thunder Bay Island Light. Part of the wreck is still above water, and diving portions are generally 30 to 40 feet underwater. 45°08.08N/W83°09.35W

2 *Lucinda Van Valkenburg*
The *Valkenburg* collided with the propeller *Lehigh* on June 1, 1887, in heavy fog. The only one hurt was the cook, who was injured slightly—though the owner must have suffered a headache when he realized the 128-foot schooner was not insured. Resting 70 feet below, about two miles from Thunder Bay Light, the centerboard still stands, and the hull is partly collapsed. 45°03.26N/83°10.03W

3 *D.R. Hanna*
The 532-foot steel steamer encountered another steamer, the *Quincy A. Shaw,* on May 16, 1919, in calm seas. As they were about to pass, the *Shaw* gave a single blast on her whistle—signaling a port-to-port passage—and began to cut across the bow of the *Hanna,* which was carrying $2 million in cargo. Too late to reverse, the two collided and the *Hanna* slowly filled by the bow, eventually rolling and turning turtle; the Shaw suffered comparatively minor damage. The *Hanna* now rests upside-down 135 feet below and about six miles from Thunder Bay Island Light. 45°04.95N/83°65.05W

4 *Monohansett*
On November 23, 1907, flames swept through the old 165-foot wooden steamer. Fire pumps doused her from midnight until 9 a.m., but after the fire reignited that afternoon, the steamer ended up on the bottom, about 20 feet below near the dock on the southwest coast of Thunder Bay Island. The wreckage is broken in three large pieces, and parts of the boat's sides and much of the machinery are still present. 45°01.89N/83°11.86W

5 *P.H. Birckhead*
Consumed in flames, the 156-foot wooden steam barge drifted clear of the dock and sank in shallow water on September 30, 1905. The badly broken wreck can be found 12 feet down, less than a mile from the south breakwall of the Thunder Bay River. 45°02.98N/83°25.87W

6 *Oscar T. Flint*
On November 25, 1909, the captain of the 218-foot wooden steamer awoke to find thick smoke pouring into his cabin. He escaped the roaring flames with only the clothes on his back and a large fur coat. A surprisingly large part of the hull is intact, and the limestone cargo is still evident. Marine life has settled into the wreckage, about four miles from the mouth of the Thunder Bay River. 45°01.45N/83°20.67W

7 *Barge No. 1*
The barge reportedly broke in two under the strain of rolling seas on November 8, 1918. The cargo—200 crates of live chickens—washed ashore on the beaches of Thunder Bay, prompting the Coast Guard to throw a party in honor of the "survivors." The wreckage, located almost seven miles from Thunder Bay Marina Harbor, still has its side walls and planking present 45 feet below. 45°00.84N/83°18.08W

8 *Grecian*
On June 7, 1906, the 296-foot *Grecian* ran ashore in thick fog. Her badly cut bottom was patched so she could be towed to Detroit, but she ran into more rough weather, filled and settled 100 feet below. She was broken amidships with her forward deck collapsed. The three stern levels are intact and penetrable, allowing divers to view the machinery and small artifacts. 44°57.99N/83°11.83W

9 *Monrovia*
Due to a combination of miscommunication and inexperience, the 536-foot, 12,000-ton *Royalton* smashed into the 430-foot, 6,700-ton *Monrovia*'s port side as they crossed paths on June 25, 1959, in thick fog. The *Monrovia* immediately began to take in water and came to rest 140 feet below, upright and largely intact, about 13 miles from Thunder Bay Island. 44°58.88N/82°55.28W

10 *Nellie Gardner*
Experts think that this, the oldest of the bunch, is the wreck of the *Nellie Gardner,* a three-masted, 565-ton schooner lost in 1883. The wreck is located 18 feet below, its sidewalls and part of the starboard rail still intact. 44°53.67N/83°19.46W

Harrisville, Michigan

Lat 44°39.68
Lon 83°16.91

Whether you are approaching from the north or south, keep at least two miles offshore to avoid shoals, especially around Sturgeon Point. The southern entrance to the harbor, shown on some older charts, no longer exists. Enter from the east, rounding the northern tip of the main breakwall, which is clearly marked with a flashing green light **A**. The tip of the western breakwall **B** carries a flashing red light. The Harrisville Harbor of Refuge **C** is a sizable, full-service marina that can accommodate up to 75 vessels. Fuel and repairs are available. The protected area near the tip of the piers is a popular anchorage.

Distances from Harrisville (statute miles)
Alpena: 33 N
Tobermory, Ontario: 94 NE
Oscoda/Au Sable: 17 S
East Tawas: 34 S
Bay City: 86 S

Lake Huron

Harrisville, Michigan
Sandy beaches and great salmon fishing await.

With a population of about 550 in summer, this town is charmingly cozy. Unlike many of Michigan's small coastal towns, Harrisville lacks the scars typical of a lumbering past. The shady streets and older homes are reminiscent of gentler days, and the town has a full range of shops and services. As you stroll through town, be sure to stop at **Daisy's**, a shop filled with antiques and collectibles that also features delicious homemade treats, including fudge. The newer **Harbortown Market** on Main Street houses a variety of shops specializing in art, quilts, jewelry and other gifts.

The entrance to the **Harrisville Harbor of Refuge**, on the north side of the breakwall, is not readily visible to yachts making landfall. The south harbor entrance has been closed in. Well protected from all directions, the harbor has plenty of room to anchor out. Mooring buoys are assigned by the harbormaster, and up to 75 spaces are available for transient use. Gas and diesel are sold here, and pumpout is available. The grounds are spacious and well-maintained, with grills and tables for cooking out, a playground and a sandy beach for swimming. In season, chinook salmon weighing up to 30 pounds have been caught inside the harbor.

Approaching from the north, vessels should steer clear of the shoals off Sturgeon Point: the reefs ringing Harrisville have sunk their share of ships through the years. In 1866, Sturgeon Point, named for the large number of the caviar-supplying fish that once spawned there, was chosen as the location for a lighthouse. Construction of the tower was completed in 1869, with the light becoming operational in 1870. To this day, captains rely on the light to guide their course. In 1876 the life-saving station was built south of the tower.

HARRISVILLE CHAMBER OF COMMERCE

Local events

May
Keerl's Korner Fishing Tournament

July
Community-wide Garage & Sidewalk Sale

Depot Days

Lincoln Lions Tournament for Salmon & Lake Trout

Kids' Fishing Day

July
Craft Show at the Craftmaker's Cabin

4th of July Celebration

Duck Race

Antique and Collectibles Show

50s Dance

Sand-Sculpting Contest

October
Lighthouse Festival

The **Sturgeon Point Lighthouse**, lightkeeper's quarters and tower survived through the years and have been restored. The lightkeeper's house is now a museum, open to the public from Memorial Day weekend through the Lighthouse Festival on the first weekend in October. There's a gift shop, too.

The **Old Bailey School**, an original one-room schoolhouse, is a reminder of simpler times. Built of Norway pine in 1907, the school is still furnished with items used during that time. It served the community until the 1940-41 school year.

Also worth a side trip is **Cedarbrook Trout Pond**, located two and a half miles north of Harrisville near the light. The state's first licensed trout farm, it was built in 1950 to take advantage of the cold, highly oxygenated water that flows from springs on the sandy bluffs. While most of the thousands of trout raised here are shipped off to restock lakes and streams, Cedar Brook does maintain two ponds for fishing and will even supply tackle.

Where to Eat

Coffee Talk Café 989-724-6236
This downtown spot serves a light breakfast; lunch options include rollups, Mexican grilled cheese, caesar salad and pastries. Dine inside or out.

Harbortown Pizza 989-724-5000
Grab a tasty slice or two just a few blocks west of the marina.

Old Place Inn 989-724-6700
Lunch and dinner daily with homemade soups, specials and a full menu and bar.

More Info

Emergency 911
Alcona Health Center 989-724-5655
Alpena General Hospital 989-356-7390
Chamber of Commerce 800-432-2823
www.huronshoreschamber.com
Harrisville State Park 989-724-5126
Richards Pharmacy 989-724-5178
Sturgeon Point Lighthouse Museum 989-724-6297

62 LAKELAND BOATING · LAKE HURON CRUISING GUIDE

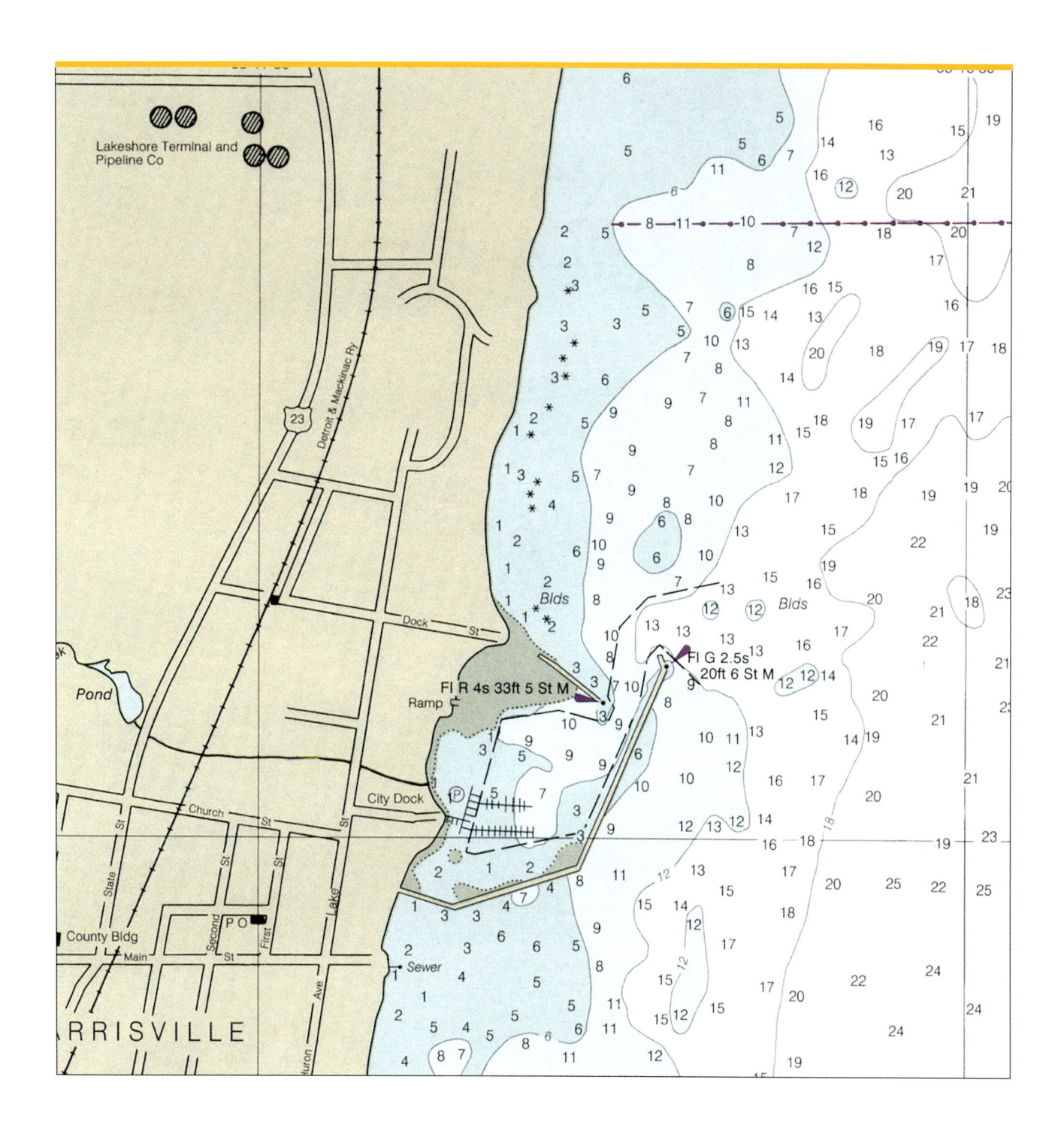

Facilities information subject to change. We suggest you call ahead.

	Monitors VHF Channel	Transient Slips Available	ALTERNATE MOORING: Wall mooring / Rafting allowed	Maximum LOA	Minimum Depth at Dock	Power (amperage or volts)	HOOKUPS: Water / Cable TV	FUEL: Gas / Diesel / Pumpout	BASICS: Heads / Showers / Laundry	AMENITIES: Swimming pool / Whirlpool / Rec area / Grills / Picnic tables / Dog walk	Take Reservations	Take Credit Cards	Haulout (Capacity in tons or feet)	REPAIRS: Mechanical / Electronics / Fiberglass / Woodworking / Sails / Canvas	CONVENIENCE: Ship's store / Ice / Convenience foods & beverages
Harrisville Harbor of Refuge 1 Harbor Dr. 989-724-5242 Harrisville, MI 48740	9 16	75	WR	75'	5'	30 50	W	GDP	HS	RGPD	N	Y		ME	I

LAKELAND BOATING · LAKE HURON CRUISING GUIDE 63

Oscoda & Au Sable, Michigan

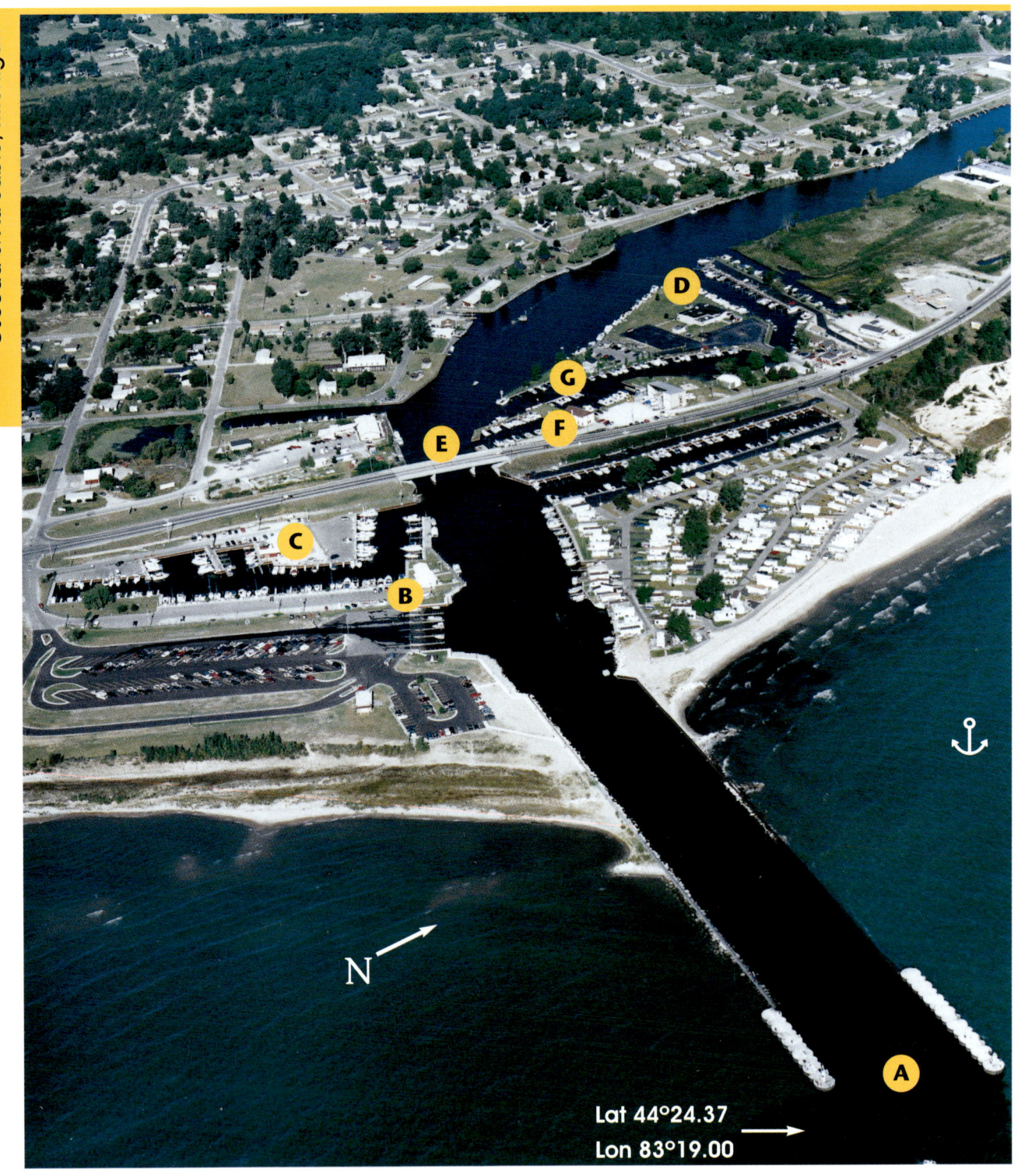

Lat 44°24.37
Lon 83°19.00

As you approach Oscoda, keep at least two miles offshore to avoid hazards. Make the turn to enter the harbor from due east. The harbor entrance **A** is between two parallel piers carrying flashing red and green lights. The Main Pier Marina Association **B** operates about 30 transient slips on the south bank just inside the harbor. Bunyan Town Marina **C** offers dockage but limited services. The Oscoda Yacht Club **D** occupies a gray building to starboard past the highway bridge **E** but does not cater to transients. The full-service Northeast Michigan Marine **F**, on the north side of the river, can handle transients up to 50 feet. Next door, Fellows Marina **G** offers dockage and a full range of services.

LAKELAND BOATING · LAKE HURON CRUISING GUIDE

LAKE HURON

Oscoda & Au Sable, Michigan
Sister cities on a famed trout-fishing river.

On opposite banks of the Au Sable River, Oscoda and Au Sable offer excellent fishing, hunting, canoeing and world-class golf. Though legally they are distinct entities, in terms of geography and spirit, these pleasant resort towns are inseparable.

The Au Sable was once a major logging highway, carrying logs down to the waiting mills. In a 30-year period, enough lumber was cut here to encircle the globe 20 times. Today, thanks to an enormous reforestation effort, the river flows through 427,000 acres of the **Huron National Forest**, established in 1909. The towering statue of Paul Bunyan that stands in an Oscoda park commemorates the area's heritage as timber country. Bunyan's tall tales were written down by Oscoda newspaperman James MacGillivray in 1906. MacGillivray's home still stands.

Where to Eat

Au Sable Inn & Finish Line Sports Bar 989-747-0305
Great food and spirits.

Charbonneau's 989-739-5230
This friendly food emporium is determined to see that no one goes home hungry. Located to port on the upriver side of the highway bridge, there is docking and a deck for alfresco dining.

The Party & Food Center 989-739-2091
This shop will deliver their special Philadelphia-style cheese steaks and hoagies, as well as pizza.

Wiltse's Brew Pub & Family Restaurant 989-739-2231
The onsite brewery produces a variety of ales and beers, as well as Lumberjack root beer. Steaks and ribs are a house specialty.

More Info

Emergency 911
Hospital 989-362-3411
Chamber of Commerce 989-739-7322
www.oscoda.com
Iosco Transit 989-362-8101
Ace Hardware 989-739-2041
Gilbert's Drug Store 989-739-7585

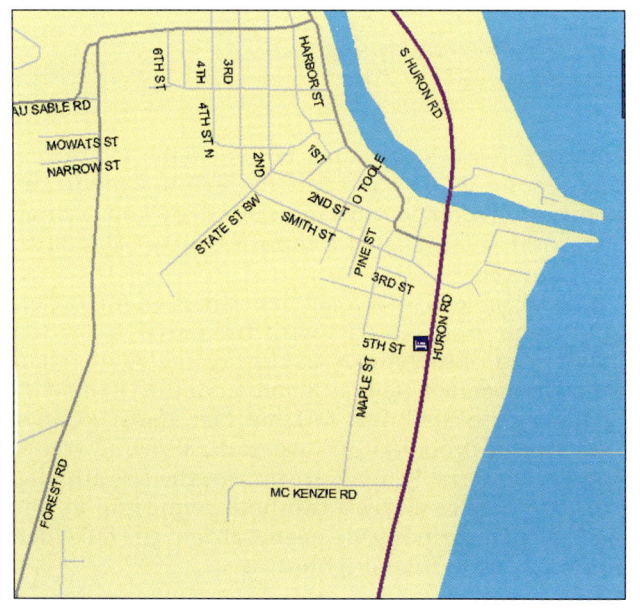

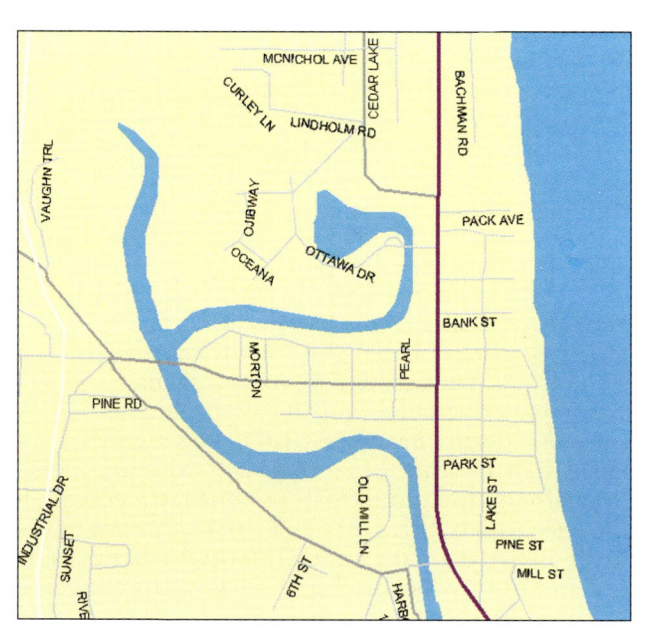

LAKELAND BOATING · LAKE HURON CRUISING GUIDE

Local events

May
Sunrise Concert Series

June
Art on the Beach

July
July 4th Parade & Fireworks

GLPAA Regional Fishing Tournament

Concerts on the Beach

Annual Au Sable River International Canoe Marathon

Gagaguwon Traditional Powwow

August
Concerts in the Park

Au Sable River Canoe & Clean

On July 11, 1911, a fire wiped out what was left of the area's forests. The lumberjacks moved out and the farmers who took their place soon discovered that the soil around Saginaw Bay was incredibly fertile.

For the cruiser heading south, the towns mark the division between Saginaw Bay and Lake Huron; the official demarcation is Au Sable Point, located about five miles to the south. Au Sable has little to offer in the way of shops, but Oscoda can be reached by taxi or dinghy. Head upriver past the second highway bridge and land on the starboard bank.

The entrance to the river is marked by two water towers and two parallel piers with flashing red and green lights mounted on towers at the entrance. Below the first highway bridge (with 23-foot clearance) on the south bank is the **Main Pier Marina Association**, which can accommodate sail- as well as powerboats, with 6 to 8 feet of water at the dock. Next door is **Bunyan Town Marina**, a fishing tackle store in a large red building with docking.

On the far side of the bridge are two marinas and the **Oscoda Yacht Club**, an attractive one-story gray-sided building on the starboard bank of the river past the highway bridge. A spacious lawn with picnic tables flanks the river. **Northeast Michigan Marine**, a full-service facility on the north side of the river, can handle transients up to 50 feet in length. Next door is **Fellows Marina,** which offers picnic tables and gas grills, as well as a fish-cleaning and -freezing service. **Fellows Bait & Tackle Shop** is the best-stocked tackle store within 50 miles. There's plenty of live bait, along with rod and reel service and daily fishing reports.

In Oscoda, two antique shops, **Hobart's Antiques** and **Wooden Nickel Antiques**, are worth a visit. North of Oscoda in nearby Greenbush is the **Great Lakes Marine Art Gallery**, which offers prints and paintings of ships, as well as half-hull models. Oscoda has a scenic public park and beach, volleyball and basketball courts and playground equipment for youngsters.

The **Art on the Beach Festival** has been growing for more than 15 years. More than 200 exhibitors come from around the United States on the last weekend in June to display their arts and crafts. The **Au Sable River International Canoe Marathon** takes place at the end of July, beginning at dusk in Grayling and ending 120 miles and some 14 hours later in Oscoda. The second leg in the canoeing Triple Crown, the event brings competitors and spectators from around the world and is the occasion for the Au Sable River Days festival.

The Au Sable River is a major destination for anglers who come to match wits with the native trout and coho. A list of the numerous fishing guides and charters that operate here is available from the chamber of commerce, located next to Main Pier Marina. Canoe liveries will drop you off upstream, allowing you to paddle comfortably downstream. For the less athletic, a paddlewheeler makes a two-hour round trip up the Au Sable. This trip is truly memorable in the fall when the leaves are at their brightest.

A Very Tall Tale

Long before tall-tale-teller James MacGillivray began pecking away on his typewriter, lumberjacks had been spinning yarns about a clever, ax-wielding giant as they sat around fires and pot-bellied stoves.

The mythic Paul Bunyan is said to have been born in Maine to parents of average size. Within a week, the child weighed 100 pounds and sported a curly black beard. Not knowing what to do with their behemoth son—who caused earthquakes when he walked and tidal waves when he waded—the Bunyans sent Paul off to an uninhabited part of the woods.

When Paul grew up, the loneliness of his isolation haunted him. But he found a lifelong companion during a particularly violent storm in the Winter of the Blue Snow. When Paul heard a muted mooing coming from the depths of a snowdrift, he reached into the densely packed drift and pulled out a mammoth baby ox that was tinted the same blue as the snow. Paul named his new friend Babe the Blue Ox. Legend has it that Babe weighed more than the combined weight of all the fish that ever got away.

Paul invented logging in response to the people's need for wood to build churches, ships, bridges and pencils. He could fell hundreds of trees with one swoop of his ax. Paul started a logging camp, enlisting 1,000 of the brawniest lumberjacks around. When drinking water for the workers became scarce, Paul and Babe dug ponds to quench the men's thirst. Today, we call these ponds the Great Lakes.

Distances from Oscoda/Au Sable (statute miles)
Presque Isle Harbor: 92 N
Harrisville: 17 N
East Tawas/Tawas City: 17 S
Port Austin: 30 SE

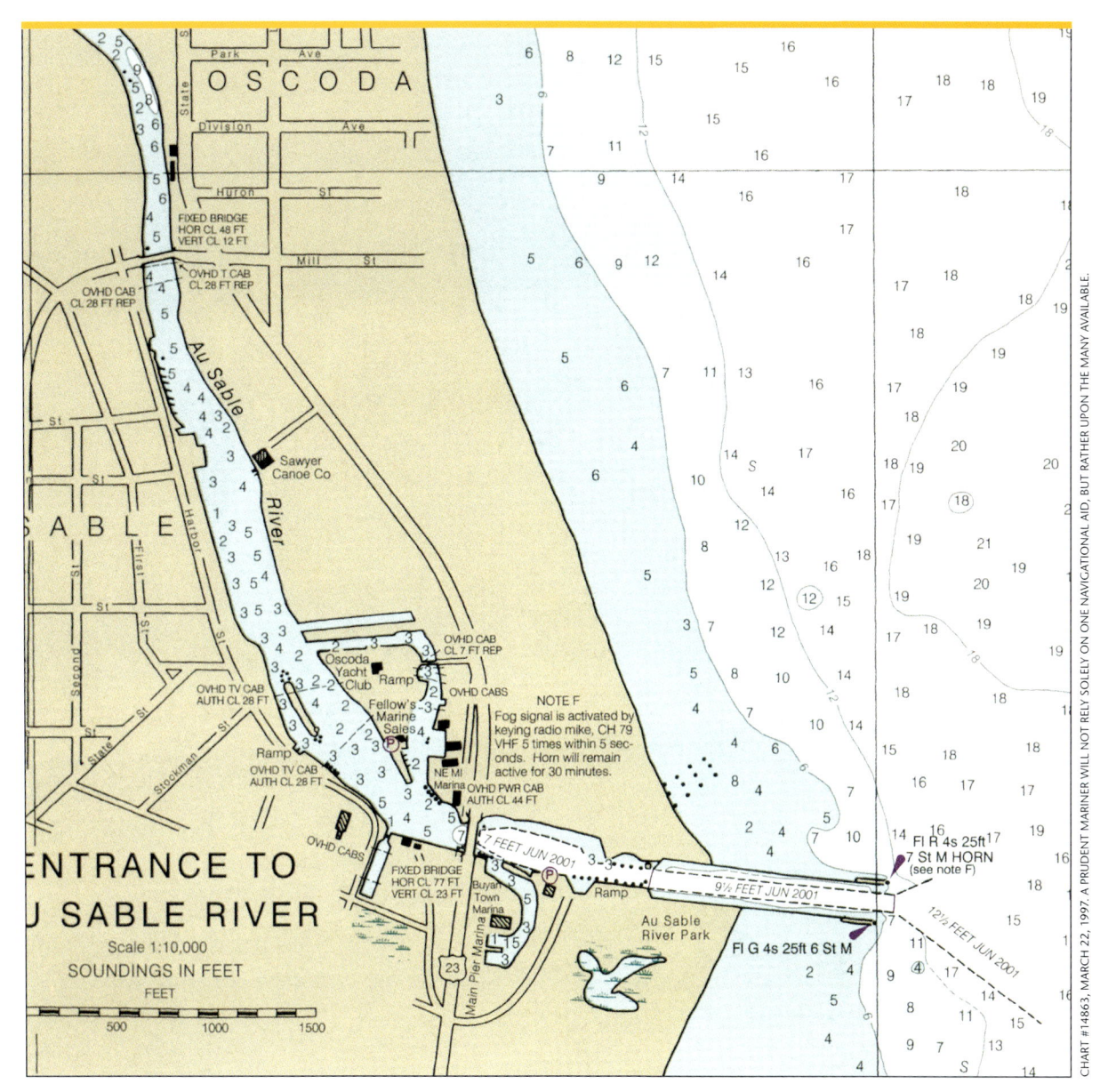

Facilities information subject to change. We suggest you call ahead.

			Monitors VHF Channel	Transient Slips Available	ALTERNATE MOORING: Wall mooring, Rafting allowed	Maximum LOA	Minimum Depth at Dock	Power (amperage or volts)	HOOKUPS: Water, Cable TV	FUEL: Gas, Diesel, Pumpout	BASICS: Heads, Showers, Laundry	AMENITIES: Swimming pool, Whirlpool, Rec. area, Grills, Picnic tables, Dog walk	Take Reservations	Take Credit Cards	Haulout (Capacity in tons or feet)	REPAIRS: Mechanical, Electronics, Fiberglass, Woodworking, Sails, Canvas	CONVENIENCE: Ship's store, Ice, Convenience foods & beverages
C	Bunyan Town Marina 989-739-2371	4494 North U.S. 23 Oscoda, MI 48750			3		W	2'6"			H		P	Y	Y		IC
G	Fellows Marina 989-739-2744	440 S. State St. Oscoda, MI 48750	16	7	WR	60'	4'	30	W	GDP	HSL	GPD	Y	Y	10	MEFC	SIC
B	Main Pier Marina Association 989-739-8530	4498 North U.S. 23 Oscoda, MI 48750	16	30	Y	65'	6'	30 50	W		HSL	RGPD					I
F	Northeast Michigan Marine 989-739-4411	470 S. State St. Oscoda, MI 48750	9,16 68,72	10	WR	50'	5'	30	W	GDP	HS	GPD	Y	Y	15	MF	SI

LAKELAND BOATING · LAKE HURON CRUISING GUIDE **67**

East Tawas & Tawas City, Michigan

This view looks northeast along the Michigan coast. Tawas Bay Condominium Marina **A** is a private facility that usually has slips for transients. The East Tawas State Dock **B**, Jerry's Marina, Inc. **C** and the Tawas Bay Yacht Club **D** are on Tawas Point, as is the Tawas Bay Lighthouse **E**. (See close-up photos on the following pages.) The area sheltered by the curving arm of Tawas Point is a popular spot to drop anchor.

Distances from East Tawas/Tawas City (statute miles)
Alpena: 66 N
Harrisville: 40 NNE
Bay City: 51 S
Port Austin: 30 SE
Harbor Beach: 60 SE

LAKE HURON

East Tawas & Tawas City, Michigan
Sister hamlets grace Michigan's Cape Cod.

The adjoining villages of East Tawas and Tawas City owe their existence to the track crews of the Detroit and Mackinac Railroad, who built a rail line over what had long been considered impenetrable swamp. Today, over a century later, beautiful beaches and excellent fishing make the towns well worth a visit.

Coming from the north, be sure to clear the shoals off Tawas Point, sometimes referred to as the Cape Cod of Michigan, by honoring buoy R"2".

The point, formed by the currents of Lake Huron—which drop sand rounding the point and increase in speed entering Tawas Bay—only emerged as dry land in the last century. The small island off of the point is just the latest evidence of this phenomenon. A wreck some 2,000 yards off the point is believed to be the three-masted schooner *Kitty Reeves*, which went down in a November gale in 1870 with a cargo of copper valued at $250,000, which has never been salvaged.

The large **East Tawas State Dock** is well-protected from any swell. The two-story **Holiday Inn** on the shore is easily visible from all approaches, as are the two high, charted radio towers. The entrance to the basin is on the east side and is protected from all but south-southwest winds.

For a more rural setting, turn to the east once you have cleared buoy R"4" and make for the private channel that leads to **Jerry's Marina, Inc.** in the bight of Tawas Point. The marina has a Travelift and a fuel dock. The **Coast Guard Station**, a white building with a red roof to the east of the marina, was originally built in 1876. It is the last surviving example of the first series of lifesaving stations built on the Great Lakes.

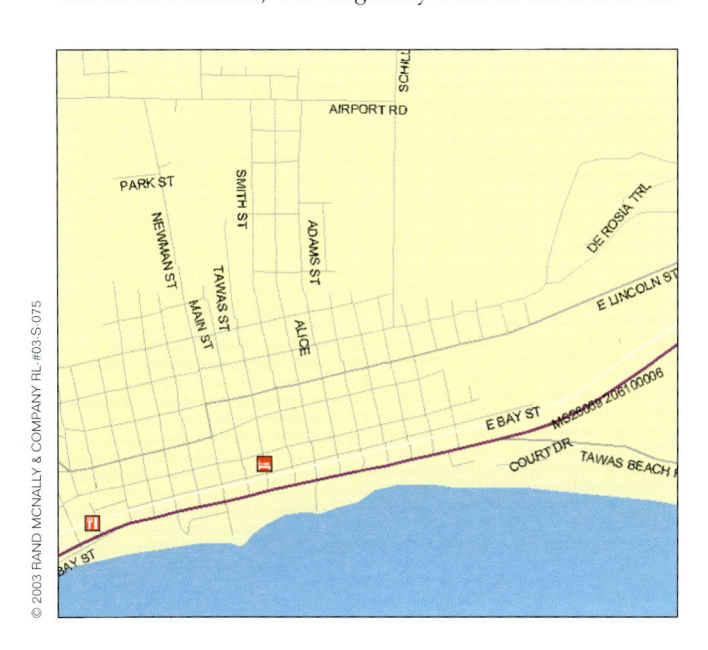

Between the marina and the Coast Guard Station is the **Tawas Bay Yacht Club,** which does not take transients. Tawas Point boasts a historic lighthouse and a state park. The brick **Tawas Point Lighthouse**, built in 1876, is 16 feet in diameter and 67 feet high. Its original lens, which housed an oil lamp,

More Info

Emergency 911
Tawas St. Joseph Hospital 989-362-3411
Chamber of Commerce 800-558-2927
www.tawas.info
Sunrise Taxi 989-362-8000

Where to Eat

Baywalk Restaurant 989-362-5201
Famous for its broasted chicken.

Champs Food & Spirits 989-362-8080
A short walk away, across from the city park. Featuring steak, seafood and sandwiches.

Chum's Bar 989-362-8601
Breakfast, lunch and dinner.

G's Pizzeria & Deli 989-362-8659
Across the street from the Holiday Inn, this eatery has free delivery.

The Perfect Pickle 989-362-8482
Home of the deep-fried pickle and pickle soup, as well as daily lunch specials.

Pier 23 989-362-8856
Dine on the patio deck overlooking the bay.

Sonny's on the Bay 989-362-6378
Home-style cooking with indoor and outdoor seating.

Whitetail Café 989-362-1090
Known for its homemade soups and coneys. Open for breakfast and lunch.

Local events

May
Shoreline Arts and Crafts Show

June
Michigan Free-Fishing Weekend

Tawas Point Celebration Days

Summer Arts and Crafts Show

July
Independence Day Grand Parade and Fireworks

Summerfest

Sidewalk Sale Days

August
Christmas Railroad Days in Summer

Tawas Bay Waterfront Fine Art Festival

Tawas Bay Antique Market Show & Sale

Blues by the Bay Festival

July
Labor Day Arts and Crafts Show

Lat 44°16.58
Lon 83°29.08

Located at the head of Tawas Bay, the East Tawas State Dock **B** offers transient dockage and a ramp **F**. On approach, look for the charted radio towers **G** and two-story Holiday Inn **H**.

Lat 44°15.22
Lon 83°26.97

Jerry's Marina, Inc. **C** occupies two basins in the northeast corner of Tawas Bay and offers transient slips, a ramp **I**, fuel and limited repairs. Tawas Bay Yacht Club **D**, next to the Coast Guard Station **J**, is a private facility that does not cater to transients. Protected by Tawas Point, the northeast corner of the bay is a popular anchorage. The Tawas Point Lighthouse **K** can be seen in the distance.

---

was replaced with the present glass lens, which was made in Paris in 1880. At 200,000 candlepower, the oscillating white light is illuminated at four-second intervals and is visible for 16 miles.

A red light showing 90 degrees inside Tawas Bay warns mariners off the shoals to the west of the point. The park here is noted for its abundance of fauna and flora—in particular its wild strawberries, which make an incomparable jam. It is also one of the world's most spectacular stopovers for migrating birds.

The town of East Tawas runs directly inland from the State Dock. The quaint, turn-of-the-century village of low brick buildings gives way after a few blocks to green lawns and tidy houses. The town has a full complement of shops, including a bookstore, to meet all your provisioning needs.

The **Iosco County Historical Museum** is located on Bay Street in a house built in 1903 by James D. Hawks, the first president of the Detroit and Mackinac Railroad.

To the southwest, Tawas City has limited facilities. **Tawas Bay Condominium Marina** has a protected basin

70 LAKELAND BOATING · LAKE HURON CRUISING GUIDE

with a stylish two-story gray clubhouse and can handle boats up to 60 feet.

Anglers should consider this a must-visit port of call. Thanks to its unique location between the cool waters of Lake Huron and the warm waters of Saginaw Bay, Tawas Bay is renowned for the variety of its catch. The Tawas River kicks off the season with a good steelhead run, followed by smelt and walleye. The king and coho salmon fishing is top-rated, while the shore along Tawas Point is noted for huge northern pike.

The **Tawas Bay Summer Arts and Crafts Show**, which features an assortment of collectibles, is held in June. Another big art show, the annual **Tawas Bay Waterfront Fine Art Festival**, is held the first weekend in August, followed by an antique show and sale the next weekend.

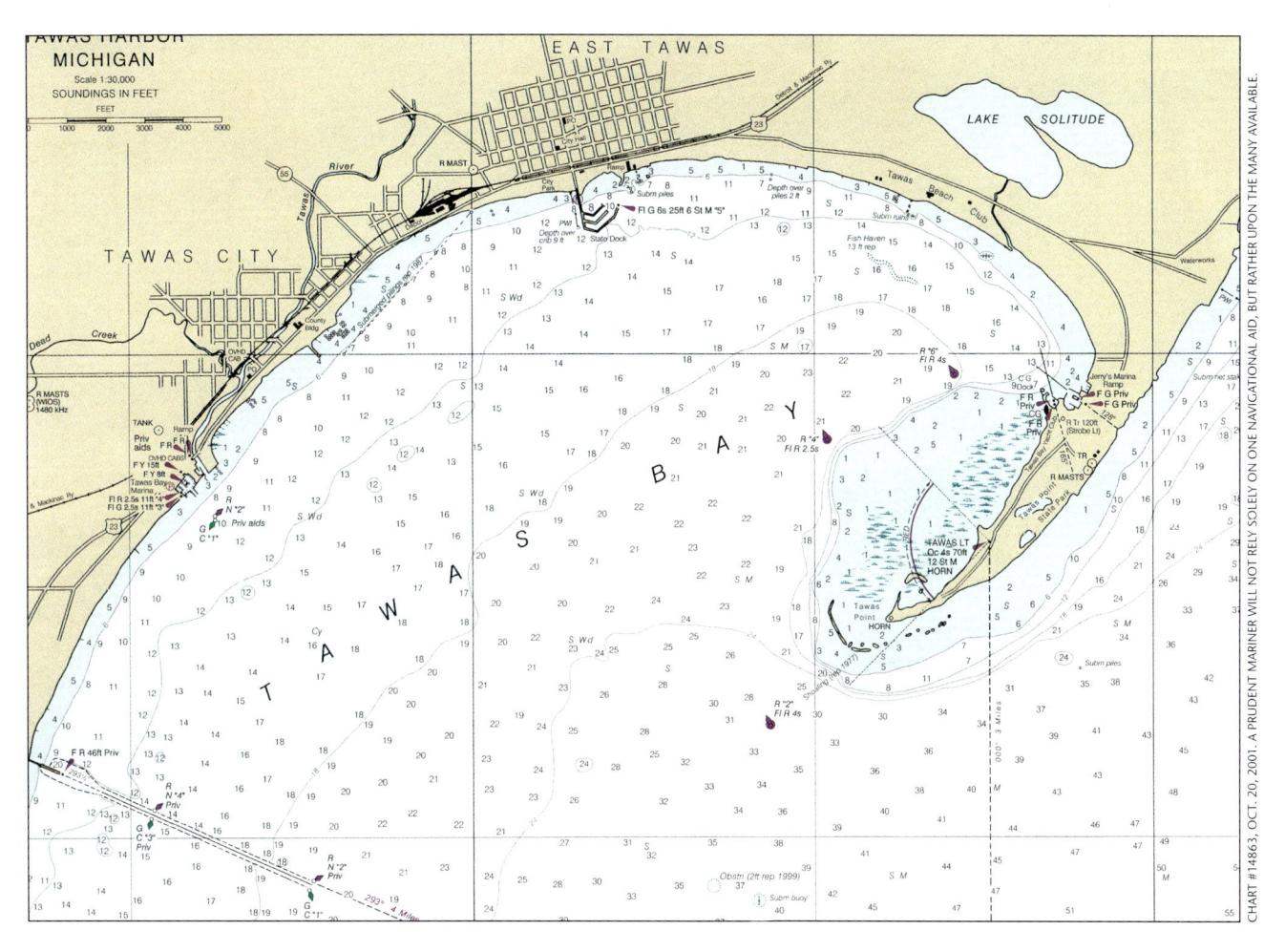

Facilities information subject to change. We suggest you call ahead.

			Monitors VHF Channel	Transient Slips Available / Wall mooring	ALTERNATE MOORING: Rafting allowed	Maximum LOA	Minimum Depth at Dock	Power (amperage or volts)	HOOKUPS: Water Cable TV	FUEL: Gas Diesel Pumpout	BASICS: Heads Showers Laundry	AMENITIES: Swimming pool Whirlpool Rec area Grills Picnic tables Dog walk	Take Reservations	Take Credit Cards	Haulout (Capacity in tons or feet)	REPAIRS: Mechanical Electronics Fiberglass Woodworking Sails Canvas	CONVENIENCE: Ship's store Ice Convenience foods & beverages
B	East Tawas State Dock 989-362-2731	686 Tawas Beach Rd. East Tawas, MI 48730	9	115	WR	100'+	5'	30 50	W		GDP	HS	PD	Y	Y			
C	Jerry's Marina, Inc. 989-362-8641	542 Tawas Beach Rd. East Tawas, MI 48730	68	20	W	40'	3'	30	W		GP	HS	PD	Y	Y	15	MEFW	SIC
A	Tawas Bay Condominium Marina 989-362-3595	939 S. U.S. 23 Tawas City, MI 48764-0300	16	12		60'	6'	30 50	WC			HS	SRGPD	N	Y			IC

LAKELAND BOATING · LAKE HURON CRUISING GUIDE 71

Au Gres, Michigan

Lat 44°01.10
Lon 83°41.00

When approaching Au Gres, keep at least one mile off of Point Au Gres and Point Lookout. Look for the flashing red light that marks the channel entrance. Parallel jetties with red and green flashing lights **A** mark the entrance to the Au Gres River. The Au Gres Yacht Club Marina **B**, to port about a quarter-mile upriver from the mouth, sells fuel and occasionally has dockage for members of affiliated clubs. Transients should continue on to the Au Gres State Dock **C**, about one and a half miles upriver to port, which provides overnight dockage and fuel. Harbor Town Marina **D**, just opposite the State Dock, does not cater to transients. Inland Marine **E**, which is next to Harbor Town, does some repairs and offers transient slips.

Distances from Au Gres (statute miles)
Tawas: 20 N
Bay City: 26 S
Caseville: 33 E
Port Austin: 51 NE

72 LAKELAND BOATING · LAKE HURON CRUISING GUIDE

LAKE HURON

Au Gres, Michigan
Set up camp by this beautiful bay.

Situated on the bay formed by Point Au Gres to the south and Point Lookout to the north, Au Gres is more a resort locale than a town, boasting pristine beaches and good fishing for perch and walleye.

Approach with care, keeping at least one mile off the points; the coastline from Oscoda south to here is low with a sandy beach, making landfall difficult. The entrance to the Au Gres River is marked with parallel jetties with red and green flashing lights.

The **Au Gres Yacht Club Marina**, located a quarter of a mile in from the mouth of the river, offers up to six transient spaces to boaters who belong to an affiliated yacht club. The **Au Gres State Dock** is one and a half miles upriver on the port side. Across the river from the state dock are **Inland Marine, Inc.**, which offers repairs, boating supplies and some slips, and **Harbor Town Marina**, which does not cater to transients.

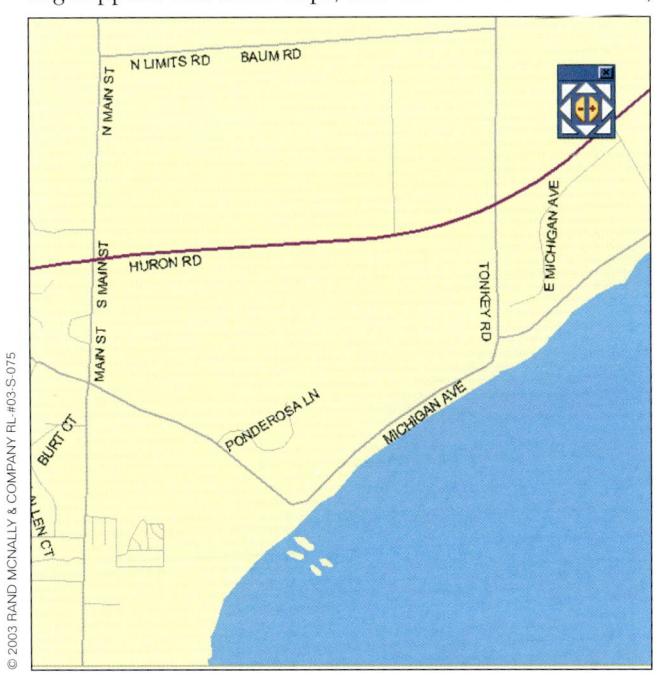

The city itself is on U.S. 23 a short distance from the marinas. Next to the state dock, a city-run campground offers more than 119 campsites with full hookup, picnic areas, baseball diamonds, playgrounds and tennis and basketball courts. A nearby water park features a large slide, a go-kart track, mini-golf and a trout pond.

More Info

Emergency 911
Tawas St. Joseph Hospital 989-362-3411
Chamber of Commerce 989-876-6688
www.augres.com
Au Gres City Campground 989-876-8310
Au Gres Drug Store 989-876-8899
K&C Sales (hardware) 989-876-7816

Where to Eat

Dunleavy's Eatery and Pub 989-876-7022
Catch a bite of their famed fresh fish.

H&H Bakery and Restaurant 989-876-7144
Great pizza and fresh bread.

LAKELAND BOATING · LAKE HURON CRUISING GUIDE

Au Gres offers motels, groceries, ice cream shops and restaurants. Golfers can test their luck on one of the area's four golf courses, and history buffs will want to visit the **Arenac County Historical Society Museum**, housed in an 1883 Methodist church. Exhibits include Native American artifacts and an old-fashioned schoolroom.

Local events

July
AuGres Car Cruiser

August
Venetian Day

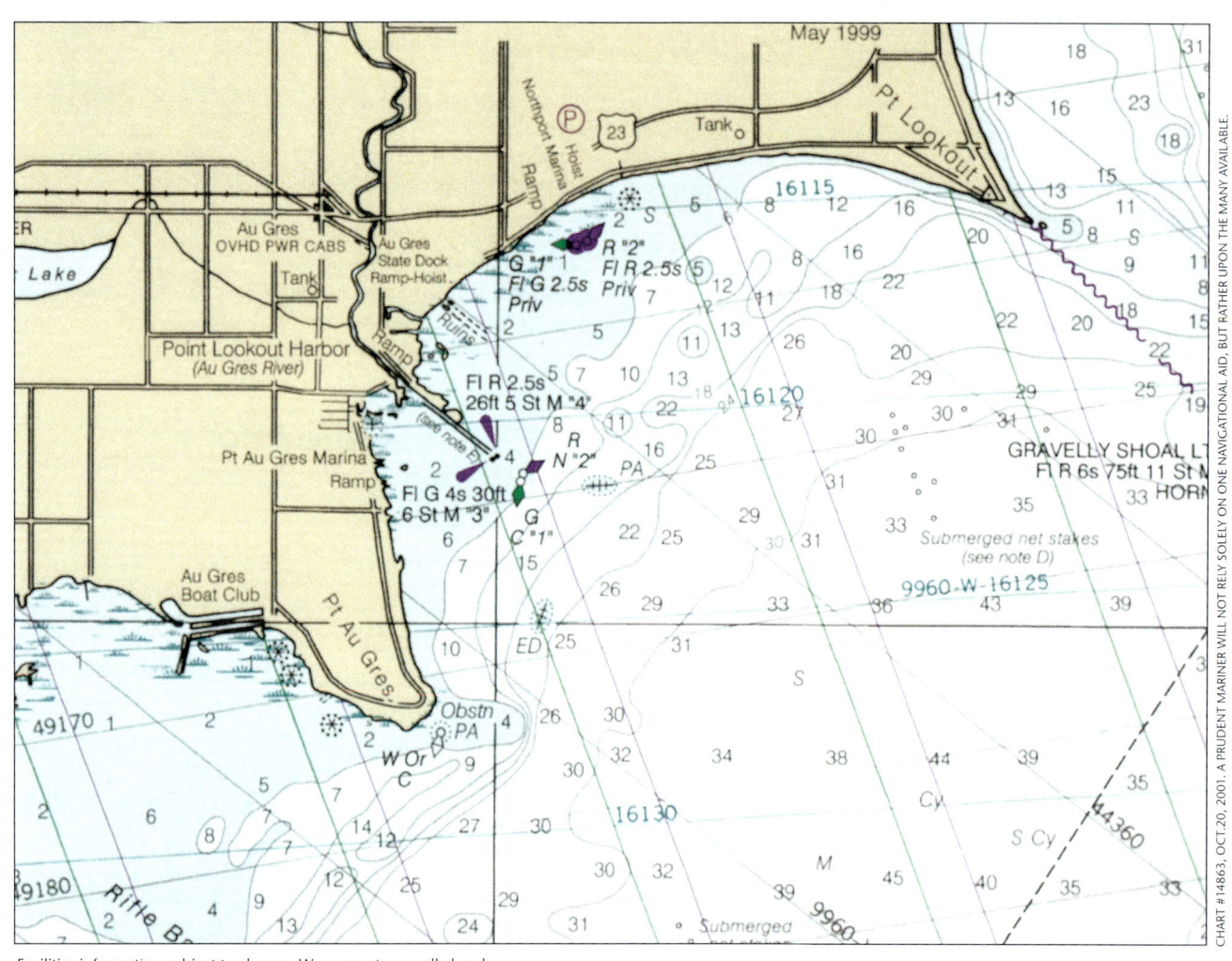

Facilities information subject to change. We suggest you call ahead.

			Monitors VHF Channel	Transient Slips Available	ALTERNATE MOORING: Wall mooring / Rafting allowed	Maximum LOA	Minimum Depth at Dock	Power (amperage or volts)	HOOKUPS: Water / Cable TV	FUEL: Gas / Diesel / Pumpout	BASICS: Heads / Showers / Laundry	AMENITIES: Swimming pool / Whirlpool / Rec area / Grills / Picnic tables / Dog walk	Take Reservations	Take Credit Cards	Haulout (Capacity in tons or feet)	REPAIRS: Mechanical / Electronics / Fiberglass / Woodworking / Sails / Canvas	CONVENIENCE: Ship's store / Ice / Convenience foods & beverages
C	Au Gres State Dock 989-876-8729	201 Water St. Au Gres, MI 48703	9 16	30	R	55'	3'5"	30,50 110	W	GP	HS	GPD		N	Y		
B	Au Gres Yacht Club Marina 989-876-8155	3135 Midshipman Drive Au Gres, MI 48703	16	6		40'	4'	30	W	GDP	HS	SPD		Y	Y	15	MEFWC	SIC
E	Inland Marine Inc. 989-876-7185	333 S. Main St. Au Gres, MI 48703	9	Y	W	40'	5'	30	W					N	Y	8	MF	S

74   LAKELAND BOATING · LAKE HURON CRUISING GUIDE

STORMY WEATHER

News Flash
Don't let lightning catch you unawares.

"The most important safety rule about lightning and boating is simple: Get off the water immediately," says James Lubner, water safety specialist for the University of Wisconsin Sea Grant.

Each year, some 400 people in the United States are struck by lightning while outside, with about 80 people killed and many more suffering permanent disabilities. Lightning is capable of blowing out the bottom of boats and has caused millions of dollars in damage to navigational equipment.

Outdoors is the most dangerous place to be in a storm, and on a small craft on the water is arguably one of the most dangerous places to be outdoors. If you see lightning or dark clouds or you hear thunder, or if marine warnings are in effect, do not venture out on the water unless you're absolutely confident that your boat can be navigated safely under the conditions.

If you're already on the water when a storm breaks and you can't head back to port, put on a life jacket and prepare for rough seas. Lightning will hit the tallest object in a boat, so stay low or go belowdecks. If you are exposed, squat down in the center of the boat, away from metal hardware, and cover your head with your hands; this position diffuses the impulse of the lightning bolt. Do not huddle in a group.

To minimize the potential for strike damage, quickly lower or remove the radio antenna and any other protruding devices that aren't part of your lightning protection system. Switch off all nonessential electronics, shut off electrical system breakers providing power to nonvital systems and disconnect power cables to bracket-mount electronics. Obviously, flush-mounted electronics won't be as easy to disconnect.

Keep away from metal objects that are not grounded to the protection system. Don't touch more than one grounded object at the same time or you may become a shortcut for electrical surges passing through the protection system. Stay well away from the lightning-conducting system components and be sure that you aren't zapped by leaning up against the refrigerator or stove.

As every boater knows, it's vitally important to stay alert when on the water, even when the weather seems fair. At least 10 percent of lightning occurs when there are no visible clouds in the sky, and most people who are struck by lightning aren't even in the rain. Many deaths from lightning occur ahead of the storm, because people wait until the last minute before seeking shelter. Heavy static on your AM radio can be an indication of nearby thunderstorm activity. Pay attention to any steady increase in wind or sea, decreasing visibility such as from fog, and any increase in wind velocity in the opposite direction of a strong tidal current. Your hair might also stand on end.

A lightning strike can tear through a boat's electronics and AC/DC electrical systems, destroying all kinds of equipment at the ends of power lines—stereo, lighting, fire extinguisher and air conditioner controls, engine instruments and senders, switch and breaker panels. The most effective lightning protection system depends on a good-sized ground plate that is made specifically for this purpose and mounted below the waterline. These systems typically utilize very heavy-gauge wire to lead a strike from the highest point on the boat directly to the underwater plate in as straight a line as possible. Surge or spike protectors in electrical equipment power lines are worth considering to help protect your appliances or electronics. If you suffer a direct hit, you won't escape undamaged, but near misses can be made less destructive if you follow these safety measures.

For more information, head to nws.noaa.gov/warnings.html and www.safeboatingcouncil.org.

Bay City, Michigan

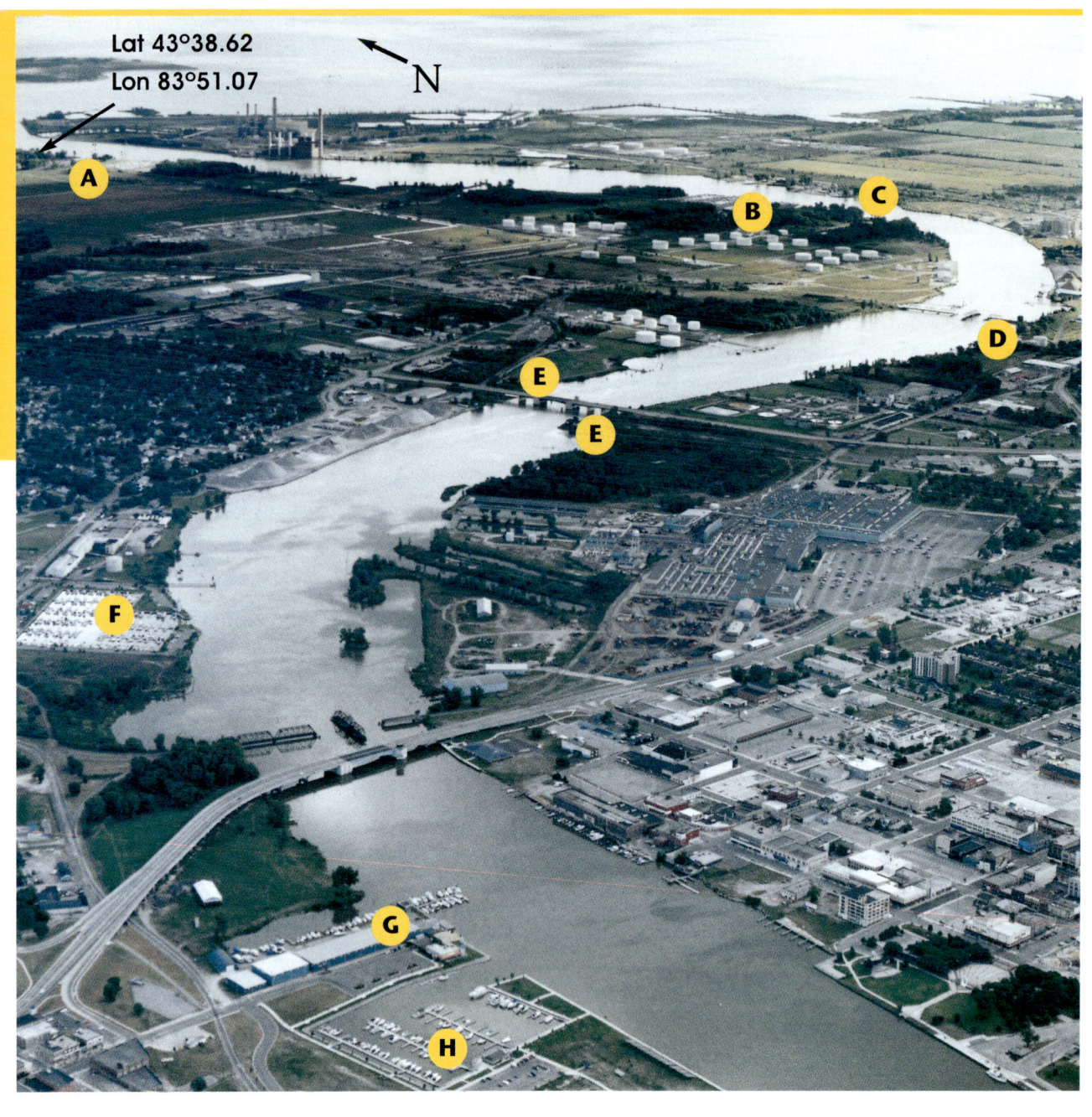

Lat 43°38.62
Lon 83°51.07

From Gravelly Shoal Light, set a course of 212.75°T to arrive at the well-marked 14½-mile channel leading to the mouth of the Saginaw River. The Bay City Yacht Club **A** is to starboard just inside the river mouth. This private club does not cater to transients. Just upriver to starboard, the Bay Harbor Marina **B** offers full services. On the opposite bank, Saginaw Bay Yacht Club **C** is private and rarely has room for transients. Farther upriver, just beyond the bascule bridge—which has a clearance of 20-plus feet and usually stays open, unless a train is coming—there's a public launch ramp **D**. A pair of new launch ramps **E** has also been added on either side of Independence Bridge, which has a clearance of 30 feet. Situated at the second bend, Wheeler Landing **F** now offers transient dockage. After passing under the second railroad bridge, which is usually open, and Liberty Bridge, with a clearance of 23 feet at the middle of the bridge in low water, you'll find Pier 7 Marina **G**, which offers fuel and full services. Transient slips are available next door at the municipal Bay City Liberty Harbor Marina **H**.

Distances from Bay City (statute miles)
Harrisville: 86 N
East Tawas: 52 N
Au Gres: 26 N
Caseville: 51 NE
Port Austin: 61 NE

LAKE HURON

Bay City, Michigan
A former shipping hub turns into a music mecca.

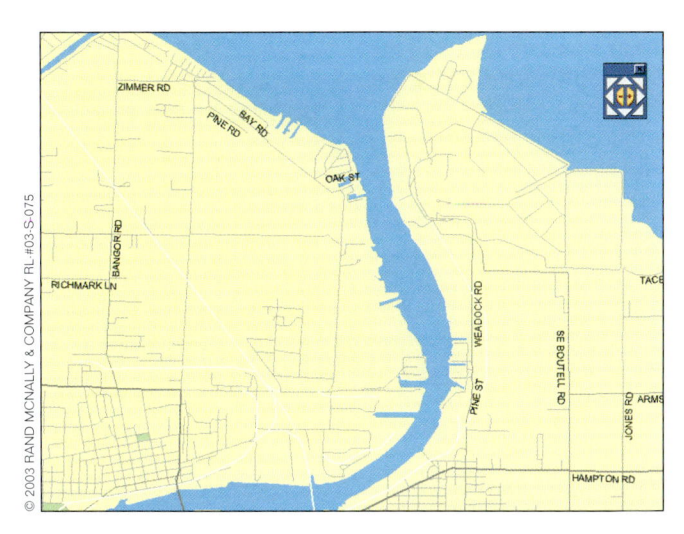

The wetlands of Saginaw Bay have been home to Native Americans for thousands of years. Bay City's more modern history dates back to 1837, when the dense forests of the area, combined with the ease of transport afforded by the Saginaw River, drew an army of loggers, speculators and camp followers. Today the downtown waterfront is undergoing a renaissance, with the historic brick buildings of Midland Street boasting a wide spectrum of shops and restaurants ranging from casual to elegant. The area has also become a mecca for music lovers: Nightclubs and pubs offer everything from rock and country music to karaoke and piano bars.

Once one of the busiest ports on the Great Lakes, Bay City used to see a thousand schooners a month pass upriver to load. Millions of tons of Great Lakes and ocean cargo still pass through the port each year. It was also a major shipbuilding center through World War II, thanks to the pioneering efforts of the Defoe shipyards, which built many of the Navy's destroyers. Defoe pioneered the technique of

More Info

Emergency	911
Bay Regional Medical Center	989-894-3000
Bay Area Convention & Visitors Bureau	888-BAY-TOWN www.tourbaycitymi.org
Chamber of commerce	989-893-4567
Greater Bay Area Cab	989-894-1118

Continuing upriver, you'll find Brennan Marine **1**, a huge full-service marina offering fuel, dockage with full hookups and repair service.

LAKELAND BOATING · LAKE HURON CRUISING GUIDE

Where to Eat

The Lantern Restaurant 989-894-0772
The area's only outdoor bar and grill features live entertainment and docking.

River Rock Café 989-894-5500
Specialties include steaks, perch and sandwiches.

Old City Hall—The Chambers 989-892-4140
Casual, fine dining in a historic downtown building.

O'Hare Bar & Grill 989-893-5181
On Midland Street, this claims to be the "home of the best bar burger."

O Sole Mio Restaurant 989-893-2371
A short walk from the river, this place has been serving fine Italian food since 1951.

Tommy V's Café & Pizzeria 989-895-3500
Dig in to Chicago-style deep-dish pizza served in a cast-iron pan.

---

the Saginaw River, which offers a fuel dock and on-site repairs. **Wheeler Landing,** at the second bend, has water and cable TV hookups and a beautiful swimming pool, and **Bay City Liberty Harbor Marina,** farther upriver, can take cruisers up to 104 feet long. Two bridges farther upriver is the full-service **Brennan Marine.** Fuel and repairs are also available at **Pier 7 Marina. The Lantern** restaurant, located across the river from Pier 7 and Bay City Liberty Harbor, also offers dockage to patrons.

As the waterfront along the eastern side of the river undergoes a major renewal, new shops and restaurants are coming into the area. One of the coolest destinations is the **Bay City Antiques Center,** which occupies an entire city block on North Water Street between Third and Fourth streets. The center is housed in a three-story building that was built in 1867 during the

Local events

June
Free Fishing Festival
Riverside Art Festival
St. Stanislaus Polish Festival
Dobson Industrial River Roar

July
Fireworks Festival

August
Labadie Pig Gig Ribfest

September
River of Time Living History Encampment

---

building hulls upside-down and launching them by rolling them over into the water.

Access is easy for cruising yachtsmen, as there are several facilities in the area. Popular options for transients include **Bay Harbor Marina,** near the mouth of

---

PIER 7 MARINA

- Ships Store-Parts & Accessories
- Beer • Wine • Snacks
- Mercury Outboards
- Boston Whaler
- Mercury Inflatables
- Certified Mercury & Mercruiser Technicians
- Complete Fiberglass Repairs & Refinishing
- Custom Wood Work & Restoration
- Gas Dock
- 35 Ton Lift - Deep Water
- Brokerage Service
- Open 7 Days (April thru November)
- Located on Saginaw River
- Downtown Bay City - Next to Hooters
- Transient Available Next Door at Liberty Harbor

BOSTON WHALER

MERCURY The Water Calls

989-894-9061 • Fax: 989-894-0424
963 E Midland Street, Bay City, MI 48706
www.pier7marina.com

---

78 LAKELAND BOATING · LAKE HURON CRUISING GUIDE

TOM KAEKEL

rough and rowdy days of the lumber boom. The building is largely intact, with original floors and ceilings. Formerly a hotel known as the Campbell House, it now houses 200 vendors offering 53,000 square feet of antiques for sale.

While the grand homes of the lumber barons along Center Avenue are striking, the most impressive historical site is **City Hall**. This Romanesque stone edifice dominates the town with its 125-foot-high clock tower. A climb of 68 steps affords a great view of the city and waterfront. The original wooden interior of the building, with its distinctive metal pillars, was restored in 1976.

If you love your action fast and loud, be sure to time your cruise to coincide with River Roar, Bay City's annual powerboat race, which takes place the last weekend in June. The city really decks itself out for this event, with gaily striped waterfront pavilions and nonstop entertainment. Then, a few weeks later, the city turns itself out in full force again for the three-day Fourth of July festival, which features what is billed as one of the largest fireworks displays in the state.

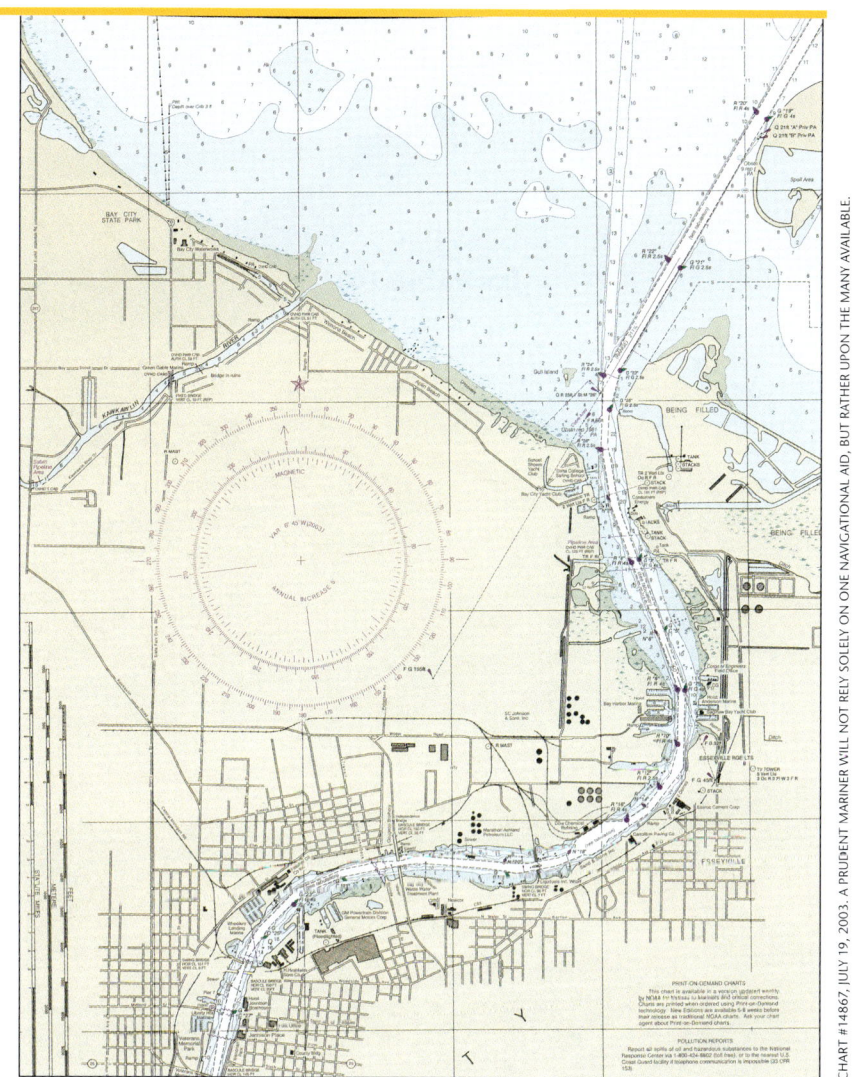

Facilities information subject to change. We suggest you call ahead.

			Monitors VHF Channel	Transient Slips Available	ALTERNATE MOORING: Wall mooring Rafting allowed	Maximum LOA	Minimum Depth at Dock	Power (amperage or volts)	HOOKUPS: Water Cable TV	FUEL: Gas Diesel Pumpout	BASICS: Heads Showers Laundry	AMENITIES: Swimming pool Whirlpool Rec area Grills Picnic tables Dog walk	Take Reservations	Take Credit Cards	Haulout (Capacity in tons or feet)	REPAIRS: Mechanical Electronics Fiberglass Woodworking Sails Canvas	CONVENIENCE: Ship's store Ice Convenience foods & beverages
H	Bay City Liberty Harbor Marina 989-894-2800	215 JFK Dr. Bay City, MI 48706	16	48	WR	104'	6'	30 50	WC	P	HSL	GPD	Y	Y			IC
B	Bay Harbor Marina 989-684-5010	5309 E. Wilder Rd. Bay City, MI 48706	16	30	WR	60'	8'	30 50	W	GDP	HS	SPD	N	Y	60	MEFW	SIC
I	Brennan Marine 989-894-4181	1809 S. Water St. Bay City, MI 48707		20		60'	5'	30	W	GDP	HS	GPD				MF	SI
G	Pier 7 Marina 989-894-9061	963 E. Midland St. Bay City, MI 48706	9 16			150'	10'		WC	GP	HS	P	N	Y	35	MEFW	SIC
F	Wheeler Landing 989-667-0030	600 Marquette Ave. Bay City, MI 48706		12		40'	8'	Y	WC	P	HSL	SRGPD	Y	N			IC

Sebewaing, Michigan

Lat 43°44.16
Lon 83°27.82

Depths here are shallow, so be sure to approach the well-marked channel leading to the town of Sebewaing from well offshore. The entrance to the Sebewaing River **A** is marked with lights on the north and south banks. Transient dockage is available at Sebewaing Harbor Marina **B**, to starboard, just before the railroad bridge **C** as you head upriver. Amenities include a gas dock, head and showers, and a repair shop. A boat ramp **D** is located nearby. Boaters should call ahead to make sure the marina is open; in past years it has had to close due to low water levels.

LAKE HURON

Sebewaing, Michigan
A sweet spot for walleye fishing.

Sebewaing is known for two things—sugar and walleye. Unfortunately, like many ports on the Great Lakes, it has also become known for low water levels in recent years. But great fishing and some unique attractions make this pleasant, turn-of-the-century farm town a sweet destination—when water levels permit.

Because the east shore of Saginaw Bay is low-lying and shallow, be sure to approach from well offshore. The dredged channel leading into the Sebewaing River is well marked. The outermost buoy, R"2", is at the dropoff. The water on either side of the channel is extremely shallow. The depth in the channel is generally 6 to 8 feet, except for a low spot of less than 4 feet at N"12", opposite a low-lying shoal on the north side.

Sebewaing Harbor Marina, an attractive facility with about 50 transient slips that can handle boats up to 50 feet, lies on the starboard side of the river just before the railroad bridge. The marina has a small marine store and sells gas. There are campfire pits for grilling out, and live bait is sold. Town is just a 15-minute walk away. However, boaters are urged to call before planning a trip to this marina, which was closed in summer 2003 due to low water levels.

Sebewaing is a quintessential small town consisting of

Local event
June
Sugar Festival

several blocks of turn-of-the-century brick buildings. It is the home of **Norman's**, a work clothes emporium that claims to have the lowest prices in the country. The towers of the sugar plant dominate the skyline, yet there's a small-town, not industrial, vibe.

The **Old Heidelberg Gallery** has different shows each month, put on by the **Lake Huron Arts Council**. Take a walk down memory lane at **Antiques-by-the-Bridge**, a mall featuring collectibles dating from the 1850s to the 1940s. Upstairs, the **Antique Inn** bed and breakfast offers four rooms furnished with beautiful period pieces. For a walk down another kind of lane, head to the **Sebewaing Lanes** bowling alley. The high point of the summer is the annual three-day Sugar Festival at the end of June, featuring the a carnival, a chicken barbecue and, of course, fireworks.

Distances from Sebewaing (statute miles)
Caseville: 30 NE
Bay City: 23 SW

Where to Eat

The Lamplighter. 989-883-9224
Offering a full menu in a relaxed setting.

Peking City 989-883-3106
This authentic Chinese restaurant is a wonderful surprise in the middle of the farming heartland.

Ricky B's Restaurant & Lounge 989-883-9622
Great home cooking including daily specials, Friday night fish frys and Sunday brunch.

Sebewaing Lanes 989-883-2721
Relax at the bowling alley bar and grill.

More Info

Emergency 911
Scheurer Hospital 989-453-3223
Chamber of Commerce 989-883-2150
www.sebewaing.org

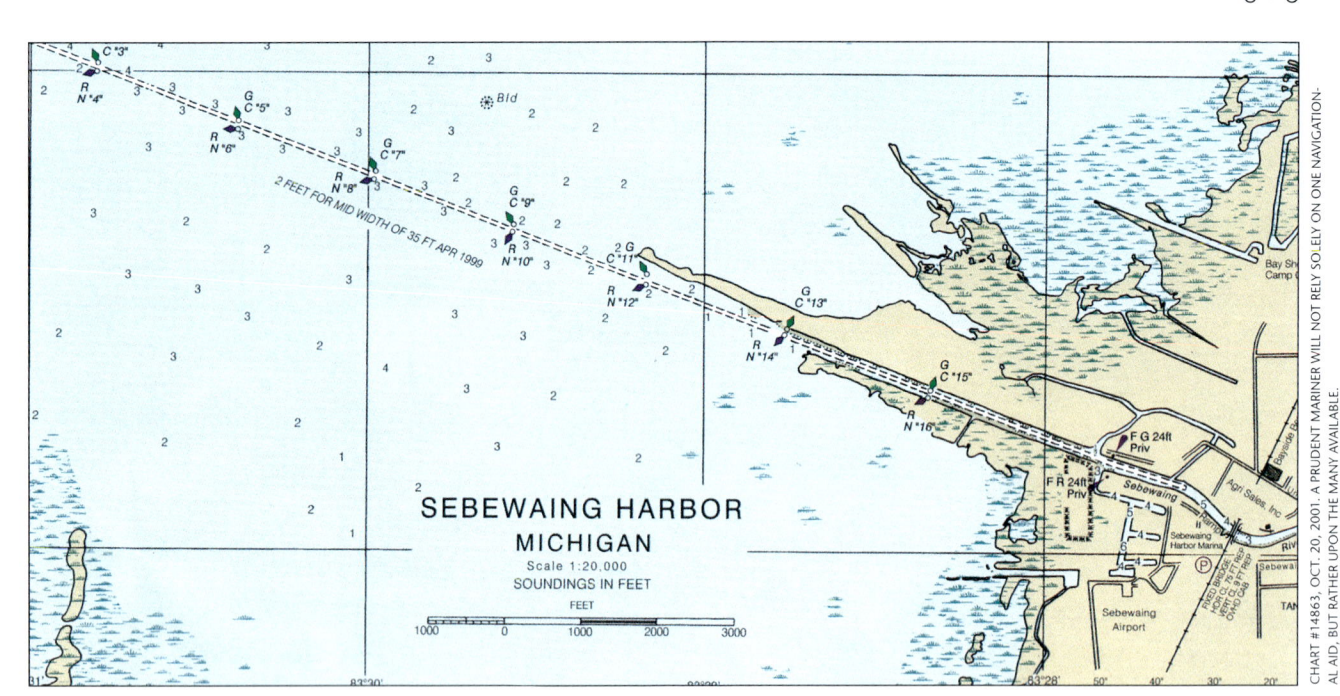

SEBEWAING HARBOR
MICHIGAN
Scale 1:20,000
SOUNDINGS IN FEET

CHART #14863, OCT. 20, 2001. A PRUDENT MARINER WILL NOT RELY SOLELY ON ONE NAVIGATIONAL AID, BUT RATHER UPON THE MANY AVAILABLE.

Facilities information subject to change. We suggest you call ahead.

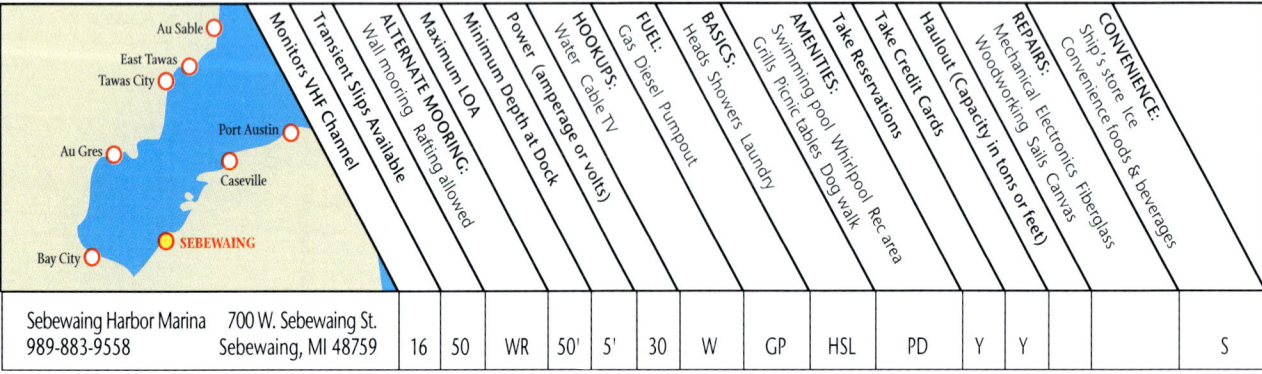

		Monitors VHF Channel	Transient Slips Available	ALTERNATE MOORING: Wall mooring / Rafting allowed	Maximum LOA	Minimum Depth at Dock	Power (amperage or volts)	HOOKUPS: Water / Cable TV	FUEL: Gas Diesel Pumpout	BASICS: Heads Showers Laundry	AMENITIES: Swimming pool Whirlpool Rec area / Grills Picnic tables Dog walk	Take Reservations	Take Credit Cards	Haulout (Capacity in tons or feet)	REPAIRS: Mechanical Electronics Fiberglass / Woodworking Sails Canvas	CONVENIENCE: Ship's store Ice / Convenience foods & beverages
B	Sebewaing Harbor Marina 989-883-9558 700 W. Sebewaing St. Sebewaing, MI 48759	16	50	WR	50'	5'	30	W		GP	HSL	PD	Y	Y			S

82 LAKELAND BOATING · LAKE HURON CRUISING GUIDE

Helicopter Rescue

Here's how to help the Coast Guard help you.

Should you ever face a major emergency aboard your vessel, you may find yourself awaiting the arrival of a Coast Guard helicopter to airlift an injured passenger to safety or provide other help.

The Coast Guard will give specific instructions for helicopter evacuation by radio. But visualizing this procedure and its logical steps right now—before an emergency happens—will prepare you to make the transfer of crew quickly and safely.

When calling to report an emergency, use precise navigational information. Reporting your position accurately will cut down on response time.

Although they are fast and agile, helicopters are delicate. Remember to lower all masts, booms, antennae, bimini and outriggers that might impede the helicopter's flight path. If there are riggings that can't be lowered, mark them with lights (at night) or flags to point them out to the pilot. Clear the cockpit and deck of all loose gear—objects left on deck, like folding chairs and papers, can be sucked into the chopper's jet engines—and keep all unnecessary personnel out of the way.

Keep as many crewmembers as possible belowdeck, and order those who stay topside to don lifejackets—the powerful gusts of wind produced by the propeller can be enough to knock crewmembers off balance and send them tumbling into the drink. Put a lifejacket on the evacuee, and secure a note to him stating his medical history, condition or difficulty, in case he loses consciousness. If the radio works, the CG radio operator will want this information. It's also a good idea to keep firefighting equipment handy, in case of an accident.

When the helicopter arrives, change course to put the wind 30 degrees off your port bow (most hoists are located on the helicopter's starboard side) and attempt to make contact with the helicopter for further instructions. Downdraft may make it difficult to control your vessel, so you will need enough speed to maintain steerage and directional control in a given sea state. Rotor noise will make it difficult to hear radio instructions, so it is a good idea to have someone standing by in the cabin with a handheld or secondary VHF to help relay radio instructions.

If the evacuation must take place after dark, light the pickup area. Be careful not to shine lights upward or at the helicopter. Bright lights can impair the pilot's vision.

A rescue device (sling or litter) will be lowered on a steel cable. Have a crewman standing by to guide the device with the attached non-conductive steadying line. Don't touch the cable or device—allow it to make contact with your vessel to discharge any static electricity generated by the helicopter's propeller. The built-up energy can produce a shock strong enough to knock an adult to the deck.

If the evacuee can move into the cockpit (or other designated hoisting area), secure him in the rescue device and signal the helicopter to hoist away with a thumbs-up sign.

If the sling or litter must be moved out of the cockpit/pickup area, disconnect it and let the hook end of the line go free. Do not, under any circumstances, attempt to move the rescue device out of the cockpit with the hoisting cable attached. And never attach the helicopter cable to the boat. A strong gust of wind could cause the chopper to lose control and plummet into the water. Move the disconnected rescue device, secure the evacuee in it, and head back into the cockpit. If the helicopter has moved away, the pilot will move back in for reconnection. Once again, allow the steel cable or hoist hook to contact your boat to dissipate static electricity. Refasten the hook to the rescue device and signal the helicopter to hoist away.

Knowing what to do in case of emergency on the water will help you and your crew by helping out the Coast Guard—and could even save a life.

Caseville, Michigan

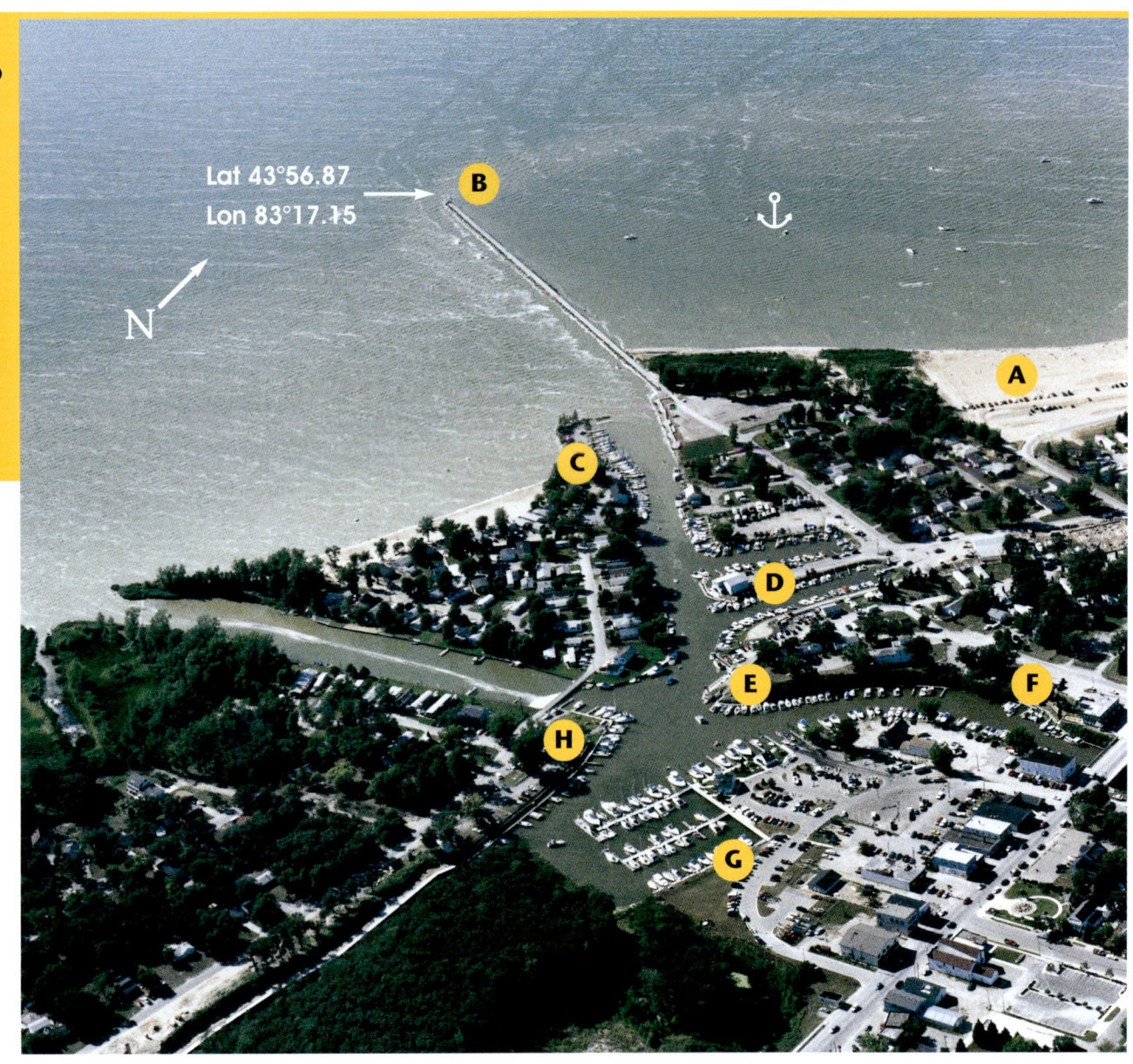

Lat 43°56.87
Lon 83°17.15

If approaching Caseville from Tawas Bay, stay clear of Charity Island. The waters off the beautiful Caseville County Park Beach **A** are a popular anchorage. Caseville's dredged entrance channel is marked with unlit buoys. A flashing green light **B** marks the end of the jetty. Hug the breakwall to avoid the shallow water to the south of the entrance. Huron Yacht Club **C**, which is private and doesn't cater to transients, is to starboard just inside the river mouth. Upriver to port, friendly Hoy's Saginaw Bay Marina, Inc. **D** offers slips, fuel and repairs in its two basins. The Caseville Resort and Marina **E** is a new year-round condo community. Next to it is Hill's Riverside Marina **F**, which offers a few transient slips, but Caseville Municipal Harbor **G**, situated at the head of the harbor, is the main attraction for transients, offering fuel and slips with full hookups. Port Elizabeth Marina and Yacht Club **H** has primarily seasonal slips, but you might find a spot here in a pinch.

Distances from Caseville (statute miles)
East Tawas: 25 NW
Port Austin: 18 E
Bay City: 49 SW

84   LAKELAND BOATING · LAKE HURON CRUISING GUIDE

LAKE HURON

Caseville, Michigan
Break out the beach blankets!

At the mouth of the Pigeon River on the inside of the Thumb, Caseville is one of Lake Huron's most charming natural harbors, offering gorgeous sunsets, miles of sandy beaches and all the recreational offerings of a full-fledged summer resort.

If you are approaching from Tawas Bay, be sure to stay clear of Charity Island. The dredged entrance channel in from the bar is well-marked with unlit buoys, and a flashing green marks the end of the jetty. In addition, there are front and rear harbor range lights. A long breakwater to the north lessens the swell from Lake Huron.

The waters off **Caseville County Park Beach** are a popular spot to anchor out and catch some rays. Just beware of the buoys that mark swimming areas. In the summer, the beach, considered one of Michigan's prettiest, is packed with locals.

The **Huron Yacht Club**, located to starboard as you enter the river, is an attractive facility favored by sailors. Continuing upriver, you will find **Hoy's Saginaw Bay Marina, Inc.** to port, followed by **Caseville Resort and Marina**, which has been turned into condos. Next door, **Hill's Riverside Marina** does repairs, offers a few transient slips and operates a bar and grill. The **Caseville Municipal Harbor** is at the head of the harbor in a scenic setting with a tree-lined bluff overlooking unspoiled marshes to the west. Across the way, you'll find the new **Port Elizabeth Marina and Yacht Club**, which sometimes has a few slips for transients.

The town is directly behind the Caseville Municipal Harbor. **Bay Liquor & Food** has beer, wine and deli sandwiches and is open from 6 a.m. to midnight. Be sure to stroll to the old Caseville **United Methodist Church**, whose spire

Lat 43°54.32
Lon 83°21.45

Beadle Bay Marina is located on the south side of Sand Point, about five miles south of the mouth of the Pigeon River.

LAKELAND BOATING · LAKE HURON CRUISING GUIDE   **85**

Where to Eat

Bay Window Restaurant 989-856-2676
Enjoy a family-style dinner followed by a movie at the attached theater.

Dufty's Blue Water Inn. 989-856-3663
Take in the live entertainment or relax on the double deck overlooking the harbor. Steaks are the specialty.

Giuseppe's Pizzeria. 989-856-2035
Featuring barbecued ribs and chicken, Italian dishes and, of course, pizza.

Hill's Riverside Bar. 989-856-9917
Great burgers and beef sandwiches at the marina.

Walt's 989-856-4020
This casual eatery is a short stroll from the municipal harbor.

More Info

Emergency 911
Scheurer Hospital 989-453-3223
Chamber of commerce 800-606-1347

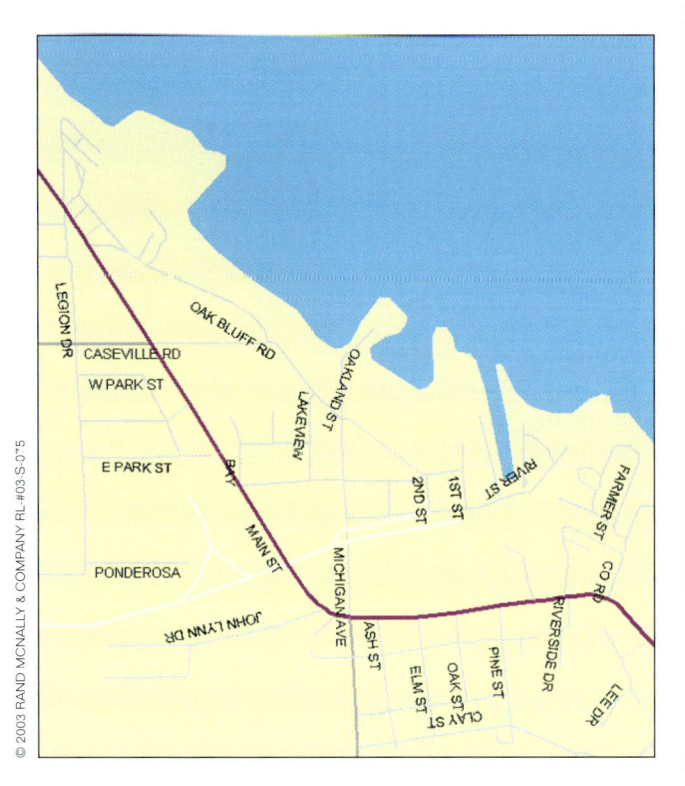

is visible from offshore. The 74-foot-high gothic steeple is a focal point for the community. The first church on the site was built in 1868 by the Reverend Manasseh Hickey and 12 settlers. The present church was built in November 1874.

Children of all ages will enjoy **Putt-Putt Golf & Games**, with its bumper boats and water slide. Opposite the yacht club is a beautiful municipal beach, with a boardwalk along the river for evening strolls.

Additional dockage is available along the lakeshore about five miles south of the Pigeon River. **Beadle Bay Marina**, nestled on the south side of Sand Point, is a first-class, family-oriented facility. **Bayshore Marina**, located in the southern cusp of Wild Fowl Bay, offers full services, too.

Anglers, take note: The Caseville area is called the perch capital of Michigan for good reason. There are 300,000 shallow, sandy acres of spawning ground in the inner bay. The outer bay, which has depths up to 90 feet and numerous reefs, is noted for its walleye, salmon and lake trout.

Local events

June
Walk Through Time Pre-Civil War Camp Re-enactment

Great Lakes Bowfishing Championship

August
Cheeseburger in Caseville

Lat 43°51.09
Lon 83°21.05

Friendly, family-operated Bayshore Marina J, at the southern cusp of Wild Fowl Bay, has ample transient dockage.

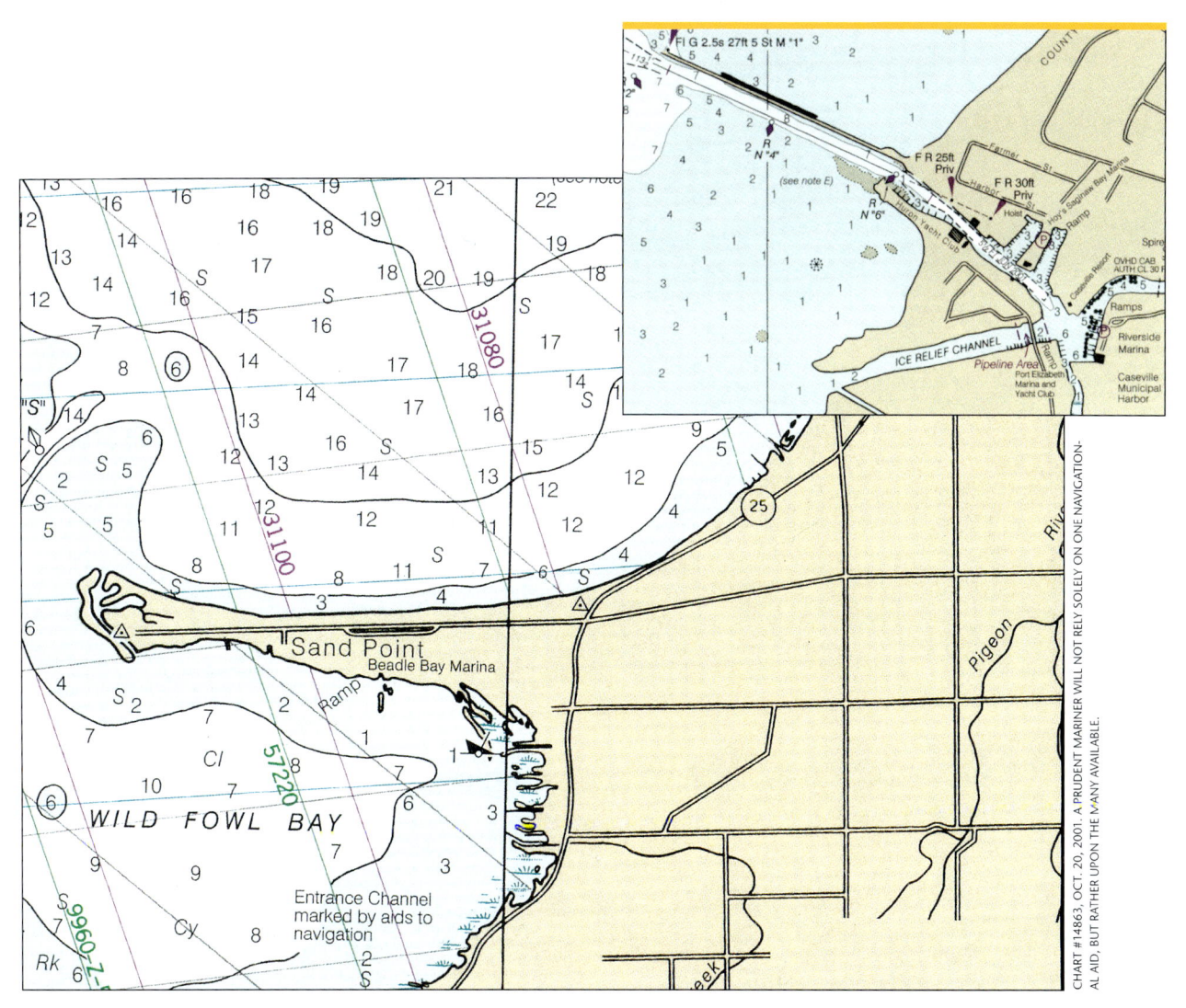

Facilities information subject to change. We suggest you call ahead.

			Monitors VHF Channel	Transient Slips Available	ALTERNATE MOORING: Wall mooring Rafting allowed	Maximum LOA	Minimum Depth at Dock	Power (amperage or volts)	HOOKUPS: Water Cable TV	FUEL: Gas Diesel Pumpout	BASICS: Heads Showers Laundry	AMENITIES: Swimming pool Whirlpool Rec area Grills Picnic tables Dog walk	Take Reservations	Take Credit Cards	Haulout (Capacity in tons or feet)	REPAIRS: Mechanical Electronics Fiberglass Woodworking Sails Canvas	CONVENIENCE: Ship's store Ice Convenience foods & beverages
J	Bayshore Marina 989-656-7191	2612 Wallace Cut Bay Port, MI 48720	16	10	W	45'	5'	30	W	GP	HS	PD	Y	Y	35	MEF	SIC
I	Beadle Bay Marina 989-856-4911	4375 Lone Eagle Trail Caseville, MI 48725	16	10	W	45'	3'	20	W	GP	HS	PD	Y	Y			IC
G	Caseville Municipal Harbor 989-856-4590	7040 Main St. Caseville, MI 48725	16	18	WR	70'	8'	30	WC	GDP	HSL	RPD	N	Y			IC
D	Hoy's Saginaw Bay Marina Inc. 989-856-4475	6591 Harbor St. Caseville, MI 48725	16	10	W	40'	5'	30	W	GP	HS	D	Y	Y	25	MFC	SIC
F	Hill's Riverside Marina 989-856-9917	6538 Main St. Caseville, MI 48725		3		46'	6'	30	W		HS	D	Y	N	14	MFWC	SC
H	Port Elizabeth Marina 989-856-8077	6635 River St. Caseville, MI 48725	16	4		48'	4'	30 110	WC		HS	GPD	Y	N	25		IC

LAKELAND BOATING · LAKE HURON CRUISING GUIDE   **87**

Port Austin, Michigan

Lat 44°03.24
Lon 82°59.62

If approaching from the east, maintain at least a 40-foot depth until you reach the No. 1 can, about a mile northwest of the reef lighthouse. Stay about a half-mile from the lighthouse. After rounding the can, take a direct south heading and motor toward the taller of two radio towers until you see the harbor entrance. Do not take a heading on the water tower. Also, do not enter from the northeast **A**, where depths are about 2 feet, but rather through the lighted entrance between the newer detached breakwall and the old breakwall **B**. Inside the breakwalls, the Port Austin State Dock **C** offers plenty of transient dockage and a boat ramp and sells fuel **D**. Chuck's Marine **E**, located about a quarter-mile up Bird Creek, does not cater to transients.

Distances from Port Austin (statute miles)
Oscoda: 30 NNW
Tawas City: 30 NW
Harbor Beach: 30 S

88 LAKELAND BOATING · LAKE HURON CRUISING GUIDE

LAKE HURON

Port Austin, Michigan
This quiet village offers big-city dining and entertainment.

Located at the tip of Michigan's Thumb, Port Austin basks in spectacular sunrises as well as sunsets over the water. With the completion of the small boat harbor in 1959—and an upgrade in 1990—the sleepy village, which encompasses all of one square mile, became a popular rest stop for cruisers traversing the Michigan coast of Lake Huron.

The entrance is well-marked by the Port Austin Reef Light, a square tan tower with a flashing white light and bright red roof. This must be kept to port to avoid the rocks between Pointe aux Barques and the tower. The entrance to Bird Creek is protected by two breakwaters. The one to the north on an east-west axis is marked by a flashing green; the breakwater to the west of the entrance lies on a north-south axis and is marked by a flashing red.

The popular, full-service **Port Austin State Dock** is located to the west of the river mouth. **Chuck's Marine**, located about a quarter-mile upstream, is equipped for repairs on all types of hulls but has transient dockage. **Thumb Marine** (989-738-5271) has been offering towing and repairs for the past 30 years.

Port Austin has plenty of provisioning stops, including a hardware store and a supermarket, along with interesting shops offering everything from folk art and curios to antiques and collectibles. The **Fisherman's Widow** has antiques and gifts with a nautical bent.

A craving for sweets can be indulged at **Finan's Store**, an old-fashioned soda fountain. **Murphy's Bakery** is renowned

More Info

Emergency	911
Huron Medical Center	989-269-9521
Chamber of Commerce	989-738-7600
	www.portaustinarea.com
Thumb Area Transit	800-322-1125

LAKELAND BOATING · LAKE HURON CRUISING GUIDE   **89**

Lat 44°03.71
Lon 82°53.20

Local events

May
Classic Car Show
Craft Show
Memorial Day Parade

June
Salmon Fishing Tournament
Pioneer Day
Tag Art Show

July
4th of July Parade

August
Sawmill Day

Located about five miles northeast of Port Austin along the shore, Grindstone City is home to two marinas, which occasionally handle transients. However, the harbor is accessible to small vessels only, and the approach is very hazardous during periods of low water. Harbor Marina **F** offers dockage, sells fuel and sometimes offers overnight stays, based on availability of seasonally rented slips. Captain Morgan's **G** is a resort with cabins, a restaurant and some transient slips.

for its doughnuts.

For fun head to the **go-cart track**, shoot a round of **mini golf**, take a **fishing charter**, sign up for **scuba lessons** or check out www.thumbtravels.com for even more ideas. Book an **evening cruise** with Capt. Fred Davis and hear tales about the Port Austin Reef and historic lighthouse. Theater fans shouldn't miss the **Port Austin Community Players.** Within the village are four parks and three designated national historic sites: the **Bank 1884,** the **Garfield Inn** and **Lake Street Manor Bed & Breakfast.** And the **Huron City Museums,** eight miles east of Port Austin, is a collection of 10 beautifully restored and furnished 19th-century buildings overlooking the lake.

Grindstone City

Located at the tip of the Thumb, Grindstone City was once a thriving locale, but its local industry, millstones, has long since gone the way of the passenger pigeon. What remains today is small, completely enclosed harbor. The entrance to the basin is marked with red and green daymarks that favor the portside jetty. **Captain Morgan's** and **Harbor Marina** occasionally have space for the odd transient boater. Captain Morgan's also offers cabins and a full-service restaurant that specializes in great seafood.

Where to Eat

The Bank 1884 989-738-5353
Located in a historic building less than a block from the harbor, this restaurant's cuisine rivals that of big-city restaurants. Dinner only; reservations suggested.

The Farm 989-874-5700
This highly rated restaurant features heartland specialties accompanied by fresh-picked vegetables. Call for pickup from the marina.

Finan's Store 989-738-8412
An old-fashioned soda fountain serving malts, banana splits, burgers, fries and chili dogs.

Garfield Inn 989-738-5254
This national historic site, dating to the 1830s, offers fine dining and lodging and is just three blocks from the harbor.

Murphy's Bakery 989-738-7192
Close to the dock, offering fresh breads, cakes, pastries, coffee and locally famous doughnuts.

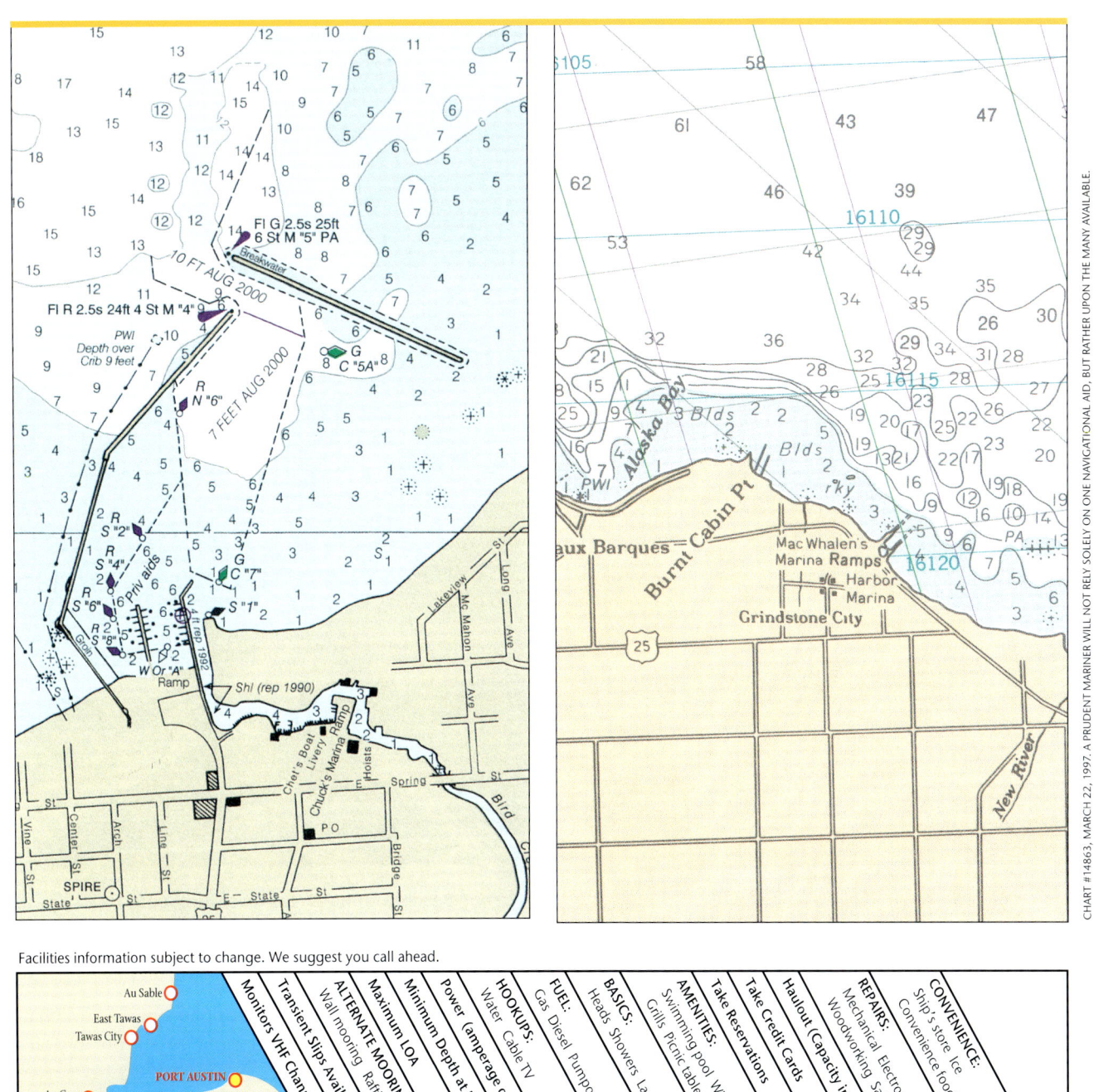

Facilities information subject to change. We suggest you call ahead.

			Monitors VHF Channel	Transient Slips Available	ALTERNATE MOORING: Wall mooring / Rafting allowed	Maximum LOA	Minimum Depth at Dock	Power (amperage or volts)	HOOKUPS: Water / Cable TV	FUEL: Gas Diesel Pumpout	BASICS: Heads Showers Laundry	AMENITIES: Swimming pool Whirlpool Rec area / Grills Picnic tables Dog walk	Take Reservations	Take Credit Cards	Haulout (Capacity in tons or feet)	REPAIRS: Mechanical Electronics Fiberglass / Woodworking Sails Canvas	CONVENIENCE: Ship's store Ice / Convenience foods & beverages
G	Captain Morgan's 989-738-6050	3337 Pointe aux Barques Rd. Grindstone City, MI 48467		2		32'	3'	30	W				GPD	Y	Y			IC
E	Chuck's Marine 989-738-2628	119 E. Spring St. Port Austin, MI 48467												Y	15	MFW	SC
F	Harbor Marina 989-738-7558	3379 Pointe aux Barques Rd. Grindstone City, MI 48467		68	4	29'	3'			G	H		D	Y	Y			IC
C	Port Austin State Dock 989-738-8712	8787 Lake St. Port Austin, MI 48467	9 16	49	W	55'	2'	30	W	GDP	HS		RPGD	Y	Y	M		IC

LAKELAND BOATING · LAKE HURON CRUISING GUIDE

Harbor Beach, Michigan

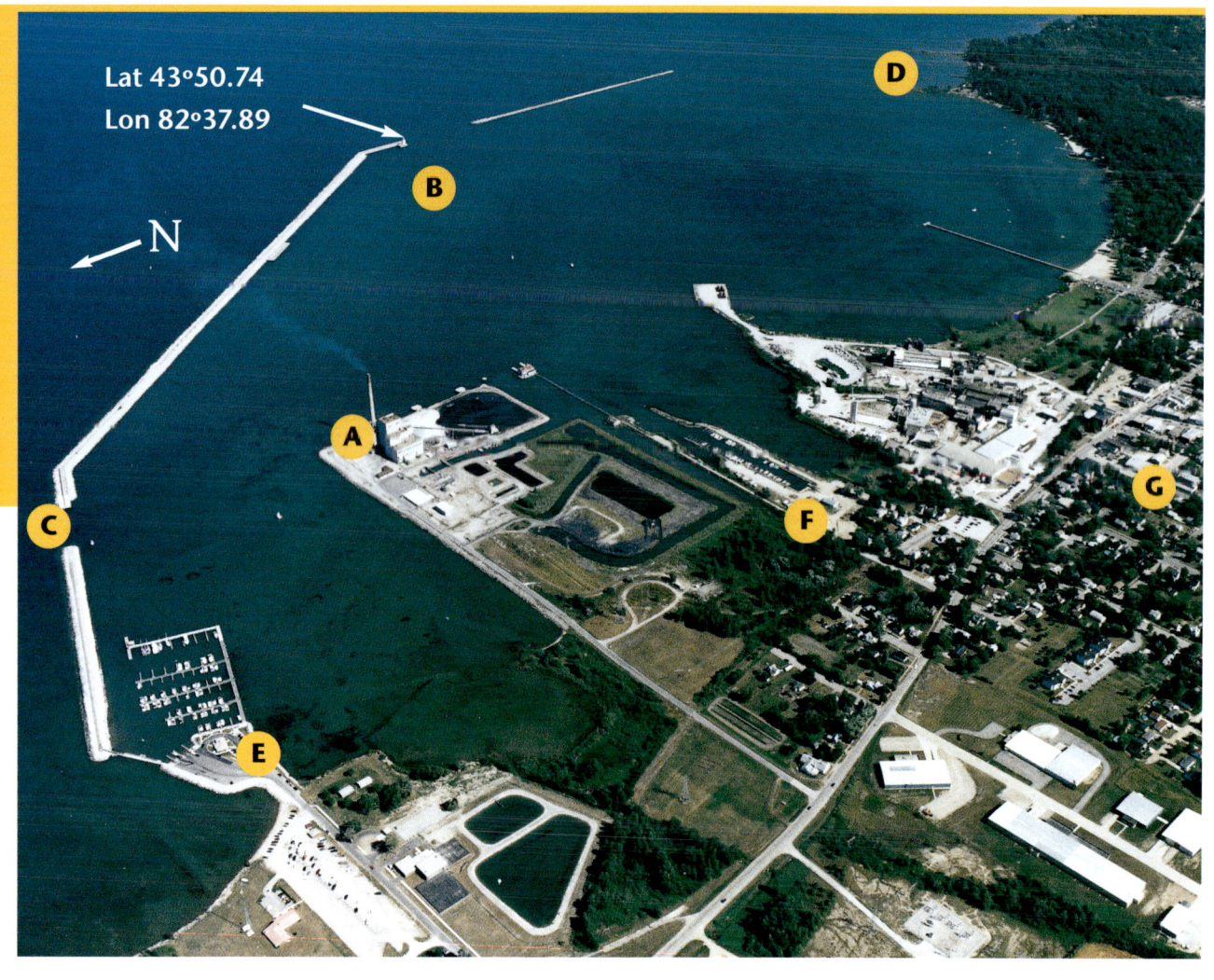

Lat 43°50.74
Lon 82°37.89

When approaching Harbor Beach, look for the strobe-topped stack of the power plant **A**. There are two entrances in the breakwater; the preferred entrance, between the south and middle breakwalls, is marked by the Harbor Beach Light **B**. Vessels with drafts less than 7 feet can enter between the northern and middle breakwalls **C**. Do not attempt to enter from the south **D**.

Harbor Beach Marina **E** can accommodate boats up to 125 feet and does minor repairs. Enter the harbor through the main entrance and bear to starboard. Follow the green spar buoys and hug the north breakwall into the basin. Off Shore Marina **F**, a popular option for shoal-draft powerboats, is located close to Harbor Beach's downtown area **G**.

Facilities information subject to change. We suggest you call ahead.

			Monitors VHF Channel	Transient Slips Available	ALTERNATE MOORING: Wall mooring / Rafting allowed	Maximum LOA	Minimum Depth at Dock	HOOKUPS: Water / Cable TV	Power (amperage or volts)	FUEL: Gas / Diesel / Pumpout	BASICS: Heads / Showers / Laundry	AMENITIES: Swimming pool / Whirlpool / Rec area / Grills / Picnic tables / Dog walk	Take Reservations	Take Credit Cards	Haulout (Capacity in tons or feet)	REPAIRS: Mechanical / Electronics / Fiberglass / Woodworking / Sails / Canvas	CONVENIENCE: Ship's store / Ice / Convenience foods & beverages
E	Harbor Beach Marina 989-479-9707	1 Ritchie Dr. Harbor Beach, MI 48441	16	74	WR	125'	7'5"		30 50	W	GDP	HS	PD	Y	Y	M	IC
F	Off Shore Marina 989-479-6064	5 Lytle Ave. Harbor Beach, MI 48441	16 68	20		35'	4'		20	W	GP	HS	PD	Y	Y		SIC

LAKE HURON

Harbor Beach, Michigan
Superlative fishing in the largest manmade harbor on the Great Lakes.

When Congress authorized the construction of a harbor of refuge at Sand Beach in 1873, it wasn't long before the town was the proud home of the largest manmade freshwater port in the United States—and had changed its name to Harbor Beach. The expansive **Harbor Beach Marina** can dock a 1,000-foot ore carrier, making this one of the few harbors that can accommodate superyachts.

The port has become one of the most popular cruising stops on Lake Huron and has been recognized as a good layover for north-south shipping since the harbor was constructed. It's also a favorite among anglers for its charterboat, breakwall and shore fishing.

Thanks to the power plant stacks that can be seen from more than 30 miles away on a clear day, it is an exceptionally easy landfall. The U.S. Coast Guard station to the north of town has a tall radio antenna. Another lofty radio tower is located inland about 500 yards. Give the shoreline a wide berth to avoid the prevalent shoals.

There are two harbor entrances in the breakwater. Harbor Beach Light marks the main entrance to the east and is the preferred route. Inside, cruisers can anchor out or take a slip at Harbor Beach Marina. Shoal-draft powerboats can dock to the south of the power plant at **Off Shore Marina.**

The town is located right behind the harbor and has much to offer, including a **beach, nature trails,** several **parks,** two nice **museums** and a **theater.** Those docked at Harbor Beach Marina will have a 15-minute hike to downtown, but the marina operates a courtesy shuttle van, and there is a bike path as well.

The **Grice Museum,** a gothic stone structure that's listed in the national registry, houses numerous local artifacts and includes an old-fashioned schoolroom. Another interesting museum is the former **home of Frank Murphy,** who was governor of Michigan, governor-general of the Philippines and a U.S. Supreme Court justice.

The **Harbor Beach lighthouse,** located on the north side of the main harbor entrance, was built in 1885. Its original fourth-order Fresnel lens, made in Paris in 1884, is now in the Grice Museum. Harbor Beach is also home to the first U.S. Lifesaving Station on the Great Lakes, built in 1881, which is still operating.

Visitors can find more information at the welcome center, located on South Huron Avenue next to the Murphy house. And those planning to be in the area the third weekend in July won't want to miss the annual **Maritime Festival,** which features lectures and exhibits on Harbor Beach's maritime history, along with entertainment, car and boat shows, a fish fry and games for the kids.

More Info
Emergency	911
Harbor Beach Community Hospital	989-479-3201
Chamber of Commerce	989-479-6477
Thumb Area Transit	800-322-1125

Where to Eat
Ernesto's Pizza 989-479-9013
Famous "secret recipe" pizza and other specials.

**Harbor Light Inn Restaurant
& Cocktail Lounge** 989-479-6005
Dine in a pleasant, informal atmosphere.

Harbor Beach Anchorage 989-479-9494
A local favorite renowned for its broasted chicken by the bucket and ice cream.

Distances from Harbor Beach (statute miles)
Mackinaw City: 180 NW
Presque Isle: 112 N
Harrisville: 64 N
Tobermory, Ontario: 107 NNE
Port Sanilac: 29 S
Port Huron: 61 S

CHART #14862, APRIL 13, 2002. A PRUDENT MARINER WILL NOT RELY SOLELY ON ONE NAVIGATIONAL AID, BUT RATHER UPON THE MANY AVAILABLE.

LAKELAND BOATING · LAKE HURON CRUISING GUIDE   93

Port Sanilac, Michigan

The approach to Port Sanilac is relatively free of hazards. Enter the harbor **A** from the south. The ends of both breakwalls are marked with lights. Anchor in the harbor or dock at Port Sanilac Harbor Commission's slips **B**, where there's power, gas and diesel, and pumpout. A ramp **C** is located nearby. There are some additional slips at Port Sanilac Marina **D**, which has a ship's store but no longer sells gasoline. Uri's Landing **E**, an upscale restaurant, overlooks the lake from the second floor of the marina building.

Lat 43°25.82
Lon 82°32.09

Distances from Port Sanilac (statute miles)
Harbor Beach: 29 N
Tobermory, Ontario: 155 NNE
Kincardine, Ontario: 79 NE
Goderich, Ontario: 52 ENE
Lexington: 13 S
Sarnia, Ontario: 37 S

LAKE HURON

Port Sanilac, Michigan
A historic port offering relaxation and wreck-reation.

The oldest recreational harbor in the state, Port Sanilac is sometimes overlooked in favor of nearby Port Huron, and cruisers heading north often speed by as they shoot for the tip of the Thumb and the Saginaw Bay crossing. But little-known Port Sanilac is one of the most historic and beautiful ports on the coast.

Settled in 1848 and known then as Bark Shanty Point, Port Sanilac was one of the first boom towns in the Thumb. It was named for Sanilac, a Wyandot chief who gained a reputation as a great warrior in the conflict between the Iroquois and the Wyandots. "Sanillac," a poem published in 1831 by Maj. Henry Whiting, immortalized Sanilac's love for a Native American maiden named Wona.

Entrance to the harbor, which was built in 1951, is to the south. Well-protected by two jetties, it is not visible from directly offshore. The red and green entrance lights, however, can be seen, so be sure not to try to make the entrance from the east.

The state dock, **Port Sanilac Harbor Commission**, is located in the northern bight of the harbor and offers most services. Next door to the south is **Port Sanilac Marina** (www.portsanilac-marina.com), which is a dealer for MerCruiser and Yanmar. The marina has some transient slips but no longer sells gas. The attractive two-story marina building houses a ship's store on the ground floor and **Uri's Landing** restaurant above. A town park overlooking the harbor has grills and picnic tables.

Port Sanilac boasts more than 40 centennial buildings. Of particular interest to cruisers is the historic 1886 brick lighthouse with its octagonal tower overlooking the basin to the south. The **Sanilac County Historical Museum** is housed in the old **Loop-Harrison House**, which was built in 1872 for the town's first doctor, Joseph Loop. Displays include Victorian-era furnishings and an old-time medical office. The shopping district, one block inland and a few blocks to the south of the harbor, is a magnificent stretch of turn-of-the-century buildings. **Raymond Hardware** is an

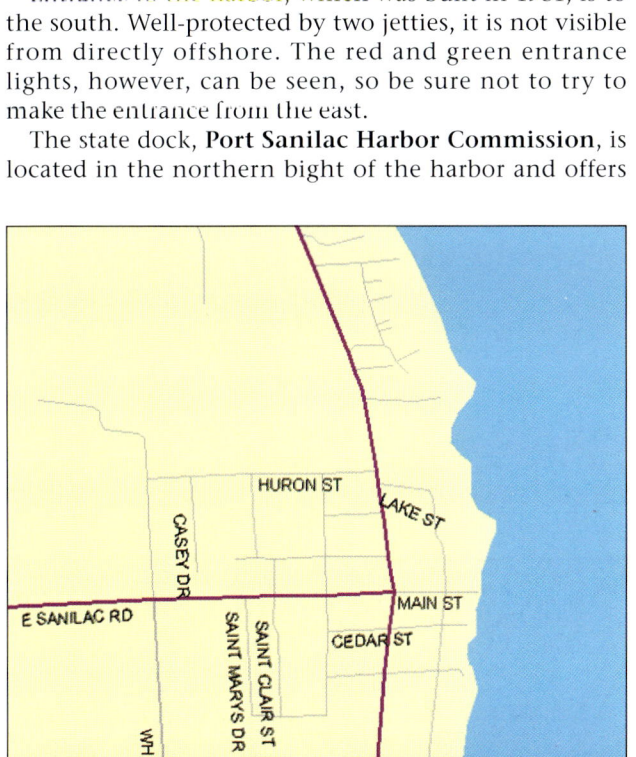

More Info

Emergency	911
Coast Guard	810-984-2602
Port Sanilac Village Offices	810-622-9963

Where to Eat

Bellaire Lodge & Motel 810-622-9981
This century-old Victorian house is home to what many say is the best fish fry in the Thumb. The glass-enclosed porch overlooks the gardens.

Mary's Diner. 810-622-9377
Head here for hearty home-cooked fare and breakfast.

Uri's Landing 810-622-9470
Located above the chandlery in Port Sanilac Marina, Uri's has a great view of the lake and is open for lunch and dinner, along with breakfast on weekends.

LAKELAND BOATING · LAKE HURON CRUISING GUIDE

authentic historic site as well as a working store. Also worth a visit is **Liberty Rose Antiques**.

For an evening's fun, be sure to attend an evening performance of the **Barn Theater**, which is, in fact, held in a barn. The town is also known for its annual summer festival in July. For more adventurous recreation, don a mask and fins and explore one of the 100 wrecks in the **Sanilac Shores Underwater Preserve** through **Four Fathoms Diving** (810-622-3483), or spend the day fishing for salmon and trout with the charter fleet.

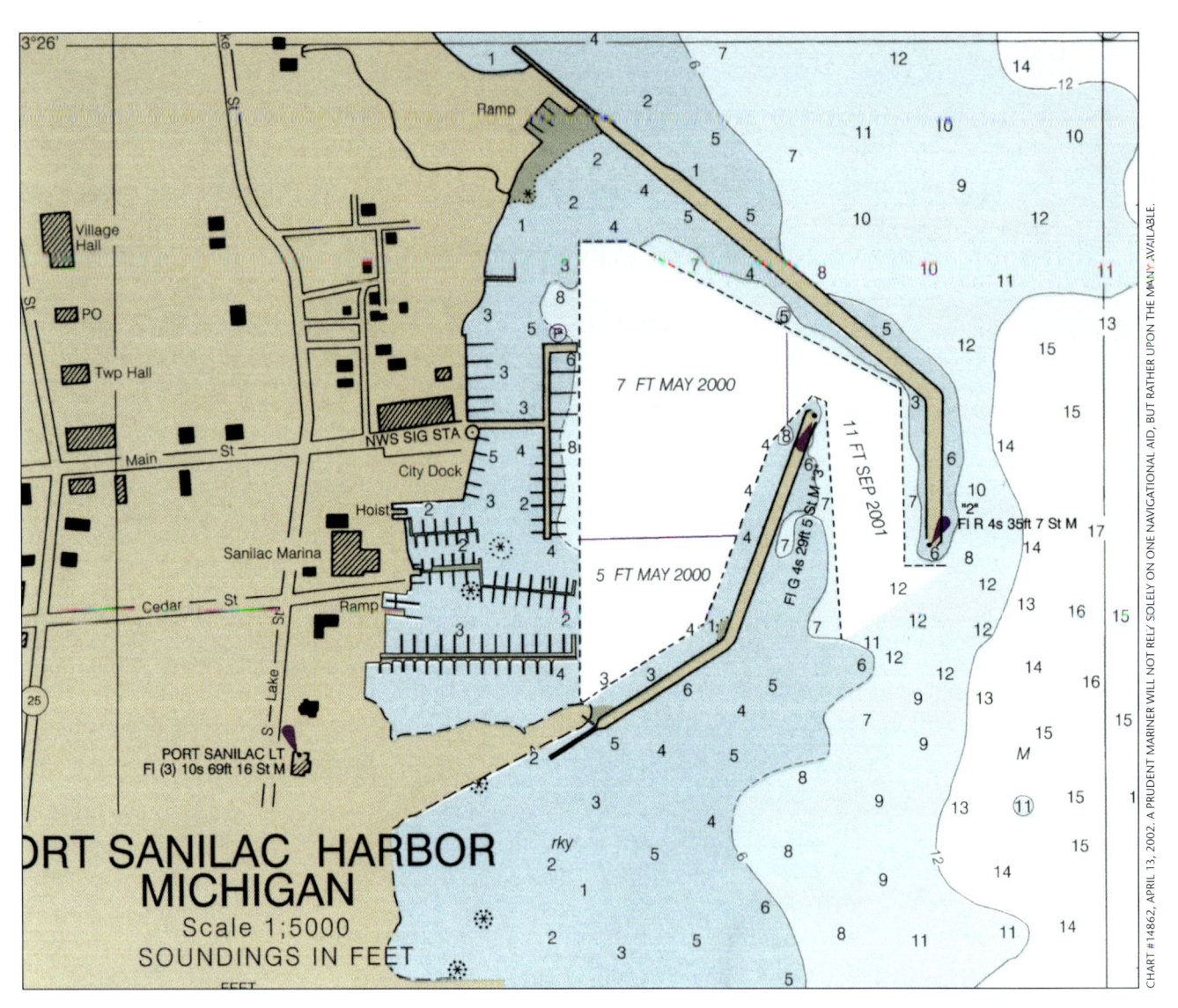

Facilities information subject to change. We suggest you call ahead.

		Monitors VHF Channel	Transient Slips Available	ALTERNATE MOORING: Wall mooring / Rating allowed	Maximum LOA	Minimum Depth at Dock	Power (amperage or volts)	HOOKUPS: Water, Cable TV	FUEL: Gas, Diesel, Pumpout	BASICS: Heads, Showers, Laundry	AMENITIES: Swimming pool, Whirlpool, Grills, picnic tables, Dog walk, Rec area	Take Reservations	Take Credit Cards	Haulout (Capacity in tons or feet)	REPAIRS: Mechanical, Electronics, Fiberglass, Woodworking, Sails, Canvas	CONVENIENCE: Ship's store, Ice, Convenience foods & beverages
B	Port Sanilac Harbor Commission 7376 Main St. 810-622-9610 Port Sanilac, MI 48469	16	40	WR	110'	10'	30 50	W		GDP	HS	GPD	N	Y			I
D	Port Sanilac Marina, Inc. 7365 Cedar St. 810-622-9651 Port Sanilac, MI 48469	16	20		60'	5'	30	W		HS		Y	Y	27	MF	SIC

96   LAKELAND BOATING · LAKE HURON CRUISING GUIDE

Web Updated Daily!

A ONE-TWO PUNCH FOR SELLING BOATS!

1. Lakeland Boating Magazine
2. www.lakelandboating.com

The Premier Online Boat Search for the Great Lakes

- Buyers Can Search by Length, Manufacturer, Year and Price
- Reach More Than 4.5 Million Potential Buyers
- Easy to Use, Easy to Order, Easy to Update

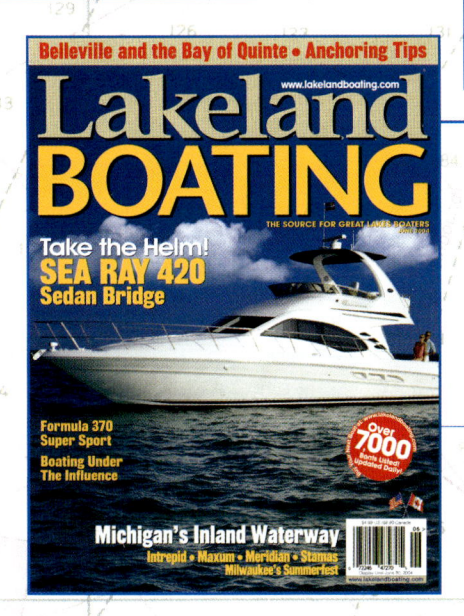

With more than **4.5 million boaters** on the Great Lakes and over **54,000 *LB* readers**, it's clear that the best way to sell your boat is to advertise online at www.lakelandboating.com.

2 Easy Ways to Order:
- **Online:** www.lakelandboating.com
- **By Phone:** 800-892-9342

Lexington, Michigan

The waters around Lexington are mostly free of hazards. Keep your eyes peeled for the lights—red on the right and green on the left—that mark the entrance through the breakwaters **A**, entering from the south. Dockage is available at the full-service Lexington State Dock **B**. Oldford's Marina **C** is a private facility that accepts transients based on availability. The area east of the docks is a popular anchorage. Though the channel was recently dredged, many boaters find space to beach themselves on the 3-foot-wide sandbar **D** along the east breakwall. There's a beach **E** with a playground **F** just north of the harbor.

Lat 43°16.00
Lon 82°31.37

Facilities information subject to change. We suggest you call ahead.

			Monitors VHF Channel	Transient Slips Available	ALTERNATE MOORING: Wall mooring Rafting allowed	Maximum LOA	Minimum Depth at Dock	Power (amperage or volts)	HOOKUPS: Water Cable TV	FUEL: Gas Diesel Pumpout	BASICS: Heads Showers Laundry	AMENITIES: Swimming pool Whirlpool Rec area Grills Picnic tables Dog walk	Take Reservations	Take Credit Cards	Haulout (Capacity in tons or feet)	REPAIRS: Mechanical Electronics Fiberglass Woodworking Sails Canvas	CONVENIENCE: Ship's store Ice Convenience foods & beverages
B	Lexington State Dock 810-359-5600	7411 Huron St. Lexington, MI 48450	16	108	R	100'	5'	30 50	W	GDP	HS		RPD	Y	Y			IC
C	Oldford's Marina 810-359-5410	7412 Huron St. Lexington, MI 48450		3		45'	6'	30 50	W		HSL		RGPD	Y	Y			IC

98 LAKELAND BOATING · LAKE HURON CRUISING GUIDE

LAKE HURON

Lexington, Michigan
Make this 'berry' attractive port your first stop.

After the great storm of 1913 destroyed its harbor and docks, Lexington almost faded into insignificance. But today the quaint Victorian town, with its well-protected harbor, offers a relaxing alternative to bustling Port Huron and industrialized Sarnia, Ontario, just 18 miles to the south.

Once known as "Berry Town," Lexington was founded in 1855 and quickly boomed as a farming and shipping port. The town boasted an organ factory, a flour mill, fishing docks, a brewery and six saloons. A favorite summer resort since the early 1990s, it remains a picturesque community, with four homes on its trim main street listed in the National Register of Historic Places.

The harbor of refuge is protected by overlapping jetties and a southern entrance; take care if arriving after dark not to come in on a westerly heading. The town is also identifiable by a large silver water tower to the south.

The **Lexington State Dock**, with more than 100 slips available, sells gas and diesel and provides pumpout as well as bathroom and shower facilities. The adjacent **Oldford's Marina** occasionally offers transient dockage. The two marinas are just a short distance from town and adjacent to a pretty town park with picnic tables, barbeque grills and a playground. One mile west of town, the public **Lakeview Hills Country Club & Resort** has a nice 36-hole course. Those inclined to underwater adventures can check out the three wrecks off the harbor.

Grocery and **hardware** stores are within walking distance of the marinas, along with **antique shops, gift stores** and **restaurants**. Stop in at the beautifully restored **Charles H. Moore Public Library**, next to the village hall on Huron Street. Built in 1859, this imposing structure boasts a stun-

Local events

May
Memorial Day Parade

June
Spring Crafts Show

July
Patriotic Parade
Independence Day Festival
Black River Art Show
Music in the Park

August
Fine Arts Street Fair
Blue Water Longest Antique Show
Thumb Folk Fest

Distances from Lexington (statute miles)
Harbor Beach: 41 N
Port Huron: 21 S
Grand Bend, Ontario: 38 E
Goderich, Ontario: 52 E
Tobermory, Ontario: 168 NNE

ning interior with stained-glass windows and a graceful staircase.

The **public boat launch** stays busy with anglers fishing for trophy salmon and trout.

More Info

Emergency	911
Port Huron Hospital	810-987-5000
Lexington Business Association	810-359-5175
Lexington Tourism	www.lexingtonmichigan.com
City Cab	810-984-4109

Where to Eat

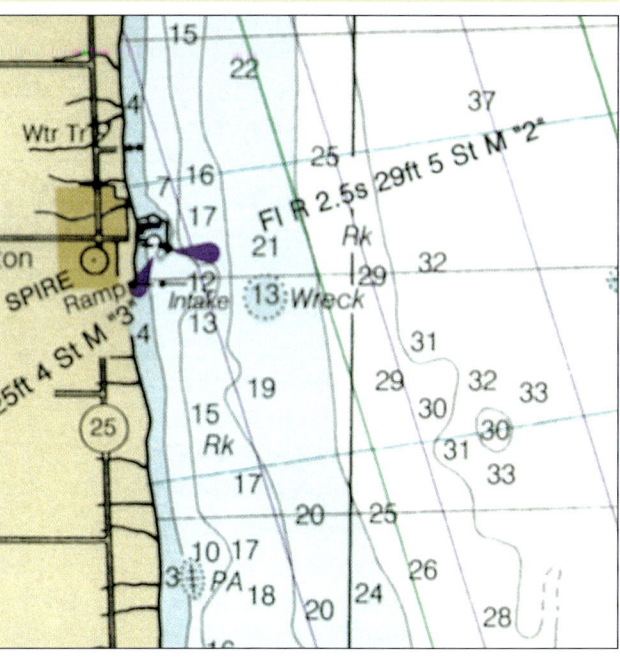

Cadillac House 810-359-7201
Stop in to try their famous broasted chicken or the Saturday surf and turf special.

Wimpy's 810-359-5450
This diner is paradise for the breakfast lover.

100   LAKELAND BOATING · LAKE HURON CRUISING GUIDE

Yacht Clubs
Clubs can offer great perks, but cruising is the best fellowship.

Many yacht clubs across the Great Lakes have limited dock facilities, if any at all. Clubs that belong to the association of yacht clubs are, in theory, supposed to offer reciprocal privileges to affiliated transient members. Some are not only open to all transients but are the preferred place to stay because of their superior accommodations. Others are strictly private and closed to visitors.

Yacht clubs often have limited transient space, and you may find that your needs are best met at one of the marinas that cater especially to transients. On the other hand, if you already belong to a club for the social activities it offers, you will probably be able to share a good deal of this camaraderie with members of other clubs that you may visit in your travels. Many clubs are big on interclub sailboat racing, for example, and welcome competition from outside. Others organize cruises so members with similar interests can travel as a group.

Numerous rendezvous are held during the season, many sponsored by manufacturers. Others are organized by type: trawler gatherings, wooden boat shows or annual poker runs for high-performance boats. These, too, have a clubby feeling.

Keep in mind, though, that the very act of cruising creates a kind of fellowship that transcends any individual club membership or activity. Once you experience the fun of meeting new people, swapping stories and exchanging addresses and phone numbers, you will come to realize that cruising the Great Lakes is the ultimate yacht club. Anyone can join, and everyone is welcome.

Know Your Pollution Rules and Regs
It *is* easy being green.

Discharge of Oil
The Federal Water Pollution Control Act prohibits the discharge of oil or oily waste into or upon navigable waters and contiguous zones of the United States if such discharge causes a film or sheen upon, or discoloration of, the surface of the water, or if it causes a sludge or emulsion beneath the surface of the water. Violators are subject to a penalty of $5,000. Boats 26 feet in length and longer with enclosed engine compartments are required to carry a 5-inch-by-8-inch placard with this warning in machinery spaces or near bilge controls, and smaller boats should do so as well.

Dumping of Garbage
The Refuse Act of 1899 prohibits the throwing, discharging or depositing of any refuse—including trash, oil and other liquid pollutants—into navigable U.S. waters. Penalties are steep: up to a $25,000 civil fine, $50,000 criminal fine or five years imprisonment for each violation. The Act to Prevent Pollution from Ships, modified by the Protocol of 1978 (Marpol 173/78), prohibits plastic trash. A Marpol placard must be carried on some boats. Additionally, a Vessel Waste Management Plan placard must be posted on boats 40 feet or more in length.

Marine Sanitation Devices
Federal law prohibits the overboard discharge of untreated sewage anywhere on inland waters, which includes the Great Lakes, and within three miles of the coasts. To protect the waters we love, make sure you have a holding tank that can be pumped out at a dockside facility. MSD installations with both overboard and holding-tank capabilities must have their Y-valves "permanently closed" against overboard discharge everywhere on the Great Lakes. This means the valve must be sealed or padlocked.

Port Huron, Michigan & Sarnia, Ontario

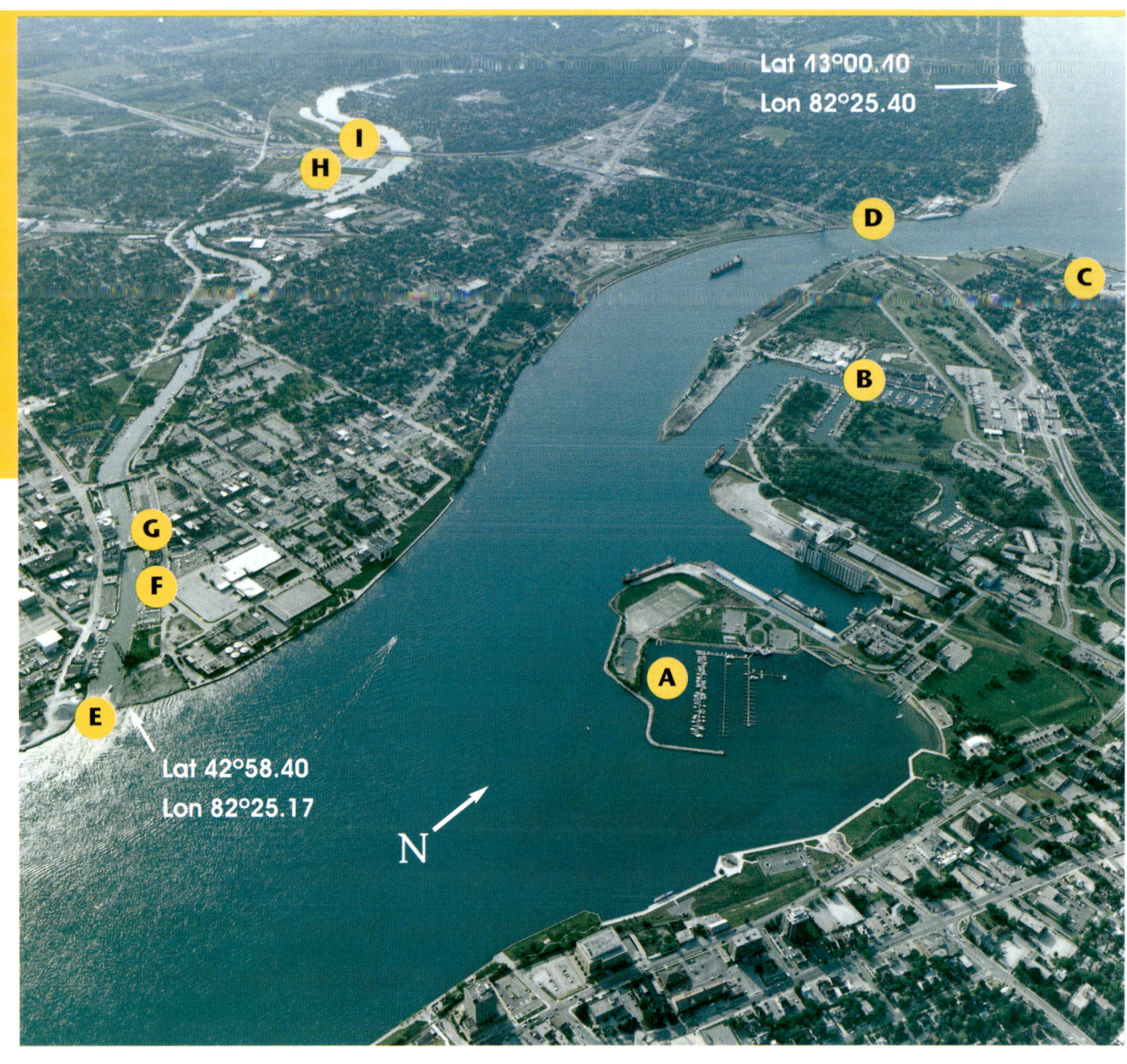

Lat 43°00.40
Lon 82°25.40

Lat 42°58.40
Lon 82°25.17

The St. Clair River is home to several marinas, including Sarnia Bay Marina **A**, where there's dockage and fuel; Bridgeview Marina **B**, a full-service facility; and Sarnia Yacht Club and Lake Huron Yachts **C** (for members of affiliated clubs), in a large sailing basin just beyond the Blue Water Bridge **D**. The Black River **E** flows into the St. Clair less than a mile south of the Blue Water Bridge, which has an ample 135-foot clearance. The Port Huron Yacht Club **F** grants dockage to sailors from recognized clubs. Just past the club are the municipal docks **G** at Fort, Quay and River streets. A slip at any of these facilities provides immediate access to town. Four bridges span the river in the next mile and a half. The large municipal Port Huron Water Street Marina **H**, located two miles upstream, will accept transients if the other facilities are full. Bridge Harbour Marina **I**, just north of the municipal marina, offers slips, along with a heated swimming pool, hot tub and other amenities.

Distances from Port Huron and Sarnia (statute miles)
Harbor Beach, Michigan: 66 N
Port Sanilac, Michigan: 37 N
Port Franks, Ontario: 36 NE

Grand Bend, Ontario: 40 NE
Goderich, Ontario: 62 NE
Tobermory, Ontario: 161 NE

LAKE HURON

Port Huron, Michigan & Sarnia, Ontario

Sister cities see you off on your Huron adventure.

Straddling the St. Clair River where it flows out of Lake Huron, the towns of Port Huron and Sarnia provide cruisers a last shore leave before setting out into the wide-open expanses of the second-largest Great Lake. Those approaching from Lake Huron will see the Blue Water Bridge from miles away—a welcome sight for anyone who has weathered a hard northerly.

Of the two small cities, Port Huron, on the west bank, is the more convenient for boaters. The town, which bills itself as "the Maritime Capital of the Great Lakes," boasts shops of all descriptions just a short walk from the marinas.

On the south bank of Port Huron's Black River, gas and marine services are available at **Black River Marine**. Opposite is the brand-new building of the **Port Huron Yacht Club**, which accepts sailors from recognized yacht clubs. To the west of the club are several municipal facilities, including the **Port Huron Municipal Marina**. Next door to that is **Bridge Harbour Marina**, which offers a number of transient slips and full amenities, including laundry, a clubhouse and a heated swimming pool and hot tub.

Sarnia's marine facilities are a longer walk from downtown, but all the marinas can be reached by public transportation or taxi from the main strip. As long as a northerly isn't kicking up Lake Huron, the **Sarnia Yacht Club** and **Lake Huron Yachts**, for sailboats

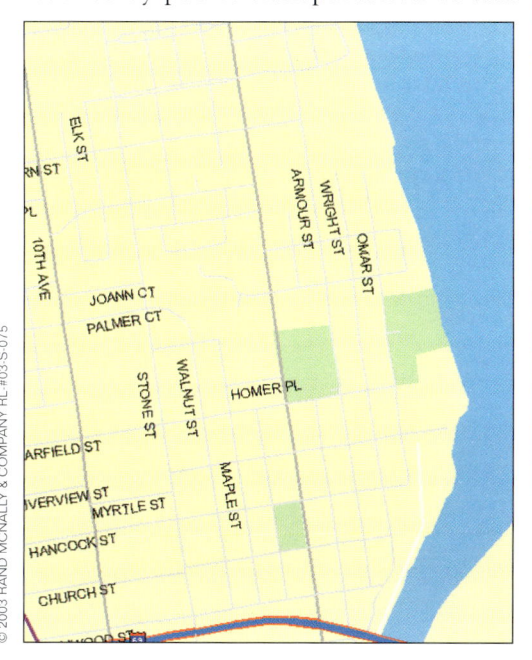

More Info

Emergency	911
Port Huron Hospital	810-987-5000
Lambton Hospital Group	519-464-4500
Tourism Sarnia-Lambton	800-265-0316
The City of Sarnia	www.city.sarnia.on.ca
Bluewater Area CVB	800-852-4242
The City of Port Huron	www.porthuron.org
City Cab	810-984-4109

This view of Port Huron up the Black River shows the Port Huron Water Street Marina **H** and the Bridge Harbour Marina **I**. Sarnia **J** is visible in the background across the St. Clair River, with Sarnia Bay Marina **A** on the river and the Sarnia Yacht Club's sailing basin **C** on Lake Huron.

LAKELAND BOATING · LAKE HURON CRUISING GUIDE   **103**

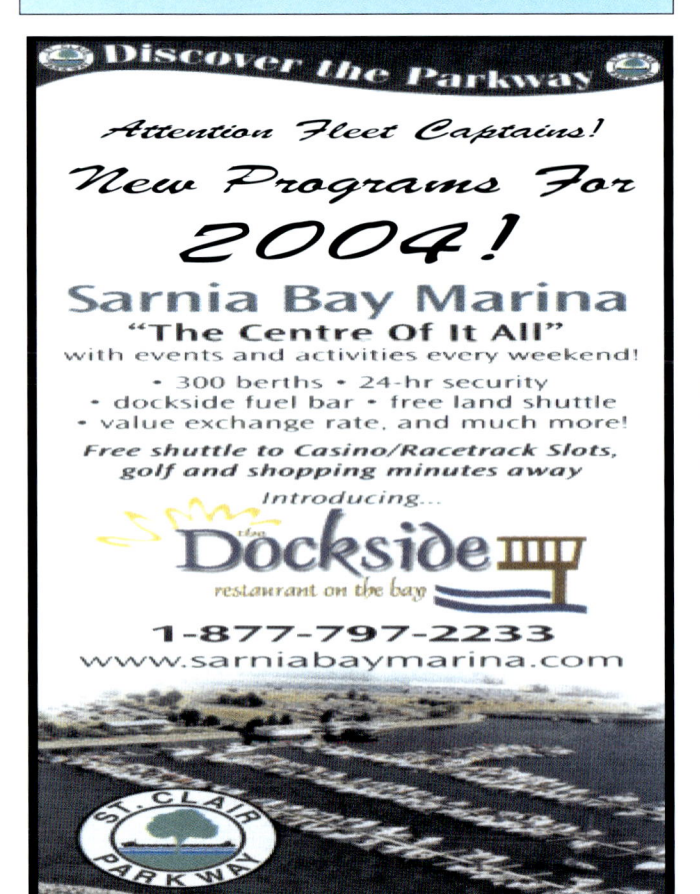

only on Huron's Point Edward offer the prettiest stopovers on the Canadian side. Otherwise, the first marina as you head downriver is the **Bridgeview Marina**, entered to the south of Bay Point. The pool here is not available to transients.

Farther downriver, **Sarnia Bay Marina** is entered at buoy RG "AHG," just to the north of the Black River. All of the Sarnia marinas offer convenient access to the St. Clair River and Lake Huron, and you don't have to wait on any lift bridges. Sarnia Bay Marina is the closest facility to downtown and is adjacent to **Centennial Park**. There's an on-site restaurant and two more across the street. A complimentary shuttle service takes you around on weekends during the summer.

Among the interesting sights in Port Huron is the *Huron* **Lightship Museum**, located in Pine Grove Park. This vessel, built in 1920, was stationed in Lake Michigan and later Lake Huron until 1970. The 82-foot-

Local events

May
Blue Water Classic Salmon Tournament

July
Port Huron to Mackinac Race

Antique & Classic Car Parade & Show

Blue Water Brass Drum & Bugle Corp Show

Southside Blues and Jazz Fest

Carnival

International Day Parade

Blue Water Volleygrass Tournament

Yale Bologna Festival

4-H County Fair

August
Blue Water Offshore Power Boat Races

Tally Ho Inn Classic Car Show

Blue Water Indian Celebration Powwow

Blue Water Area Antique/Yard Sale Trail

Rumble in the Park

Art in the Park

Antique and Classic Boat Show

Where to Eat

The Bridge Tavern 519-336-3143
Near the Sarnia Yacht Club, this tavern is a local institution, thanks to its delicious fish and chips.

The Fogcutter 810-987-3300
Atop the People's Bank building in Port Huron. Great food and view. Try the Swiss onion soup.

The Ivy Dining Room 810-984-8000
Gourmet sandwiches, seafood and steaks. Located in Port Huron's Thomas Edison Inn.

Rizzo's 519-337-7944
Good, wholesome food in a simple setting in Sarnia. An easy walk from your boat.

high **Fort Gratiot Lighthouse**—Michigan's oldest surviving lighthouse—to the north of the Blue Water Bridge is also worth a visit.

To enjoy a festive atmosphere and a thrilling sight, schedule your cruise for July to coincide with the start of the famous **Port Huron to Mackinac sailing race**, which takes place either the weekend before or the weekend after the Chicago to Mackinac race. Be sure to make your dock reservation early: This is the one event of the year that leaves the marinas packed to overflowing.

Sarnia boasts a challenging public golf course as well as the beautifully landscaped Centennial Park, with more than an acre of flower gardens. For sheer exuberance and a taste of old world culture, don't miss the annual **Sarnia Highland Games** in August.

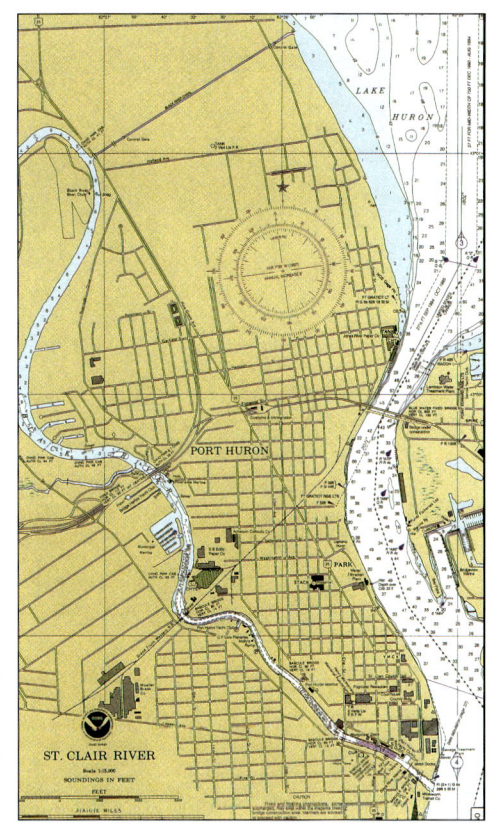

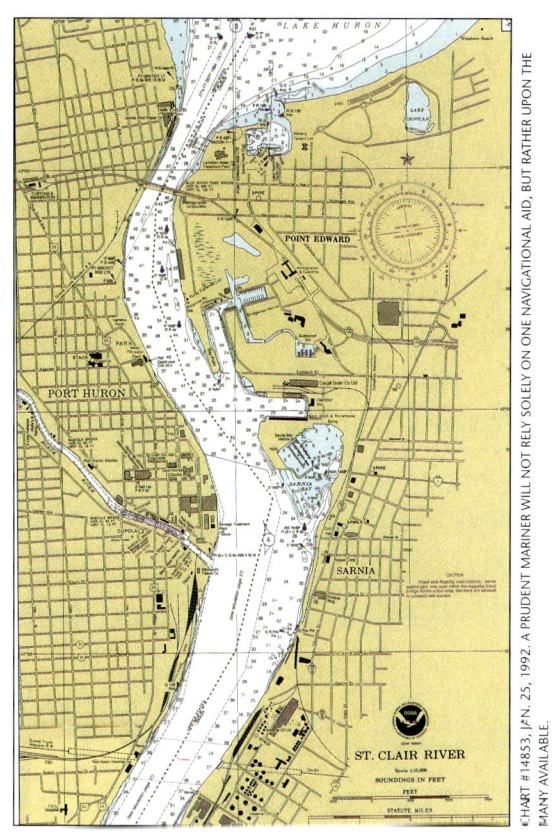

Facilities information subject to change. We suggest you call ahead.

	Marina	Address	Monitors VHF Channel	Transient Slips Available	ALTERNATE MOORING: Wall mooring / Rafting allowed	Maximum LOA	Minimum Depth at Dock	Power (amperage or volts)	HOOKUPS: Water / Cable TV	FUEL: Gas / Diesel / Pumpout	BASICS: Heads / Showers / Laundry	AMENITIES: Swimming pool / Whirlpool / Rec area / Grills / Picnic tables / Dog walk	Take Reservations	Take Credit Cards	Haulout (Capacity in tons or feet)	REPAIRS: Mechanical / Electronics / Fiberglass / Woodworking / Sails / Canvas	CONVENIENCE: Ship's store / Ice / Convenience foods & beverages
I	Bridge Harbour Marina 810-982-2492	2200 Water St. Port Huron, MI 48060	9 16	50	W	70'	8'	30 50	WC	P	HSL	SWRGPD	Y	Y		C	IC
B	Bridgeview Marina 519-337-3888	1 Marina Rd. Sarnia, ON N7T 7J7	68	20		72'	7'	30 50	WC	GDP	HSL	SPD	Y	Y	75	MEFWSC	SIC
H	Port Huron Water Street Marina 810-984-9745	2021 Water St. Port Huron, MI 48060	16	275		40'	5'	30 50	WC		HSL	GPD	N	Y		MEFWC	IC
F	Port Huron Yacht Club 810-985-9424	212 Quay St. Port Huron, MI 48060	16	2	WR	35'	7'	30	W	P	HS	GPD	N	N			IC
G	River Street Municipal Dock 810-984-9746	525 River St. Port Huron, MI 48060	16	65	WR	135'	7'	30 50	WC	GDP	HS	PD	N	Y			IC
C	Lake Huron Yachts 519-336-5575	1241 Sandy Lane Sarnia, ON N7V 4J9	16	6	WR	50'	7'	30	W	P	HS	RPD	Y	N	5		
A	**Sarnia Bay Marina** 519-332-0533	**97 Seaway Rd.** Sarnia, ON N7T 8E6	68	160	WR	110'	6'	30	W	GDP	HSL	GPD	Y	Y			SIC
C	Sarnia Yacht Club 519-332-6779	1220 Fort St. Point Edward, ON N7V 1M2		Y	WR	45'	7'	110	W	GDP	HS	RGPD	Y	Y			IC

Port Franks, Ontario

Lat 43°14.01
Lon 81°54.25

The marinas and the town itself **A** are all located on the south side of the Au Sable River. As you enter, hug this shore to avoid shallow water. The first marina inside the entrance is the Rivermouth Marina **B**, which has a limited number of transient slips and full facilities, including a bar and grill. Next door, Kennedy's Landing **C** is a private facility and does not offer services to transients. The next facility upriver with transient docking is Seven Winds Marina **D**. The Port Franks Conservation Area **E** maintains a large facility on the peninsula past the town. Lighthouse Marine **F** is a private facility that doesn't cater to transients. Boondocks Marina **G**, which is to the east past the conservation area, offers transient slips as well as gas and pumpout.

LAKE HURON

Port Franks, Ontario
Stopping in at this lovely spot is a natural.

The proximity of Port Franks to the St. Clair River—just 31 miles—might tempt those headed south to skip this scenic port, but they'd be missing out on a chance to enjoy beautiful dunes and beaches and perhaps get a glimpse of some rare birds and plants.

Cruisers headed north can hug the low-lying coast with its sandy shoreline. On a calm day, you can anchor out and swim or dinghy ashore to stretch your legs. But if the wind is blowing from the north, this can become a dangerous lee shore. And watch out at all times for the fish traps.

Port Franks is located on the west branch of the Au Sable River, about five miles east of Kettle Point. A sea buoy marks the western end of this rocky reef, which extends out two miles and breaks heavily in a northerly gale. From Kettle Point to Port Franks, the coast turns rocky. The only visible landmark is a 207-foot tower, which is located about three and a half miles east-northeast of the entrance.

The entrance, marked by a green buoy with a flashing green light, is kept dredged to a minimum depth of 4 to 5 feet, according to *Sailing Directions*, a publication of the Canadian Hydrographic Service. Locals say the depth is usually closer to 8 feet, except after a northeaster. The deepest water is generally found on the south side of the entrance, opposite the sandy point on the far shore. The river is reported to have depths of 3 to 8 feet for about five miles.

The marinas are all located along the south bank of the river. The other bank is largely undeveloped and ideal for exploring by dinghy. **Rivermouth Marina**, **Seven Winds Marina**, **Port Franks Conservation Area** and **Boondocks Marina** all offer dockage to transients. Gas is available at Boondocks and Rivermouth, which also has a Travelift for boats up to 25 tons. The **Rivermouth Marina Bar & Grill** overlooks the lake.

The town itself is also located on the south shore of the river. Visitors will find that it offers the necessities—including two stores selling groceries and other essentials. The town, which is within walking distance of the docks, is laid out in a grid consisting of just eight streets running north to south and four running east to west. Port Franks also has a church and a post office, but no liquor store, pharmacy or bank.

Nature lovers will find much here to see and do. **Pinery Provincial Park**, one of the largest forests in southwest Ontario, extends along the shore for more than four miles northeast of Port Franks and offers nature trails, picnic areas and 1,000 campsites. The **Port Franks Forested Dunes Important Bird Area**, meanwhile, boasts dramatic dunes and a surprisingly intact woodland ecosystem.

Where to Eat

Harbourside Family Restaurant 519-243-2461
The Alaskan pollack and halibut are a great deal, and if you're lucky, there'll be pan-fried pickerel.

MacPherson's Restaurant 519-243-2990
This family-oriented place prides itself on home-cooked meals ranging from pasta to roast beef.

Rivermouth Marina Bar & Grill 519-243-3636
Grab a drink or something to eat while enjoying the view of the lake and river.

More Info

Emergency 911
Strathroy Middlesex
General Hospital 519-235-2700
Grand Banks
Chamber of Commerce 519-238-2001
Tourism Sarnia-Lambton 800-265-0316
Pinery Provincial Park 519-243-2220
Lakeshore Taxi 519-238-5266

You'll find more than 320 bird and 800 plant species among the forests and wetlands, including tulip trees and endangered Pitcher's thistles and bluehearts, the endangered Acadian flycatcher and the threatened hooded warbler, redheaded woodpecker and red-shouldered hawk.

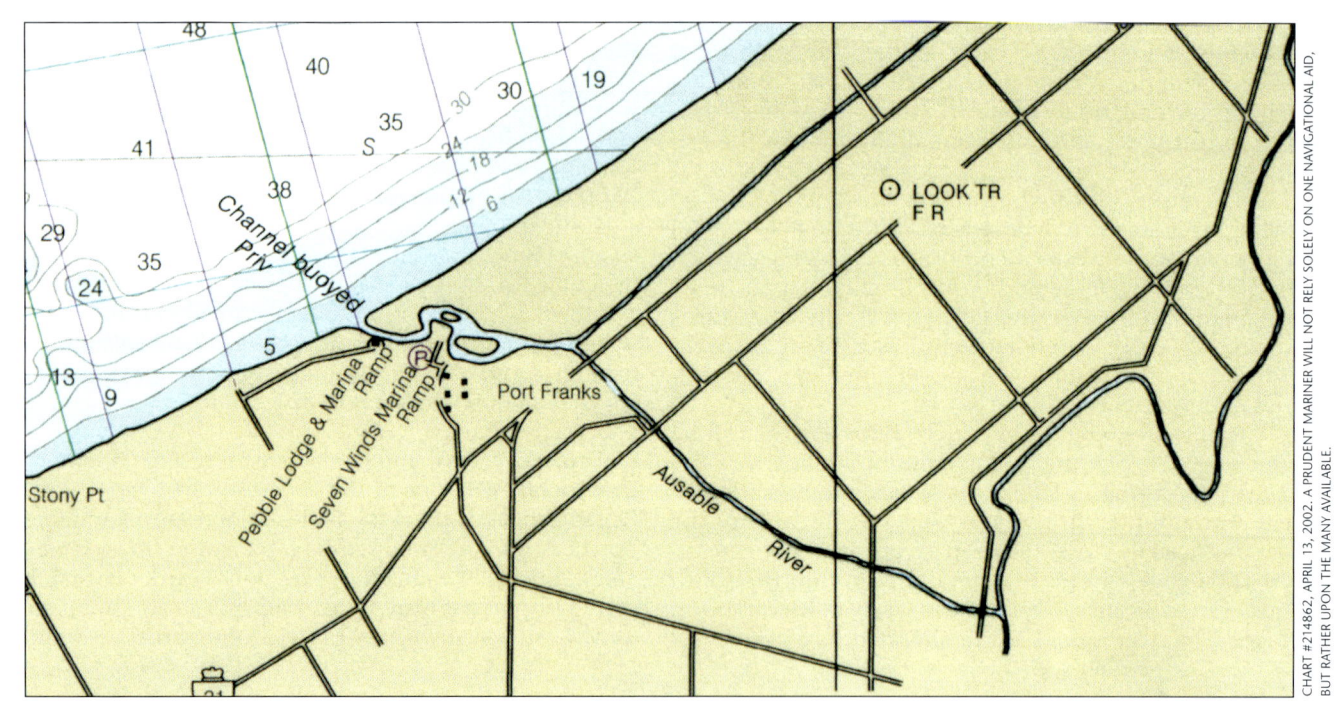

Facilities information subject to change. We suggest you call ahead.

	Marina	Address	Monitors VHF Channel	Transient Slips Available	ALTERNATE MOORING: Wall mooring / Rafting allowed	Maximum LOA	Minimum Depth at Dock	Power (amperage or volts)	HOOKUPS: Water / Cable TV	FUEL: Gas / Diesel / Pumpout	BASICS: Heads / Showers / Laundry	AMENITIES: Swimming pool / Whirlpool / Rec area / Grills / picnic tables / Dog walk	Take Reservations	Take Credit Cards	Haulout (Capacity in tons or feet)	REPAIRS: Mechanical / Electronics / Fiberglass / Woodworking / Sails / Canvas	CONVENIENCE: Ship's store / Ice / Convenience foods & beverages
G	Boondocks Marina 519-243-3636	7625 Biddulph St. Port Franks, ON N0M 2L0	68	5	WR	44'	5'	15	W		GP	HS	D	Y	Y		IC
E	Port Franks Conservation Area 519-243-2354	7574 Biddulph St. Port Franks, ON N0M 2L0	68	2		40'	3.5'	20	W			HS	RGPD	Y	Y		I
B	Rivermouth Marina 519-243-3636	10072 Poplar Ave. Port Franks, ON N0M 2L0	68	8	R	45'	5'	20	W		GP	HS	GPD	Y	Y	25	IC
D	Seven Winds Marina 519-243-1663	7562 Biddulph Street Port Franks, ON N0M 2L0		2	W	50'	4'	30	W			HS	GPD	Y	N		I

108 LAKELAND BOATING · LAKE HURON CRUISING GUIDE

LEAP OF FAITH

Going the Distance
It takes a lot of nerve to cross the lake for the first time.

JEAN LIZMORE

Moving out of the sight of land—especially for the first time—can be nerve-wracking. Those with nerves of steel and 30-knot cruisers who cross the Great Lakes without batting an eye may scoff, but many consider such an undertaking to be a major leap of faith. If you're toying with the idea of going across, here are a few points to consider.

Weather

In terms of miles, crossing the lake can be as adventurous as you care to make it. It can range from a relatively short shot—say, from Port Franks to Lexington—to a long-angle route such as Rogers City to Tobermory. But whatever your route, bad weather and high seas surely have to rank No. 2 on the list of things that can quickly spoil a cruise—right behind a fire at sea! The right day to cross can be determined by matching your cruising speed with the best possible weather forecast. If your crossing time will only be a few hours, any day that the waves look good to you and the weather suggests no serious changes will work. If, on the other hand, you are running a 6-knot sailboat or a 10-knot trawler and you will be a half-day or more on the open water, start early on a day when you won't be fighting head seas or passing through a storm front.

Navigation

In the old days (B.E., or before electronics), finding your port of choice on the other side was a scary experience. Hitting the other shore was not that big of a trick, but knowing whether to turn north or south after you got there was a different matter. Holding a course when you're out of sight of land is the boater's equivalent of flying through darkness, relying solely on instruments.

Today, with GPS and loran, the trip is no different than any other, except for the faith factor. When the shoreline disappears and there is nothing but water in every direction, you have to believe in those little liquid crystal numbers flashing on the screen.

Traffic

If you are accustomed to running the shoreline, you are also used to heavy traffic. Most offshore traffic is heading north or south, so you can easily anticipate it. During a lake crossing, you will be heading at right angles to the north/south traffic, especially the big boats and freighters. Figuring the angles on these close encounters is trickier than swerving right to avoid a head-on confrontation along the coast. Stay alert, especially in poor visibility.

Psychology

Crossing the lake for the first time is a major cruising experience. Once you get this trip behind you and duly noted in your log, your confidence level and your enjoyment of cruising will increase at least tenfold.

Many cruisers opt to make their crossings with another boat, on the assumption that there is safety—or at least courage—in numbers. This is good psychology, but bad logic. We know of one ocean-going 30-foot sailboat that offered to shepherd a little 19-foot outboard-powered sailboat across. The big inboard ended up with a vapor-locked engine and a tow from the outboard! Eventually, everyone did arrive safely but have yet to stop telling this cautionary tale.

Having a story, though, is the ultimate goal of boating that keeps most of us coming back for more. Every serious cruiser should look at every new port as the next worthwhile challenge—even if it is across the lake and out of sight.

Grand Bend, Ontario

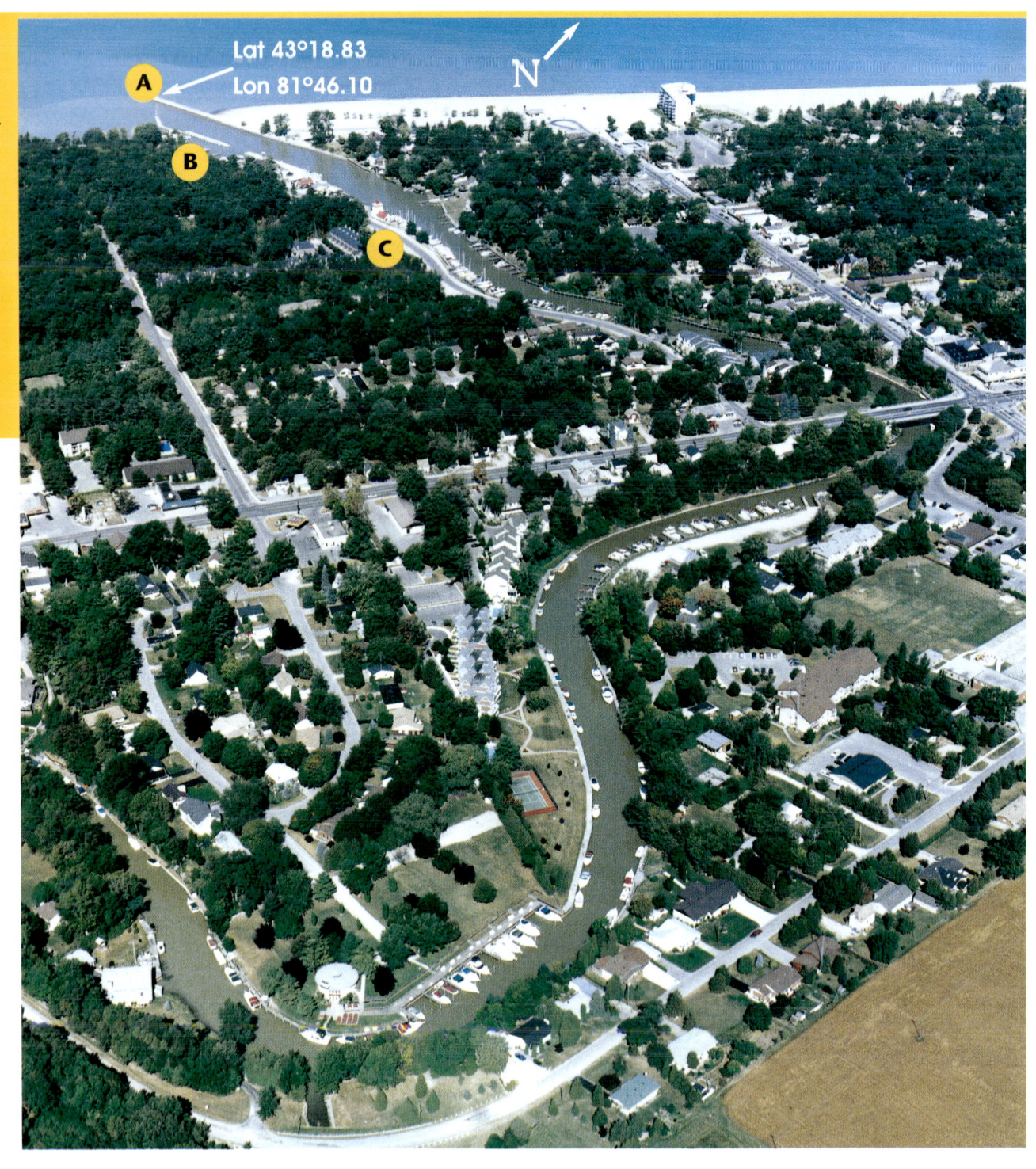

Lat 43°18.83
Lon 81°46.10

When approaching Grand Bend, beware the shallows around Kettle Point Shoal to the south and Dewey Point to the north. Enter the harbor between the two parallel piers painted day-glo orange. The north jetty **A** holds a flashing green light, the south jetty a flashing red light. This is a shoal draft harbor, so avoid entering during heavy seas. Inside the mouth of the river and immediately to starboard is Lambton Shores **B**, a popular spot for transient boaters. Just beyond is Grand Bend Yacht Club **C**, a private facility that normally doesn't cater to overnighters.

LAKE HURON

Grand Bend, Ontario
Rustic tranquility and an amazing beach.

Located on the north branch of the Au Sable River, Grand Bend is a small, tranquil village that blooms in the summer as its sidewalks fill with savvy tourists who know where to find a beautiful sandy beach and great shopping. The short stretch of Main Street that runs from the highway down to the beach is lined with shops of all description, and the town and its environs boast lots of attractions for cruisers with children.

Grand Bend is easy to spot from the water. One of the best landmarks is a large multi-story concrete building about one and a half miles north of the harbor entrance. There are also tall red and white TV towers to the east and south of the entrance. Both towers carry red air obstruction lights. There is a four-second isophase flashing green light 28 feet high on the end of the north pier and a flashing red light on the south pier. At night, the lights of the main street are easy to spot just to the north of the harbor mouth.

When entering the channel from the northwest after a period of heavy winds, sound your way in carefully, as silting can occur rapidly. Expect breaking seas in the shallower waters on both sides of the channel.

Once inside, the first facility you'll see is **Lambton Shores**, situated on the starboard bank. Depths of 5 to 7 feet are reported at the dock, where all services are available. You can't miss the next facility to starboard, the **Grand Bend Yacht Club**—the clubhouse is modeled after North Channel lighthouses like the one on Strawberry Island. They do not cater to transients directly but have an arrangement with the harbor office; the yacht club will fill any empty slips if the municipal ones are all taken. This happens quite often on holiday weekends.

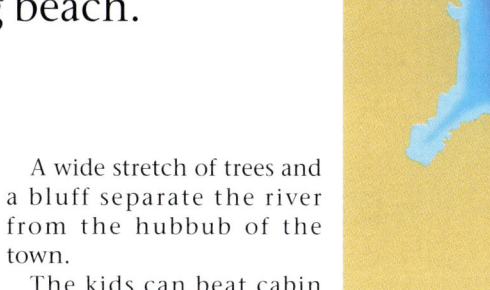

A wide stretch of trees and a bluff separate the river from the hubbub of the town.

The kids can beat cabin fever with a trip to **Sam's Playing Field**, a miniature golf course and arcade on Main Street, or to the nearby **batting cage**. Ride out to **Pinery Provincial Park** on the safe bike path or check out **Pineridge Zoo,** located a couple miles south of town on Highway 21. If you're looking for a little greased lightnin', Grand Bend boasts a fully sanctioned quarter-mile drag strip named **Thunder by the Beach**. The action is hot and loud Saturday evenings and Sundays at the facility, less than two miles south of town.

Despite its carnival atmosphere, Grand Bend does offer a number of more subdued sights and activities. Heading the list is the **Huron Country Playhouse**, which offers a changing playbill of professional stock in

Where to Eat

Aunt Gussie's 519-238-6786
Country cooking—fish and chips, shepherd's pie and pecan tarts—served up in a casual setting.

Colonial Hotel Dining Room 519-238-2371
Family dining just a short walk from the beach, featuring fresh Lake Huron fish.

Lakeview Café. 519-238-2622
Fine dining with a great view of the water. Try the rack of lamb and don't skip dessert!

Oakwood Inn Dining Room 519-238-2324
Delicious cuisine in a rustic log-cabin atmosphere. The Saturday and Sunday night buffets are popular, as is Sunday brunch, so make reservations.

More Info

Emergency 911
South Huron Hospital 519-235-2700
Tourism Sarnia-Lambton 800-265-0316
www.grandbendtourism.com
Chamber of Commerce 519-238-2001

LAKELAND BOATING · LAKE HURON CRUISING GUIDE

a charming converted barn just two miles east of town. The **Lambton Heritage Museum** features exhibits relating to pioneer days. Golfers are welcome at the **Oakwood Golf & Country Club**, located a quarter-mile north of town. The facility was recently renovated and now features a spa and fitness center, along with a 1920s-era restaurant and pub.

Distances from Grand Bend (statute miles)
Sarnia: 40 SW
Bayfield: 17 N
Goderich: 30 N
Lexington, Michigan: 38 W
Tobermory: 134 NNE

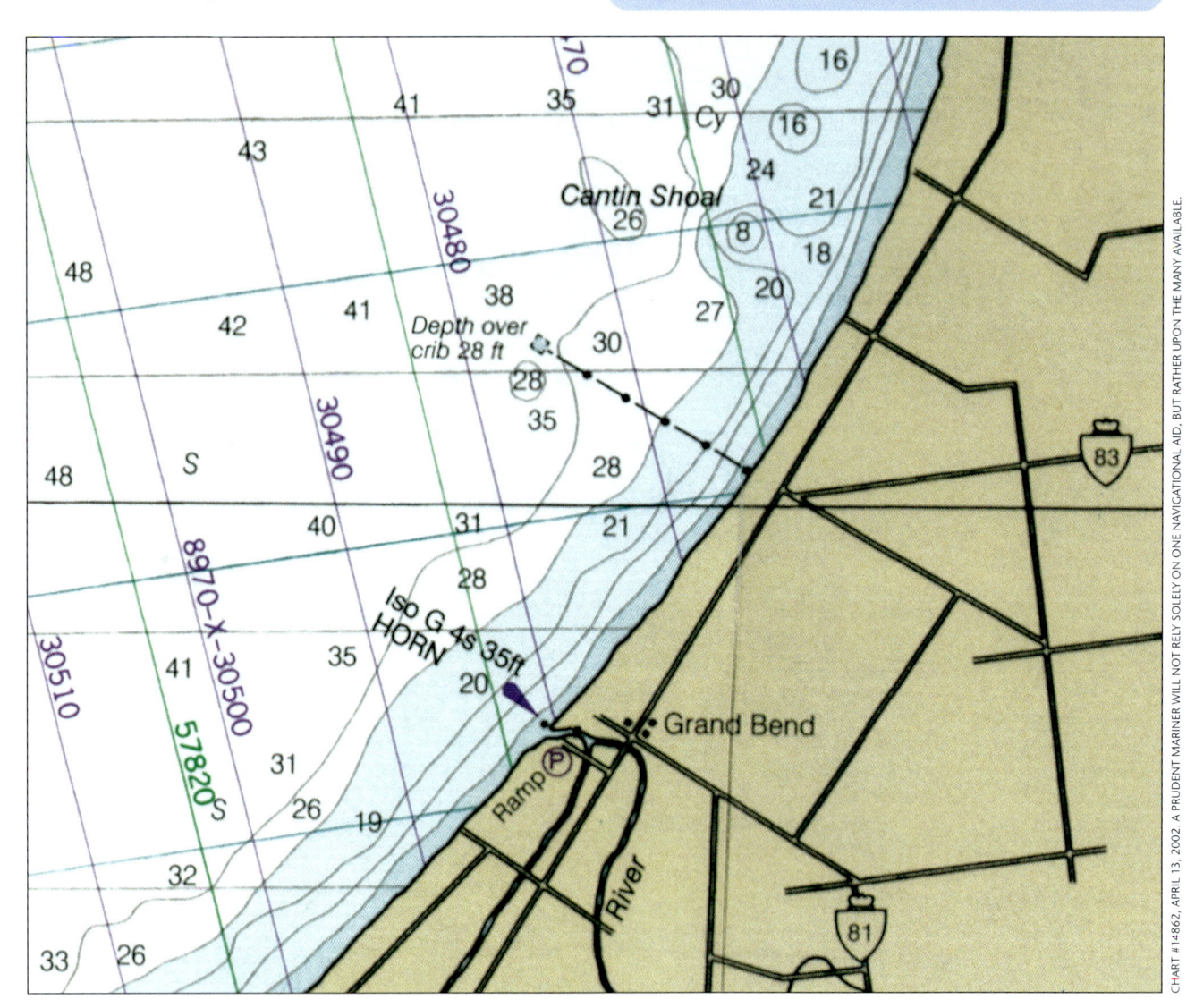

Lambton Shores 519-238-2000	P.O. Box 340 Grand Bend, ON N0M 1T0	68	80	WR	80'	5'	30	W	GDP	HSL	RGPD	Y	Y	IC

Facilities information subject to change. We suggest you call ahead.

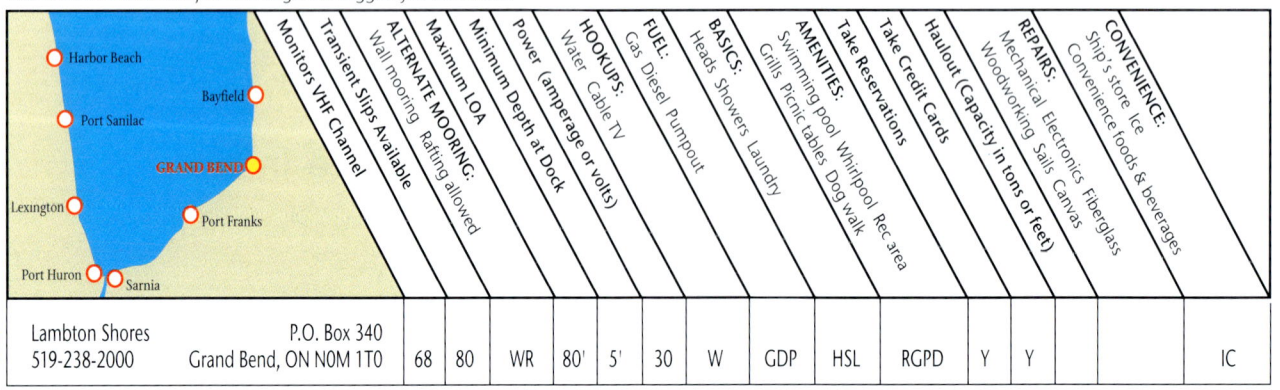

A Bone-Chilling Scenario
Boost your chances of surviving hypothermia.

Even if you're wearing a lifejacket that keeps your head above water, if you spend more than 50 minutes in 50-degree water, you have just a 50-50 chance of survival, according to the U.S. Coast Guard. Your chances slim to approximately one in 99 after three hours. Higher water temperatures improve your odds, which are also affected by factors such as your size, health, what you're wearing and how fit you are.

If your boat goes down, float quietly or huddle with others rather than exerting yourself. If possible, pull yourself up onto a floating object—an overturned hull, dinghy, life raft, or rescue platform—to get partially or totally out of the water. Water conducts heat away from your body roughly 25 times faster than air does.

After retrieving a victim from the water, remove wet clothes and wrap the person in blankets. Apply water bottles heated to body temperature to areas that lose a lot of heat, such as the head, neck, chest and groin. Warm, noncaffeinated drinks are good, but avoid alcohol, hot drinks, massage and exercise at all costs. Be prepared to administer CPR.

HYPOTHERMIA CHART

If water temperature (F) is:	Exhaustion or unconsciousness in:	Expected length of survival is:
32.5 degrees	Less than 15 minutes	Less than 15-45 minutes
32.5-40 degrees	15-30 minutes	30-90 minutes
40-50 degrees	30-60 minutes	1-3 hours
50-60 degrees	1-2 hours	1-6 hours
60-70 degrees	2-7 hours	2-40 hours
70-80 degrees	3-12 hours	3 hours-indefinitely
More than 80 degrees	Indefinitely	Indefinitely

Stormy Weather
This checklist will help you ride it out.

When rough weather is expected along your route, it's a good idea to postpone your departure from safe harbor. If you're already under way and a storm is brewing, reverse course if at all possible. If you can't, you have two options: put as much distance as possible between you and the storm center or give yourself plenty of sea room and prepare to ride it out. If you're sailing, reduce sail well before the arrival of the storm.

As you prepare for heavy weather, make sure you do the following:

❏ Check all through-hull openings. Close all seacocks, except those needed for engine and bilge pumps.

❏ Stow loose items in lockers and close securely.

❏ Lock the refrigerator door.

❏ Make sure that each crewmember is wearing raingear and a PFD. Harnesses are de rigeur for offshore sailboats.

❏ Establish a watch system to allow each crewmember sufficient rest.

❏ Secure or remove anchors stored on bow rollers.

❏ Disconnect the anchor chain and mechanically attach it to the deck pipe cap.

❏ Double the lashings on any deck- or transom-stored dinghy.

❏ Remove any small outboard motors, sailboards, cooking grills, fishing rods, fenders and other gear deployed along rails. Store everything below.

❏ Remove dorade vent cowls on sailboats. Turn all other ventilators away from the wind direction.

❏ Latch and lock all hatches, especially those smaller ones used solely for ventilation.

❏ Install companionway hatch boards on sailboats. Rig storm sails.

❏ Latch and lock all deck and cockpit lockers.

❏ Turn on navigation lights and make sure they're operational.

Bayfield, Ontario

Lat 43°34.17
Lon 81°42.55

On approach, look for the high clay bluff just south of the mouth of the Bayfield River. Green range lights on the north side of the entrance **A** lead into the harbor on a bearing of 76°T. Most transients tie up at the expansive Harbour Lights Marina, which occupies the four basins **B** along the river's north bank from the mouth to the highway bridge. This full-service marina offers myriad posh amenities, including a swimming pool, tennis courts and a restaurant. Just inside the river mouth to starboard, the Village of Bayfield Marina **C** offers dockage and fuel. Farther upstream, South Shore Marina **D** offers dockage for boats up to 32 feet.

Distances from Bayfield (statute miles)
Sarnia: 61 SW
Goderich: 14 N
Tobermory: 119 NNE

114 LAKELAND BOATING · LAKE HURON CRUISING GUIDE

LAKE HURON

Bayfield, Ontario
Step back into a charming, restful past.

When the railroad bypassed Bayfield in the 1800s, it actually did the town a favor—from today's perspective, at least. Having escaped the blight of industrialization that is so much a part of Great Lakes history, Bayfield evolved into a quiet lakeside haven. It also happens to be the largest pleasurecraft harbor on the Ontario side of Lake Huron.

Bayfield, with its grassy walkways and huge trees, is the epitome of old-fashioned charm. It boasts many examples of early Ontario architecture, including two fabulous inns with double verandas and many quaint homes and cottages. Main Street is designated under the Ontario Heritage Act and features unique shops and galleries, as well as wonderful restaurants, unmarred by franchise stores, fudge shops or fast-food chains.

Just 50 miles from the Blue Water Bridge, Bayfield is an easy first day's stopover for cruisers approaching from Lake St. Clair and its environs. The rhumbline course from the mouth of the St. Clair River takes you well offshore, so be prepared to steer an accurate compass course, or make sure your loran or GPS is in working order.

The mouth of the Bayfield River is all but invisible until you approach the outer buoy. The town is located behind the high clay bluff on the south side. Green range lights on the north side of the entrance bearing 76°T lead into the harbor. The front light is on a 16-foot-high mast with a white daymark with a fluorescent-orange vertical stripe; the rear light is 20 feet high on a similar daymark. The south side of the entrance is marked by a flashing red light on a white square tower with a red and white daymark.

Harbour Lights Marina offers many services at its facility, which runs along the north bank of the Bayfield River from the river

Bayfield, Ontario

BAYFIELD AREA CHAMBER OF COMMERCE

LAKELAND BOATING · LAKE HURON CRUISING GUIDE **115**

Where to Eat

Admiral Bayfield Restaurant 519-565-2326
Housed in a converted 1840 general store, Admiral Bayfield starts the day with a hearty breakfast and winds up with dinner accompanied by cool jazz.

Harbour Restaurant 519-565-2554
Family dining at its best. Sit on the patio that overlooks the marina. Weekend entertainment.

Little Inn of Bayfield Restaurant 519-565-2611
This historic inn offers great food in a homey setting. Have a drink by the grand piano in the parlor before dining on the veranda. Sunday brunch is a standout.

King's Bakery and Tea Room 519-565-2859
The place to go for baked goods. It also serves light lunches in the indoor café or scenic outdoor area.

The Red Pump Restaurant 519-565-2576
This historic Main Street establishment is consistently rated one of Ontario's top restaurants. Eat outside or pop into the cozy pub for a nightcap.

Local events

June
Bayfield Sail & Canvas

July-August
Bayfield Historic District walking tours (every Saturday)

August
Annual Bayfield Antiques Show & Sale

Bayfield Fall Fair & Parade

---

sis is on artistry, craftsmanship and design—not on brand names or labels. **Marten Arts Gallery** displays the works of more than 80 artists and craftsmen from all over Canada. **Micheline's Jewelry Boutique** has one-of-a-kind pieces in silver and gold. Clothes hounds should head for the fashion emporium that is part of the **The Red Pump Restaurant**. This store attracts shoppers from all over Ontario for its selection of designer clothes (the owner picks them up at European fashion shows), sold at fabulous prices. And the attached restaurant? Among the best you'll find anywhere.

Just outside town, on Highway 21, you'll find the **Bluewater Golf Course** (519-482-7197). And east of Bayfield, on Bayfield River Road, is something truly unique: **Folmar Windmill**, North America's only working wind-driven saw and grist mill.

mouth to the highway bridge. It can handle yachts up to 65 feet, and the channel carries depths of at least 5 feet up to the fuel dock on the north side. The marina has a swimming pool and tennis courts. The courteous staff also operates an antique fire truck to shuttle boaters to and from town. Hop aboard for a one-of-a-kind ride!

Bayfield River Cottage Colony & Marina is primarily a vacation community. Transient boaters can occasionally find an open slip at the development's 26-slip marina.

On the south side of the river, just inside the entrance, is the attractive **Village of Bayfield Marina**, which offers gas, diesel, pumpout and some repairs. This is a good spot for transient dockage, with town just a short walk up the hill. Just before the highway bridge on the starboard bank is **South Shore Marina**, which can handle boats up to 32 feet.

Huron County was originally settled by immigrants from Germany, Ireland, Holland and Britain. Many were responding to ads placed in newspapers by the Canada Company back in 1821. These industrious settlers cleared the forests and built farms, many of which are still inhabited by their descendants. Huron remains the most agriculturally productive county in Ontario. Bayfield was established in 1830 and two years later, the **Little Inn of Bayfield** opened its doors as a stage stop. Still open today, it houses one of the best restaurants in town.

Shoppers will love Bayfield for its wide range of unique antique stores and craft shops. Here the empha-

More Info

Emergency 911
Chamber of commerce 866-565-2499

"A Safe and Welcoming Marina"

- Deep Water
- Sail
- Power
- Laundry
- Pool
- Tennis Courts
- Private Beach
- Washroom Facilities
- Shuttle Service
- Laundry
- Clubhouse
- Grassy Docking Areas

VISA MasterCard

Harbour Lights

- VISITORS WELCOME

Office: 519-565-5150
Fax: 519-565-2078
Gas Bar: 519-565-2149

E-Mail: harbourlights@tcc.on.ca
www.harbourlightsmarina.on.ca

Situated along the North Shore of the Bayfield River

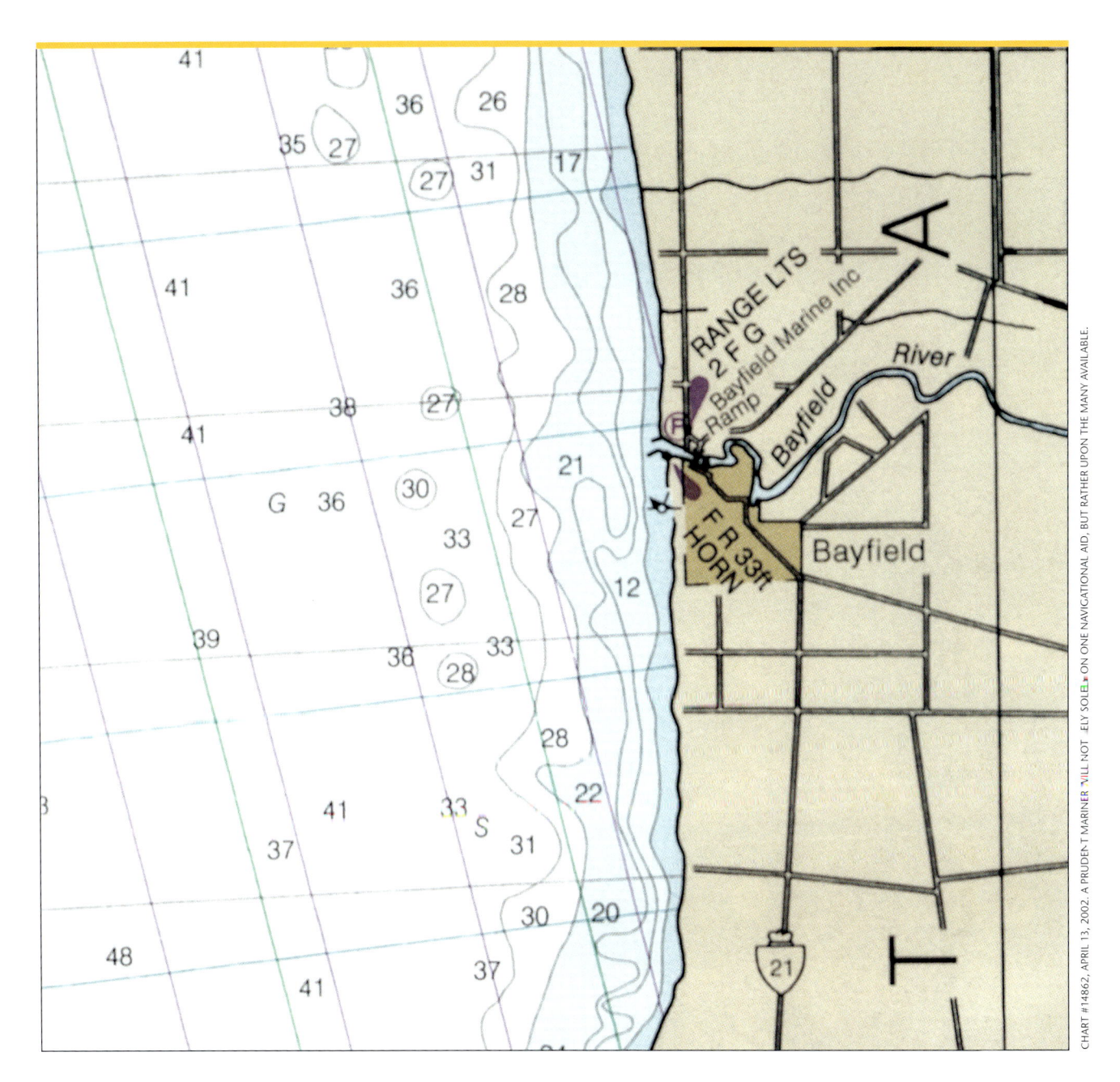

Facilities information subject to change. We suggest you call ahead.

			Monitors VHF Channel	Transient Slips Available	ALTERNATE MOORING: Wall mooring / Rafting allowed	Maximum LOA	Minimum Depth at Dock	Power (amperage or volts)	HOOKUPS: Water / Cable TV	FUEL: Gas / Diesel / Pumpout	BASICS: Heads / Showers / Laundry	AMENITIES: Swimming pool / Whirlpool / Rec area / Grills / Picnic tables / Dog walk	Take Reservations	Take Credit Cards	Haulout (Capacity in tons or feet)	REPAIRS: Mechanical / Electronics / Fiberglass / Woodworking / Sails / Canvas	CONVENIENCE: Ship's store / Ice / Convenience foods & beverages
B	Harbour Lights Marina 519-565-2149	27 Charthouse Hill Bayfield, ON N0M 1G0	68	70	W	65'	5'	30 50	W	GDP	HSL	SRPD	Y	Y	20	MEFWSC	SIC
D	South Shore Marina 519-565-2110	Highway 21 & Mill Road Bayfield, ON N0M 1G0	16 68	5	W	32'	6'	15	W		HS	PD	Y	N			SC
C	Village of Bayfield Marina 519-565-2233	Long Hill Road Bayfield, ON N0M 1G0	16 68	25	WR	125'	6'	20,30 50	W	GDP	HS	PD	Y	Y		ME	IC

LAKELAND BOATING · LAKE HURON CRUISING GUIDE

Goderich, Ontario

When approaching Goderich, look for the storage buildings **A** and grain elevators **B** of the Sifto Salt Mine. Look for the brilliant white pier lights at the outer end of the first set of breakwalls **C**. Goderich has two harbors with separate entrances. To enter the main harbor, proceed on range, passing Sifto to port. Snug Harbour Marina **D** is immediately to port. To enter the north harbor, keep north of the long breakwall **E** to enter the mouth of the Maitland River. Maitland Inlet Marina is **F** to port just inside the river. Bear starboard to enter Maitland Valley Marina **G**. The Menesetung Bridge and walking trail **H** cross overhead at the bend in the river. Take a hike or play a round at the nearby golf course **I**. Relax on one of the town's beaches **J**, **K** and enjoy a stroll along the boardwalk **L**. Visit the maritime museum **M**, the lighthouse upon the bluffs **N** and the octagonal Courthouse Square, which forms the center of the town's shopping district. This is easily accessed by the stairs to town **O**.

118 LAKELAND BOATING · LAKE HURON CRUISING GUIDE

LAKE HURON

Goderich, Ontario
It's hip to be...octagonal?

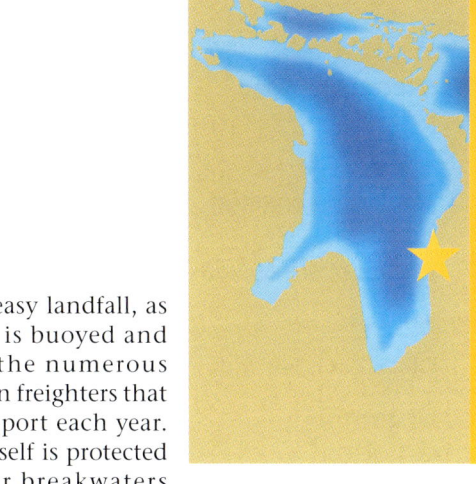

There are many sides to Goderich, both literally and figuratively. Its unique octagonal layout is highlighted by historic buildings and breathtaking gardens, earning it the moniker "the prettiest town in Canada." But Goderich is also the largest town on the Canadian coast of Lake Huron and home to the world's largest salt mine. This multifaceted existence just makes it a more fascinating place than your average port of call. With three beautiful beaches connected by an old-fashioned boardwalk, and a unique shopping district that encircles the central park, you'll never run out of interesting things to do.

Goderich began life as a lumber, fishing and transportation center. But it was salt—accidentally discovered in 1886 by oil prospectors—that made it the distinctive port we know today. The Sifto Salt Mine, which descends more than a quarter of a mile down and three and a half miles out under the lake, produces the majority of the salt mined in North America and is the largest working salt mine in the world. The port is also accentuated by tall grain elevators situated on the periphery of the main harbor that can be seen from well offshore.

This is an easy landfall, as the channel is buoyed and marked for the numerous lake and ocean freighters that frequent the port each year. The harbor itself is protected by two outer breakwaters with lights. Range lights on North Pier bearing 86°T and daymarks bearing an orange vertical stripe lead through the channel between the breakwalls. **Snug Harbour Marina**, the municipal facility, is located on the west wall next to the salt storage silo. The retired lake freighters at the head of the harbor are used for grain storage.

A more scenic setting can be found just north of the harbor at one of the two marinas on the Maitland River. A flashing white light is situated at an elevation of 30 feet on the outer end of the boulder breakwater, which forms the south side of the river entrance. **Maitland Inlet Marina**, formerly Homan's Inlet Marina, is immediately to port as you enter the river. Located behind a mile of private sand beach, this marina has mostly seasonal slips; boaters are advised to go its sister marina, **Maitland Valley Marina** (www.maitlandmarina.on.ca), which is just up the channel in a protected, surge-free basin. Once you enter the channel, proceed upriver, keeping the breakwall to starboard. Navigate between the buoy markers, which mark depths of more than 9 feet. The marina's large, modern facility offers a wide variety of services, including a clubhouse, pool and free transportation to town. Maitland Valley also features excellent fishing right from its banks, golf and tennis next door at the **Maitland Golf & Country Club**, and numerous hiking and biking trails.

Only a few feet east of the marina entrance is the gateway to the river valley. Take in the view from the

Where to Eat

Bailey's 519-524-5166
Casual fine dining in the square. Dine in your Dockers and Topsiders but get treated like royalty as you savor the scrumptious offerings served up by the British-trained chef. His wife is the ultimate hostess. Reservations are advised.

Benmiller Inn 519-524-2191
A short cab ride from Goderich, in a resort that has been restored to reflect its origins as an 1800s-era mill. The fare ranges from old-style Canadian to international. Prepare to be pampered here.

Captain Fats 519-524-9211
Certainly nothing fancy—just fresh Lake Huron fish. Stop in and take some back to the boat or eat on the patio. Located on the south dock.

Paddy O'Neil's 519-524-7337
On the historic Square in the bright and airy Hotel Bedford. A bar and eatery.

Robindale's Fine Dining 519-524-4171
The ultimate in upscale cuisine, served up in an 1800s Victorian mansion. You'll have that "at home" feeling while enjoying delectable food.

LAKELAND BOATING · LAKE HURON CRUISING GUIDE   119

Local events

June
Massed Pipe Band Festival

July
Kinsmen Summerfest Weekend

Festival of Arts & Crafts

August
Celtic Roots Festival

September-October
Marine Heritage Festival

700-foot Menesetung Walking Bridge, a converted railway trestle bridge that spans the Maitland River and forms part of a hiking trail.

The heart of Goderich is laid out with a courthouse and a park in the center and eight avenues radiating out from there. Some 80 shops and restaurants ring the park, comprising a unique shopping area known as "the Square." A short walk away are such noteworthy landmarks as the **Huron Historic Gaol**, a jail built in 1839—also in the shape of an octagon—and considered an architectural masterpiece. The facility, which was in use until 1972, housed the courts and council chambers, as well as the governor's house. Today it is a museum and National Historic Site. The **Huron County Museum** contains more than 10,000 artifacts chronicling the pioneering activities of the area. The small **Marine Museum** is housed in the wheelhouse of the *S.S. Shelter Bay*, docked in the harbor. Its exhibits chronicle the killer storm of 1913, which took the lives of 150 seamen in just 48 hours. The historic **Bedford Hotel** occupies one block of the octagon.

Another attraction is the live theater. See what's on at the **Goderich Little Theatre**, housed in the **Livery**, a heritage building in the Square. Photography and history buffs shouldn't miss the **Reuben R. Sallows Gallery**, located in the library on Montreal Street. The permanent collection features the work of Goderich native Sallows, a renowned photographer who traveled throughout Canada in the early 20th century.

More Info

Emergency 911
Alexandra Marine
& General Hospital 519-524-8323
Tourism Goderich 800-280-7637
www.town.goderich.on.ca
Goderich Taxi 519-524-6594

Distances from Goderich (statute miles)
Sarnia: 61 SW
Port Sanilac, Michigan: 52 WSW
Kincardine: 33 N
Tobermory: 100 N

Maitland Valley MARINA

...in beautiful Goderich!
We feature a full service marina which appeals to both seasonal and transient dockers!

Our vast amenities include...
- Clubhouse • 30/50 Amp Power • Pool • Transportation
- 1,200 ft. of private beach • Tennis & Golf Course (next door)
- Reservations accepted.

For More Information Call... **519-524-4409**

P.O. Box 175, 100 North Harbour Road, West,
Goderich, Ontario, Canada N7A 3Z2
www.maitlandmarina.on.ca

Deep Water Dockage!
Monitor Channel 68 -
"Maitland Valley"

Use our convenient fax...
519-524-2301

Goderich is...
- 50 Nautical Miles to Port Huron
- 90 Nautical Miles to Detroit
- 38 Nautical miles to Port Sanilac

LAKELAND BOATING · LAKE HURON CRUISING GUIDE

GODERICH HARBOU[R]

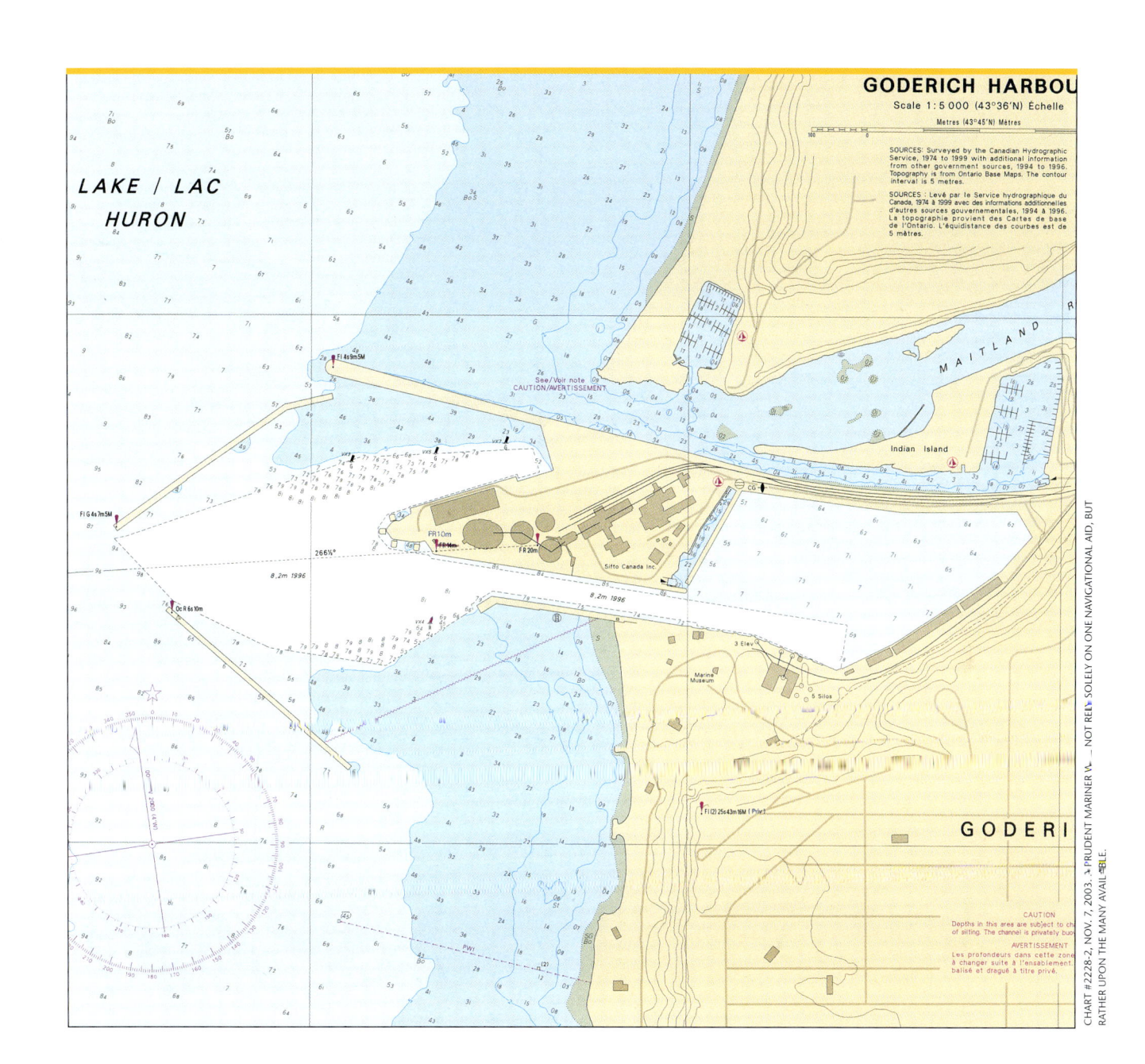

Scale 1:5 000 (43°36'N) Échelle

SOURCES: Surveyed by the Canadian Hydrographic Service, 1974 to 1999 with additional information from other government sources, 1994 to 1996. Topography is from Ontario Base Maps. The contour interval is 5 metres.

SOURCES: Levé par le Service hydrographique du Canada, 1974 à 1999 avec des informations additionnelles d'autres sources gouvernementales, 1994 à 1996. La topographie provient des Cartes de base de l'Ontario. L'équidistance des courbes est de 5 mètres.

Facilities information subject to change. We suggest you call ahead.

	Marina	Address	Monitors VHF Channel	Transient Slips Available	ALTERNATE MOORING: Wall mooring / Rafting allowed	Maximum LOA	Minimum Depth at Dock	Power (amperage or volts)	HOOKUPS: Water/Cable TV	FUEL: Gas/Diesel/Pumpout	BASICS: Heads/Showers/Laundry	AMENITIES: Swimming pool/Whirlpool/Rec area/Grills/Picnic tables/Dog walk	Take Reservations	Take Credit Cards	Haulout (Capacity in tons or feet)	REPAIRS: Mechanical/Electronics/Fiberglass/Woodworking/Sails/Canvas	CONVENIENCE: Ship's store/Ice/Convenience foods & beverages
F	Maitland Inlet Marina 519-524-4409	Rural Route 5 Goderich, ON N7A 3Y2	68		WR	100'	10'	30 50	W	GDP	HS	RGPD	Y	Y	25	MEF	I
G	Maitland Valley Marina 519-524-4409	100 N. Harbour Rd. W. Goderich, ON N7A 3Z2	68	50	WR	100'	10'	30 50	W	GDP	HS	SRGPD	Y	Y	25	MEF	I
D	Snug Harbour Marina 519-524-8813	North Harbour Road Goderich, ON N7A 4C7	68	10	W	100'	25'	30 50	W	P	HS	PD	Y	N			

LAKELAND BOATING · LAKE HURON CRUISING GUIDE

Kincardine, Ontario

Lat 44°10.70
Lon 81°38.63

When approaching Kincardine, take care to avoid the shallow waters around Point Clark, 10 miles to the southwest. Beware of the unlit detached breakwater **A** about 800 feet northwest of the entrance—during high water it is sometimes partially submerged. The harbor entrance is marked by parallel piers. The south pier **B** has a new tri-color sector light. A gray condominium building **C** provides a good landmark on approach. The municipal Kincardine Marina **D** lies directly to starboard just inside the basin. The harbor was expanded in 1999-2000, with two sets of floating docks replacing the old west wall docks. A double wide launch ramp was also added in the southwest corner.

Distances from Kincardine (statute miles)
Sarnia: 94 SW
Port Sanilac, Michigan: 79 SW
Goderich: 33 S
Port Elgin: 26 NNE
Tobermory: 67 N

LAKE HURON

Kincardine, Ontario
Kick up your heels in this bonny port.

In Kincardine, the influence of Scottish culture is evident everywhere—from the haunting strains of the phantom bagpiper to the annual Scottish Festival and Highland Games. The bustling little town, which boasts a variety of shops and restaurants, is located a block from the marina along a stretch of sandy shoreline that extends miles in either direction.

Knechtels Food Market, located on Queen Street within walking distance of the marina, will deliver to boats. The government-operated liquor and beer stores are on Broadway Street, and fresh produce and other tasty treats can be purchased every Monday from Victoria Day to Labour Day at the **Flea and Farmers Market**.

This small town of 7,000 traces its history to 1848, when Allan Cameron and William Withers landed at the frozen mouth of the Penetangore River. Cameron built a hotel and Withers, his brother-in-law, constructed a dam and a sawmill. Hospitality and small industry have been the town's hallmarks ever since.

The entrance to the man-made harbor is between two parallel stone piers; the south pier has a new tri-color sector light. Entering the harbor itself is straightforward, but watch out for a detached breakwater located about 800 feet to the northwest. It has no lights, and during high water it may be partially submerged. Cruisers heading north to Kincardine should stay well offshore from Point Clark to the harbor to avoid shoals and reefs. Clark Reef, which has a charted 3-foot spot, is marked by a buoy.

The **Municipality of Kincardine Marina** is the only facility in the harbor, which was expanded in 1999-2000. It now boasts two new sets of floating docks, which replace the old west wall docks, and a double-wide launch ramp in the southwest corner.

The most scenic feature of the harbor is the lighthouse on the north bank. An 80-foot-high octagonal white frame building with a red lanternhouse and balcony, it went into service in 1881, and the light, which can been seen for 21 miles, is still in use. It is maintained for the Canadian Coast

MARGE BEAVER

Where to Eat

Brasserie **519-396-6000**
Offering elegant cuisine in a lovely century-old building with a beautiful lake view. Smoke-free.

The Erie Belle **519-396-4331**
British to the core. Wash down your fish and chips with an imported ale. Sit on the lakeview patio or get it to go.

**The Harp and Whistle
Irish Pub & Restaurant** **519-396-9133**
An extensive seafood menu, including the house specialty, fish and chips. Darts and live entertainment complete the pub atmosphere.

Hawg's Breath Saloon **519-396-6565**
Their slogan is "Never a boar." Great lunches and daily specials.

Kincardine, Ontario

Local events

May
Fish Kincardine Derby

June
Driftwood Festival

June through Labour Day
Pipe Band Parades

July
Scottish Festival & Highland Games

July through August
CFPS Chantry Chinook Classic Fish Derby

August
Summer Music Festival
Massed Bands Festival

More Info

Emergency 911
Kincardine & District
General Hospital 519-396-3331
Kincardine Visitor Information 866-546-2736
www.kincardine.net
Kincardine Taxi Service 519-396-3411

LAKELAND BOATING · LAKE HURON CRUISING GUIDE   **123**

Guard by the **Kincardine Yacht Club**, which operates out of the marina. The **Kincardine Lighthouse Museum** houses old photographs, glass negatives and artifacts from Kincardine's seafaring past. Climb the tower for a spectacular view.

A fascinating shore excursion is a visit to **Bruce Power**, located between Kincardine and Port Elgin. This is one of the largest nuclear power plants in the Western world. While tours are currently not given, the plant does offer information sessions. Weekend visitors to Kincardine should to hike up to **Victoria Park** to hear the Kincardine Scottish Pipe Band, which has been entertaining townspeople since 1909. On Saturdays in July and August, the ensemble leaves the park at 8 p.m., marching up Queen Street and back for a concert that begins at 9 p.m. Another pleasant diversion is the **Bluewater Summer Playhouse**, which stages performances of light musical comedy in a cabaret setting Tuesdays through Sundays.

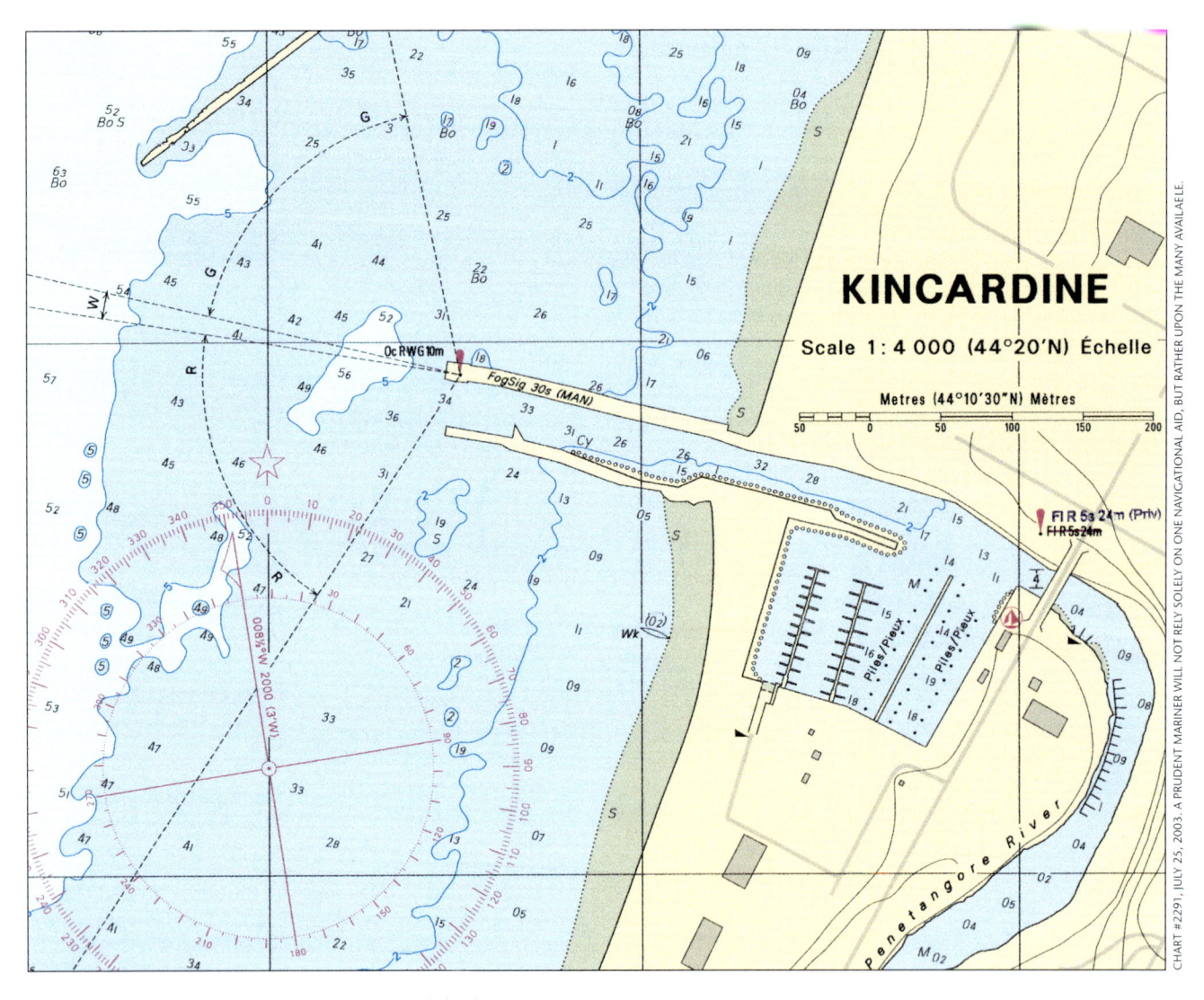

Facilities information subject to change. We suggest you call ahead.

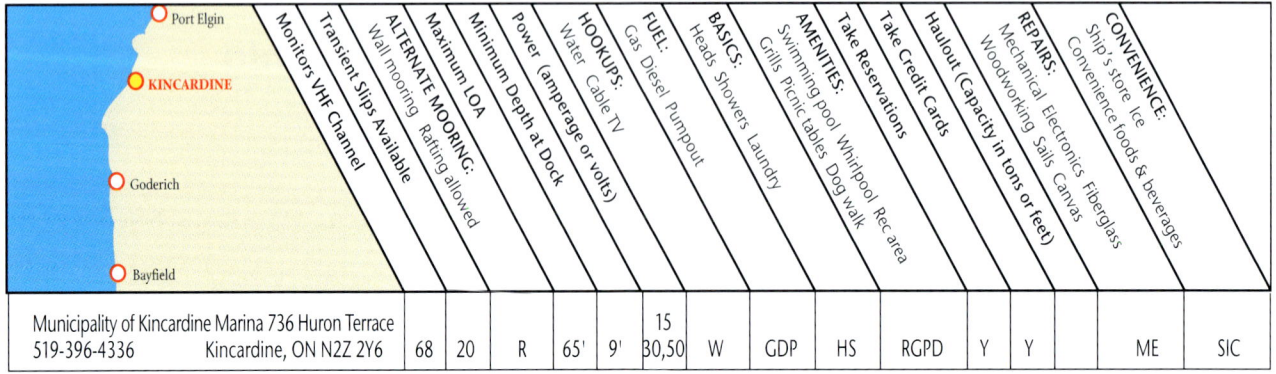

	Monitors VHF Channel	Transient Slips Available	ALTERNATE MOORING: Wall mooring Rafting allowed	Maximum LOA	Minimum Depth at Dock	Power (amperage or volts)	HOOKUPS: Water Cable TV	FUEL: Gas Diesel Pumpout	BASICS: Heads Showers Laundry	AMENITIES: Swimming pool Whirlpool Rec area Grills Picnic tables Dog walk	Take Reservations	Take Credit Cards	Haulout (Capacity in tons or feet)	REPAIRS: Mechanical Electronics Fiberglass Woodworking Sails Canvas	CONVENIENCE: Ship's store Ice Convenience foods & beverages
Municipality of Kincardine Marina 736 Huron Terrace 519-396-4336 Kincardine, ON N2Z 2Y6	68	20	R	65'	9'	15 30,50	W		GDP	HS	RGPD	Y	Y		ME	SIC

S.O.S.

Abandon Ship!
It's always best to be prepared for the worst.

It's more than just a Scout's motto: being prepared for emergencies is an absolute necessity when you head out onto the water. In a real-life game of *Survivor*, it can do more than get you to the next round — it can save lives.

If you venture offshore more than 10 miles, you should consider purchasing a self-inflating life raft—and know how to use it. The best models come with features such as canopies for protection from the elements, water ballast systems for improved stability, and survival kits packed inside. Inflatable rescue platforms are not life rafts but are particularly well suited for near-coastal boating.

Any life raft worth having will easily inflate by integral gas cylinder and come packed with paddles, a sea anchor and line, a repair kit and a manual pump. Have your raft inspected and repacked annually and you'll never have to wonder whether it will work as the water climbs above your ankles.

A preassembled kit bag, preferably one that floats, stored in an easily accessible locker (away from engine compartments and galleys, where most fires occur) can save your life should you have to abandon ship. The kit should contain, at a minimum: a manually activated EPIRB, a knife, a bailer, two waterproof flashlights, a first-aid kit, a variety of flares in waterproof containers, watertight food rations, a can opener, a signalling mirror, chemical light sticks and spare flotation cushions or PFDs. A waterproof VHF handheld radio is also helpful. And if you are heading out to sea, be sure to include watertight receptacles of fresh water and a drinking cup.

"Prepare to abandon ship!" means just what it says—*prepare*. The strongest crew member should stand by the life raft and await the command to deploy it. He or she should have a sharp knife, a second tether (if the first one parts, the life raft will be difficult to recover) and the abandon ship kit.

Conventional wisdom says that you only get into a life raft when it is necessary to step up into it—that is, when it's no longer possible to stay aboard your vessel. But staying with a floating and potentially salvageable boat, even if you're tethered nearby in the life raft, makes you that much more visible to helicopters participating in an air search. Just be ready to cast off should the boat start to sink.

Abandon ship procedures
- Put on PFDs.
- Try to quickly obtain a position fix.
- Switch the VHF to channel 16 and transmit a distress call.
- Turn on EPIRB.
- Try to save the boat.
- Inflate the raft and tether it to the boat.
- Get survival kit and gather food and water if possible.
- Place the kit, food and water in the raft and climb aboard.

A distress call
In a distress call, you should mention the following:
- Boat's name and radio call sign
- Boat's position
- Boat's problem (engine trouble, sinking, etc.)
- What assistance is needed
- Number of people on board
- Length, model and color of boat

Repeat several times if the distress call does not produce a response. Also, tune into channels that you think might be busy and issue a distress call. If you abandon ship, tape down the microphone button on a channel other than 16. This will give rescuers a signal to home in on.

Signaling
Visible signs are important for directing your rescuers. At night, you should use rockets and flares. On a clear night, parachute signals can be launched to an altitude of about 1,000 feet and can be seen from more than 35 miles away. Rockets are visible from 20 miles away, flares from five miles away. During the day, a signal mirror can be flashed toward the shoreline or in the air to grab the attention of pilots. The flash from a signal mirror is visible from 35,000 feet. Smokescreens and dyes can also be used on clear days.

Port Elgin, Ontario

Lat 44°26.69
Lon 81°24.37

The Bruce Power nuclear facility, 12 miles south of the harbor, provides a good landmark. Shoals extend offshore **A**, so plot a course to the buoy marking Logie Rock before setting a course of 109°T into the harbor. Stay inside the channel to avoid shallow water. The Y-shaped breakwaters **B** carry lights on masts on their outermost projections. Port Elgin Harbour **C** offers transient dockage and sells fuel. On Sunday evenings you can enjoy outdoor concerts featuring a variety of styles at the bandshell on the nearby beach **D**.

LAKE HURON

Port Elgin, Ontario
Unrivaled beaches and matchless sunsets.

Port Elgin bills itself as "the town of maples." *National Geographic* named it "the world's sunset capital." But the summer resort boasts more than just gorgeous trees and spectacular sunsets—a full range of shops, eateries and services are ready to cater to the transient cruiser.

The manmade harbor, formed by breakwaters and retaining walls, offers good protection from all except strong west and northwest winds that might cause a surge. The outer breakwaters have been rebuilt in the shape of a Y to reduce the surge effect. Range lights atop white daymarks with orange vertical stripes in a line bearing 109°T lead into the harbor entrance. An additional light mounted atop the outer range light is visible from all directions. The north and south breakwaters carry privately maintained lights on masts.

Coming from the south, the huge Bruce Power nuclear plant, between Douglas and Macpherson points about 12 miles south of Port Elgin, is a conspicuous landmark. Shoals extend well offshore. To be safe, plot a course to the buoy marking Logie Rock, which is about three miles offshore and slightly to the north of the 109°T course into the harbor. This buoy makes a good turning point when entering or leaving the harbor. But be sure to come in on the proper course—there is a large shoal with a depth of one foot situated north of the entrance channel. The sandy beach south of the breakwaters is the only distinguishing landmark.

Port Elgin Harbour, the only game in town, has dockage space for up to 25 boats and can accommodate vessels up to 140 feet. The marina sells gas and diesel fuel, as well as

Local events

May
Museum Month at the Bruce County Museum

Natural Farmers Market
Saturdays at the Wellness Centre.

June
Up with the Birds Hike at Macgregor Point Provincial Park

July
Lobster Picnic

Antique Show and Sale

July through August
Flea Markets
Every Wednesday at the beach.

Murder Mystery Ghost Hike

August
Fireworks

More Info

Emergency	911
Grey Bruce Health Services	519-797-3230
Port Elgin Chamber of Commerce	800-387-3456
	www.sunsets.com/portelgin
Port Elgin Taxi	519-832-2024

LAKELAND BOATING · LAKE HURON CRUISING GUIDE   **127**

Where to Eat

Andre's Swiss Country Dining **519-832-2461**
Try the sampler platter, which comes overflowing with varieties of schnitzel and other delicacies.

Lakefield Bistro **519-389-5775**
Excellent lake fish and local farm produce.

Lord Elgin **519-832-2224**
Offers a full seafood menu, but it's the famed fish and chips that draws fans from miles away. Takeout, too.

Our House Family Restaurant **519-832-2216**
Famous for its charbroiled steaks and prime rib Fridays.

Queen's Bar and Grill **519-832-2041**
Pub fare, pool and live music in a historic building.

---

ice and some snacks. **United Marine** makes boat calls for maintenance. **Sunshine Laundromat** is located nearby.

The town itself, about a mile up the hill behind the harbor, has an interesting history. The remains of a Nodwell Indian village dating back to the 14th century lie buried beneath the town's **Participark**. Unfortunately, a restoration of the site is incomplete due to lack of funds. Port Elgin, incorporated in 1874, was named for a governor-general of Canada, James Bruce, who was the eighth Earl of Elgin and Kincardine. The town's sugar maples, planted by one of the founding fathers, Sam Bricker, line Green, Mill and Elgin streets, which lead to the harbor.

Port Elgin's six miles of sandy beaches are connected to the marina by a paved walkway. Just behind the beach is **the Station**, which sells snacks and souvenirs and is the embarkation point for a real **24-gauge steam train** that makes a one-mile trip along the harbor several times daily in July and August.

The **Peter Sheeler Gallery** is owned and run by this local artist, who is renowned for his wildlife paintings and shore scenes. **Gallery Visions** on Green Street offers a terrific selection of original watercolor paintings of the Bruce shoreline. For a different shopping experience, stay for a Wednesday and take in the weekly **flea market**, which has up to 90 vendors set up along the water. Sunday evenings during the summer, head to the bandshell on the beach for music to suit all tastes—from rock and country to Big Band and classical.

Distances from Port Elgin (statute miles)
Sarnia: 117 S
Bayfield: 72 S
Goderich: 60 S
Kincardine: 22 S
Port Sanilac, Michigan: 155 SSW
Tobermory: 78 N

Facilities information subject to change. We suggest you call ahead.

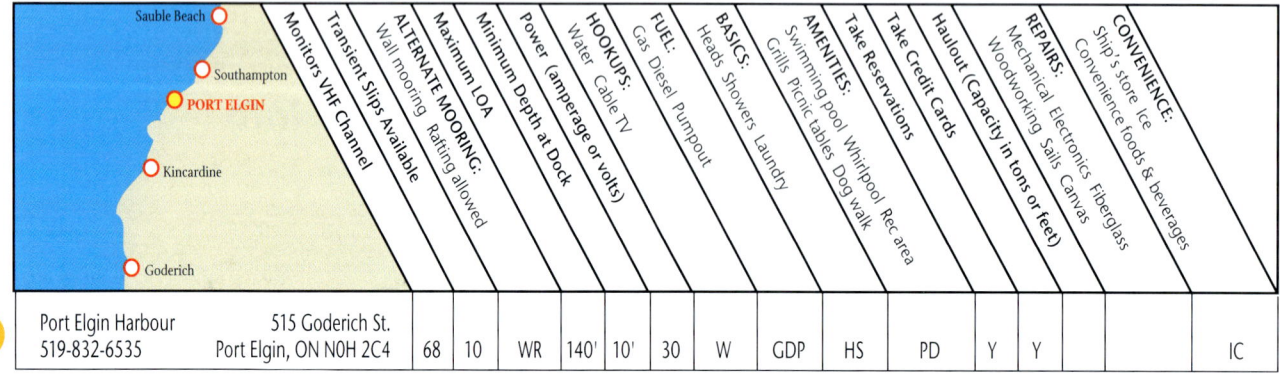

	Monitors VHF Channel	Transient Slips Available	ALTERNATE MOORING: Wall mooring, Rafting allowed	Maximum LOA	Minimum Depth at Dock	Power (amperage or volts)	HOOKUPS: Water, Cable TV	FUEL: Gas, Diesel, Pumpout	BASICS: Heads, Showers, Laundry	AMENITIES: Swimming pool, Whirlpool, Rec area, Grills, Picnic tables, Dog walk	Take Reservations	Take Credit Cards	Haulout (Capacity in tons or feet)	REPAIRS: Mechanical, Electronics, Fiberglass, Woodworking, Sails, Canvas	CONVENIENCE: Ship's store, Ice, Convenience foods & beverages
Port Elgin Harbour 519-832-6535 / 515 Goderich St. Port Elgin, ON N0H 2C4	68	10	WR	140'	10'	30	W		GDP	HS	PD	Y	Y		IC

128 LAKELAND BOATING · LAKE HURON CRUISING GUIDE

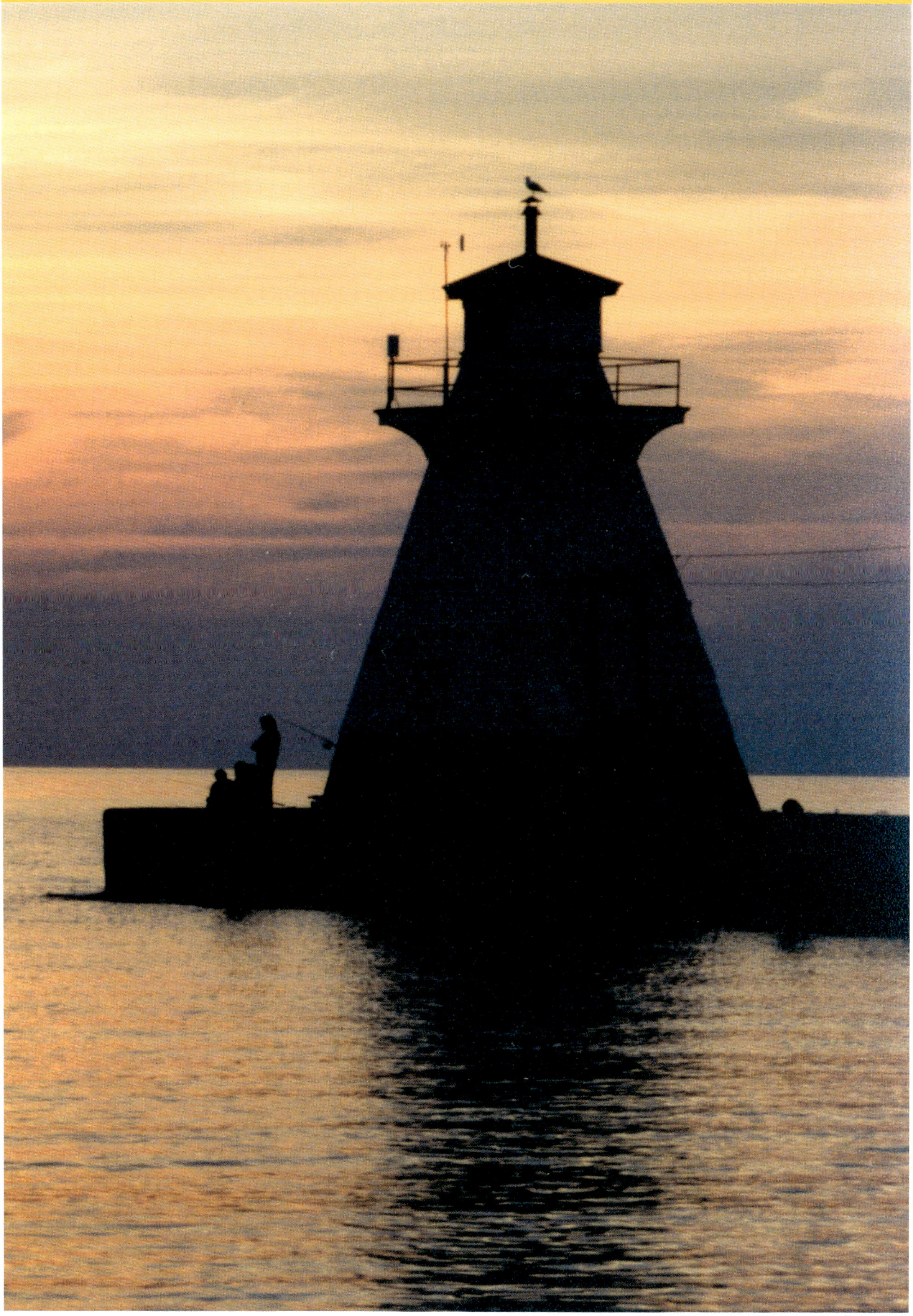

PORT ELGIN CHAMBER OF COMMERCE

Southampton, Ontario

Lat 44°29.40 Lon 81°24.15

Lat 44°30.10 Lon 81°22.55

Cruisers approaching from the south can pass between Southampton and nearby Chantry Island **A**. The island was formerly connected to the mainland by a now-submerged breakwall. Buoys mark a 450-foot gap **B** in the ruined breakwall where cruisers can make safe passage. Use the Saugeen River Turning Light Buoy to line up for the entrance channel. The tip of the northern pier carries a flashing red atop a 30-foot tower **C**. A town dock **D** just inside the river to starboard is available for bare-bones dockage. Anchoring is available near the Southampton Yacht Club **E**.

Where to Eat

Duffy's Famous Fish & Chips. 519-797-5972
Proprietors Ernie and Brenda Duff will serve you an unforgettable fish dinner at this High Street hangout.

Grosvenor's 519-797-1226
Housed in the old railway station, this is perhaps Southampton's best restaurant. Serving innovative Canadian cuisine, including fresh local fish and game and homemade pastas.

Highview Restaurant. 519-797-2199
A block from the beach, it opens at 8 a.m. every day. Charbroiled steaks are a speciality.

Walker House 519-797-2772
Up the hill from the harbor, this large white building has a full menu and live entertainment on Thursday nights.

LAKE HURON

Southampton, Ontario
A popular migration stopover for birds—and cruisers, too.

Founded as a fishing and trading outpost in 1848, Southampton is the oldest port on the Bruce coastline, so it's not surprising to find it a quiet retreat filled with the historic homes of early mariners. Though not exactly geared toward transient cruisers, the town nevertheless offers many interesting things to see and do, including a first-rate museum.

The approach to Southampton must be made with care because of the shoals between Port Elgin to the south and the Lee Bank to the north. Between 1848 and 1948, the surrounding waters claimed 50 ships. Coming from the south, cruisers with confidence in their navigational skills can pass inside of Chantry Island, which is connected to the mainland by an underwater shoal and the remains of a breakwater, now in ruins and largely submerged. A 450-foot gap marked by buoys is the only safe passage through.

Boats coming from the north (or from any direction when seas are running or visibility is poor) must be careful to steer well clear of the shoals and to use the Saugeen River Turning Light buoy (VJ"3") to line up for the entrance channel. From there, line up the Saugeen range lines on a bearing of 95°T. The outer end of the north pier shows a flashing red atop a 30-foot tower with an orange vertical daymark; the inner light is shown from a 31-foot white tower with an orange daymark. A foghorn on the front light structure sounds one blast every 20 seconds, as it has for most of this century.

There is no marina catering to transients here, but cruisers can tie up just inside the river mouth along the wall on the south side. Facilities here consist of a public washroom and a pay telephone, as well as a boat launch. During summer months, most of the available berthing is taken by commercial fishing tugs, but there is usually one spot at the end reserved for use by the **Southampton Yacht Club**.

You can anchor off of the yacht club, but the 275-foot **wooden dock** on the north side of the river is usually reserved for club members. If someone is away, the club will try to find you a spot. There are no facilities here.

Ashore, more than 100 homes of early mariners—captains, fishermen and ship builders—still stand, and a **walking tour** map is available from the tourism office in the old town hall.

The striking **town hall**, with its four-sided clock tower that was built in 1910, is worth a look itself. But perhaps the best reason to stop at Southampton is the **Bruce Coast Gallery**—also known as the Shipwreck Gallery—of the **Bruce County Museum & Archives**. The colorful marine history of the Bruce is told through exhibits on shipbuilding, shipwrecks, rescues, lighthouses and maritime notables. A military

Local events

May
Victoria Day Fireworks

May – September
Chantry Island Lighthouse Excursions
Cruise Night
Antique, custom and classic cars.
Huron Fringe Birding Festival
Up with the Birds Hike

June
Shoreline Artists Studio Tour

June – August
Friday Fun Nights
Fairy Lake Bandshell Concerts

July
Canada Day Celebrations

August
Lantern Festival
Saugeen First Nation Powwow

gallery chronicles the county's involvement in conflicts from the Fenian raids of the 1860s to World War II. Historical and imaginative programs are offered for young and old. The historic **Art School & Gallery** offers summer lessons for artists of all calibers given by well-known instructors.

Chantry Island, a migratory bird sanctuary, is home to a lighthouse that was built in the 1850s and is now being restored. The island protects the wide beaches of the mainland from storm waves, and over the past several thousand years, a unique dune complex has developed here. The Chantry Dunes, at the southern end of Southampton's waterfront, provide a trail system that allows visitors to enjoy this sensitive ecosystem without disturbing it.

In town, there are 13 tennis courts available for use, along with four excellent public golf courses within a 15-minute drive. Check out the well-groomed greens and beautiful clubhouse of the **Lawn Bowling Club**, which hosts several tournaments and jitneys every year.

Every Sunday evening at 7:30, head to **Fairy Lake** to enjoy the Music by the Lake series. Other big events include the Canada Day Celebrations & Fireworks and the Arts & Crafts Show.

After seeing the exhibits on shipwrecks, you can take your boat, if you dare, and dive to the nearby wrecks. Or try your hand angling for the steelhead that make the Saugeen River famous.

More Info

Emergency	911
Grey Bruce Health Services	519-797-3230
Chamber of Commerce	888-757-2215
	www.sunsets.com/southampton
Weiss Taxi	519-797-2443

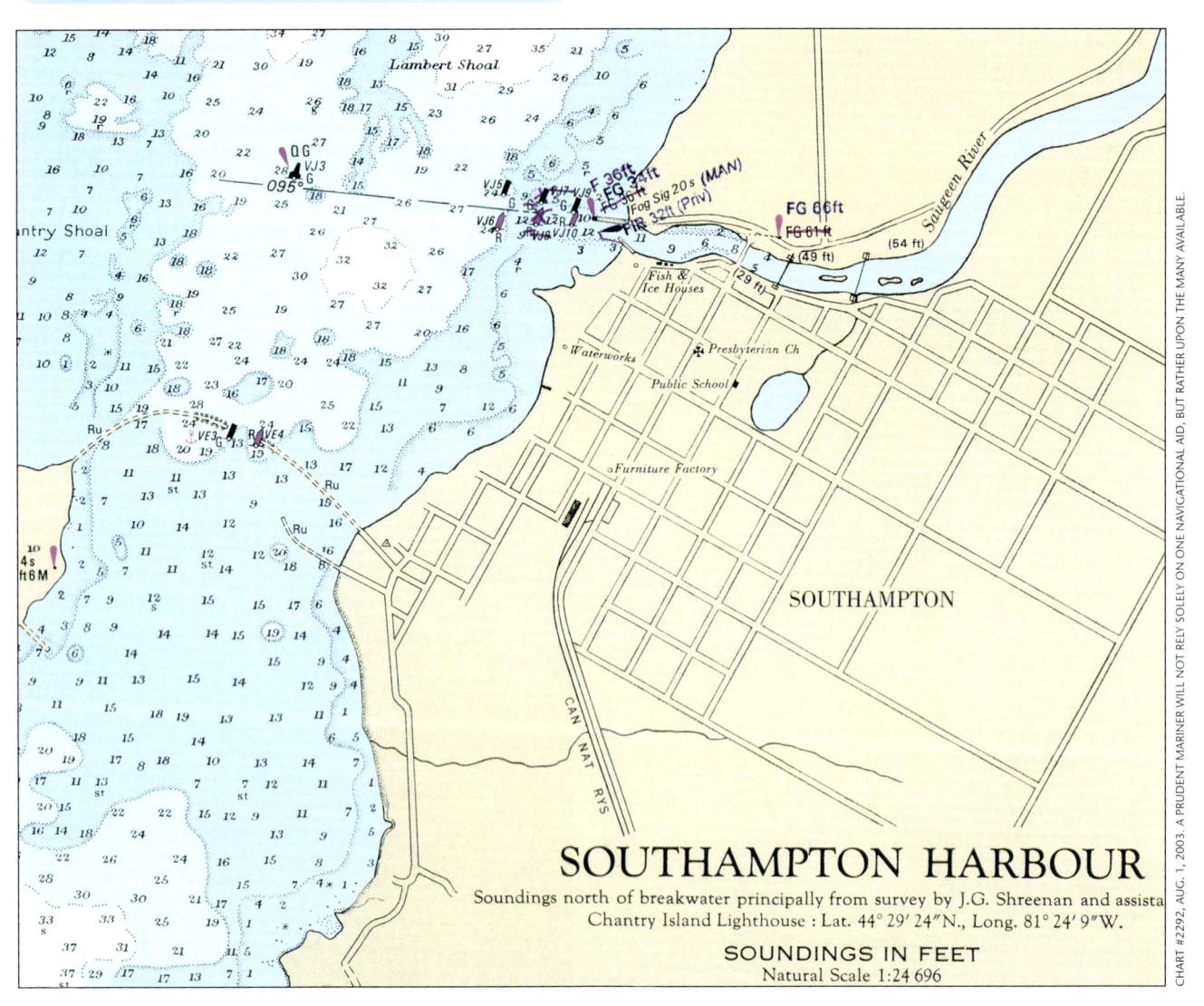

Distances from Southampton (statute miles)
Port Elgin: 5 S
Sauble Beach: 14 N
Stokes Bay: 32 N
Tobermory: 67 N

SOUTHAMPTON HARBOUR
Soundings north of breakwater principally from survey by J.G. Shreenan and assista
Chantry Island Lighthouse : Lat. 44° 29' 24"N., Long. 81° 24' 9"W.
SOUNDINGS IN FEET
Natural Scale 1:24 696

LAKELAND BOATING · LAKE HURON CRUISING GUIDE

The West Coast of the Bruce Peninsula, Ontario

The shoal waters of the rugged Stokes Bay area keep all but the bravest cruisers away. Even the locals are ever-wary. This is a good place to be in the prop-shaft repair business. The first landfall is the town of Sauble Beach. Follow the buoyed channel to the entrance of the Sauble River **A**. There's a public launch ramp **B** just inside the river to starboard.

Two marinas are located on the river's south shore. Kit-Wat Motel Marina **C**, located at the second bend in the river, has a restaurant, gas dock and occasional transient dockage. Sauble River Marina **D** offers additional dockage just upriver. Don't forget to visit the beautiful sandy beach **E** for which this town is named, just south of the river mouth.

134   LAKELAND BOATING · LAKE HURON CRUISING GUIDE

Lake Huron

The West Coast of the Bruce Peninsula, Ontario
Sauble Beach, Oliphant and Stokes Bay.

The jagged coast that extends from Southampton to Tobermory is one of the most colorful parts of Lake Huron, but the rocks and shoals can scare cruisers off. Even in the recent era of low water, however, several ports in this area can be reached in good weather—and are worth the effort.

The Fishing Islands extend from Chief's Point north for about 10 miles past Oliphant. From there to Cape Hurd at the top of the Bruce, the waters are a jumble of islands, shoals and rocks, as the Bruce's limestone backbone fractures into Lake Huron.

Heading north, the first possible landfall is the Sauble River, which enters the lake north of the town of Sauble Beach. The buoyed channel leads to two marinas on the south shore. The first is the **Kit-Wat Motel Marina**, with a restaurant, transient docking and gas. One-tenth of a mile upriver is the **Sauble River Marina**, with gas and services. Across the street is **Joseph's Market & Ice Cream Parlor**, open seven days a week.

Going north, you pass Oliphant and Little Red Bay, two small boat harbors that are extremely difficult to enter without local knowledge. If you want to explore, contact **Murdoch McKenzie Marine** in Oliphant.

Next you'll come to Stokes Bay, which is entered

North of the Sauble River, you'll find Oliphant and Little Red Bay, two small boat harbors that require local knowledge to successfully navigate. Murdoch McKenzie Marine **F** in Oliphant occupies two basins, offering transient dockage and selling gas. There is additional dockage (but no services) and another fuel dock at an unstaffed government dock **G** immediately to the south of Murdoch McKenzie. Supplies are available just up the street.

Lakeland Boating · Lake Huron Cruising Guide   **135**

Where to Eat

The Copper Kettle 519-592-5666
Across the street from the general store in Stokes Bay, this attractive restaurant is abloom with flowers and serves a full menu in a casual setting.

Joseph's Ice Cream Parlor 519-422-1041
Sauble Beach's stop for cool treats.

Kit-Wat Motel Marina 519-422-1282
A family restaurant with a full menu ranging from hot beef to salads. They also have a patio overlooking the water.

Tamarac Island Inn 519-592-5810
This green building, located on Tamarac Island south of the government wharf in Stokes Bay, has a small boat dock and is an easy landmark. Call ahead to let them know you're coming.

mysterious Indian graveyard was established here long before the coming of the white man, but locals will not divulge its exact location for fear that souvenir hunters will dig up arrowheads and other artifacts.

Distances from Sauble Beach (statue miles)
Southampton: 14 S
Oliphant: 7 N

from Oliphant
Sauble Beach: 7 S
Red Bay: 6 N

from Red Bay
Oliphant: 6 S

from Stokes Bay
Southampton: 32 S
Tobermory: 35 N

between Lyal Island, marked by a flashing four-second light, and Greenough Point to the north. Coming from either direction, do not shoal to less than 10 fathoms until the Stokes Bay range is in line. The buoyed channel is marked by the Stokes Bay entrance light buoy VK. Come in on a bearing of 69°T until Mad Reef is abaft the port, then turn northward to the anchorage between Greenough and Ferguson points and enjoy the solitude. Or proceed on a northeasterly course to the head of Stokes Bay. The **Stokes Bay Government Dock** is on the west side. Water levels are currently quite low in the area, so call ahead.

Head to tranquil Tamarac Island, which was the site of the last existing sawmill in the area. Here you'll find **Tamarac Island Inn**, a quiet facility with a friendly staff. They offer transient dockage with water hookup and a tavern and restaurant that maintains an antique feel. If you want transportation to the shopping area of town, they'll provide it.

The **Stokes Bay General Store** is less than half a mile up the Stokes River. A

More Info

Emergency 519-793-3332
Lion's Head Hospital 519-793-3424
Bruce County Tourism Office 800-268-3838

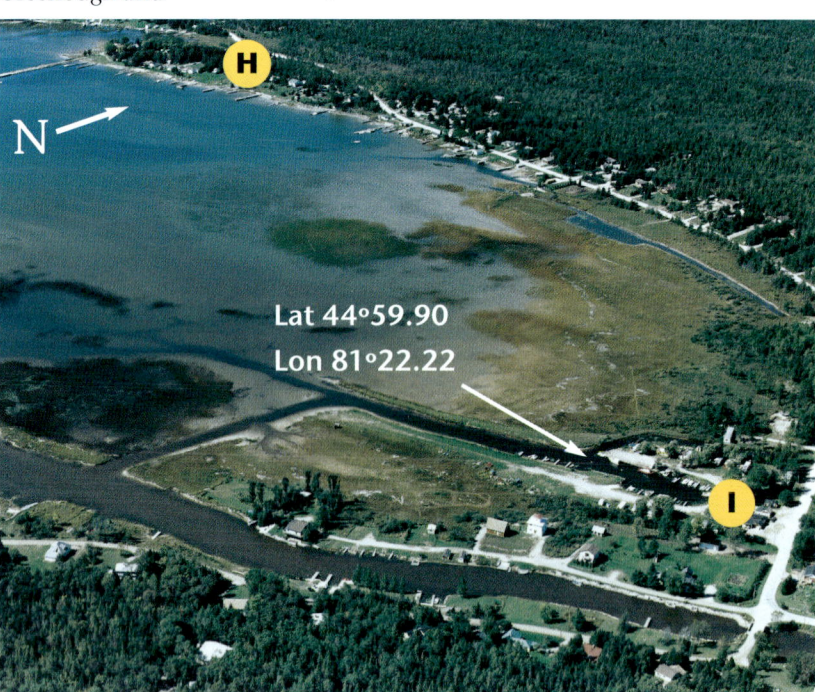

Lat 44°59.90
Lon 81°22.22

Local events

July
Stokes Bay Summer Fest
Sauble Beach Canada Day
Fireworks & Concert
Heritage Cruisers Antique Car Show

July-August
Sauble Sandpipers Craft Show

August
Sandfest
Sauble Speedway Stock Car Show

An anchorage can also be found in Stokes Bay. Just west of Lyal Island, look for the flashing red buoy light. Follow the marked channel on a bearing of 69°T. Turn to port to find some great anchorages between Ferguson Point and Greenough Point. Or bear northeast to Stokes Bay Government Dock **H**. Call ahead, however, because water levels can be quite low. Stokes Bay Marina **I** has recently experienced low water levels, and locals do not know when or if it will reopen. Tamarac Island Inn is a quarter-mile south of the government dock.

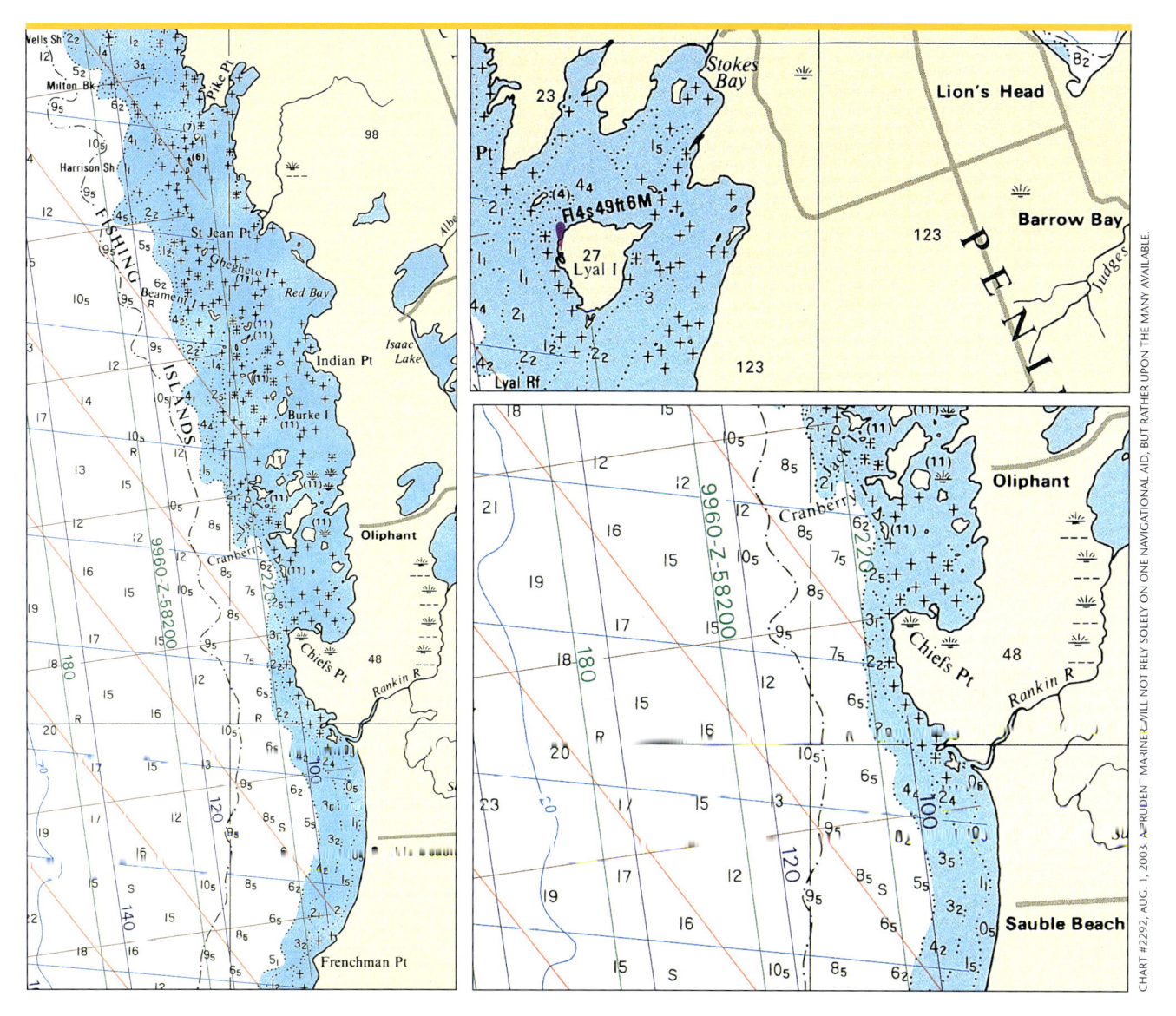

Facilities information subject to change. We suggest you call ahead.

		Transient Slips Available	Monitors VHF Channel	ALTERNATE MOORING: Wall mooring / Rafting allowed	Minimum Depth at Dock	Maximum LOA	Power (amperage or volts)	HOOKUPS: Water / Cable TV	FUEL: Gas / Diesel / Pumpout	BASICS: Heads / Showers / Laundry	AMENITIES: Grills / picnic tables / Dog walk / Swimming pool / Whirlpool / Rec area	Take Reservations	Take Credit Cards	Haulout (Capacity in tons or feet)	REPAIRS: Mechanical / Electronics / Fiberglass / Woodworking / Sails / Canvas	CONVENIENCE: Ship's store / Ice / Convenience foods & beverages
C	Kit-Wat Motel Marina 71 Sauble Falls Rd. 519-422-1282 Sauble Beach, ON N0H 2G0	2			28'	4'			G				GP	N	Y	IC
F	Murdoch McKenzie Marine Bay Street 519-534-1017 Oliphant, ON N0H 2T0	20			25'	4'			G	H			PD	Y	Y	
D	Sauble River Marina & Lodge 18 Marina Ave. 519-422-1762 Sauble Beach, ON N0H 2G0	4		W	28'	3'				H				Y	Y	C
H	Stokes Bay Government Dock Government Road 519-592-5828 Stokes Bay, ON N0H 2M0	12		WR	100'	8'		W		H			PD	Y	N	I
	Tamarac Island Inn 240 Tamarac Rd. 519-592-5810 Stokes Bay, ON N0H 2M0	9		WR	48'	2'6"	15	W		HSL			GPD	Y	Y	IC

LAKELAND BOATING · LAKE HURON CRUISING GUIDE

South Baymouth, Ontario

Lat 45°33.58
Lon 82°00.47

Lat 45°33.46
Lon 82°00.80

The approach to South Baymouth, the northern terminus of the *Chi-Cheemaun* ferry route, is simple—provided you approach on the range and respect the "big canoe," which arrives four times daily from Tobermory. South Bay Marina **A**, the only public docking facility, is immediately west of the ferry terminal **B**. South Bay **C**, a 16-mile inlet, offers good anchorages and fine fishing.

Facilities information subject to change. We suggest you call ahead.

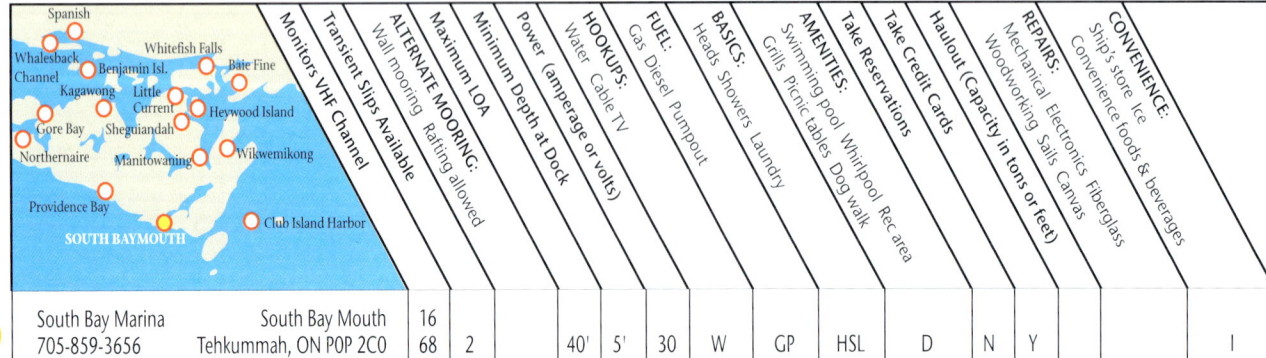

| A | South Bay Marina 705-859-3656 | South Bay Mouth Tehkummah, ON P0P 2C0 | 16 68 | 2 | | 40' | 5' | 30 | W | GP | HSL | D | N | Y | | I |

138 LAKELAND BOATING · LAKE HURON CRUISING GUIDE

LAKE HURON

South Baymouth, Ontario
Pull in to a 'ferry tale' town.

South Baymouth may be the home port of the *Chi-Cheemaun* ferry, affectionately known as the "big canoe," but smaller craft will feel equally at home in the deep, hazard-free harbor. Wind-bent evergreens, seagulls and the island's last operational foghorn enhance this 100-year-old fishing port's timeless atmosphere.

The slips at **South Bay Marina**, with 5 feet of water, can accommodate as many as 65 craft, though only a handful of slips are set aside for transients. Power and water are available at most. Showers, washrooms and laundry facilities are provided. Pumpout and gas, but no diesel, are available at the service dock, and propane is sold in town. Boats in need of repairs would be better served elsewhere, as there is neither haulout nor a resident mechanic. In a pinch, however, repairs can be arranged. The eastern shore, incidentally, is part of the **Wikwemikong Unceded Indian Reservation**.

The tiny village of South Baymouth has limited offerings but makes for a pleasant stroll. There's a small **grocery store** and several dining options, including a **fish and chip shack** and a restaurant with a patio facing the ferry docks. Bicycles can be rented from **Weatherock**, a resort located a short hike up the road. The **lighthouse**, located just beyond the marina, can be reached within minutes. While in South Baymouth, don't be surprised if the foghorn lets loose. And keep your ears cocked for the warning blast of the approaching ferry. With its immense jaws, which open both fore and aft to take on and disgorge passengers, it is well worth checking out.

Distances from South Baymouth (statute miles)
Tobermory: 28 SE
Rattlesnake Harbour: 15 E
Providence Bay: 17 NW
Presque Isle Harbor, Michigan: 84 SW
DeTour Village, Michigan: 85 W

More Info
Emergency 705-859-3155
Mindemoya Hospital 705-377-5311
Little Current Tourist Information 705-368-3021

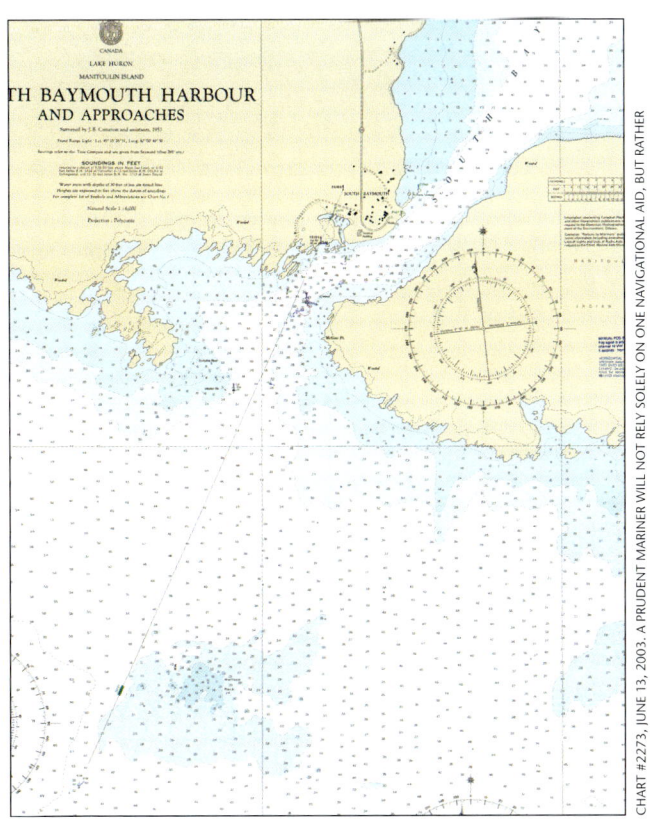

Sightseeing Tours
A four-hour bus tour of island sights such as Bridal Veil Falls, Ten Mile Point scenic lookout and Indian craft shops departs from the South Baymouth ferry docks at 1:15 p.m. Monday through Saturday in July and August.

Where to Eat
Family Brown Restaurant 705-859-3481
The dining room features locally caught whitefish, steak, pasta and a salad bar. The coffee shop next door serves light entrées, sandwiches and salads. Famous for their Sunday buffet.

Providence Bay, Ontario

Lat 45°39.52
Lon 82°16.16

Cruisers who have braved the big waters of Lake Huron to reach Providence Bay will be welcomed by a remote light at Providence Point followed by a second red marker light at the end of the town pier **A**. The government-owned Providence Bay Marina **B** has the harbor's only docks. A short hike up the shore is one of northern Ontario's best sand beaches **C**. A boardwalk leads to the village of Providence Bay **D**, about three-quarters of a mile away.

Facilities information subject to change. We suggest you call ahead.

		Monitors VHF Channel	Transient Slips Available	ALTERNATE MOORING: Wall mooring, Rafting allowed	Maximum LOA	Minimum Depth at Dock	Power (amperage or volts)	HOOKUPS: Water, Cable TV	FUEL: Gas, Diesel, Pumpout	BASICS: Heads, Showers, Laundry	AMENITIES: Swimming pool, Whirlpool, Rec area, Grills, Picnic tables, Dog walk	Take Reservations	Take Credit Cards	Haulout (Capacity in tons or feet)	REPAIRS: Mechanical, Electronics, Fiberglass, Woodworking, Sails, Canvas	CONVENIENCE: Ship's store, Ice, Convenience foods & beverages
B	Providence Bay Marina Manitoulin Island 705-377-7225 Providence Bay, ON P0P 1T0	68	1	WR	80'	17'			P	HS	D	Y	N	N	M	I

140 LAKELAND BOATING · LAKE HURON CRUISING GUIDE

LAKE HURON

Providence Bay, Ontario
Set your course for a beach that's just divine.

For those who welcome the open challenge of Lake Huron and don't expect every amenity from their stopover, "Prov," as the locals affectionately call it, has much to offer—in particular its long, sandy beach, regarded as one of northern Ontario's finest. With no serviced ports nearby but South Baymouth, this small village is a rustic, relaxing spot.

If you head west from here to Mississagi Strait and Meldrum Bay, or southwest to Presque Isle or Rogers City, there are only two "ports in a storm": Great Duck Island and Burnt Island Harbour. If you choose either of these routes, proceed carefully with a good chart in hand.

Providence Bay Marina, situated on the eastern shore of the bay, generally has one or two of its 29 slips available to transients, though it's a good idea to call ahead. The horseshoe-shaped bay, with an average depth of about 25 feet, affords a good selection of anchorages but is vulnerable to a southerly blow. The biggest drawback to Providence Bay as a port of call is the lack of facilities at the marina: fuel, propane and pumpout must be ordered from town, and there is no power or water to the docks. Showers, washrooms and a fish-cleaning station are provided at the marina building.

The stroll into town is a little over a mile, but it's a pleasant one. Visitors can follow the long, snaking boardwalk or chart their own trail among the dunes. The **Harbour Centre**, located right on the beach, houses an interpretive center and a sandwich bar. In town are a few places to eat, including **The School House Restaurant**, which specializes in locally raised meat and produce and fresh lake fish. The weathered cedar building is somewhat hidden on a side street, but it's worth seeking out. A **coin-operated laundry** can be found in the village, but there's no liquor store or bank, and only a limited supply of groceries. Providence Bay hopes to be home to a full-service marina someday soon. In the meantime, with its sand beach, cedar-shake buildings and rich history as a logging port, Prov still has a lot to offer.

Local event
August
Providence Bay Fair

Where to Eat

Lake Huron Fish and Chips 705-377-4500
Tasty whitefish, great burgers and fresh-cut fries.

The School House Restaurant 705-377-4055
Everything is made from scratch, including the stuffed chicken with spinach and shallots and the Belgian chocolate almond gâteau. Fresh seafood, too.

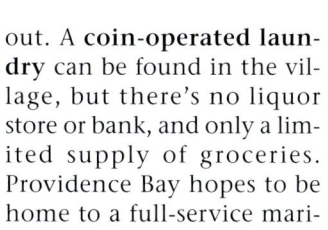

Distances from Providence Bay (statute miles)
South Baymouth: 17 SE
Tobermory: 45 SSE
Presque Isle Harbor, Michigan: 68 WSW
DeTour Village, Michigan: 85 W

More Info
Emergency 911
Mindemoya Hospital 705-377-5311
Township of Central Manitoulin 705-377-5726
www.manitoulin-island.com

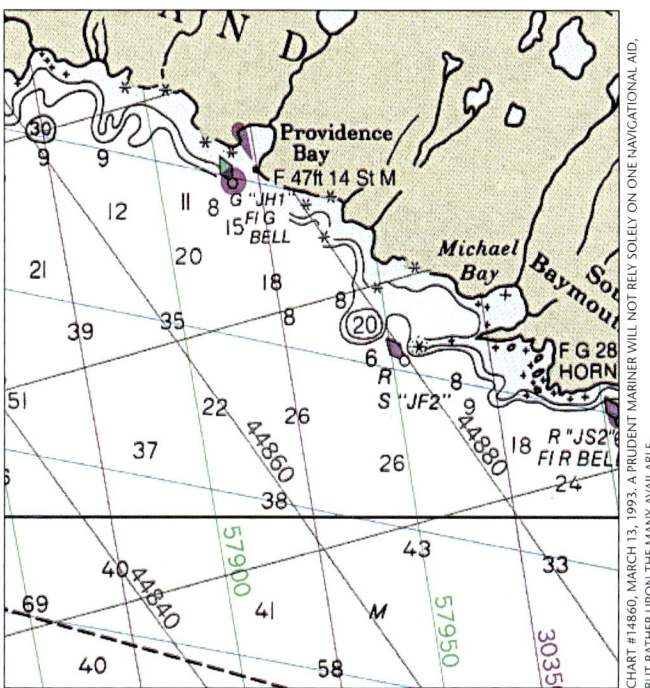

LAKELAND BOATING · LAKE HURON CRUISING GUIDE

LAKE HURON

Les Cheneaux Islands
A wonderland of clear waters and quiet coves.

Off the southeastern coast of Michigan's Upper Peninsula some 15 miles from Mackinac Island, the Les Cheneaux are an archipelago of 36 islands along 12 miles of Lake Huron shoreline. But that's just the textbook definition. It's really a kind of cruising paradise, with crystal-clear waters, emerald coves and gaily hued boathouses. Loons squawk and swoop overhead as St. Martin's Reef Light winks on the horizon.

This area was originally home to a French mission before emerging as a thriving lumbering, fishing and hunting outpost. By the 1890s it had become a classy summer resort area. Including the towns of Cedarville and Hessel, the Les Cheneaux stretch some 25 miles from St. Martin's Point in the west to Beaver Tail Bay in the east. Navigating these waters is effortless—most boaters use the western entrance to the island chain as they arrive from Mackinac Island, and Coast Guard buoys are supplemented by locally installed aids to navigation that clearly mark all underwater hazards.

Boaters should be aware, however, that buoyage is inbound from the east and also inbound from the west as far as Cedarville, where buoy colors switch sides!

Contrary to popular belief, Les Cheneaux doesn't translate as "the channels"—in fact, the word *cheneaux* doesn't even exist in French. Perhaps the best translation is this: a delicate conference of rustic simplicity and posh civility, where the growl of snazzy mahogany runabouts ricochet around the ancient forests that touch the waterline.

Distances from the Les Cheneaux Islands (statute miles)

From the west entrance:
Mackinac Island: 12 SW
Cheboygan: 24 S

From the Government Island entrance:
DeTour Passage: 18 E
Cedarville: 8 NW

Cedarville, Michigan

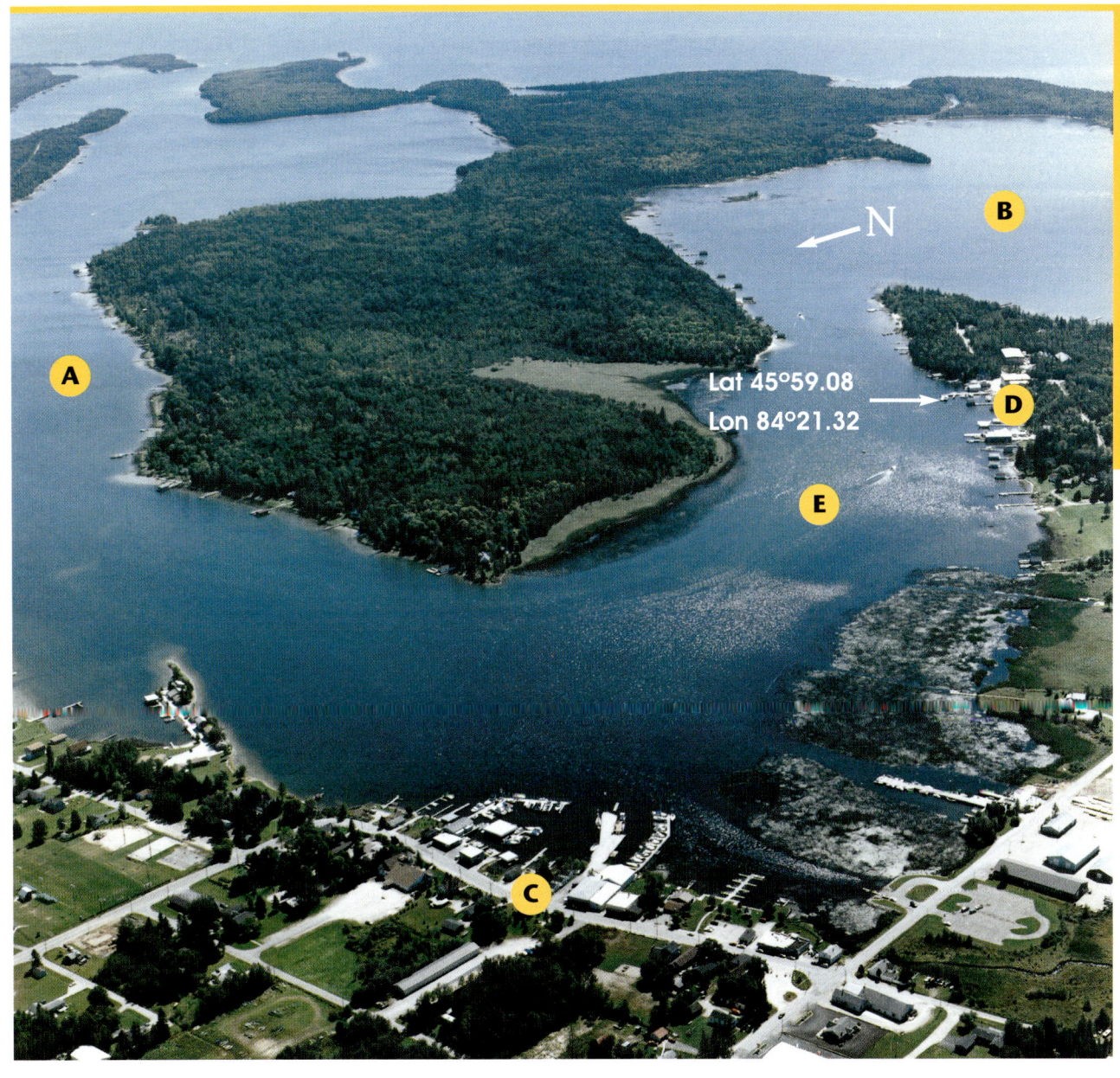

Lat 45°59.08
Lon 84°21.32

Approach Cedarville from either of the two channels formed by protective LaSalle Island. From DeTour and North Channel, you would take the eastern channel **A**. From Hessel, Mackinac or the west, the western channel **B** is most convenient. About 20 transient slips are available at Cedarville Marine **C**; Viking Boat Harbor, Inc. **D** has four. Both facilities do repair work and sell fuel and supplies.

Distances from Cedarville (statute miles)
DeTour Village: 26 E

Hessel: 7 W
Mackinaw City: 30 SW

LAKELAND BOATING · LAKE HURON CRUISING GUIDE   **143**

LAKE HURON

Cedarville, Michigan
Make your way—carefully!—
to this unspoiled vacation paradise.

Cedarville and its sister city of Hessel may have started out as lumbering and fishing towns, but it wasn't long before vacationers discovered the Les Cheneaux area and made it their own. Summer residents began flocking to the two towns in the late 1800s, taking the train to Mackinaw City and catching the steamboat service over to Les Cheneaux.

Antique boats and marine artifacts from those early resort days are exhibited at the excellent **Les Cheneaux Historical Maritime Museum**, housed in a 1920s boathouse in Cedarville. Classic Chris-Crafts are a highlight of the museum. Other aspects of the area's history—including Native American artifacts and exhibits detailing the lives of the loggers and other early settlers—are documented at the **Les Cheneaux Historical Museum**, also in Cedarville.

Tucked back into a deep bay along the convoluted Upper Peninsula coast and protected by LaSalle Island, Cedarville is not particularly well situated as a transient stopover. It can be a bit tricky to get into, especially with the low water of recent years, but if you pay attention to the marked channel and your charts (and show no more than a couple of feet), you should have no problems.

Cedarville Marine provides clean showers and floating docks with space for about 20 transients. Overnight fees are reasonable. **Viking Boat Harbor, Inc.** offers fuel and repair service and about four transient slips.

Most of the resorts in the islands are on the water and many have private marinas, but they do not usually cater to transients. In a pinch—such as during the Les Cheneaux Antique Wooden Boat Show in August—the very hospitable **Les Cheneaux Welcome Center** can often help work something out with a local resort so you can use their facilities.

In both Hessel and Cedarville, dockage availability depends on the weather. Many boaters lay over in one of these two cities until the weather is good enough to cross the 20 miles of open water to DeTour Passage and entry to the North Channel. In fair weather, transient slips are plentiful.

Take a stroll around town before stopping in at one of the town's restaurants. Two **hardware stores**, a **laundromat** and a **convenience store** are also within walking distance.

Unlike many other areas of Michigan, the islands and the rest of the 80-mile northern Lake Huron shoreline have been the focus of the **Nature Conservancy** in partnership with land owners from Hessel and Cedarville

More Info

Emergency	911
Mackinac Straits Hospital	906-643-8585
Les Cheneaux Welcome Center	906-484-3935

www.lescheneaux.org

Where to Eat

Ang-Gio's..................906-484-3301
A homey and attractive place just up the road from the marina. Featuring Italian and American dishes and some of the best pizza in town.

Pammi's Restaurant............906-484-7844
The nautical theme of this charming bistro extends to the menu—locally caught fish is the specialty. Don't miss the corn chowder.

Pine Cone...................906-484-3413
A good place for a quick bite. Featuring a full takeout menu that includes burgers and coney dogs.

east to DeTour. More than 60 rare plant and animal species thrive here.

Creekside Herbs & Art, an artists' colony hidden in the forest, can be tricky to reach from the water but is well worth the effort. You'll find sculptures, blown glass, a boutique, an ecotourism company and an amazing herb garden.

Local events

July
4th of July

Michigan's Great Outdoors Culture Tour

August
Les Cheneaux Antique Wooden Boat Show and Festival of Arts

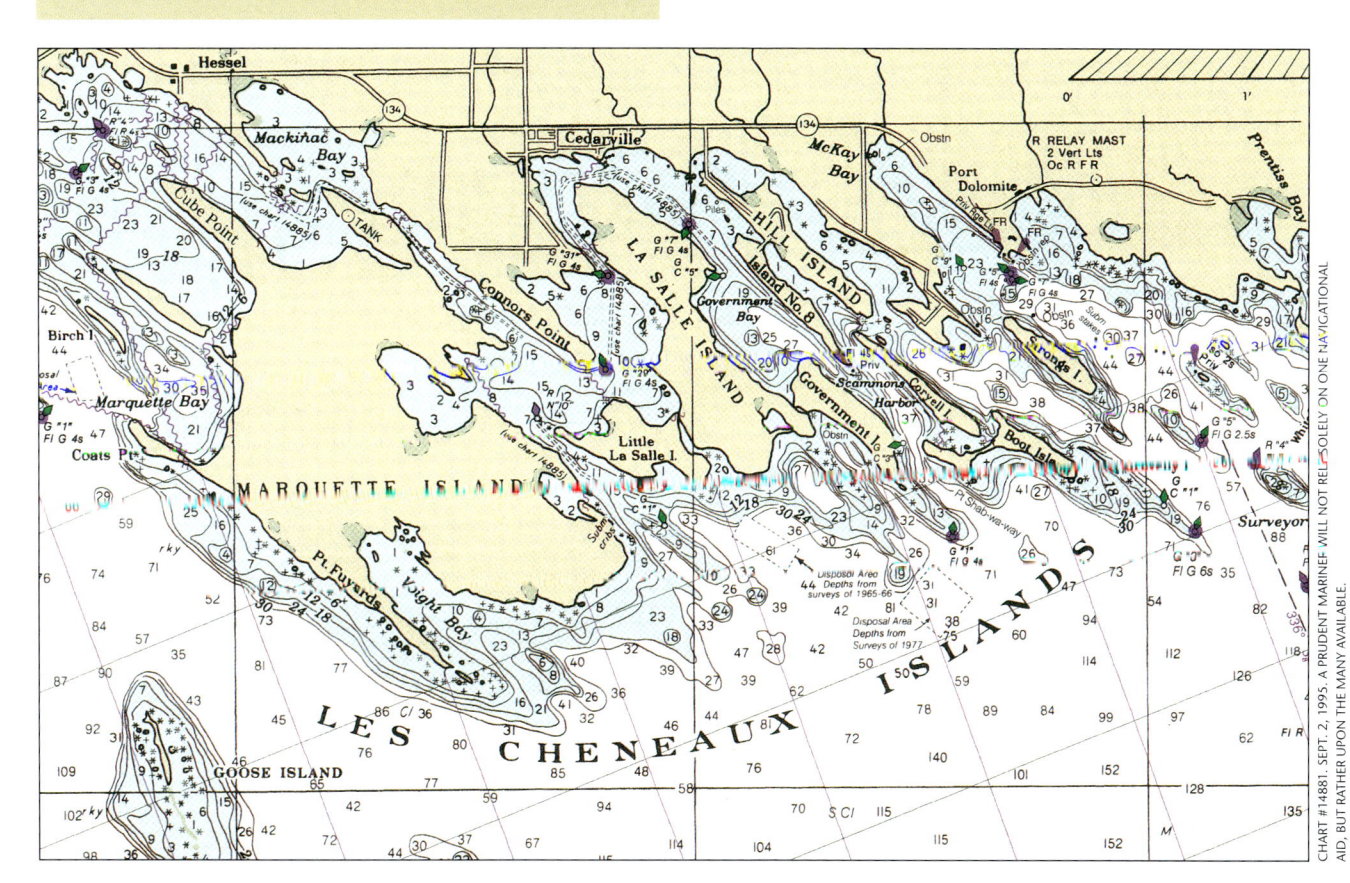

Facilities information subject to change. We suggest you call ahead.

			Monitors VHF Channel	Transient Slips Available	ALTERNATE MOORING: Wall mooring Rafting allowed	Maximum LOA	Minimum Depth at Dock	HOOKUPS: Power (amperage or volts) Water Cable TV	FUEL: Gas Diesel Pumpout	BASICS: Heads Showers Laundry	AMENITIES: Swimming pool Whirlpool Rec area Grills picnic tables Dog walk	Take Reservations	Take Credit Cards	Haulout (Capacity in tons or feet)	REPAIRS: Mechanical Electronics Fiberglass Woodworking Sails Canvas	CONVENIENCE: Ship's store Ice Convenience foods & beverages
C	Cedarville Marine 906-484-2815	100 Hodeck St. Cedarville, MI 49719	16	20		100'	4'	30 50	W	G	HS	G	N	Y	20	MEFW	SI
D	Viking Boat Harbor, Inc. 906-484-3303	1121 Islington Road Cedarville, MI 49719	16	4		60'	4'	20	W	GDP	HS		Y	Y	50	MEFWSC	SI

LAKELAND BOATING · LAKE HURON CRUISING GUIDE   **145**

Hessel, Michigan

Lat 46°00.12
Lon 84°25.52

The harbor at Hessel is uncomplicated. Coast Guard buoys are augmented by buoys put out by the local islands association. Approaching from the west entrance, bear starboard after passing Goat Island **A** and Haven Island **B** and continue southeast along Hessel Point. First you'll find the private docks of Wilson Cottages **C**. Just beyond is the full-service Hessel Marina **D**, which is your best bet for transient space. This is a great place to dock and stock up at the Hessel Grocery **E** or dine at the Hessel Bay Inn **F**. The city also operates a double ramp **G** here. The popular Hessel Home Bakery is just up the road. Next is the E.J. Mertaugh Boat Works, Inc. **H**, a private marina and full-service repair shop that is famous as the first Chris-Craft dealer in the country. Downtown Hessel is only a short block away. Continuing up the coast, you'll find the private docks of the Loreli Cottages **I** and the Lindberg Cottages **J**.

Distances from Hessel (statute miles)
Mackinac Island: 14 SW
St. Ignace: 17 SW
Cedarville: 7 E

146 LAKELAND BOATING · LAKE HURON CRUISING GUIDE

LAKE HURON

Hessel, Michigan
A mecca for vintage boat lovers.

Hessel is just a short run from the action at the Straits of Mackinac, and for cruisers making their way there in easy stages, or for those headed for the North Channel, the port is a favorite stopping place. Its rugged coastline and offshore islands provide a teasing taste of things to come as boaters head into unspoiled waters. Hessel is a rare port whose personality and reputation far outweigh its size.

Hessel Bay runs northwest to southeast for three and a half miles, with mid-channel depths from 8 to 18 feet. The town, settled in 1884, was named after the area's first postmaster, a Swedish fellow named John Erickson. His last name was changed to Hessel (after his home town in Sweden) when he immigrated to the United States—apparently there were too many John Ericksons here already.

Hessel captivates its guests with the genuine nautical charm of the **E.J. Mertaugh Boat Works.** Like a village suspended in time, this marina radiates a special nostalgia for wooden boats and craftmanship. History is preserved in slips lined with immaculately reconditioned speedboats. Definitely stop in for a visit, but note that they do not cater to transients (though they do sell gas and diesel). They do all repairs and have a haulout capacity of 65 tons. The best bet for cruisers looking for an overnight berth is the full-service **Hessel Marina,** with 36 transient slips.

A flock of Chris-Crafts and other classics migrates to the area the second weekend in August for the annual Les Cheneaux Antique Wooden Boat Show. About

Where to Eat

Hessel Bay Inn 906-484-2460
Right across from the marina, this place has a reputation for serving the best fish dinners in the area. The lakefront patio offers great views.

Hessel Harbor Market & Deli 906-484-2435
A century-old general store stocked with gourmet items, deli treats and an extensive wine selection.

Hessel Home Bakery 906-484-3412
A local favorite that will satisfy even the largest appetite for fresh-baked goodies. Their pasties (meat pies) and garlic toast are legendary.

Islander Bar 906-484-3359
Mingle with the friendly locals at this landmark pub.

E.J.Mertaugh BoatWorks
Est. 1925

BOSTON WHALER • JC PONTOON • EVINRUDE Johnson GENUINE PARTS
MERCURY • MerCruiser STERN DRIVES & INBOARDS The Only Logical Choice. • BOMBARDIER E-TEC

- Full service Boat Yard with 2 locatons to service you
- Specializing in wood boat restoration as well as outboard/IO/Fiberglass, Gel Coat repairs.
- Inside/outside/heated/cold/storage and shrink wrap
- Hessel Location - travellift to 60 ton
- Detour Location - travellift to 20 ton
- New and used boat and motor sales and brokerage
- Ships Store / Gas and diesel fuel
- Operate DNR Detour Harbor Marina

Member MBIA   Member ABYC

Detour: 906-297-6888 • Hessel: 906-484-2434
P.O. Box 40 Hessel, MI 49745
Email: ejmert@lighthouse.net

www.ejmertaughboatworks.com

LAKELAND BOATING · LAKE HURON CRUISING GUIDE   **147**

9,000 people arrive to stroll the docks, drool at the boats and chat about restoration techniques. Anyone with an all-wooden boat can enter. Awards are given in 15 classes, including for best restoration job and best boat in the show.

If you dock in Hessel, all services are exceptionally close. **Hessel Grocery**, just across the street from the marina, provides all the necessities. **Hessel Bay Inn**, located adjacent to the harbor, serves up delicious blackened whitefish and other specialties. The enticing aroma of **Hessel Home Bakery** permeates the forest up on M-134, just a few blocks from the docks. An 18-hole championship golf course stretches out across from the airport. Nearby **Kewadin Casino** is popular with visitors looking to try their luck.

The streets of Hessel are lined with a variety of quaint gift shops, bait shops and cabin rental services. **Lindberg Cottages** seems to be a favorite.

Local events

July
Fourth of July

August
Les Cheneaux Antique Wooden Boat Show and Festival of Arts

Salmon Derby

More Info

Emergency 911
War Memorial Hospital 906-635-4460
Les Cheneaux Welcome Center. 906-484-3935

www.lescheneaux.org

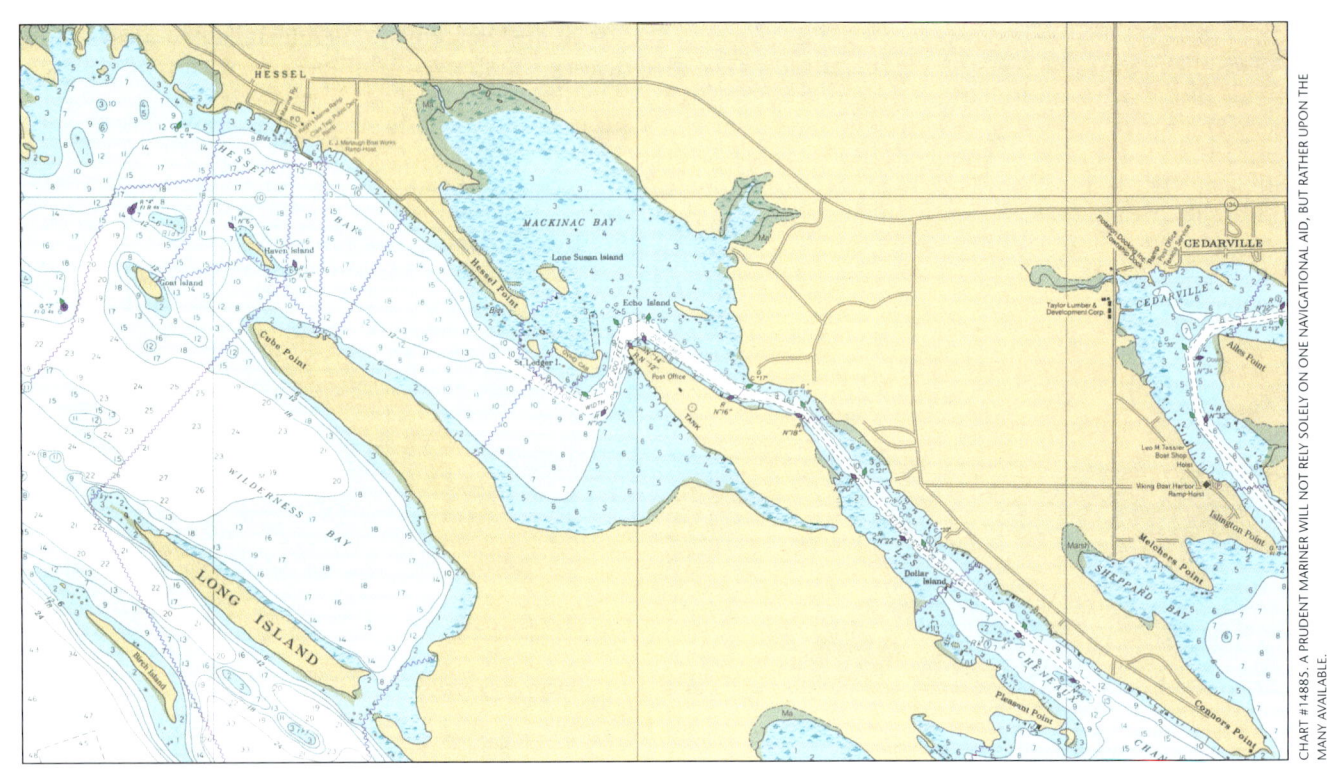

Facilities information subject to change. We suggest you call ahead.

			Monitors VHF Channel	Transient Slips Available	ALTERNATE MOORING: Wall mooring, Rafting allowed	Maximum LOA	Minimum Depth at Dock	Power (amperage or volts)	HOOKUPS: Water, Cable TV	FUEL: Gas, Diesel, Pumpout	BASICS: Heads, Showers, Laundry	AMENITIES: Swimming pool, Whirlpool, Rec area, Grills, Picnic tables, Dog walk	Take Reservations	Take Credit Cards	Haulout (Capacity in tons or feet)	REPAIRS: Mechanical, Electronics, Fiberglass, Woodworking, Sails, Canvas	CONVENIENCE: Ship's store, Ice, Convenience foods & beverages
D	Hessel Marina 906-484-3917	P.O. Box 367 Cedarville, MI 49719	9 16	36	WR	40'	7'	15 20	W	P	HSl	GP		N	Y			C
H	E.J. Mertaugh Boat Works 906-484-2434	296 Hessel Point Rd. Hessel, MI 49745	16	N			6'			GD					Y	65	MEFWSC	S

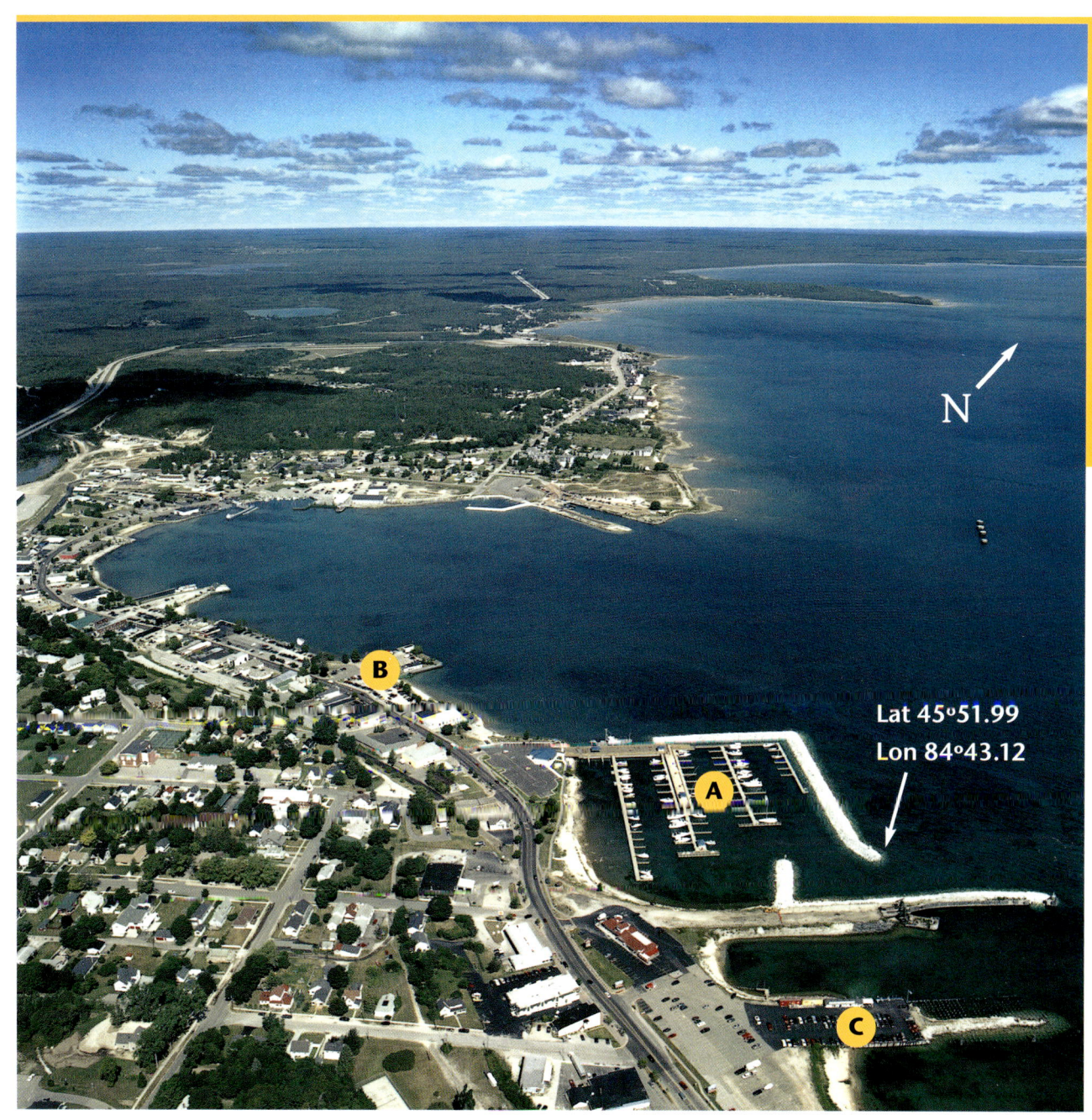

St. Ignace, Michigan

St. Ignace, with its brand-new, state-of-the-art marina **A**, nestled between the busy ferry docks **B** and **C**, is a historic tourist outpost at the northern end of the Mackinac Bridge.

Lat 45°51.99
Lon 84°43.12

Facilities information subject to change. We suggest you call ahead.

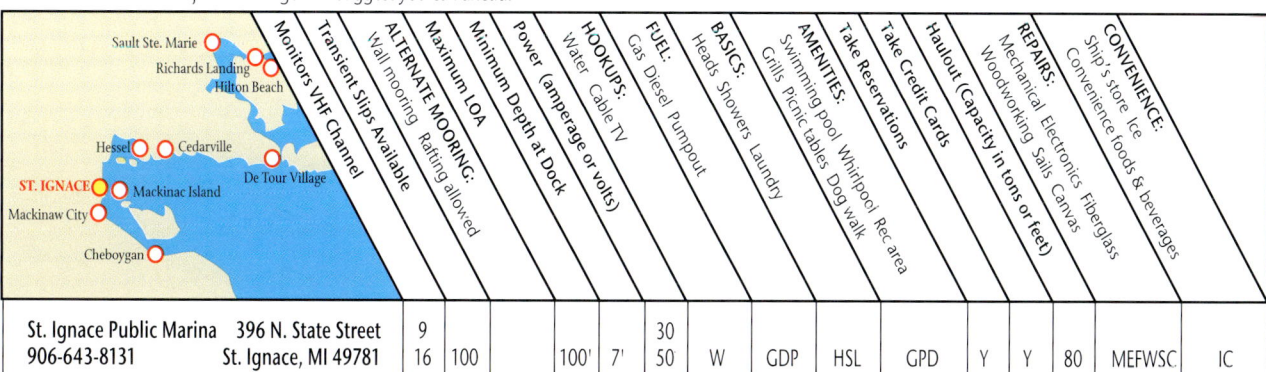

		Transient Slips Available	Monitors VHF Channel	ALTERNATE MOORING: Wall mooring, Rafting allowed	Maximum LOA	Minimum Depth at Dock	Power (amperage or volts)	HOOKUPS: Water, Cable TV	FUEL: Gas, Diesel, Pumpout	BASICS: Heads, Showers, Laundry	AMENITIES: Swimming pool, Whirlpool, Rec area, Grills, Picnic tables, Dog walk	Take Reservations	Take Credit Cards	Haulout (Capacity in tons or feet)	REPAIRS: Mechanical, Electronics, Fiberglass, Woodworking, Sails, Canvas	CONVENIENCE: Ship's store, Ice, Convenience foods & beverages
A	St. Ignace Public Marina 396 N. State Street 906-643-8131 St. Ignace, MI 49781	9 16	100		100'	7'	30 50	W	GDP	HSL	GPD	Y	Y	80	MEFWSC	IC

LAKE HURON

St. Ignace, Michigan
Rediscover the charms of this former trading outpost.

Anchoring the mighty Mackinac Bridge at its northern end, St. Ignace is a small city with a long history. Founded in 1671 by famed Jesuit priest and voyageur Père Jacques Marquette (but named for St. Ignatius of Loyola, the founder of his order), the city and its harbor have long hosted explorers, traders and travelers.

After years of planning, the old bare-bones **St. Ignace Public Marina** has been transformed into a 136-slip state-of-the-art facility that offers full services, including pumpout, restrooms, showers, laundry, grills and picnic tables. The marina is just two miles south of the **General Utility Airport**. The three ferry docks serving Mackinac Island make the port a very busy place.

The Museum of Ojibwa Culture showcases the native people who lived on the site before the coming of the Europeans and their place in the 17th-century fur trade.

The downtown business district follows the shoreline of Lake Huron's Moran Bay and is lined with restaurants, gift shops and fudge stores. There's also a **coin-operated laundry**, a **bank/ATM**, a **pharmacy**, a **bakery**, a **grocery store** and **marine-supply providers** within walking distance of the marina. Along the waterfront **Huron Boardwalk**, open-air exhibits provide details of the area's rich history.

Bayside Live!, a free weekly entertainment series, is held Thursdays at the new marina pavilion. Each Tuesday evening in July and August, Arnold Line offers its **Vesper Cruise** through the Straits of Mackinac and under the

Where to Eat

B.C. Pizza 906-643-0300
"Best Choice Pizza" will deliver pizza, subs, salads, pasta and more.

Driftwood Sports Bar & Grill 906-643-9133
Breakfast, lunch and dinner. Fresh fish, steaks and salads.

Galley Restaurant & Lounge 906-643-7960
Famous for its whitefish and great prime rib.

Mackinac Grille & Waterfront Pub 906-643-7482
A popular place for Sunday brunch or nightly entertainment.

Marina Pub 906-643-0556
Daily specials and entertainment overlooking the marina and Mackinac Island.

Village Inn 906-643-9511
Planked whitefish, steaks, homemade pastas.

150 LAKELAND BOATING · LAKE HURON CRUISING GUIDE

Local events

June
Kids' Fishing Day

Antiques on the Bay Car Show

St. Ignace Car Show & Down Memory Lane Parade, Cruise Night and Oldies Concerts

July
July 4th Street Dance, Parade, Community Picnic and Fireworks

Red Hacker Basketball Tournament

July-August
Bayside Live!

Straits of Mackinac Vesper Cruises

August
Straits of Mackinac Underwater Treasure Hunt

St. Ignace Salmon Derby

September
Michinemackinong Powwow

Mackinac Bridge Walk

Mackinac Bridge. **Kewadin Casino**, three miles north of the marina, offers a free shuttle service.

If you're up for a little exercise, head to **Castle Rock**, a 200-foot rock formation that rewards those energetic enough to climb the stairs to the top with a spectacular view of the St. Ignace area.

Throughout the years, many ships have fallen victim to the treacherous waters of the straits, providing sunken treasure for scuba divers. The 148-square-mile **Straits of Mackinac Underwater Preserve** holds more than 15 shipwrecks. The premier wreck of the Straits Preserve is the Sandusky, a 100-foot wooden brig that sank in 1856. The brig is well-preserved and is among the few Great Lakes wrecks that boast a carved figurehead. Discovered in 2002, the William Young is a 148-foot wooden three-masted schooner/barge that sank in 1866. A more recent wreck is the Cedarville, a 588-foot self-unloading freighter that sank when it collided with the Norwegian vessel Topdalsfjorfd. Wrecks are buoyed and loran and GPS coordinates are available.

Distances from St. Ignace (statute miles)
Mackinaw City: 7 S
Cheboygan: 20 SE
Beaver Island: 47 W
Mackinac Island: 6 E
Hessel: 17 NE
DeTour Village: 44 ENE

More Info

Emergency 911
Mackinac Straits Hospital 906-643-8585
Mackinac County Airport 906-643-7165
Chamber of Commerce 800-338-6660
www.stignace.com
Arnold Mackinac Island Ferry 800-542-8528
Shepler's Mackinac Island Ferry 800-828-6157
Star Line Mackinac Island Ferry 800-638-9892

Welcome To The New St. Ignace City Marina

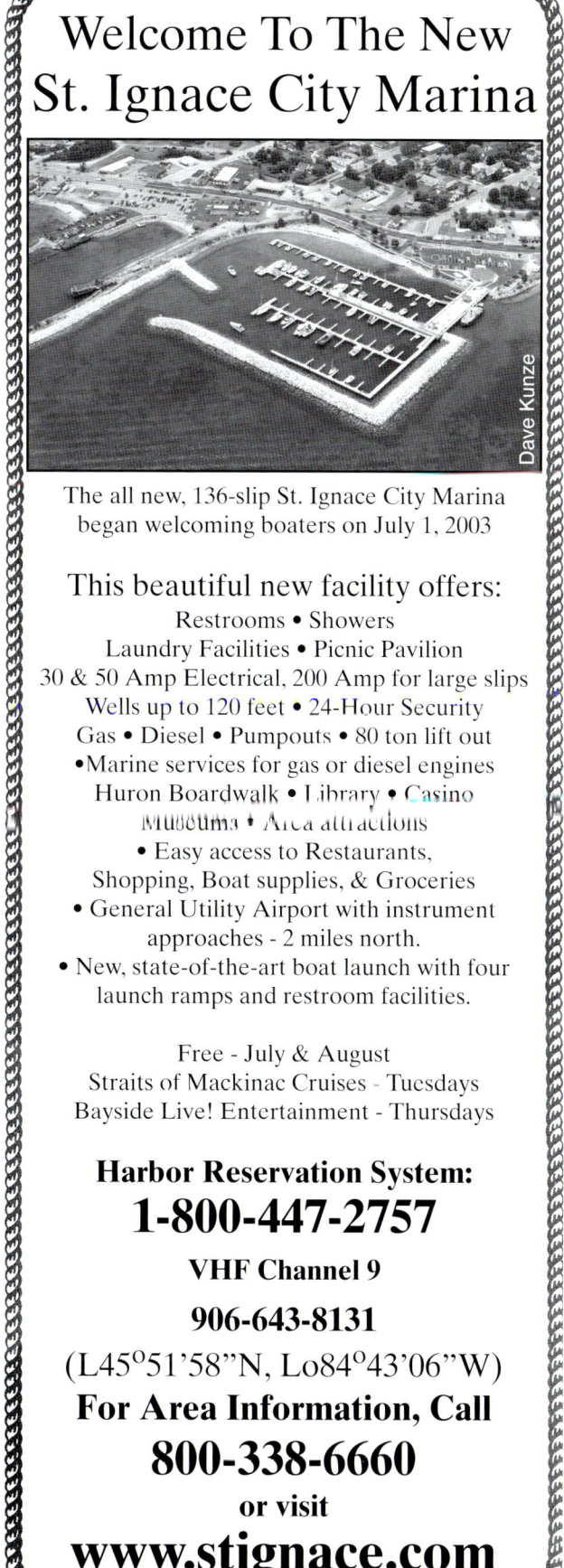

The all new, 136-slip St. Ignace City Marina began welcoming boaters on July 1, 2003

This beautiful new facility offers:
Restrooms • Showers
Laundry Facilities • Picnic Pavilion
30 & 50 Amp Electrical, 200 Amp for large slips
Wells up to 120 feet • 24-Hour Security
Gas • Diesel • Pumpouts • 80 ton lift out
• Marine services for gas or diesel engines
Huron Boardwalk • Library • Casino
Museums • Area attractions
• Easy access to Restaurants, Shopping, Boat supplies, & Groceries
• General Utility Airport with instrument approaches - 2 miles north.
• New, state-of-the-art boat launch with four launch ramps and restroom facilities.

Free - July & August
Straits of Mackinac Cruises - Tuesdays
Bayside Live! Entertainment - Thursdays

Harbor Reservation System:
1-800-447-2757
VHF Channel 9
906-643-8131
(L45°51'58"N, Lo84°43'06"W)

For Area Information, Call
800-338-6660
or visit
www.stignace.com

Section Three

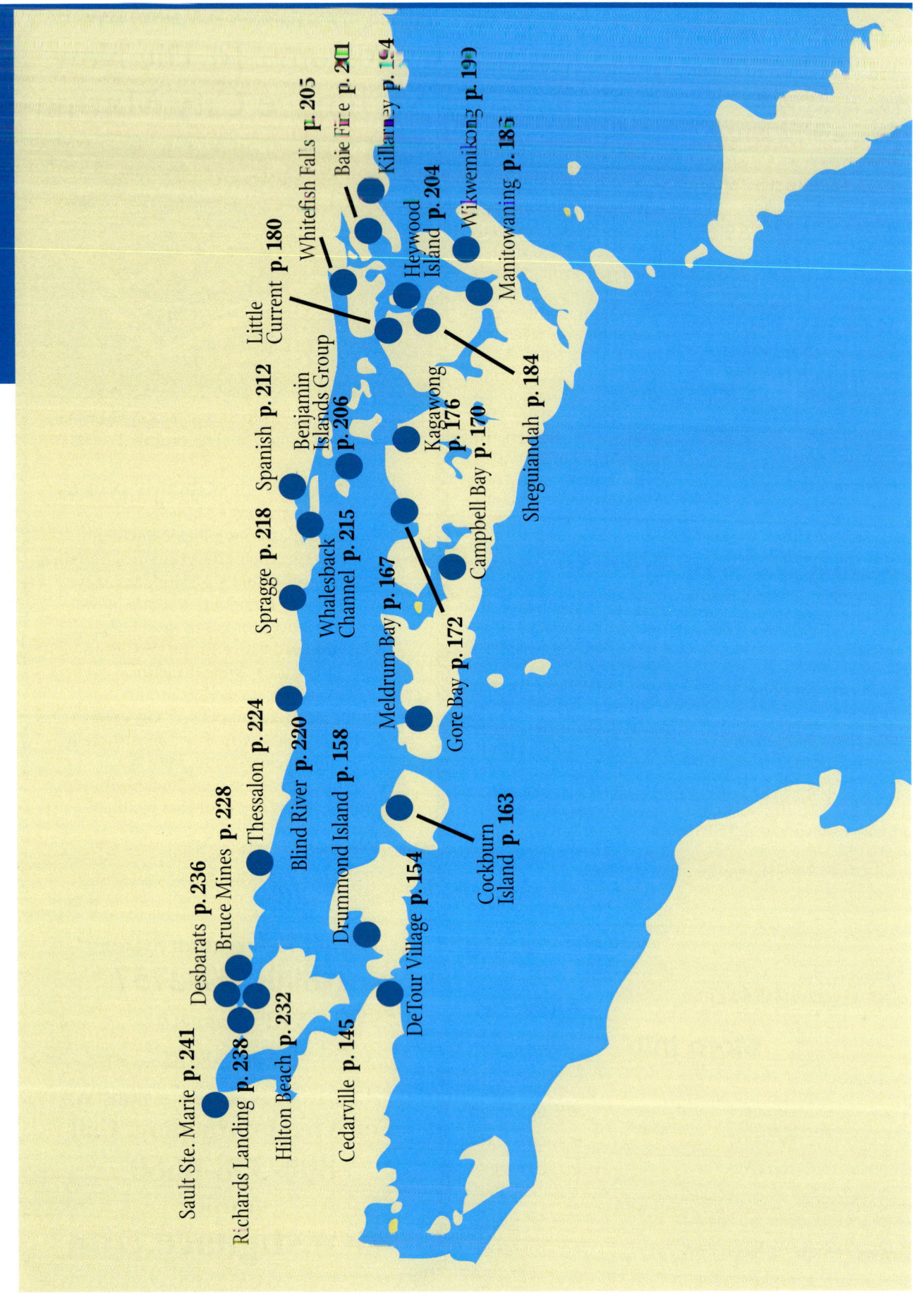

- Whitefish Falls **p. 205**
- Baie Fine **p. 201**
- Killarney **p. 194**
- Heywood Island **p. 204**
- Wikwemikong **p. 199**
- Manitowaning **p. 185**
- Little Current **p. 180**
- Benjamin Islands Group **p. 206**
- Kagawong **p. 176**
- Sheguiandah **p. 184**
- Spanish **p. 212**
- Campbell Bay **p. 170**
- Spragge **p. 218**
- Whalesback Channel **p. 215**
- Meldrum Bay **p. 167**
- Gore Bay **p. 172**
- Thessalon **p. 224**
- Blind River **p. 220**
- Drummond Island **p. 158**
- DeTour Village **p. 154**
- Cockburn Island **p. 163**
- Sault Ste. Marie **p. 241**
- Desbarats **p. 236**
- Bruce Mines **p. 228**
- Richards Landing **p. 238**
- Hilton Beach **p. 232**
- Cedarville **p. 145**

152 LAKELAND BOATING · LAKE HURON CRUISING GUIDE

North Channel: Pristine Beauty
Cruisers and trailerboaters alike can seek out secluded spots.

The words "North Channel" conjure up visions of white quartzite mountains, forested hillsides, countless islands, secluded anchorages, quaint ports of call and sparkling waters. Long the province of the cruising boater, all this is now available to the trailerboater as well, with many fine launch sites throughout this 3,100-square-mile area.

Boaters can enter these breathtaking cruising grounds from the northwest at Sault Ste. Marie and the St. Joseph Channel, or they can come in from the southwest at DeTour and Drummond Island through Potagannissing Bay or south of St. Joseph Island via the well-marked Canadian shipping channel. From the south, cruisers take False DeTour Channel between Drummond Island and Cockburn—also the boundary between the United States and Canada—or the Mississagi Strait between Cockburn and Manitoulin Island. From the east, the starting point is Killarney.

Well into the 19th century, water remained the main link between North Shore communities, as distances were great and the terrain too difficult to warrant a link by road. All that changed in the early 1960s, when the trans-Canada highway was completed and a road was built across the tundra from Sudbury to Killarney. Today the water is primarily the domain of fishermen, cottagers and cruising boaters.

The mystique of the North Channel lies not only in its fabled history and breathtaking natural beauty, but in the fact that nearly 400 years after the arrival of Samuel de Champlain and the voyageurs, the pristine character of the cruising mecca remains virtually unchanged. May it forever remain so.

DeTour Village, Michigan

Lat 45°59.74
Lon 83°53.95

Situated on the west flank of DeTour Passage, the village of DeTour marks the southern entrance from Lake Huron **A** to the St. Marys River. At DeTour Harbor Marina **B**, the village's only transient marina, cruisers can choose from a total of 80 slips. The ferry dock **C** sends boats back and forth to Drummond Island on the hour; it cannot be used as a tie-up spot. The private docks operated by the Fogcutter restaurant **D** are located just adjacent to the ferry dock. The town's main shopping district **E** is just a few blocks away.

Distances from DeTour Village (statute miles)
Mackinac Island: 39 W
Les Cheneaux Islands: 6 W
Drummond Island: 8 NE
Meldrum Bay: 42 W
Sault Ste. Marie: 48 N

154 LAKELAND BOATING · LAKE HURON CRUISING GUIDE

THE NORTH CHANNEL

DeTour Village, Michigan
Make this detour into a destination.

Although the fur-trading voyageurs preferred to use the main route across the North Channel to Sault Ste. Marie, today DeTour Passage is a well-trafficked channel and the natural choice for anyone heading north from Lake Michigan or southwestern Lake Huron.

From DeTour, located at the mouth of the St. Marys River, boaters can either continue up the St. Marys River or make a straight run for the North Channel.

When approaching from the south, boaters should watch for the DeTour Reef Light, located about a mile offshore on the DeTour Reef. Nearer the harbor, the DeTour Village water tower should also be conspicuous. The entrance to **DeTour Harbor Marina**, the village's only facility serving transient boaters, is at the south end of a long breakwall, signaled by a green and red can. There are some shallow areas within the harbor, but all are clearly marked with orange and white buoys.

DeTour Harbor is a state-owned facility operated locally by **Mertaugh Boat Works,** a full-service boatyard. The marina offers about 80 slips, 90 percent of which are set aside for transients. Thirty-amp electricity and some 50-amp service are provided to the docks, as is water. Depth at the overnight slips ranges from 7 to 10 feet. Gasoline, diesel and pumpout are available at the service dock. Other amenities include showers, restrooms and ice, as well as a ship's store for basic supplies. There is an excellent launch ramp just south of the harbor.

Although the marina has no haulout lift or service shop, there are mechanics in town who will come to

LAKELAND BOATING · LAKE HURON CRUISING GUIDE   **155**

Where to Eat

DeTour Village Inn 906-297-2881
A bar serving standard pub fare: burgers, hot sandwiches, and fish and chips.

Dockside Restaurant 906-643-7911
The menu ranges from seafood to steak, but the specialties are whitefish and walleye.

Fisher's Restaurant 906-297-2801
Open for breakfast, lunch and dinner, featuring whitefish, turkey and burgers.

Fogcutter Bar & Grille 906-297-5999
Sports bar and deck dining overlooking the St. Marys River. Salad bar and the self-declared world's best whitefish.

Fortino's Pizzeria 906-297-3051
A short walk from the state dock, Fortino's serves up hot pizza and cold ice cream cones.

Mainsail Saloon and Restaurant . . . 906-297-2141
Full-service dining, homemade soups, whitefish, steak, ribs and chicken.

---

your boat. A towing service is also available. The nearby Mertaugh Boat Works offers haulout, repairs and a mobile mechanic.

For laundry, it is a short hike to the **Sudzy Wash and Dry laundromat**, located just behind the marina to the north. Groceries, beer and liquor are available about a block to the south at **Sune's Food Center**. An excellent place to stock up on charts or boating supplies is **North Country Sports**, at the corner of Ontario and Elizabeth streets. This sporting goods store carries a full line of charts; supplies such as fenders, lifejackets and cleaners; and a selection of bait, tackle and other fishing equipment.

Herring fishing is usually in full swing right around the Fourth of July and lasts for about six weeks. Walleye, pike and bass are also popular catches. Salmon fishing begins the first week in July and usually runs through August and into the fall.

There are a number of dining opportunities in DeTour, but one in particular stands out: **Fogcutter Bar & Grille**, right beside the ferry docks. Fogcutter features a complete lunch and dinner menu, with whitefish cooked any way you want as the house speciality.

Other dining options include **Fortino's Pizzeria**, a restaurant and ice cream parlor located north of the marina, and the **Mainsail Saloon and Restaurant**, which is south of the marina. On Ontario Street is **Fisher's Restaurant**, a half-mile walk from the marina. The **Garage Coffee Shop**, which features local art and pottery, is also on that block. Drinks can be had at the **DeTour Village Inn**, just across from the sporting goods store.

The **DeTour Passage Historical Museum** is right next to the ferry dock on Elizabeth Street. Among the artifacts on display is the Fresnel lens from the original DeTour lighthouse, which was shut down when the new one was built offshore. A **bank** with a cash machine is located on the same block.

DeTour is a small village, with modest frame houses, most now re-sided with vinyl, lining its streets. As a provisioning stop en route to the North Channel, or to Mackinac and points farther south, it is a great place to pull in. While in DeTour, you can also enjoy a unique vantage of the immense freighters beginning to nose up the St. Marys River to Sault Ste. Marie. Take a walk to the **DeTour Botanical Garden** at the south end of town, where hundreds of species of flowers and trees grow along the shore.

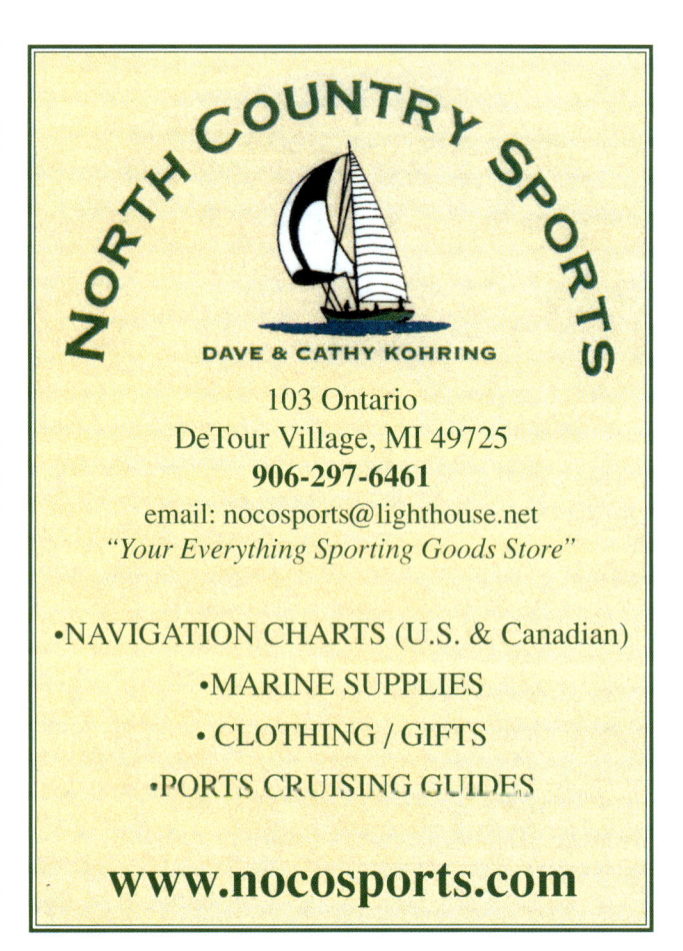

More Info

Emergency 911
DeTour Village Medical Center 906-297-3204
Chamber of Commerce 906-297-5987
Eastern U.P.
Transportation Authority 906-632-2898

Local events

June
Mininising Honoring Our Children Powwow

July
4th of July Celebration

NORTH COUNTRY SPORTS

DAVE & CATHY KOHRING

103 Ontario
DeTour Village, MI 49725
906-297-6461
email: nocosports@lighthouse.net
"Your Everything Sporting Goods Store"

- NAVIGATION CHARTS (U.S. & Canadian)
- MARINE SUPPLIES
- CLOTHING / GIFTS
- PORTS CRUISING GUIDES

www.nocosports.com

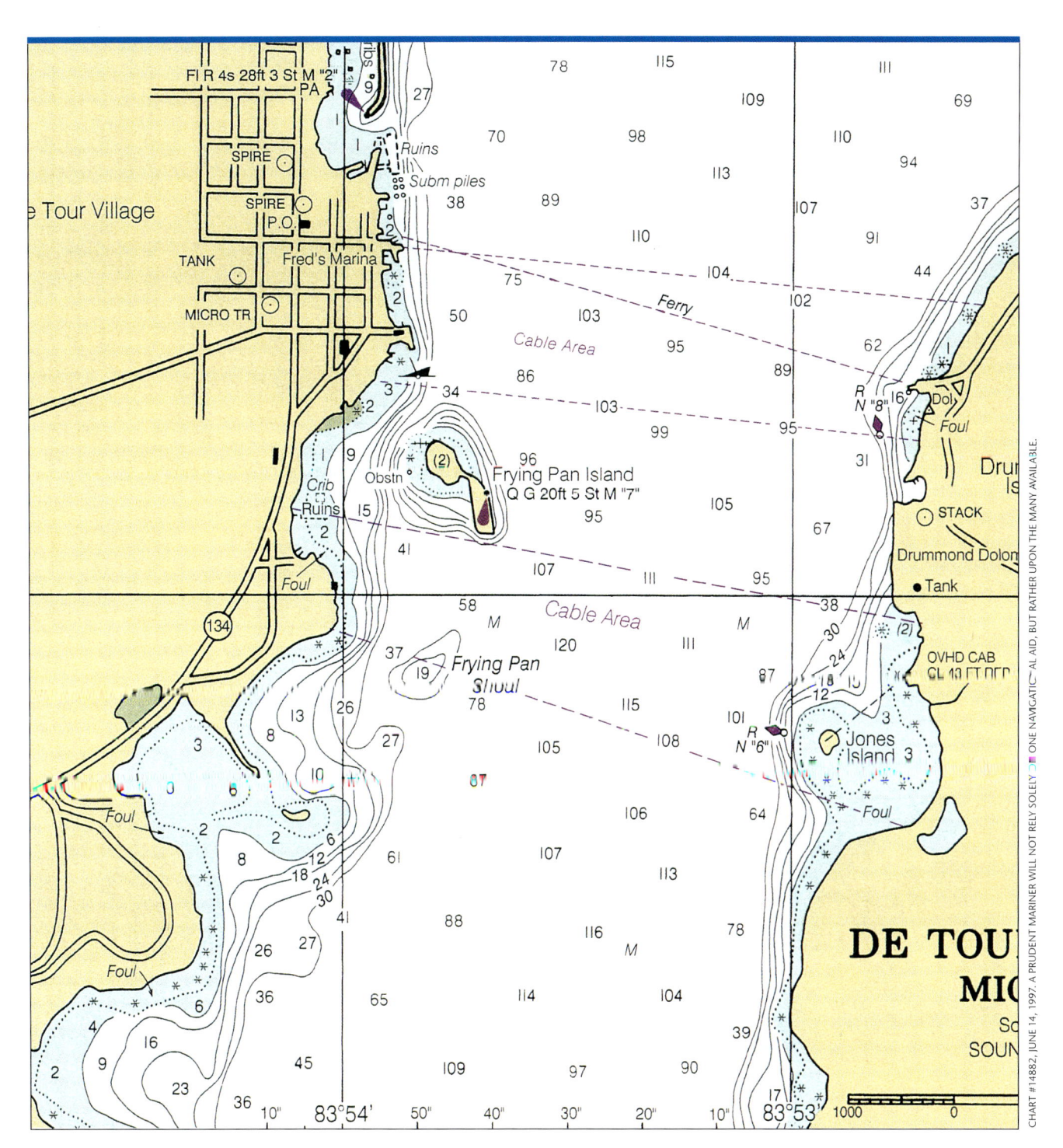

Facilities information subject to change. We suggest you call ahead.

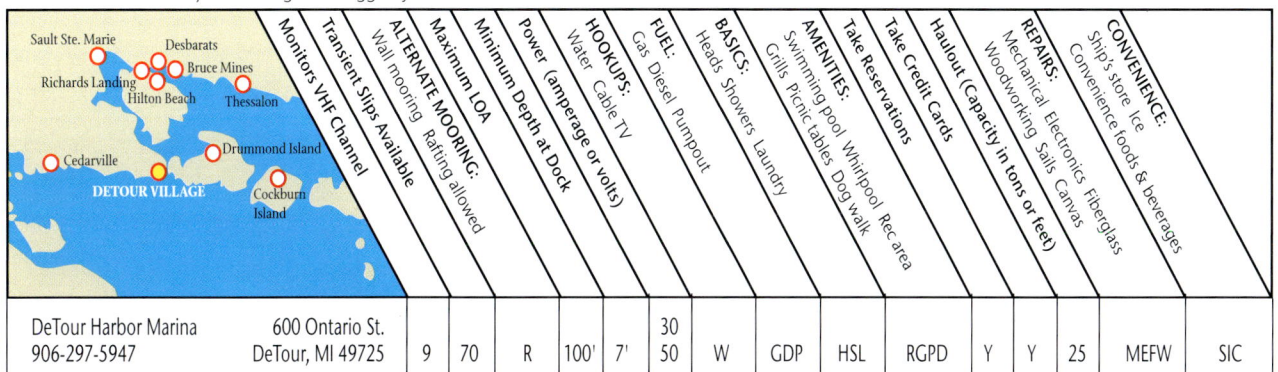

		Monitors VHF Channel	Transient Slips Available	ALTERNATE MOORING: Wall mooring Rafting allowed	Maximum LOA	Minimum Depth at Dock	Power (amperage or volts)	HOOKUPS: Water Cable TV	FUEL: Gas Diesel Pumpout	BASICS: Heads Showers Laundry	AMENITIES: Swimming pool Whirlpool Rec area Grills Picnic tables Dog walk	Take Reservations	Take Credit Cards	Haulout (Capacity in tons or feet)	REPAIRS: Mechanical Electronics Fiberglass Woodworking Sails Canvas	CONVENIENCE: Ship's store Ice Convenience foods & beverages
B	DeTour Harbor Marina 906-297-5947	600 Ontario St. DeTour, MI 49725	9	70	R	100'	7'	30 50	W		GDP	HSL	RGPD	Y	Y	25	MEFW	SIC

LAKELAND BOATING · LAKE HURON CRUISING GUIDE **157**

Drummond Island Michigan

Lat 46°01.45
Lon 83°44.88

Drummond Island Yacht Haven **A**, a private marina that accepts transients, is located at the south end of Potagannissing Bay, on the northern shoreline of Drummond Island. The small village of Drummond **B** is about two miles away. There is a launch ramp at the marina and another in the village **C**.

Distances from Drummond Island (statute miles)
Mackinac Island: 47 W
DeTour Village: 8 SW

Sault Ste. Marie: 42 N
Meldrum Bay: 34 E

THE NORTH CHANNEL

Drummond Island, Michigan

Get back to nature at this quiet port.

The rustic little village of Drummond makes a perfect base for exploring Drummond Island, which offers a bit of history, plenty of wildlife and some world-class golfing.

Drummond Island Yacht Haven is positioned on the very edge of the North Channel cruising area, eight miles from DeTour and only a mile and a half from the popular anchorage of Harbor Island. It's a great place to rest up before crossing the channel, or to check back in if you're returning to the United States, because owner Dennis Bailey also acts as a U.S. Customs official.

Approaching from the north, you can pass either to the east or west of Propeller Island. In the first instance, follow the green navigation buoys south and then west, staying just east of Bald and Rogg islands. If you come around the other side, skirting Standerson and Harbour islands, mind both Harbour Island Reef and Cheney Shoal, keeping them to starboard. Coming from DeTour Passage, pass south of both Gull and Wreck islands.

The marina, conspicuous because of its large American flag, will take as many as 30 boats at its permanent dock, which kinks around on the inside of the point. Cruisers can count on 10 feet of water here. At the marina's outer dock, boats up to 126 feet long can be accommodated. Both 50- and 30-amp power are provided, as are water, pumpout, gasoline and diesel. A laundromat and shower building are located right beside the docks. Yacht Haven also maintains a well-stocked marine store that includes charts, offers scuba tank refills and rents bicycles as well as boats. Both gas and diesel mechanics are on hand in case any repairs are needed. Yacht Haven also has a 75-ton Travelift.

Those interested in playing **The Rock**, **Woodmoor Resort**'s famous five-star golf course, should radio the marina, which will set up a tee time for you. Woodmoor, a full-service resort, will send a vehicle to pick you up. Yacht Haven can provide you with local transportation, but you might want to rent a car or

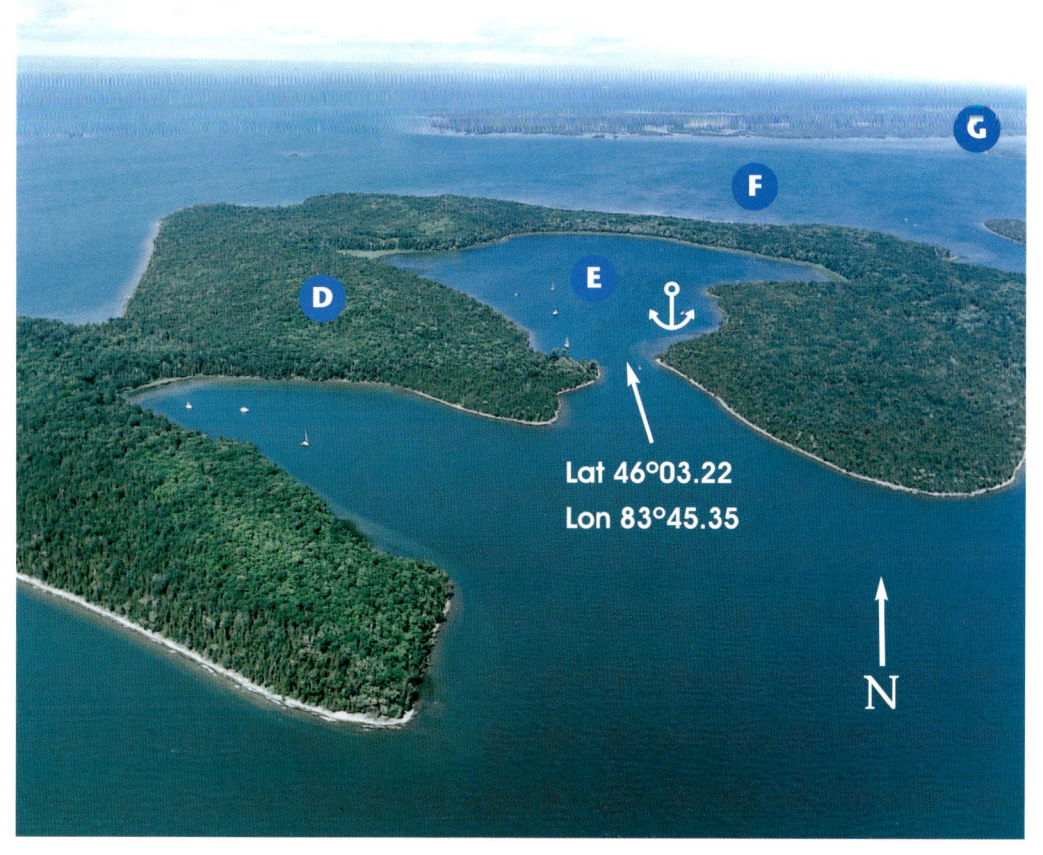

About a mile and a half directly north of Drummond Island Yacht Haven lies Harbor Island **D**, with its very popular and well-protected anchorage **E**. Beyond that lies Potagannissing Bay **F** and the North Channel **G**.

Lat 46°03.22
Lon 83°45.35

LAKELAND BOATING · LAKE HURON CRUISING GUIDE   **159**

Where to Eat

Bayside Dining 800 999 6343
A gourmet restaurant in a rustic island setting at Drummond Island Resort. Featuring great views, a dock and an extensive wine list.

**The Gourmet Galley and
Port of Call Restaurant** 906-493-5507
The Gourmet Galley is modeled after a New York deli, with sandwiches and specialty foods from all over the world available for delivery. The Port of Call features a different international theme daily. Seafood, steaks and vegetarian dishes are also on the menu.

The Northwood 906-493-5282
Stop in for burgers and sandwiches. Also known for delicious fresh whitefish and all-you-can-eat broasted chicken. Open April through October.

Pins Bar & Grill 906-493-1014
Located in the Drummond Island Resort, this casual eatery has a big-screen TV, a bar and eight bowling lanes. It serves breakfast, lunch and dinner.

More Info

Emergency 906-493-5555
Drummond Island Clinic 906-493-5221
Chamber of Commerce 906-493-5245

---

bike because the loosely arranged village of Drummond is about two miles away. At the "four corners" you'll find a **bank**, a small **supermarket** that carries alcohol, a **hardware store** and a **takeout restaurant**. **The Northwood** offers fine sit-down dining; another quarter-mile east you'll find the **Gourmet Galley and Port of Call Restaurant**, a one-of-a-kind, all-in-one New York-style deli, 1950s-style ice cream parlor, international restaurant and provisioning store that sells beer, wine and ethnic foods.

Much closer to the marina lies the **Drummond Island Historical Museum**. Handsomely constructed from local logs, this building replaced the original museum, which collapsed under heavy snow a few years ago. The good news is that most of the artifacts survived, including some relics from Fort Drummond, the last British holding in the United States.

For other forms of unhurried amusement, you might test the waters of Potagannissing Bay for smallmouth, perch, pike or walleye. Hike the mile-long **Heritage Hiking Trail** in Drummond Island Township Park, visit the **Maxton Plains** wildlife reserve or simply stroll along one of Drummond's quiet roads among the rustic houses, sturdy oaks and towering white pines.

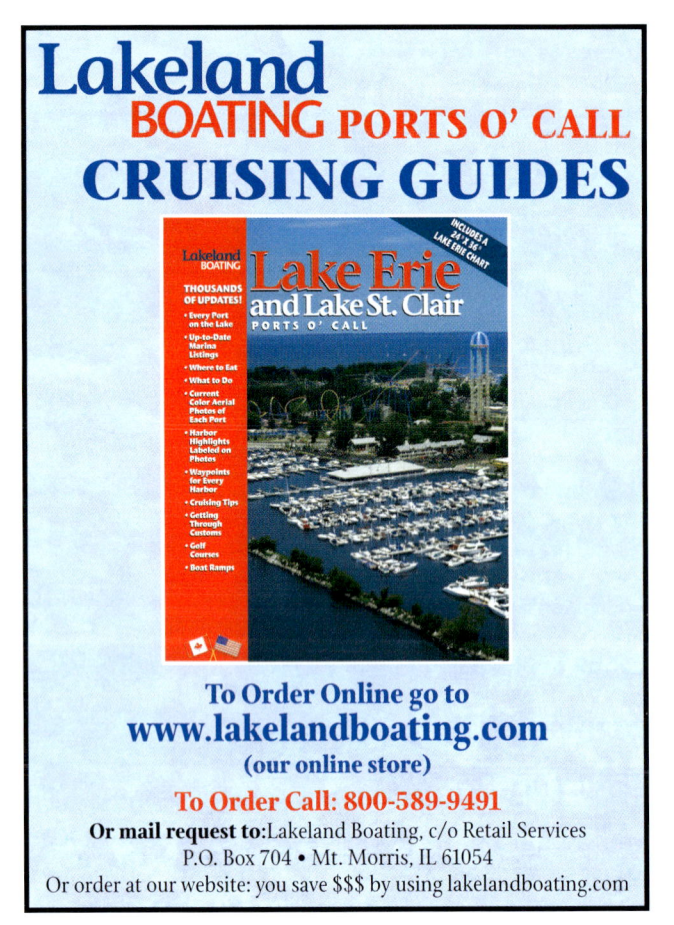

Lakeland
BOATING PORTS O' CALL
CRUISING GUIDES

To Order Online go to
www.lakelandboating.com
(our online store)

To Order Call: 800-589-9491

Or mail request to: Lakeland Boating, c/o Retail Services
P.O. Box 704 • Mt. Morris, IL 61054
Or order at our website: you save $$$ by using lakelandboating.com

DRUMMOND ISLAND
The Best Overnight Experience in the North Channel

Drummond Island Yacht Haven

Gasoline, Diesel, Oil • Indoor Heated Storage
Marine Supply Store • 75 Ton Travel Lift
Mechanic • U.S. Customs • Showers • Scuba Refills
Laundromat • Boston Whaler, Pursuit & Lund
Swimming Beach • 20 Waterfront Cottages

Call for a free chart with marked routes:
800-543-4743
www.diyachthaven.com

Drummond Island Resort and Conference Center

Bayside Dining:
Gourmet dining on the waterfront, CIA Executive Chef, extensive wine menu and incredible cuisine.

Pins Bar & Grill & Bowling Alley:
Casual Food, Salad Bar, Pizza, Full Menu.

The Rock:
Championship golf on one of Michigan's top courses.

The Cedars Sporting Clay Range:
15 unique situational shooting stations, 50 or 75 shot rounds, guns & shells available.

www.drummondisland.com
Call for more information and tee times.
800-999-6343

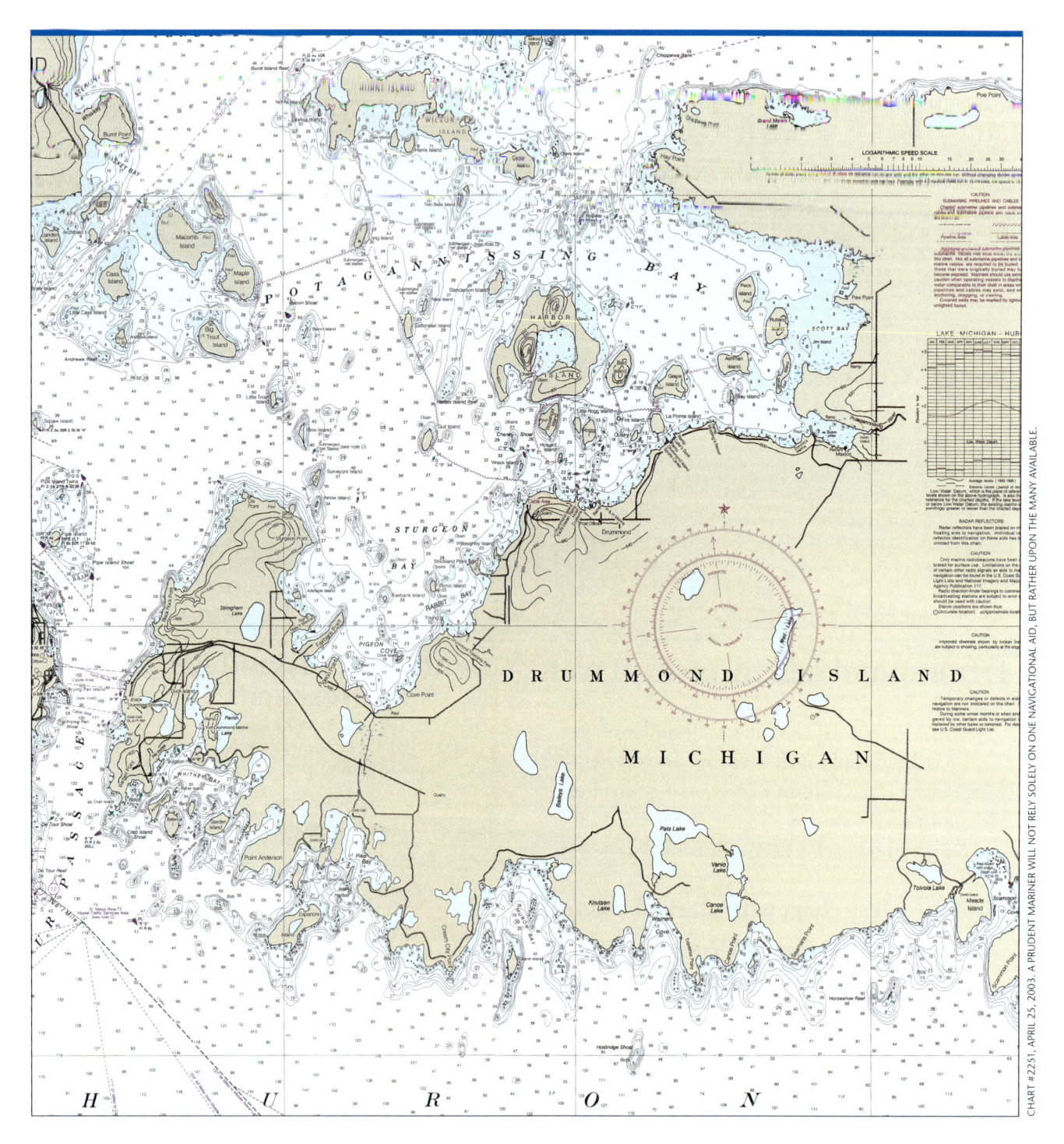

Facilities information subject to change. We suggest you call ahead.

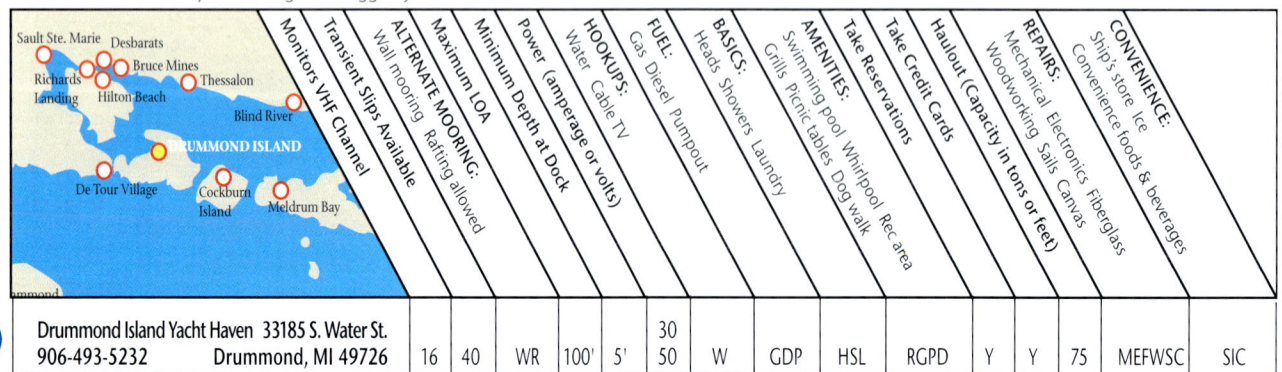

	Monitors VHF Channel	Transient Slips Available	ALTERNATE MOORING: Wall mooring Rafting allowed	Maximum LOA	Minimum Depth at Dock	Power (amperage or volts)	HOOKUPS: Water Cable TV	FUEL: Gas Diesel Pumpout	BASICS: Heads Showers Laundry	AMENITIES: Swimming pool Whirlpool Rec area Grills Picnic tables Dog walk	Take Reservations	Take Credit Cards	Haulout (Capacity in tons or feet)	REPAIRS: Mechanical Electronics Fiberglass Woodworking Sails Canvas	CONVENIENCE: Ship's store Ice Convenience foods & beverages
A Drummond Island Yacht Haven 33185 S. Water St. 906-493-5232 Drummond, MI 49726	16	40	WR	100'	5'	30 50	W		GDP	HSL	RGPD	Y	Y	75	MEFWSC	SIC

162 LAKELAND BOATING · LAKE HURON CRUISING GUIDE

Cockburn Island, Ontario

Tolsma Bay, on the northeast shore of Cockburn Island, has two government docks, both of which can be used by the transient cruiser. The one that points north **A** has been barricaded off by the federal government. The other **B**, which once welcomed passenger ferries, angles east and offers more shelter for boats up to 70 feet. The docks do not offer any amenities, though there is a dockmaster. The village of Tolsmaville is now almost a ghost town.

Distances from Cockburn Island (statute miles)
Meldrum Bay: 11 E
Gore Bay: 44 E
Thessalon: 23 N

THE NORTH CHANNEL

Cockburn Island, Ontario

A gentle breeze from the past.

Today Cockburn Island, home to the formerly bustling port of Tolsmaville, is such a well-kept secret that not even directory assistance can find it. The island (pronounced "co-burn") is something of a forgotten link in the chain of limestone islands that shields the North Channel from Lake Huron.

In the early part of the century, Tolsmaville was a busy town, providing shiploads of lumber from its numerous mills. As recently as 1964, Cockburn Island was still populous enough to warrant a weekly visit from the ferry *Normac*. Today, however, there are just a few year-round residents left on the island, and the fading houses are used mostly by summer vacationers, if at all. For the North Channel cruiser, Tolsmaville will offer nothing in the way of supplies or services, but as a bad-weather refuge—or a thought-provoking side trip—it will more than suffice.

Defined by False Detour Channel to the west and Mississagi Strait to the east, Cockburn Island can be accessed from Lake Huron, although most boaters will be more likely to light upon the harbor from the North Channel side. Meldrum Bay, the nearest serviced port, is about 11 miles to the east; Blind River and Thessalon are 17 and 23 miles north, respectively, on the mainland. At this port, the cruiser can count on about 12 feet of water at the **Tolsmaville Government Docks**; a dockmaster is employed during the summer, but not at all times. A radio telephone and outhouses are located a short distance from the docks.

Otherwise, there are no services available on Cockburn Island. The community of Tolsmaville lacks even a general store. What it does provide is a rare glimpse into a nearly forgotten era, with plenty of old buildings and strange-looking jalopies (most still miraculously in use) to ponder. And if you want an oral account of the village's heyday, a number of friendly summer residents (many of them former permanent residents) are usually happy to oblige.

More Info

Emergency 911
Cockburn Island
Municipal Office 705-844-2289

CABLES

Submarine and overhead cables may conduct high voltages and contact with or proximity to these poses an extreme danger. Mariners should not anchor close to submarine cables, however if anchors become hooked onto a cable care must be taken not to sever the cable. Cables installed since the date of publication of this edition may not be charted. Sufficient clearance must be allowed under all overhead cables. Actual clearances may differ from charted values due to changes in atmospheric conditions and / or water levels.

CÂBLES

Les câbles sous-marins et aériens peuvent conduire une haute tension; tout contact direct ou indirect peut être fatal. Les navigateurs ne devraient pas mouiller près de câbles sous-marins; cependant si les ancres accrochent un câble, il faut se soucier de ne pas rompre le câble. Il se peut que les câbles installés depuis la date de publication de cette édition ne soient pas cartographiés. Une hauteur libre suffisante doit être allouée sous tous les câbles aériens. Les hauteurs libres actuelles peuvent varier des valeurs cartographiées à cause de changements dans les conditions atmosphériques et/ou les niveaux d'eau.

COURSE

Bearings and distances of sailing courses indicated in magenta are recommended by the Lake Carriers Association and the Dominion Marine Association.

ROUTE

Les relèvements et les distances des routes de navigation indiqués en magenta sont recommandés par la Lake Carriers Association et la Dominion Marine Association.

CHART #2251, APRIL 25, 2003. A PRUDENT MARINER WILL NOT RELY SOLELY ON ONE NAVIGATIONAL AID, BUT RATHER UPON THE MANY AVAILABLE.

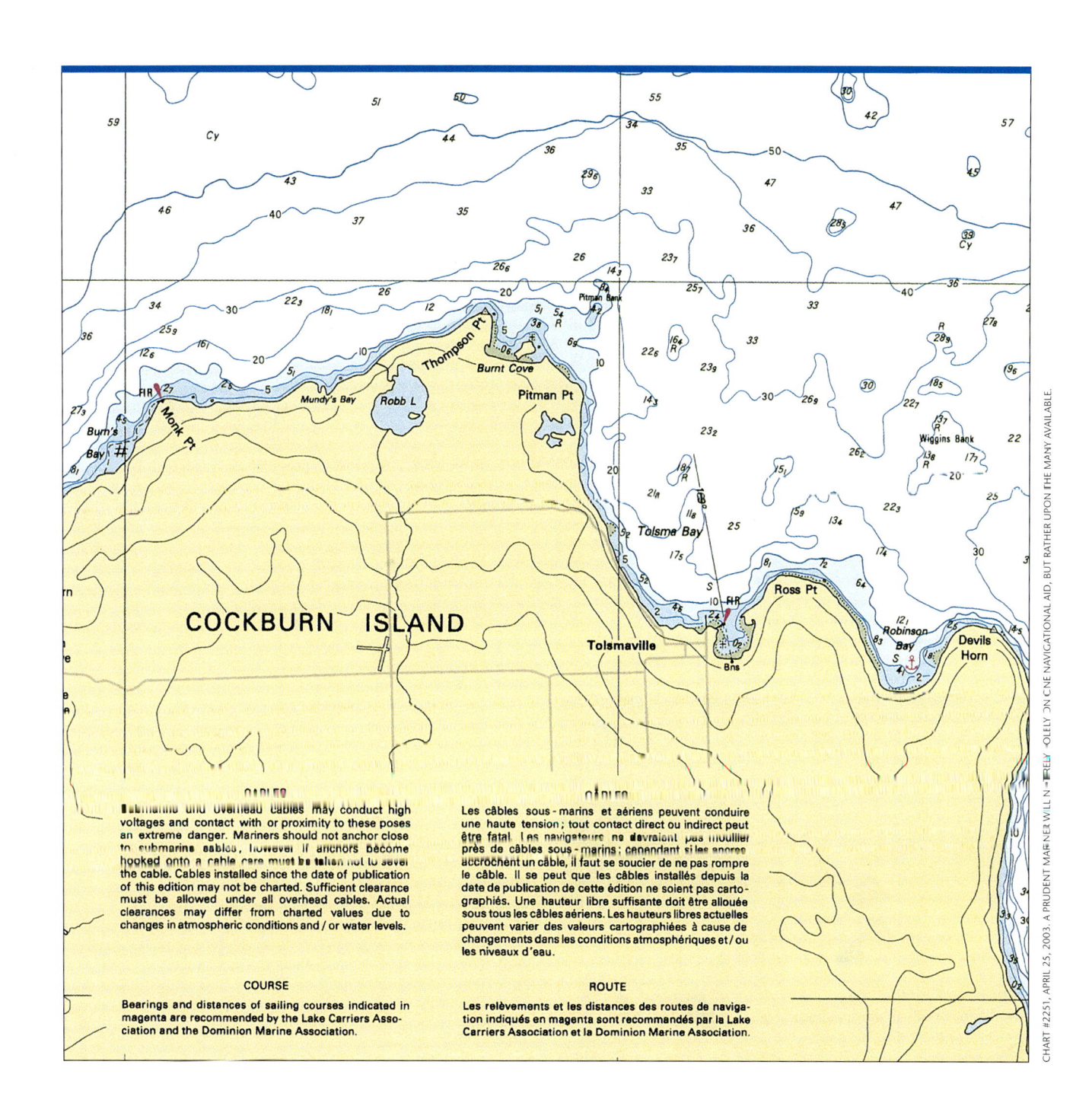

Facilities information subject to change. We suggest you call ahead.

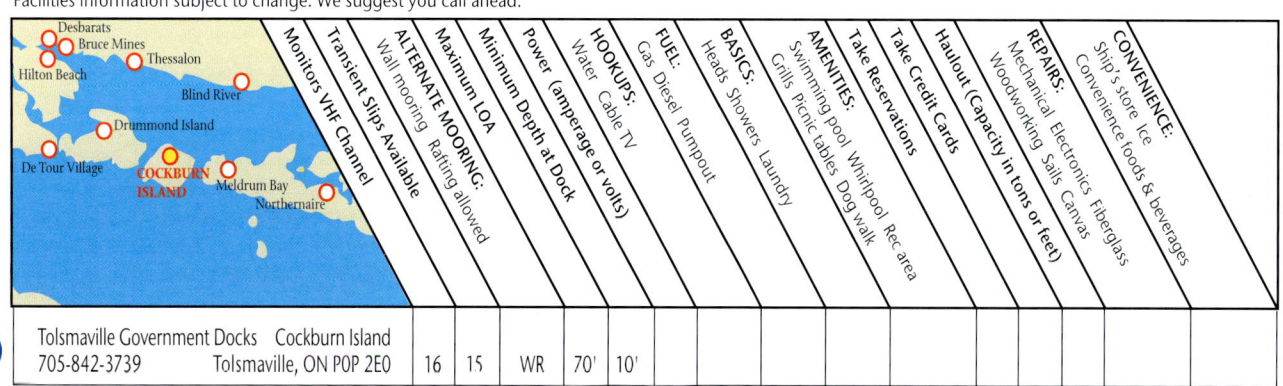

B	Tolsmaville Government Docks Cockburn Island	Monitors VHF Channel	Transient Slips Available	Alternate Mooring (Wall mooring, Rafting allowed)	Maximum LOA	Minimum Depth at Dock	Power (amperage or volts)
	705-842-3739 Tolsmaville, ON P0P 2E0	16	15	WR	70'	10'

Lakeland Boating · Lake Huron Cruising Guide   **165**

THE NORTH CHANNEL

Manitoulin Island, Ontario
Commune with the spirits on this wild isle.

Manitoulin Island is a kind of crossroads of the spirit world: ancient Indian gods and heroes dwell alongside the European explorers of folklore and history. In the woods and vales and along the sandy beaches, strands of mythology and traditional culture mingle with pioneer and maritime history and are woven into the fabric of daily life. This is evident throughout Manitoulin, from Kagawong's breathtaking Bridal Veil Falls to Wikwemikong's unceded Indian reservation, home to the descendants of the Ojibway, Odawa and Pottawotami tribes.

Drop a line into one of the island's many cerulean inland lakes, creeks and rivers—you're sure to be rewarded with a chinook salmon, rainbow trout, perch or whitefish. As you skirt the isle on your boat, keep a lookout for families of deer timidly peeking out from clusters of cedar, birch, ash and oak trees. Walk barefoot along the golden shoreline, keeping a sharp lookout for fossils. Waterfowl dot the azure skyline above the lush, rolling, daisy-strewn hills. While there are plenty of ports to pull into on Manitoulin—some of them quite bustling— much of the island is untamed wilderness. Try to spot some curious raccoons and scurrying porcupines as you twist along the numerous well-marked trails, which are ideal for hiking and exploring. More than 100 lakes, some of which have islands of their own, dot Manitoulin, the largest freshwater island in the world.

If you plan your trip well, you can attend one of the powwows held on the island. Observe the dancers and drummers as they perform their rituals in brightly colored costumes. It's a rare opportunity to witness some of the traditions that permeate this area and give it its life force. Your experience on this legendary isle is sure to be *esuna-manda*—amazing.

Meldrum Bay, Ontario

Lat 45°55.54
Lon 83°06.85

As Manitoulin Island's westernmost harbor, Meldrum Bay is the port of entry for many U.S. boaters to clear customs. The tiny community is perched on the U-shaped inlet's hilly southwestern shore. Transient dockage is available at the Meldrum Bay Marina **A** and at Whitesea Cottages & Charter **B** in the southeastern part of the bay. Sunset Grille **C**, adjacent to the docks, serves burgers, ice cream and fries to go or to eat at the outdoor tables. The Meldrum Bay Inn **D** is a great dining option located in a historic house near the marina. Meldrum Bay General Store **E** carries groceries, hardware, boating supplies, gas and liquor.

Distances from Meldrum Bay (statute miles)
DeTour Village: 58 SW
Sault Ste. Marie: 86 NW
Blind River: 20 NW

Hilton Beach: 49 NW
Gore Bay: 37 E
Little Current: 61 E

LAKELAND BOATING · LAKE HURON CRUISING GUIDE

THE NORTH CHANNEL

Meldrum Bay, Ontario
A friendly port where time stands still.

Meldrum Bay provides a quiet welcome to Manitoulin Island for boaters continuing eastward. Originally a busy commercial fishing port, Meldrum Bay's rocky and mostly unsettled shores, crying seagulls and breezy promontories today lend it a tough, lonesome beauty. Not much seems to have changed in this port since the trout fishery dried up in the 1940s, but the town actually offers boaters quite a bit in the way of services.

Meldrum Bay Marina, recognizable by a red light atop its white building, has ample dockage for up to 40 transients; 30- and 50-amp electric; plus gas, diesel and pumpout. Kayaks, canoes and small sailboats are available for rent by the hour, half-day or full day. There is a launch site at the southern tip of Point Park, facing the marina

Visitors should be aware that the long finger docks south of the government dock can get buffeted by a brisk northerly wind. Perhaps as recompense for this potential discomfort, the marina's proprietors seem to go out of their way to offer numerous small conveniences, including complimentary coffee, a dog walk area, a chandlery selling oil and charts, a barbecue grill for guests' use and laundry facilities. And you might want to stick close to the docks in the late afternoon so as not to miss the arrival of the "pie lady"—a local woman who delivers fresh baked goods to the marina every day.

Another good transient option is **Whitesea Cottages & Charter** on the southeast shore of the bay, where those looking for shore accommodations can spend a peaceful night or two. Constructed in 1989 and enlarged in 2003, Whitesea has good water depth and slips that are shielded in a manmade lagoon by a substantial breakwater. Depending on availability, the resort offers transients shower facilities in one of their well-appointed waterfront cottages. Whitesea also operates salmon charters for anyone who wants to go after chinook in the Mississagi Strait.

The **township campground** on the point has laundry facilities, and the **Meldrum Bay General Store** across the road stocks a full line of groceries, liquor, supplies, licenses and hardware. The **Meldrum Bay Inn**, an old-fashioned building with a wide, shaded porch, offers rooms as well as a restaurant. Whitefish, other Canadian entrées and desserts are among the affordable delicacies on the menu.

Manitoulin Scuba, a full-service dive shop with equipment sales, rentals and air fills, is located right at the marina just south of the fourth finger dock. Run by the owners of the local general store, it also offers introductory scuba lessons.

Those interested in Meldrum Bay's fishing heritage can pay a visit to **The Net Shed Museum**, a few paces beyond the general store. History buffs will also enjoy a visit to the **Mississagi Lighthouse**, located about seven

Where to Eat

Meldrum Bay Inn **705-283-3190**
A gourmet dining experience with an excellent wine list. The fare is good Canadian cooking with a French influence. Save room for dessert—the chocolate cake is sinful.

Mississagi Lighthouse Restaurant **705-283-1084**
Have a meal in a unique setting. The restaurant is housed in the historic lighthouse.

More Info

Emergency 911
Gore Bay Clinic 705-282-2262
Manitoulin Island
Tourism 705-368-3021

168 LAKELAND BOATING · LAKE HURON CRUISING GUIDE

miles west of Meldrum Bay on the edge of the Mississagi Strait. This 115-year-old landmark now houses a restaurant and museum. If you're feeling particularly energetic, the site is accessible by mountain bike.

If you're looking for a nearby anchorage, Vidal Bay and Newberry Cove are both options, though neither is recommended in a stiff northerly wind. Boaters should also be aware that no haulout service is provided in Meldrum Bay. Some hull and engine repairs can be done in a pinch, but if it's something that can wait, you will be better attended to in Blind River or Gore Bay, the next closest ports.

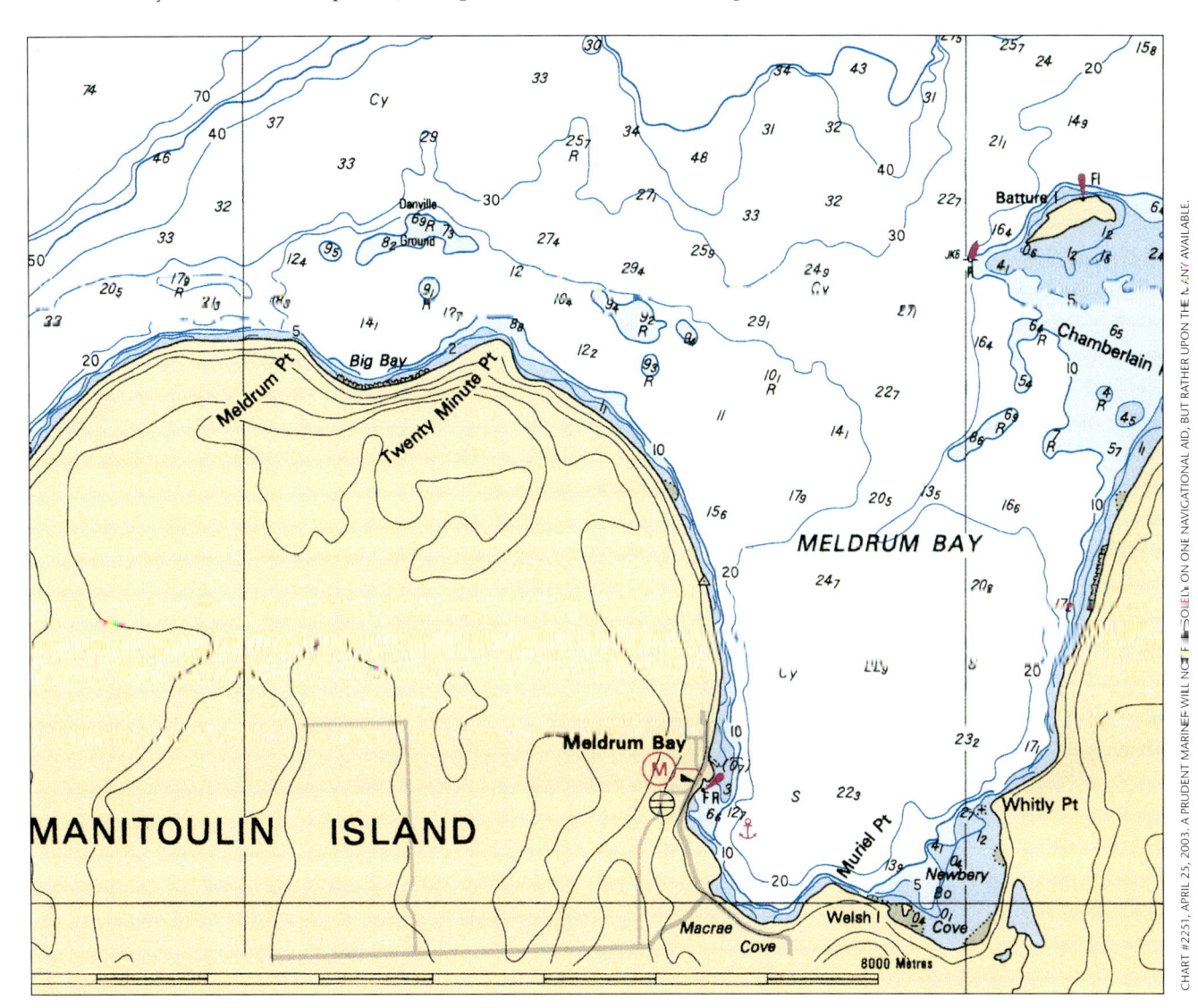

Facilities information subject to change. We suggest you call ahead.

			Monitors VHF Channel	Transient Slips Available	ALTERNATE MOORING: Wall mooring / Rafting allowed	Maximum LOA	Minimum Depth at Dock	Power (amperage or volts)	HOOKUPS: Water / Cable TV	FUEL: Gas / Diesel / Pumpout	BASICS: Heads / Showers / Laundry	AMENITIES: Swimming pool / Whirlpool / Rec area / Grills / Picnic tables / Dog walk	Take Reservations	Take Credit Cards	Haulout (Capacity in tons or feet)	REPAIRS: Mechanical / Electronics / Fiberglass / Woodworking / Sails / Canvas	CONVENIENCE: Ship's store / Ice / Convenience foods & beverages
A	Meldrum Bay Marina 705-283-3252	Highway 540 Meldrum Bay, ON P0P 1R0	68	40	R	100'	6'	30 50	W	GDP	HSL	GPD	Y	Y			I
B	Whitesea Cottages & Charter 800-732-0350	Box 18 Meldrum Bay, ON P0P 1R0	68	20		30'	8'	15 30	W	GD	HS	PD	Y	Y			IC

Campbell Bay, Ontario

Tucked in the southeast corner of Campbell Bay **A**, one of three bays in Bayfield Sound south of Barrie Island, Northernaire Lodge & Marina **B** is an idyllic—if somewhat out-of-the-way—tie-up option for North Channel travelers. Dockage is provided on the west side of the peninsula **C**, while a curving sand beach **D** stretches along its east flank. Wolsey Lake **E** is visible in the distance.

Lat 45°49.02
Lon 82°34.64

Facilities information subject to change. We suggest you call ahead.

B	Name	Location	Monitors VHF Channel	Transient Slips Available	ALTERNATE MOORING: Wall mooring, Rafting allowed	Maximum LOA	Minimum Depth at Dock	Power (amperage or volts)	HOOKUPS: Water, Cable TV	FUEL: Gas, Diesel, Pumpout	BASICS: Heads, Showers, Laundry	AMENITIES: Swimming pool, Whirlpool, Rec area, Grills, Picnic tables, Dog walk	Take Reservations	Take Credit Cards	Haulout (Capacity in tons or feet)	REPAIRS: Mechanical, Electronics, Fiberglass, Woodworking, Sails, Canvas	CONVENIENCE: Ship's store, Ice, Convenience foods & beverages
B	Northernaire Lodge & Marina 705-282-2642	Campbell Bay Bayfield Sound, ON P0P 1E0		8	R	60'	4'5"	20	W	G	HS	RPD	Y	Y			IC

170 LAKELAND BOATING · LAKE HURON CRUISING GUIDE

THE NORTH CHANNEL

Campbell Bay, Ontario
Modern comforts, vintage charms.

Cruisers seeking bright lights and fast-paced action may choose to pass by Campbell Bay, but those looking for a peaceful, secluded, low-key spot will find it's just their thing. **Northernaire Lodge & Marina**, which is located in the southeast corner of the bay, is an ideal place to stay put for a while, do some fishing or just laze around.

Hidden away on a private, wooded peninsula, Northernaire maintains 24 private cottages for its guests, most of whom arrive by car. But the lodge also offers a full-service marina, and transient boaters are more than welcome. There's space for eight to 10 boats, with water and 20-amp power hookups. Showers and washrooms are close at hand. Gas is available, although there's no pumpout or propane. Diesel can be ordered in if necessary. Check out the lodge's website at www.northernairelodge.com.

For those who want to fish, the lodge sells licenses, bait and tackle and rents fishing boats. If reeling 'em in strikes you as too much work, you can always just hit the beach.

While tied up at Northernaire, boaters can sample the cuisine offered in its cozy dining room. If you need some basic groceries, make the quarter-mile hike to **Taylors One Stop** in Evansville. Rent a car from Gore Bay if you plan more extensive touring.

Cambell Bay is one of three bays in Bayfield Sound, a large body of water lying behind Barrie Island that boasts plenty of water and many snug coves for private anchorage.

Significant Waypoint
Mouth of Campbell Bay between Cape Roberts and Jubilee Shoal:
45°58.91N/82°45.41W

More Info
Gore Bay Medical Centre. 705-282-2262
Sudbury Chamber of Commerce 705-673-7133

Where to Eat
Northernaire Lodge Dining Room . . 705-282-2642
Located inside the classic rustic lodge, constructed in 1932, the dining room offers a full menu at breakfast and two entrées to choose from at dinner.

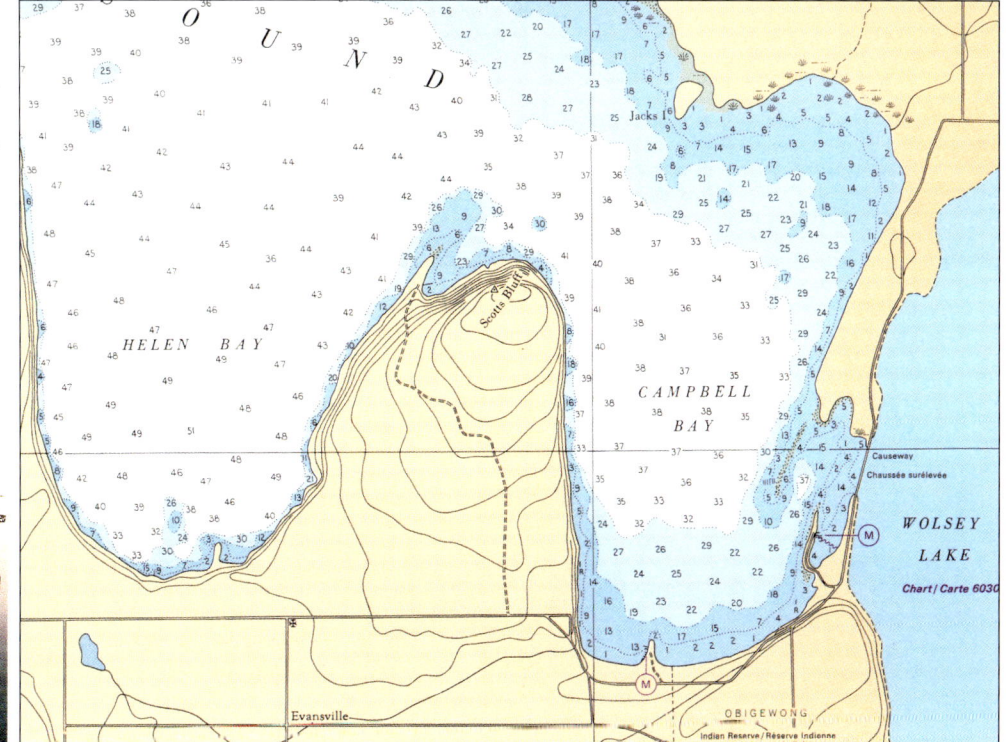

LAKELAND BOATING · LAKE HURON CRUISING GUIDE   **171**

Gore Bay, Ontario

Lat 45°55.28
Lon 82°27.51

A two-mile deep notch in Manitoulin Island's meandering north shore, Gore Bay is snug and free of hazards. Once abreast of the town point **A**, transients should see plenty of docking space available at Gore Bay Marina **B**. Canadian Yacht Charters **C** operates a substantial chandlery out of the same building that houses the marina office, while Purvis Marine Storage **D** offers haulout and some repairs. The red-roofed Gore Bay Pavilion **E**, located just south of the town docks, houses a restaurant, tourist information booth and lighthouse-style observation tower. The main street of Gore Bay **F** is just a block away, while Gordon's Lodge **G** is a short dinghy ride across the bay. The lat/lon given is just outside of the photograph.

Distances from Gore Bay (statute miles)
Killarney: 49 E
Little Current: 28 E
Kagawong: 18 E
Meldrum Bay: 37 W
Blind River: 31 N

LAKELAND BOATING · LAKE HURON CRUISING GUIDE

THE NORTH CHANNEL

Gore Bay, Ontario
Stop in for a slice of this pie-shaped bay.

Gore Bay, an easy day's run from either Meldrum Bay on Manitoulin Island's western edge or Little Current, is one of the island's busiest ports. The former freighter town, situated at the apex of a wedge-shaped inlet shielded on both sides by wooded limestone bluffs, offers a wide range of marine services and convenient provisioning options to the North Channel cruiser.

The inlet is distinguished on its northwest point by a functioning lighthouse. The approach to Gore Bay is uncomplicated; however, boaters coming from the east are advised to steer wide of the exposed clay bank located at the bay mouth, as it portends an underwater ridge. Once oriented with the range lights at the head of the bay, it is a quick, clear jaunt to the town point and the harbor entrance, which is marked by a red buoy to starboard. Boaters are asked to respect a no-wake zone after clearing the buoy.

Gore Bay Marina, whose office is located in the white building behind the point, is the best choice for a transient berth. The slips and long finger docks maintained by the town are quite roomy and can handle up to 150 boats at a time, depending on length. (Boats upwards of 120 feet can be accommodated.) At least 6 feet of water is available at the transient docks; at the fuel dock, the draft is a generous 15 feet. Service includes 30- and 50-amp power, water, gas and diesel, pumpout, washrooms and two shower facilities. Propane is not available at the dock but can be delivered on request. An excellent launch ramp is situated in the middle of the marina complex.

Provided the wind isn't out of the north, decent anchorages may also be secured offshore, where the depth is generally quite good. Boaters seeking an anchorage should be aware, however, that the sandy loam bottom can be very weedy in places.

Canadian Yacht Charters, located right next door to the town marina, offers about 10 spaces for transients and carries a good supply of charts, bait, clothing, bilge pumps and other accessories at its marine store. Any mechanical parts that are not in stock can be acquired within a day. CYC maintains an extensive charter fleet of trawlers and sailboats and rents out tennis courts, bicycles, windsurfers and waterbikes. Currently CYC has a fleet of about 25 yachts. Owners Ken and Pamela Blodgett note that they plan to add laundry facilities soon.

Purvis Marine Storage, also next door, operates a 35-ton lift and can do emergency prop and shaft work. More extensive repairs are farmed out to a widely respected local mechanic.

For the carless traveler, the beauty of Gore Bay is that virtually everything is within easy walking distance of the docks. **Bayside Restaurant and Café**, a good place to grab a gourmet meal, is located directly behind the marina. The circular wooden **Gore Bay Pavilion** is a stone's throw to the south. Artists are showcased on the main floor, while the **Rocky Raccoon Café**, a global cuisine restaurant, occupies the second. Another flight up the circular staircase leads the visitor to a windowed cupola that commands a lighthouse-keeper's view of the harbor. A sand beach and picnic area are adjacent to the pavilion.

The town's main street is one block inland. This compact shopping district includes the island's newest **grocery store**, a **beer and liquor store**, **hardware stores**, several **restaurants** (for superb food and cozy atmosphere, check

LAKELAND BOATING · LAKE HURON CRUISING GUIDE   **173**

Where to Eat

Bayside Restaurant and Café 705-282-1279
Gourmet meals, including German cuisine, located behind Canadian Yacht Charters.

Gordon's Lodge 705-282-2342
A sports lounge and fine dining restaurant featuring steaks, Manitoulin whitefish and Georgian Bay pickerel. The outstanding food is complemented by the lovely view. Call them if you need a lift.

Rocky Raccoon Café 705-282-8111
This global cuisine restaurant, located right at Gore Bay Marina, offers a special boaters' breakfast. The café also houses the local tourism info center.

Twin Bluffs Bar & Grill 705-282-2000
This award-winning restaurant features "island cuisine" prepared with fresh local ingredients.

out **Twin Bluffs Bar & Grill** on Eleanor Street), a **health and bulk food shop, drugstore, coin laundry** and **bank** with a 24-hour cash machine. Those hoping to restock the galley with fresh produce should plan their layover to coincide with a Friday morning, when the town stages its weekly **farmers market** at the **Gore Bay Arena**.

Boaters who want to stretch their legs can venture a mile or so north of the marina to visit the **Janet Head Lighthouse**, the second-oldest beacon on the island. In the other direction, a boardwalk curves around the foot of the bay, leading eventually to **Gordon's Lodge**, which features fine dining and pickup service. At the end of Dawson Street is the **Gore Bay Museum**, located in a former jail, and the town offers plenty of tree-shaded side streets to explore, most of them studded with stately limestone dwellings.

Car rentals are also available in Gore Bay, and **Jeff's Taxi and Delivery** serves most of Manitoulin. But whether you're getting ready to head out to the Benjamins or simply seeking a place to unwind and stretch your legs, the town should satisfy all of your expectations.

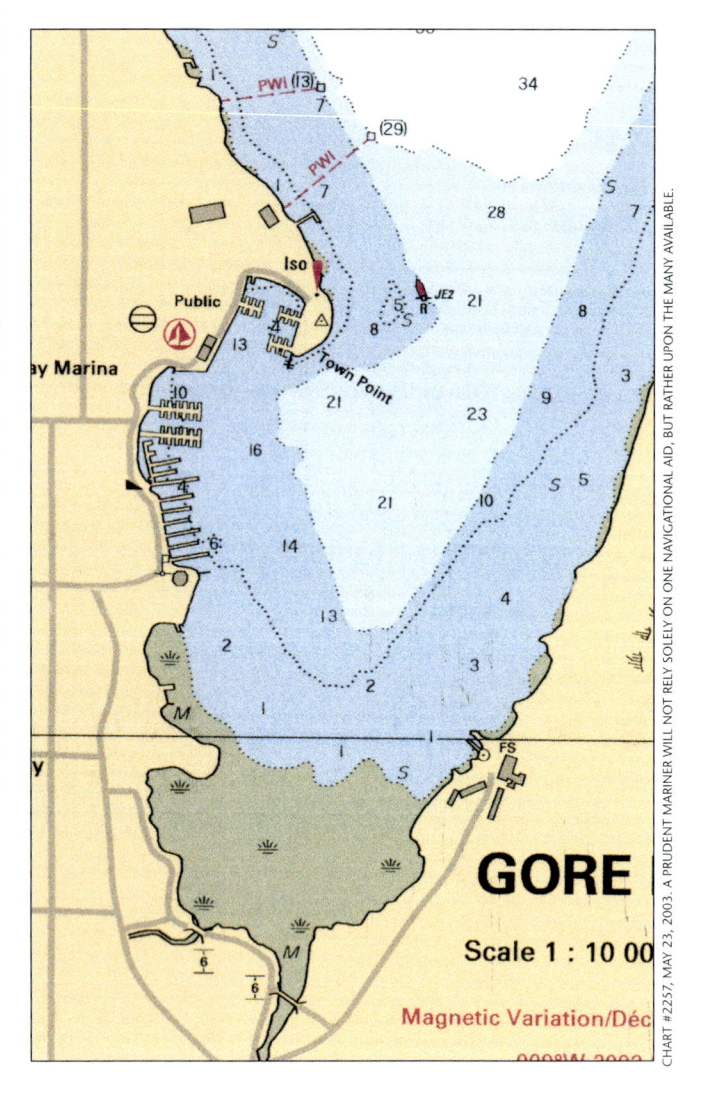

More Info

Emergency	911
Gore Bay Medical Centre	705-282-2262
Gore Bay Information Centre	705-282-3352
Jeff's Taxi and Delivery	705-377-6222

Facilities information subject to change. We suggest you call ahead.

			Monitors VHF Channel	Transient Slips Available	ALTERNATE MOORING: Wall mooring Rafting allowed	Maximum LOA	Minimum Depth at Dock	Power (amperage or volts)	HOOKUPS: Water Cable TV	FUEL: Gas Diesel Pumpout	BASICS: Heads Showers Laundry	AMENITIES: Swimming pool Whirlpool Rec area Grills Picnic tables Dog walk	Take Reservations	Take Credit Cards	Haulout (Capacity in tons or feet)	REPAIRS: Mechanical Electronics Fiberglass Woodworking Sails Canvas	CONVENIENCE: Ship's store Ice Convenience foods & beverages
C	Canadian Yacht Charters 705-282-0185	30 Water St. Gore Bay, ON P0P 1H0	16 68	10	W	100'	12'	30 50	W		HS		PD	N	Y	35	MEFWSC	SIC
B	Gore Bay Marina 705-282-2420	P.O. Box 298 Gore Bay, ON P0P 1H0	68	70	WR	120'	6'	30 50	W	GDP	HS		PD	N	Y			SI

174   LAKELAND BOATING · LAKE HURON CRUISING GUIDE

GREAT LAKES CRUISING CLUB

Join the Club
Make friends, share secrets and help promote safe boating.

One evening in 1934, seven boating friends, including an outgoing skipper by the name of Arch Gibson, were gathered around a dinner table in Chicago discussing the limitations of the charts available then and imagining how nice it would be if those who experienced distant waters could share their knowledge and keep other boaters up to date.

Soon after that, the Great Lakes Cruising Club was born.

Gibson had been in the habit of sharing his annual cruising adventures through a series of Christmas letters. With the encouragement of his friends at dinner that night, those letters evolved into a book that combined his log notes with those of other boaters. Known from the beginning as the *GLCC Log Book*, the records grew over the years from 30 to 2,500 pages. It is kept in looseleaf form so that the annual updates can be easily added; it is also now available on CD-ROM.

The book is organized by lake and connecting waterway, including sections on the Erie Canal, the St. Lawrence and the Mississippi. It covers not only the civilized harbors such as the ones in this book, but the wild and natural ones as well.

This is the unique charm of the publication: Because almost every spot has been explored by someone, even the most obscure places and little-known facts of navigation are brought to light in these pages.

But the production of the *Log Book* is only part of the Cruising Club story. Membership has grown from the original seven friends to almost 2,500. Open to everyone who is interested in cruising, it represents folks from 29 states, Canada and other foreign countries who own both power- and sailboats ranging from 10 to 125 feet.

Not only are all members encouraged to contribute fresh information to the *Log Book*, they also are invited to the traditional rendezvous held in July, usually at an exotic location. Occasionally, the club holds more than one of these rendezvous to accommodate the wide variety of cruising tastes represented.

There are dinner meetings in the off-season, usually including an informative program. The club also supports the improvement of harbors and the protection of the environment.

A vital part of the club's organization are the port captains. Now presiding over 150 ports, these key people are responsible for keeping their port information updated, as well as serving as goodwill ambassadors for current as well as prospective members. They also help plan and schedule the various meetings throughout the year.

The glue that holds the GLCC together is *Lifeline*, the club's magazine, which comes out five times a year.

Although this *Lakeland Boating* cruising guide has no formal connection with the Great Lakes Cruising Club, we agree that we all share the same overall goals: to promote both the enjoyment and the safety of boating and to encourage fellowship and support among those who venture out each season into waters both known and unknown.

Inquiries can be sent to:

Great Lakes Cruising Club
28 E. Jackson Blvd.
Suite 1300
Chicago, IL 60604
phone: 312-431-0904
fax: 312-431-0908
email: info@glcclub.com
www.glcclub.com

TOM KAEKEL

Kagawong, Ontario

Lat 45°54.57
Lon 82°15.65

The village of Kagawong lies at the end of Mudge Bay, about seven miles due south of Clapperton Island. Those approaching from the west should pay close attention to their charts, as the channel southwest of Vankoughnet and Clapperton Island is riddled with shoals. Entering Mudge Bay from the east is much simpler. Northern Marina **A**, the only choice for transients, is situated between a small functioning lighthouse **B** and a large stone building known as the Old Mill **C**. There is an excellent beach **D** adjacent to the marina. The small-craft basin **E** to the northeast is for local boaters only. There is a popular anchorage to the north in Clapperton Harbour behind Harbour Island.

Distances from Kagawong (statute miles)
Gore Bay: 18 m W

Little Current: 20 m E

Facilities information subject to change. We suggest you call ahead.

		Monitors VHF Channel	Transient Slips Available	ALTERNATE MOORING: Wall mooring / Rafting allowed	Maximum LOA	Minimum Depth at Dock	HOOKUPS: Water / Cable TV	Power (amperage or volts)	FUEL: Gas Diesel Pumpout	BASICS: Heads Showers Laundry	AMENITIES: Swimming pool Whirlpool Rec area / Grills Picnic tables Dog walk	Take Reservations	Take Credit Cards	Haulout (Capacity in tons or feet)	REPAIRS: Mechanical Electronics Fiberglass / Woodworking Sails Canvas	CONVENIENCE: Ship's store Ice / Convenience foods & beverages
A	Northern Marina 705-282-3330	P.O. Box 33 Kagawong, ON P0P 1J0	68	15	WR	65'	7'	30	W	GP	HS	PD	Y	Y			SIC

176 LAKELAND BOATING · LAKE HURON CRUISING GUIDE

The North Channel

Kagawong, Ontario
A working lighthouse shows the way.

Directly across from the docks of this charming little town are two sights that no boater will want to miss: the small **Kagawong Lighthouse**, which dates from 1894 and is still in use, and the small white **Anglican church**, whose pulpit is fashioned from the bow of a wrecked boat. Located some 18 miles east of Gore Bay, Kagawong offers most marine services and a limited selection of provisions.

Northern Marina has 7 feet of water and space for about 15 craft at its finger docks. The long outer dock has up to 25 feet of water and can take cruisers up to 150 feet. Gas, pumpout, water, 15-amp power, an excellent beach, washrooms and showers, and a launch ramp are all at hand. Diesel and propane can be secured on request. The marina also charters sailboats and houses a substantial chandlery. Government charts are available, as are Zodiac dinghies and dinghy repairs. Farquhar's ice cream is sold here as well. Minor boat and engine repairs can be attended to by **Barry Boats**, a local marine shop.

In terms of anchorages, Mudge Bay is not ideal, particularly if a strong northwest wind is blowing. If the wind is consistently out of the west, Sextant Point on the west shore will provide fine shelter; it also boasts a small beach. In fair weather, decent purchase can be found not far from the marina.

The village of Kagawong can be easily explored on foot. Main Street has a **general store** with some hardware and a modest supply of groceries, as well as liquor and beer. **Manitoulin Chocolate Works** is a purveyor of chocolate, coffee and gift baskets. A few doors up the hill, the quaint **Kagawong Stonehouse Restaurant** serves family fare. The **Old Mill building** on the waterfront was once a pulp mill and electric plant; since 1996, it has housed **Edwards Studio**, an art gallery. If you're curious about how other buildings fit into Kagawong's past, pick up the walking tour map available from the marina and let it lead you around.

Kagawong is an Ojibway word that means "where the mists rise from the falling waters," a reference to **Bridal Veil Falls**. A visit to the falls, where the Kagawong River crashes over a steep limestone bluff, is well worth the three-quarter-mile trek. You can reach the falls either by taking the nature trail that winds along the west bank of the river to the base of the falls, or by continuing up the hill to the highway, which leads to the crest of the falls. Whichever way you go, when you get there you'll have the option of cooling off behind the curtain of water or wading directly into the swirling pool at the bottom. The nearby **Manitoulin Wind & Wave** rents kayaks, canoes and paddleboats.

More Info

Emergency 911
Little Current Mindemoya Hospital 705-377-5311
Township of Billings 705-282-2611

Where to Eat

Needles Stonehouse Restaurant 705-282-2941
Home cooking—from seafood and salads to lasagna and T-bone steaks. Their specialties are locally caught whitefish and great breakfasts. Enjoy a nightcap at Stonewalls bar, located in the restaurant.

ROD MCLEOD

LAKELAND BOATING · LAKE HURON CRUISING GUIDE

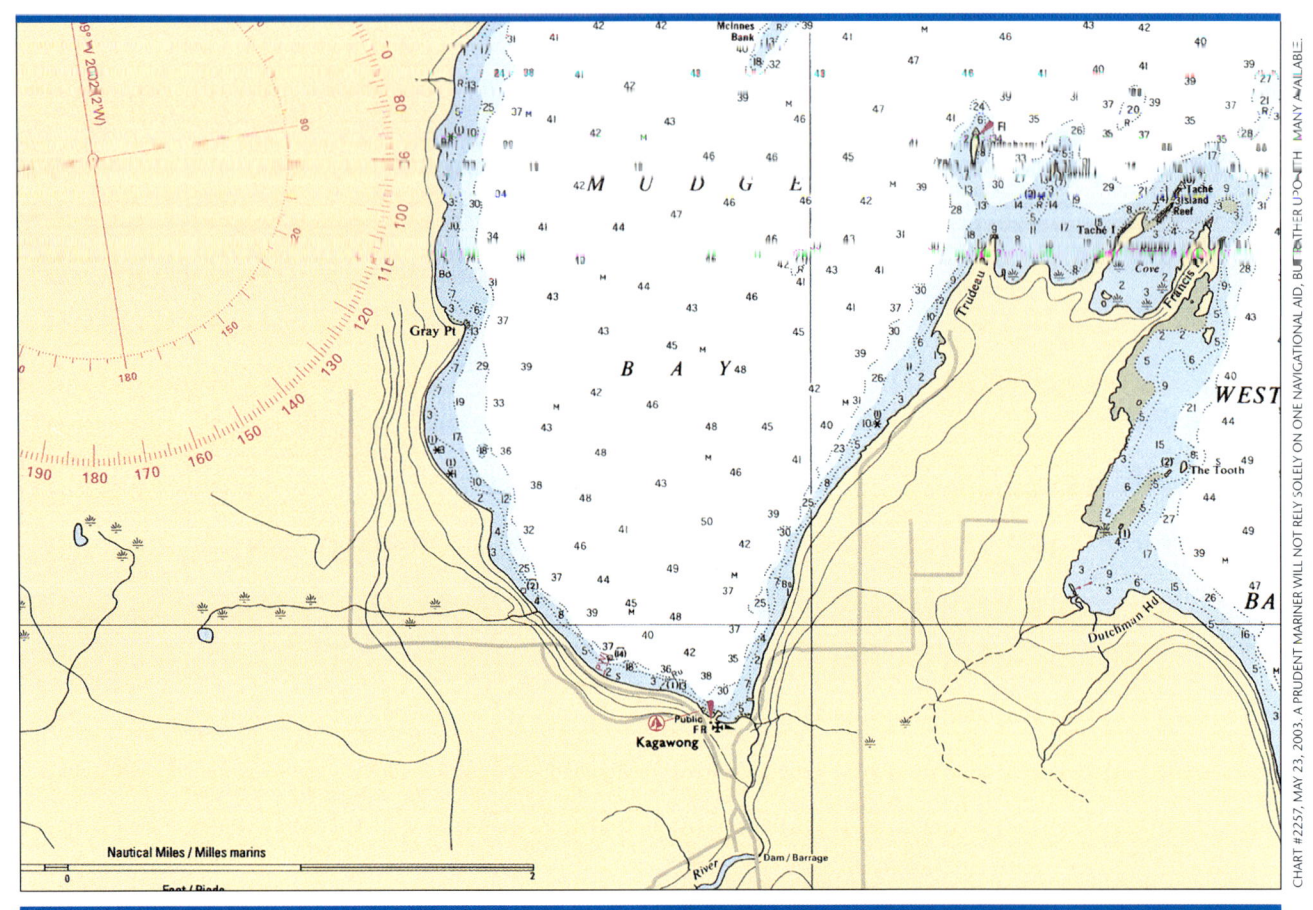

West Bay, Ontario
This quiet village offers a taste of Native American culture.

If you're in the market for Native American art but are well stocked up on all other provisions, consider a visit to the village of West Bay. The bay itself is eight miles long and six miles wide at its widest point and lies just south of the eastern approach to the Clapperton Passage leading to Kagawong, Gore Bay and points westward.

Situated at the south end of the bay, the village is the center of the largest of Manitoulin's six Indian reserves and is also home to the island's single high school. Students come by bus from as far as Wikwemikong, 40 miles to the east, and Meldrum Bay, 60 miles to the west.

West Bay has an 80-foot-long government wharf with 6 to 7 feet of water; the outer end, however, is in disrepair, and a breakwater is awash west of it. Corbier Cove, a mile north and west of the village, has good holding for anchorage but is open to the north.

The **Ojibway Cultural Foundation**, a short walk from the dock, exhibits and sells paintings, leather goods and other handicrafts by Native American artisans. Beyond the nearby Catholic church and less than half a mile from the dock is **Abby's Restaurant**, serving full home-cooked meals. A **convenience store** and **gas station** are located nearby, but no fuel is available at the dock.

On the northwest side of West Bay, Sounding Cove is protected from the north and has a good bottom for anchoring, as well as a government dock with 12-foot depth. Here you'll find **Silver Birches Resort**, which has a restaurant and snack bar; showers, water and ice are also available.

Distances from West Bay (statute miles)
Kagawong: 14 W
Little Current: 18 E

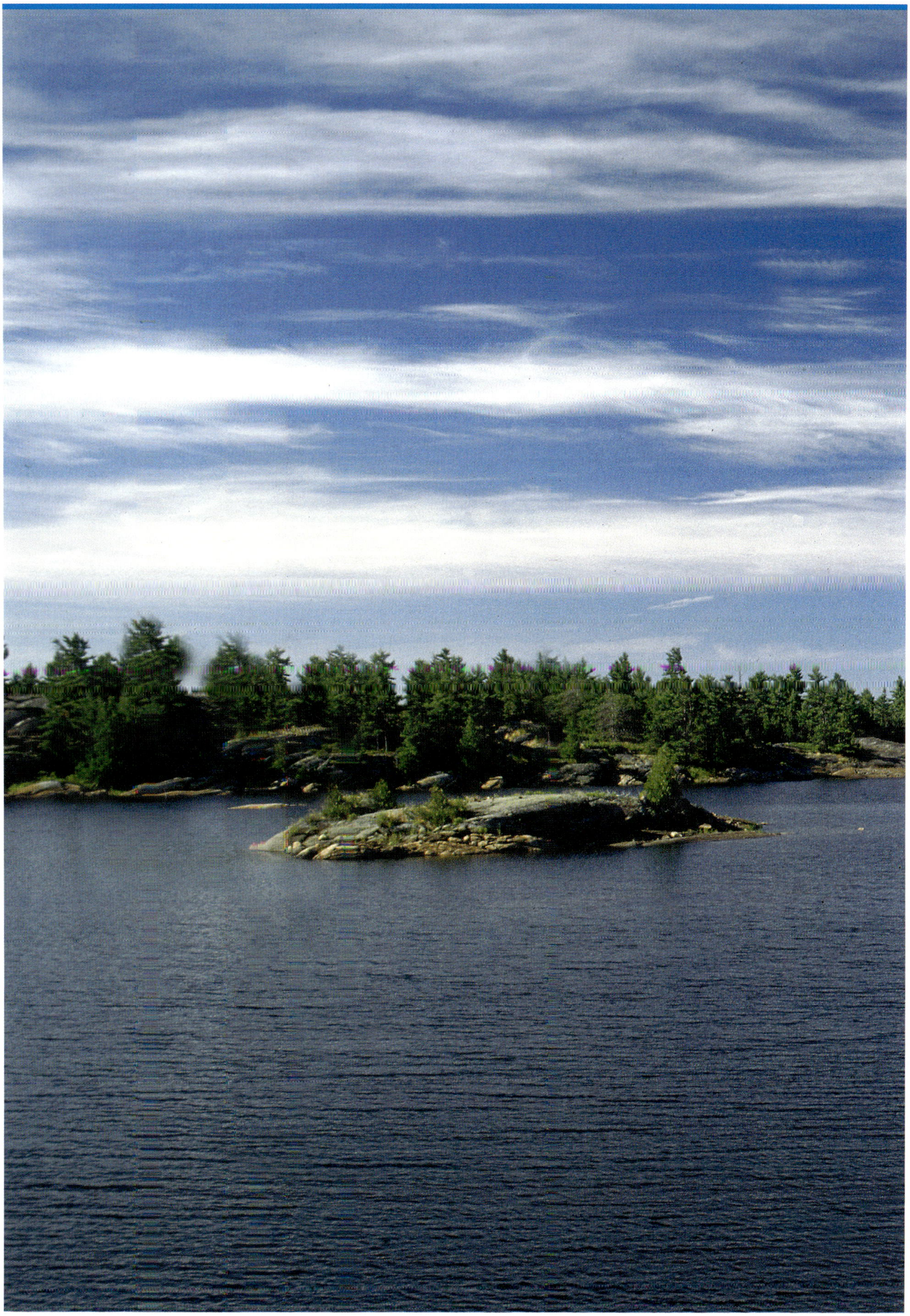

PETER MARREK

Little Current, Ontario

Lat 45°58.80
Lon 81°54.85

Little Current sits on the northeast tip of Manitoulin Island, at the narrowest point of the North Channel. The swing bridge **A** (which turns instead of lifts) opens on the hour during the summer. Closed, it has a clearance of 18 feet. Nearby, the Manitoulin Island Information Centre **B** provides a stopping point to make inquiries or to wait for the bridge to open. Gas, diesel and pumpout are all available at Wally's Dock Service **C**, which manages the wide wooden municipal dock downtown. Dockage is available at Boyle Marine **D**, which offers restrooms, showers and electrical hookups. Adjacent to Boyle's is the larger federally owned and municipally run Spider Bay Marina **E**, with numerous services available. Additional dockage is available on the east side of the island at Harbor Vue Marina **F**.

Distances from Little Current (statute miles)
Baie Fine: 16 NE
Killarney: 20 SE
Gore Bay: 30 W

180 LAKELAND BOATING · LAKE HURON CRUISING GUIDE

THE NORTH CHANNEL

Little Current, Ontario
Just about everything you need, right on the waterfront.

In Little Current, the North Channel abruptly narrows to 100 yards, making a stop in this handsome port almost inevitable for cruisers. Located midway between Killarney and the Benjamin Islands, the town is a great place for boaters to fuel up and stock up on provisions.

Situated in the natural passage between Manitoulin Island to the south and Goat Island to the north, the spot was known by the original inhabitants as Waiebijiwang—"Where the waters flow back and forth"—for its unpredictable currents. The fur-trading voyageurs, somewhat less in awe, dubbed the channel Le Petit Courant, or Little Current. Linked inextricably with the water from its earliest days as a lumber- and coal-shipping town, Little Current has never lost its marine orientation. Most of its businesses still face the waterfront, and boaters can tie up and simply cross the street to take care of the great bulk of their shopping needs.

The approach to Little Current is well-marked and easily managed, whether you are arriving from the east, following the channel between the Strawberry Island Lighthouse and Garden Island, or passing north of the Narrow Island light from the west. Cruisers reaching higher than 18 feet should try to time their arrival with the schedule of the swing bridge, which opens for 15 minutes on the hour in daylight. Be prepared for the current, which can reach up to six knots at the bridge and which can, and often does, reverse direction several times a day. To determine the direction and strength of the current, check which way and how far the channel buoys are tilted. There's a red boat-shaped current buoy located just west of the swing bridge. A no-wake speed is strictly enforced from Gibbons Point east of the bridge to the end of the buoyed channel west of Picnic Island.

For the most convenient provisioning, the best plan is to find a space along the long downtown dock. Beginning just west of the Swing Bridge, the **federal wharf** stretches more than half a mile along the shoreline to Boyle Marine; still, it tends to fill up quickly on a busy summer day. **Wally's Dock Service**, located to the east of the large red **post office** building on the waterfront, manages the dock for the town. Gas, diesel, pumpout and compressed natural gas are all available at the service dock, which has a draft of 14 feet. The office includes a marine store, which carries charts and a good selection of tackle, chemicals, rope and anchors. Wally's also collects fees for overnight dockage and maintains the shower and washroom facilities located immediately adjacent to the office. Thirty-amp power and water are available at most spots on the dock, and boaters can count on at least 8 feet of draft.

Spider Bay Marina, a second municipal facility located just west of town, offers a well-protected harbor with 150 slips, roughly half of which are set aside

Where to Eat

Anchor Inn 705-368-2023
The dinner menu changes daily and ranges from chicken cordon bleu and New York steak to ribs, veal and fresh fish. Monday nights feature live entertainment. Wing nights are Tuesday and Wednesday, and there's dancing on weekends.

China City 705-368-1234
Featuring spicy Szechwan cuisine.

G.G.'s Food Mart, Bakery & Deli . . . 705-368-2651
Come here to pack a lunch stuffed with homemade breads, items from the hot and cold deli, fried chicken and pies.

Gary's Family Restaurant. 705-368-3370
Stop off on the way to the ferry to sample the Arctic char and the Kentucky-style chicken. A great place for kids, too.

The Old English Pantry 705-368-3341
For lunch, stop in for the famous sandwiches and afternoon teas. Come back at dinner for roast beef, fresh fish, pasta and Yorkshire pudding.

Shaftesbury Inn 705-368-1945
Casual and fine dining in a historic setting just steps from the boardwalk. Features Canadian and continental cuisine.

LAKELAND BOATING · LAKE HURON CRUISING GUIDE   **181**

for overnight transient dockage. The harbor has a depth of 8 to 10 feet and offers gas, diesel, propane, pumpout, 30- and 50-amp electricity, a large picnic area, a laundromat, showers and washrooms. Downtown Little Current is only five minutes away on foot. Sail repair and electronic service are also easily arranged at Spider Bay, as are some mechanical repairs, although the marina itself offers no repair shop or haulout facility.

The friendly and hospitable **Boyle Marine** is just east of Spider Bay and just west of downtown. It operates a 30-ton Travelift and repairs shafts, props and hulls. Factory-trained technicians are on duty to handle inboard and outboard engine work. Salvage and tow calls are answered on VHF channel 68. All docks are equipped with water and power. Boyle offers slips with up to a 30-foot draft and 100-foot length on a protected system. Although the facility does not offer fuel, it has a well-stocked marine store, inside storage on site in a covered 200- by 60-foot building, and overnight service for special orders. The marina is a member of the Clean Marine Partnership and proclaims a dedication to preserving water quality.

Another ambitious service facility is located on the other side of Little Current, southeast of the swing bridge. **Harbor Vue Marina** offers extensive repairs and boasts more than 7,000 square feet of shop space. Boats up to 35 tons and 45 feet in length can be hoisted into one of Harbor Vue's spacious repair bays. The marina also maintains 55 slips with power and 8 feet of water, a quarter of which are open to transient guests. Showers and washrooms are close at hand, and security is provided around the clock. Gas is available at the service dock, and pumpout service has been added. If you want to reach town, it's a 15-minute hike along the shoreline; if that seems like too much work, the marina will happily provide transportation.

There are two nice launch ramps in Little Current: one at Spider Bay Marina, just west of town, and an enlarged double ramp at Harbor Vue, just east of town.

As the largest community on Manitoulin Island, Little Current is well-equipped to supply everything on your list. In its waterfront downtown area, nothing is more than a few paces away. The appropriately named Water Street offers a **laundromat**, a **grocery store**, **pharmacies**, two **banks** with cash machines, **gift**, **craft** and **clothing shops**, a **hardware store** and a number of dining opportunities. The stately **Shaftesbury Inn** has been renovated and reopened, providing Little Current with an upscale eatery. The restaurant in the **Anchor Inn** (which also houses a bar) and **The Old English Pantry** on Water Street are both good choices for family fare, as is **Gary's Family Restaurant**, two blocks south on Manitowaning Road. A **liquor store** is nearby; the **beer store** is a bit farther away on Meredith Street. **G.G.'s Food Mart, Bakery & Deli** is also on Meredith. Back downtown, you might want to check out **Turners' of Little Current, Ltd.** for its maps, charts and Hudson Bay blankets. The *Manitoulin Expositor*, two doors

HARBOR VUE MARINA the North Channels largest full service marina. East of the swing bridge in Little Current.

- Six service bays for inside repair
- 40 ton travel lift. •Factory trained technicians
- Hull repair for all types •Monitor VHF 68
- Prop and shaft repair •Salvage and towing •Ships store •Storage inside and out

72 Fergusons Road
Little Current, Ontario P0P 1K0
Telephone: 705-368-3212 Fax: 705-368-3379
Latitude 45-58-40N Longitude 81-54-12W

MERCURY MerCruiser
YAMAHA
VOLVO PENTA
EVINRUDE Johnson GENUINE PARTS

Just West of Downtown Little Current

Your Full Service Marina For:
- Inside/outside storage on-site
- Dockage - seasonal & overnite • Travel lift
- Certified marine mechanics • Boat & engine repairs
- Marine Supplies • Salvage & towing
- Prop & shaft repairs • Within walking distance of downtown stores & restaurants

MERCURY Outboards Set the Water On Fire.
MerCruiser STERN DRIVES & INBOARDS The Only Logical Choice.
VOLVO PENTA
Onan YANMAR

Box 487, Little Current, Ontario P0P 1K0
705-368-2239
Canadian Marinas Monitor 68
Email: boylemarine@manitoulin.net

down, is northern Ontario's oldest continually published newspaper. Its front shop carries a good selection of literary fiction.

If you want to see more of the island, a cab service is available in town. Chances are, however, that you'll be content to prowl through the town itself or dawdle along the boardwalk beside the town dock. With the tidy storefronts lining Water Street to the south and the spectacular quartzite hills of the LaCloche Mountains to the north, there is plenty of atmosphere to absorb on either side. Meanwhile, the "little" current in between promises wider stretches of water just ahead.

More Info

Emergency	888-310-1122
Manitoulin Health Center	705-368-2300
Manitoulin Island Information Centre	705-368-3021

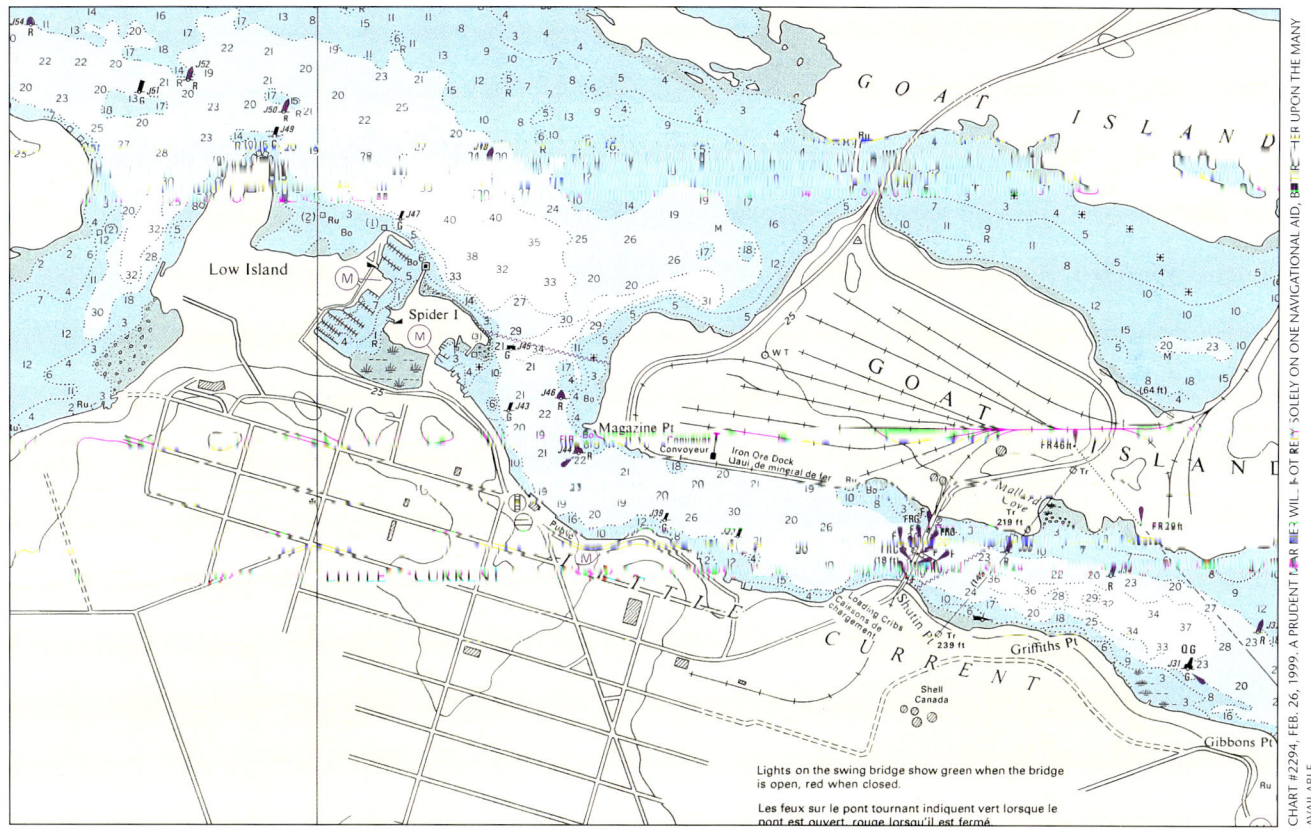

Facilities information subject to change. We suggest you call ahead.

	Marina	Address	Monitors VHF Channel	Transient Slips Available	ALTERNATE MOORING: Wall mooring, Rafting allowed	Maximum LOA	Minimum Depth at Dock	Power (amperage or volts)	HOOKUPS: Water, Cable TV	FUEL: Gas, Diesel, Pumpout	BASICS: Heads, Showers, Laundry	AMENITIES: Swimming pool, Whirlpool, Rec area, Grills, Picnic tables, Dog walk	Take Reservations	Take Credit Cards	Haulout (Capacity in tons or feet)	REPAIRS: Mechanical, Electronics, Fiberglass, Woodworking, Sails, Canvas	CONVENIENCE: Ship's store, Ice, Convenience foods & beverages
D	Boyle Marine 705-368-2239	29 Water St. Little Current, ON P0P 1K0	71	10	WR	100'	30'	30	W		HS		PD	Y	Y	30	MEFWSC	SI
F	Harbor Vue Marina 705-368-3212	72 Fergusons Rd. Little Current, ON P0P 1K0	68	10		48'	6'	30	W	GP	HS		D	Y	Y	35	MFW	SI
E	Spider Bay Marina 705-368-3148	48 Water Street Little Current, ON P0P 1K0	68	85	W	70'	8'	30 50	W	GDP	HSL		PD	Y	Y			IC
C	Wally's Dock Service 705-368-2370	32 Water St. Little Current, ON P0P 1K0		40	WR	112'	8'	30	W	GDP	HSL		P	N	Y			SIC

LAKELAND BOATING · LAKE HURON CRUISING GUIDE

Sheguiandah, Ontario

Lat 45°53.72
Lon 81°54.85

A wide nick in Manitoulin's east shore, Sheguiandah Bay opens just beyond the southern tip of Strawberry Island. Range lights at 261°T will safely lead the cruiser to the Sheguiandah Government Dock **A**, which is sheltered by O'Meara Point **B**. The floating docks here have been pulled out due to low water levels.

Distances from Sheguiandah (statute miles)
Little Current: 10 NNW

Killarney: 19 NNE

Facilities information subject to change. We suggest you call ahead.

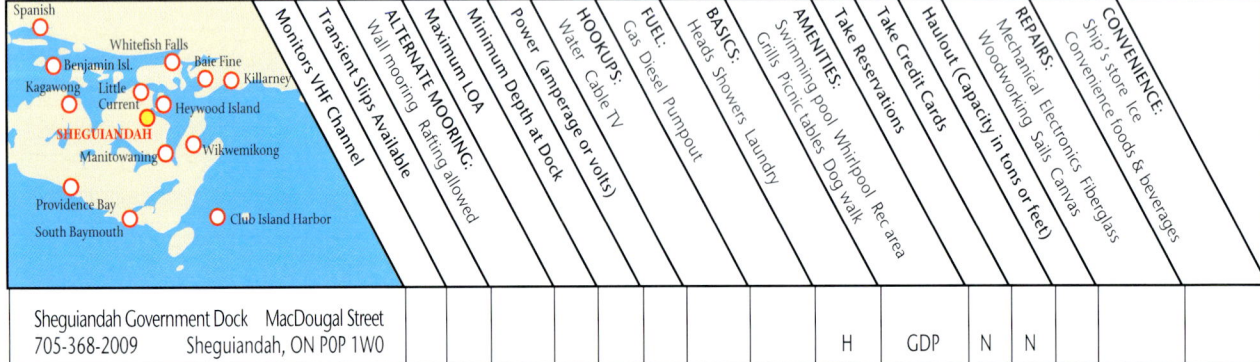

	Monitors VHF Channel	Transient Slips Available	ALTERNATE MOORING: Wall mooring Rafting allowed	Maximum LOA	Minimum Depth at Dock	Power (amperage or volts)	HOOKUPS: Water Cable TV	FUEL: Gas Diesel Pumpout	BASICS: Heads Showers Laundry	AMENITIES: Swimming pool Whirlpool Rec area Grills Picnic tables Dog walk	Take Reservations	Take Credit Cards	Haulout (Capacity in tons or feet)	REPAIRS: Mechanical Electronics Fiberglass Woodworking Sails Canvas	CONVENIENCE: Ship's store Ice Convenience foods & beverages
A Sheguiandah Government Dock MacDougal Street 705-368-2009 Sheguiandah, ON P0P 1W0									H	GDP	N	N			

184 LAKELAND BOATING · LAKE HURON CRUISING GUIDE

THE NORTH CHANNEL

Sheguiandah, Ontario
Dig the archaeological finds or take in a powwow.

Encompassing both the village at the end of the bay and the Ojibway First Nation extending to the south, Sheguiandah offers relatively few services, but is an interesting spot to poke around.

Since the municipal wharf was dismantled in 2003, the **Sheguiandah Government Dock** is the only game in town — and its floating docks have been pulled out due to low water. Near the launch ramp, there is a picnic area that includes tables, barbeque stands, rudimentary washrooms and an unusual black pebble beach. When water levels rise, the docks will be able to accommodate up to 10 boats. No fees are required here, as there are no services or security.

A boardwalk runs to the bay head, where two tourist camps can be found. **Whitehaven**, the more inland of the two, maintains a small marina for its guests and can provide gas to transient boaters if necessary. However, there is no transient dockage available. **Green Acres Camp**, about half a mile away on Highway 6, has propane, a limited supply of groceries and, occasionally, dockage for transients. The **Sheguiandah First Nation** begins just south of the store. At present, there is not much here to warrant a trip, unless your visit happens to coincide with the nation's powwow, usually held in early July.

The village of Sheguiandah does have some fascinating aspects, including a nearby **archeological dig** that suggests the area was inhabited as far back as 9,600 years ago, which would make it the oldest known settlement in North America.

The **Little Current-Howland Centennial Museum**, located within walking distance of the docks on Highway 6, has many artifacts on display, including ancient tools unearthed from the dig site. Nineteenth-century buildings still line the streets of the village, and a **replica of an old-fashioned sawmill** can be found on the banks of the stream that purls through town.

More Info

Emergency	911
Manitoulin Health Center	705-368-2300
Manitoulin Tourist Information	705-368-3021
Jeff's Taxi	705-377-6222

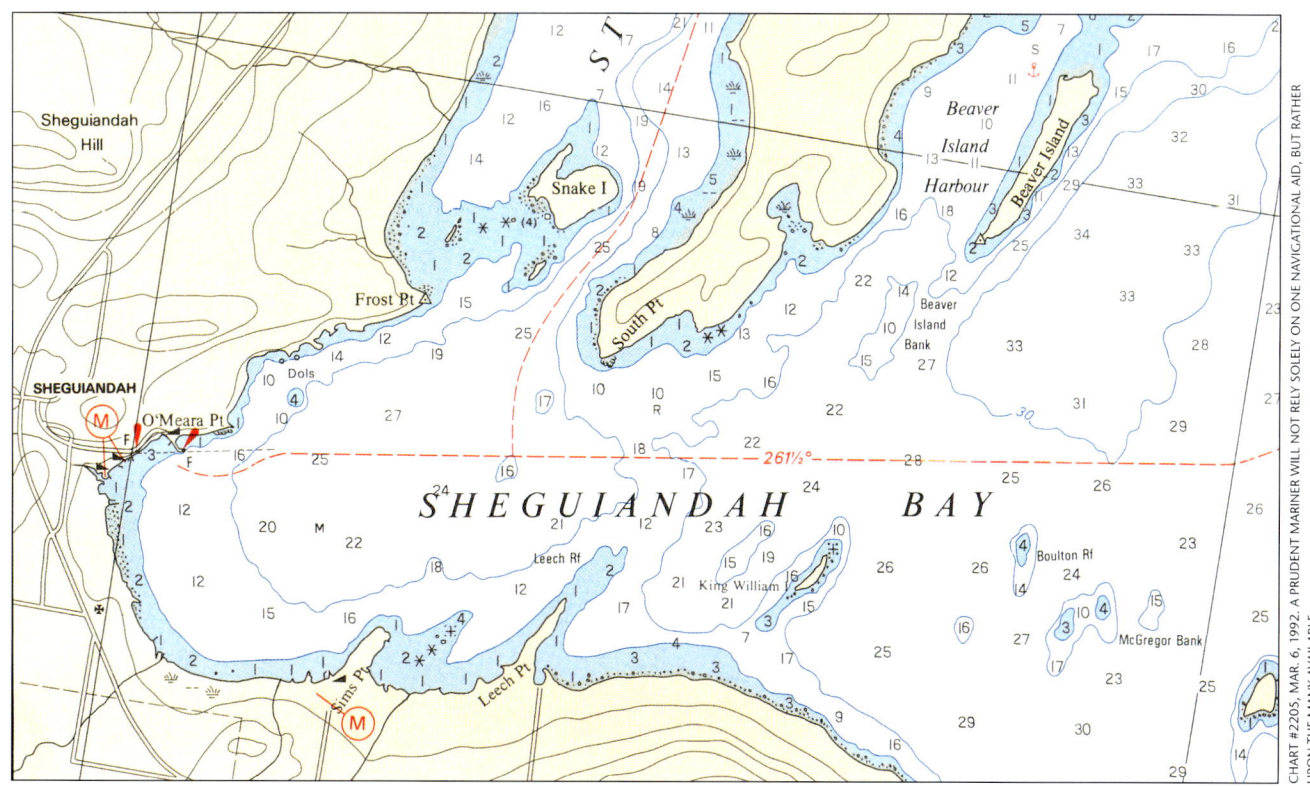

LAKELAND BOATING · LAKE HURON CRUISING GUIDE **185**

Manitowaning, Ontario

Lat 45°44.70
Lon 81°48.48

Manitowaning Bay **A**, a 10-mile stretch of deep, unobstructed water, is located directly south of Heywood Island in the North Channel. At the town of Manitowaning **B**, in the bay's southwest corner, the inlet doglegs around a natural point, creating a relatively calm space for the town docks at Bay Street Marina **C**. Your first glimpse of civilization will likely be the old lighthouse **D** with its fixed green beacon or the large yellow mill building **E**, now a historic site. The S.S. Norisle **F**, the last passenger steamer to ply the Great Lakes, is tied up alongside the marina. It, too, is now a heritage attraction. The Manitowaning Lodge Tennis & Golf Resort **G** has docking for boaters dining at the restaurant there. Anchorage protected from all but strong northerly winds can be found in the lee of Fanny Island **H**.

Distances from Manitowaning (statute miles)
Killarney: 26 NE

Little Current: 22 NW

186 LAKELAND BOATING · LAKE HURON CRUISING GUIDE

THE NORTH CHANNEL

Manitowaning, Ontario
Ships and lighthouses
and sea monsters, oh my!

The oldest European settlement on Manitoulin Island, Manitowaning boasts many examples of its marine heritage and is well worth a side trip from the more trafficked routes of the North Channel. Although its streets are lined with numerous shops, including a decent **grocery store**, Manitowaning is usually thought of more as a place to soak up atmosphere than as a full-fledged provisioning stop.

When motoring or sailing down from Little Current, exercise caution if you choose to take the inside channel of Strawberry Island. Be aware, too, that a shoal lurks off Phipps Point in the southeast part of Manitowaning Bay, a mile and a half north of town. Otherwise, the bay is exceptionally deep. According to an Ojibway legend, an underwater creature known as Mishebeshu lurks in its depths.

The only transient space is at **Bay Street Marina**, although **Manitowaning Lodge Tennis & Golf Resort**, located a short distance to the south, does have some docking and will welcome boaters who wish to dine at its gourmet restaurant—reservations are necessary. The slips of Bay Street Marina, which take up to 35 boats, are 30 feet long; most have water and 30-amp hydro. The draft is 9 feet at the fuel dock, 20 to 25 feet elsewhere. Gas, diesel and pumpout are all provided on site; shower and washroom facilities and a small beach are also close at hand. There is no propane at the marina, nor is there any haulout service. The former can be acquired with little fuss, and local mechanics can be summoned to the docks if repairs are in order. The marina does boast a fine launch ramp.

The real charm of the marina is its proximity to both the retired *S.S. Norisle*, a coal-fired ferry more than 200 feet in length, and the **Manitoulin Roller Mill**, a 19th-century grist mill that now houses a museum. Both are open to the public, and the steamer boasts the added novelty of having a fine restaurant onboard. Also nearby is the **Burns Wharf building**, which features live theater on weekends throughout the summer months.

A brief climb into town yields more dining options, a **beer and liquor store**, **hardware supplies**, **banking** (including a new **ATM**), a **grocery store** and a **coin-operated laundry**. Recent years have also seen the inclusion of an 18-hole **golf course** with clubhouse, a **medical and dental clinic**, and a **pharmacy**. The refurbished **lighthouse**, a cedar shake structure dating from 1885, is at the north end of Arthur Street, as is northern Ontario's oldest **Anglican church**. At the other end of Arthur Street is the **Assiginack Museum**, housed in an 1870s jail.

Good anchorage can be found in the lee of Fanny Island, though strong northerly winds will make even this location risky. In such weather, it is advisable to tuck yourself in at the town docks and let the nearby attractions tug you around on foot.

Where to Eat

**Manitowaning Lodge
Tennis & Golf Resort** **705-859-3136**
A must for foodies. Two superb gourmet chefs whip up a four-course menu nightly.

The Muskie Widows Tavern **705-859-1800**
Come by and relax in the rustic atmosphere of the **Muska-Lounge**. Enjoy a variety of home-cooked food while poring over the outdoor memorabilia and taxidermy displays.

More Info

Emergency 911
Little Current Hospital 705-368-2300
Municipality of Assiginack
Town Office 705-859-3196

LAKELAND BOATING · LAKE HURON CRUISING GUIDE   **187**

MANITOWANING BAY

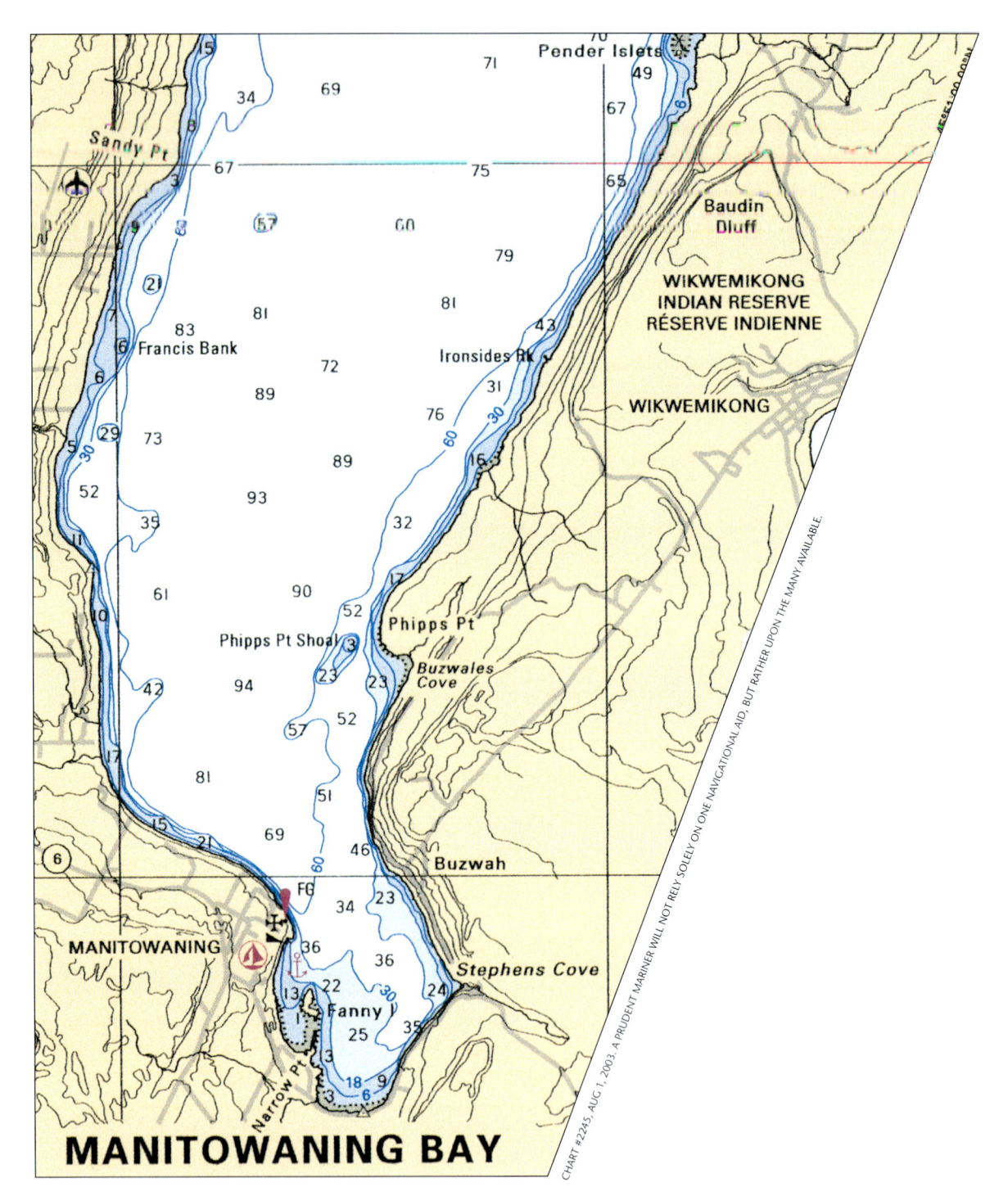

Facilities information subject to change. We suggest you call ahead.

			Monitors VHF Channel	Transient Slips Available	ALTERNATE MOORING: Wall mooring Rafting allowed	Maximum LOA	Minimum Depth at Dock	Power (amperage or volts)	HOOKUPS: Water Cable TV	FUEL: Gas Diesel Pumpout	BASICS: Heads Showers Laundry	AMENITIES: Swimming pool Whirlpool Rec area Grills Picnic tables Dog walk	Take Reservations	Take Credit Cards	Haulout (Capacity in tons or feet)	REPAIRS: Mechanical Electronics Fiberglass Woodworking Sails Canvas	CONVENIENCE: Ship's store Ice Convenience foods & beverages
C	Bay Street Marina 705-859-3700	Bay Street Manitowaning, ON P0P 1N0	68	35	R	50'	9'	30	W		GDP	HS		PD	N	Y	

188   LAKELAND BOATING · LAKE HURON CRUISING GUIDE

Lake Huron-Area Airports

Michigan

Albert J. Lindberg Airport............ 906-484-3623
Hessel

Alpena County
Regional Airport................ 989-354-2907
Alpena

Centralia/Huron Park Airport 519-228-6111
Huron Park

Cheboygan County Airport....... 231-627-5571
Cheboygan

Drummond Island Airport 906-493-5411
Drummond Island

Huron County Memorial Airport.... 989-269-6511
Bad Axe

Iosco County Airport 989-362-0052
East Tawas

James Clements Airport 989-895-8991
Bay City

Mackinac County Airport 906-643-7327
St. Ignace

Midland-Bay City-Saginaw
International Airport 989-695-5555
Saginaw

Oscoda-Wurtsmith Airport 989-739-8486
Oscoda

Pellston Regional Airport
of Emmet County 231-539-8441
Pellston

Ontario

Collingwood Regional Airport 705-445-2663
Collingwood

Goderich Municipal Airport........ 519-524-2915
Goderich

Gore Bay Manitoulin Airport......... 705-282-2101
Gore Bay

Huronia Airport 705-526-8086
Midland

Killarney Municipal Airport 705-287-2242
Killarney

London Airport.................. 519-452-4015
London

Manitoulin East
Municipal Airport............... 705-859-3009
Manitowaning

Owen Sound
Regional Airport................ 519-371-6936
Sydenham Township

Parry Sound Area
Municipal Airport............... 705-378-2897
Seguin

Sault Ste. Marie Airport 705-779-3031
Sault Ste. Marie

Sudbury Airport 705-693-2514
London

Tobermory Municipal Airport...... 519-596-2898
Tobermory

Toronto Lester B. Pearson
International Airport 416-247-7678
Mississauga

Wiarton-Keppel
District Airport 519-534-0140
Wiarton

A comprehensive list of airports in the United States can be found at www.airnav.com.

Wikwemikong, Ontario

Wikwemikong Bay Marina **A**, in the northwest corner of Smith Bay on the shores of the Wikwemikong Unceded Indian Reserve, is a state-of-the-art facility that, plagued by low water, is sitting high and dry. It is closed until high water returns. The village of Wikwemikong **B** is a little more than a mile away.

Distances from Wikwemikong (statute miles)
Killarney: 16 NE

Club Island: 23 S

Facilities information subject to change. We suggest you call ahead.

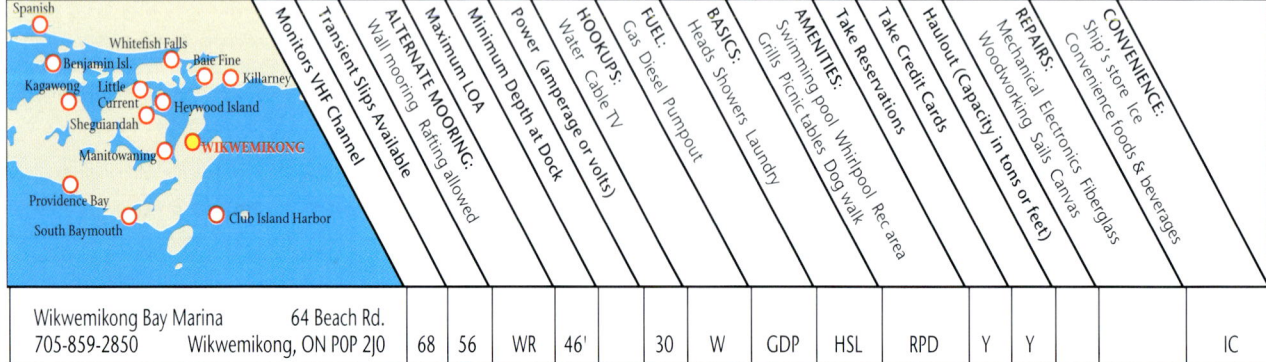

	Address	Monitors VHF Channel	Transient Slips Available	ALTERNATE MOORING: Wall mooring Rafting allowed	Maximum LOA	Minimum Depth at Dock	Power (amperage or volts)	HOOKUPS: Water Cable TV	FUEL: Gas Diesel Pumpout	BASICS: Heads Showers Laundry	AMENITIES: Swimming pool Whirlpool Rec area Grills Picnic tables Dog walk	Take Reservations	Take Credit Cards	Haulout (Capacity in tons or feet)	REPAIRS: Mechanical Electronics Fiberglass Woodworking Sails Canvas	CONVENIENCE: Ship's store Ice Convenience foods & beverages
A Wikwemikong Bay Marina 705-859-2850	64 Beach Rd. Wikwemikong, ON P0P 2J0	68	56	WR	46¹		30	W		GDP	HSL	RPD	Y	Y			IC

THE NORTH CHANNEL

Wikwemikong, Ontario
Where native traditions mingle with a Jesuit legacy.

Low water levels have recently plagued Wikwemikong, home to a state-of-the-art marina. But "Wiky," the location of the only unceded North American Indian reserve in Canada, remains a popular destination for anyone interested in native culture. Close to both Killarney and Little Current, it is an interesting departure from the crowded channels and busy sidewalks to the north. The community also offers an interesting contrast all its own: a mix of traditional Anishinabek culture and modern lifestyle.

The **Wikwemikong Bay Marina**'s 56 slips, all with power and water, are well protected by a 200-yard breakwall. However, the marina has been closed due to light water and silt buildup for several seasons. When open, the fuel dock provides gas, diesel, propane and pumpout, and the 7,200-square-foot marina building, beautifully constructed of square logs by local builders, houses showers, washrooms, a laundromat, some marine supplies and a heritage center. Security is provided around the clock.

A **beach** and swimming area are located right beside the marina, but anyone hoping to do some banking, stock up on food or see the sights will have to hike up the hill and into Wikwemikong. The marina provides a complimentary shuttle service, which comes in handy,

Where to Eat

Brook's & Dave's
Family Restaurant **705-859-2760**
Deli-style sandwiches, pizza and hot meals. Open for lunch and dinner.

Pat's Café. . . . . **No phone**
This eatery, which is open seasonally for lunch and dinner, serves up traditional native cuisine.

Patsy's Family Restaurant **705-859-2075**
Serving everything from hot sandwiches and charbroiled burgers to chicken and roast beef. Their specialty is whitefish. Open for breakfast, lunch and dinner.

More Info

Emergency 911
Wikwemikong Health Centre 705-859-3164
Wikwemikong
Heritage Organization 705-859-2385

LAKELAND BOATING · LAKE HURON CRUISING GUIDE   **191**

especially if you don't like hills.

However you choose to make the trip, a logical first stop is the **Holy Cross Mission Ruins**, located directly above the marina. This is a reminder of the Jesuit legacy in Wikwemikong. Authentic native theater performances are staged outdoors in the ruins in late July and early August.

There is a **grocery store/sweetshop** northeast of the ruins. Another grocery store can be found a mile farther west, along with a **pharmacy**, **bank** and restaurants. Marina employees also will happily direct or take visitors to local **craft shops** and **galleries**.

The Wiky marina does not yet have a boat lift or repair shop, though both are planned for the future. In the meantime, the port is slowly becoming a popular choice for cruisers, especially during the first weekend in August, when the colorful Wikwemikong Indian Days, which include a powwow, are held. Visitors are urged to book a spot at the marina in advance if planning to attend this event.

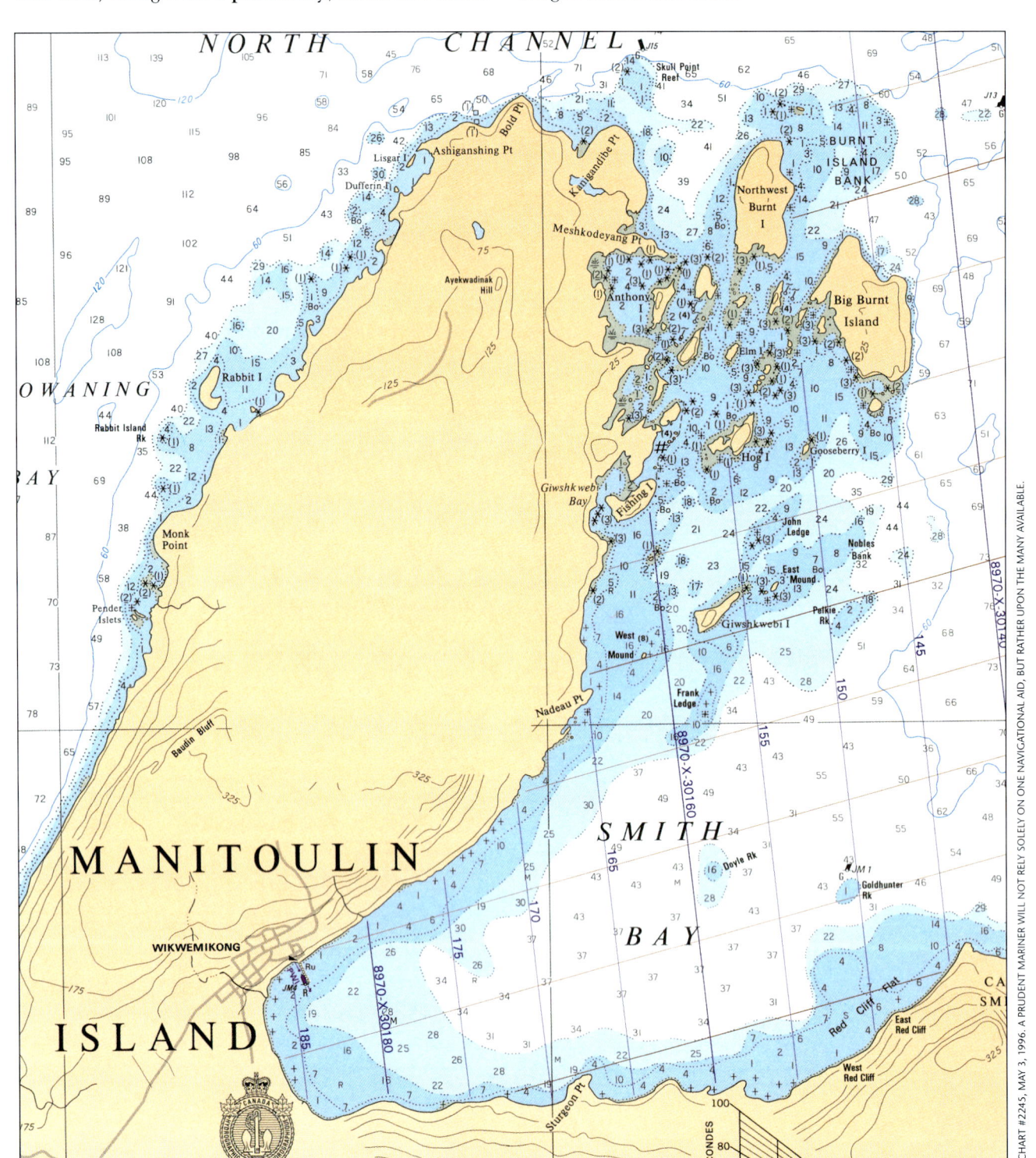

A TASTE OF NATIVE CULTURE   BY CAROLYN R. WELLS

Gathering Place
Proud traditions live on in Native American dances.

An eerie hush wafts in as an elder blesses a new drum with sweet grass. Hands rap its taut elkskin cover. Beyond the poled roof that shelters the drummers and singers, the rustle of a summer breeze harmonizes with the rush of the river that angles through the forest of birch and maple trees. The master of ceremonies announces the Grand Entry March, and the singer raises his voice in song.

Welcome to an Annual Traditional Gathering of North American Indians, in this case the Mississauga First Nation. The event—a midsummer feast of color and movement—has officially begun.

Pau-wau once meant "medicine man" or "spiritual leader," but today the term *powwow* is generally used to describe all festivities that help recall and revere the past. There are variations, but the Mississauga Traditional Gathering is representative of most powwows.

The Mississauga First Nation is part of the Three Fires, made up of the Ottawa, Potawatomi and Ojibway tribes, who moved westward from the banks of the St. Lawrence River to the forests surrounding the Great Lakes. Centuries ago, they were known as Anishinabek, or "People of the Forests," so named because their Algonquin tongue taught them to listen to the *si-si-gwa-d*—"the murmuring of the trees"—to remember the purity of man and nature and to keep the two in balance.

Each dance has a special meaning, whether signaling the rite of passage into adulthood, recognizing accomplishments and scholarship, honoring the dead, acknowledging mourners, celebrating the joy of living or giving thanks to the Creator.

The Jingle Dress Dance, the original Ojibway healing dance, is perhaps the most inspiring ritual of the afternoon. The jingle dress is sewn by the wearer, its design often inspired by a dream. The lids of snuff cans are shaped into cones and attached to the dress to make a tinkling sound that fills the air as each young dancer enters the circle.

"These women are very powerful," their mentor explains. "They have purified their bodies and received the power to aid healing."

After the dance, they form a receiving line to be congratulated by family and friends. Then the young ladies give berries, representing the first fruit, to the drummers in thanks for playing and to the spectators for their presence.

The men perform the Fancy Dance, symbolic of the hunt and scouting. A feathered headdress and bustle complete the costume. They stop dancing on the very last beat of the drum, bringing cheers from the crowd.

The Give Away Dance commemorates the loss of someone during the past year. The honor song is sung for the family as they dance, remembering their grief and love. Blankets, shawls and sometimes money taped to the twigs of tree branches are given away by the families. Relatives and friends of the departed may join the dance until the music stops.

The next several dances are more carefree. The Grass Dancers, with their stream of bright warm colors, move gracefully across the background of cool, green trees. The Crow-Hop is next, and then men and women come together, dancing the Two-Step. When candy is thrown, the rush of children's feet and their gleeful cries fill the sacred circle. The afternoon ceremony closes with the Sneak-Up Dancers giving thanks to the Creator.

JOHN DE VISSER

Killarney, Ontario

Lat 45°59.92
Lon 81°32.10

Lat 45°58.33
Lon 81°31.33

Lat 45°58.15
Lon 81°30.81

Spread along a thin, protected channel between George Island and the mainland, the community of Killarney is rich in sheltered docking opportunities. Approaching from the east, your first option for a berth will be at Killarney Mountain Lodge **A**. The next building to the west is Pitfield's General Store **B**, which provides space for fueling and stocking up, as well as for the occasional overnight guest. Herbert Fisheries **C** takes a limited number of transients on a first-come, first-served basis. Another tie-up option is George Islanders' Marina, **D**. Gateway Marine & Storage **E** always has a handful of slips set aside for transients and also owns the slips at the former Killarney Adventures Inn, now Pines Inn B&B **F**. The Sportsman's Inn **G** is a historic waterfront inn that offers still more docking opportunities on both sides of the channel. At the point, Roque's Marina **H** has up to 10 transient slips. The entrance to Covered Portage Cove **I**, a popular anchorage on Killarney Bay, is visible in the distance. Killarney Bay Inn **J** has air-conditioned rooms and a dining room overlooking the bay; those staying here can dock at Pitfield's.

Distances from Killarney (statute miles)
Little Current: 22 W
Tobermory: 46 SSW
Byng Inlet: 65 ESE

194 LAKELAND BOATING · LAKE HURON CRUISING GUIDE

The North Channel

Killarney, Ontario
A wilderness paradise with plenty of amenities.

Believe it or not, until relatively recently the mainland community of Killarney was only accessible by water—Highway 637 was finally cut through the northern woods to Killarney in 1962. Today the area is still wild and unspoiled, yet conveniently located for the transient cruiser. The peaceful strip of water on which this town is situated, Killarney Channel, acts as a kind of natural portal between the full expanse of Georgian Bay to the south and east and the lateral sprawl of the North Channel to the west.

First settled in 1820, Killarney—then known as Shebahonaning or "Narrow Channel"—has retained its focus on the water, with numerous facilities arranged neatly along the channel for the benefit of its transient guests.

Approaching from the east or south from Georgian Bay, stick to the charts, minding the numerous shoals that project from Philip Edward, Papoose and Scarecrow islands. Meanwhile, keep on the lookout for the Killarney East Lighthouse and radio tower on Red Rock Point. A light also marks the west entrance to the channel; as you approach from the west, leaving Lansdowne Channel to cross Killarney Bay, you will see it blinking at Le Haye Point.

Once in the channel, there are many choices for tying up. The historic **Sportsman's Inn**, located toward the western end of the channel, offers the most room, with 140 slips. Power and water are available at each slip, while gas, diesel, propane and pumpout are provided at the service dock. New showers and washrooms are also close at hand, as is the **Dock Shop**, which sells charts, boat supplies and scuba equipment. Fishing charters are also available. The inn itself offers rooms, a restaurant and a bar, all done up in a comfortable, rustic style. Adjoining the inn to the west is a new "VIP" dock for 100-plus-footers that offers satellite TV and all power options. Check out Sportsman's online at www.sportsmansinn.ca.

Killarney Mountain Lodge, at the other end of the channel, is an ambitious, sprawling resort with tennis courts, a heated outdoor pool, rooms and cabins, a dining room and a unique circular lounge that features live entertainment. Its docks, which range in depth from 4 to 35 feet, will handle up to 30 craft and are rigged with both 30- and 50-amp power. Gas, diesel and pumpout are all available; for showering, guests are welcome to use the lodge. Like the Sportsman's Inn, Killarney Mountain Lodge rents or charters fishing boats and maintains a marine store stocked with charts, rope and other basic supplies. Their Web address is www.killarneymountainlodge.com.

Between these two resorts lie a number of more modest options, none very far from the next. **Gateway Marine & Storage** will take up to 21 boats at its relatively new slips, seven of which are reserved for transients. Gateway has portable pumpout and supplies power to the docks. Other amenities include showers, washrooms, a laundromat and a deli-style take-out restaurant. Gateway can also lift boats up to 12 tons and 36 feet. Its marine store, located right at the docks, features a wide range of marine accessories.

Killarney Public Dock, which is managed by **Herbert Fisheries**, has space for about six craft. It's easily recognizable by the red-and-white-roofed fisheries building, which sits on the federal wharf. The docks themselves have good water depth but little else besides a nice launch ramp. Washrooms are available, but there are no showers. The real attraction of the government dock is its proximity to the renowned fishing business, which sells both fresh fillets

Where to Eat

Gateway Restaurant 705-287-2333
A full-service restaurant at the marina, open from 6:30 a.m. to 9 p.m.

Herbert Fisheries Fish Stand 705-287-2214
Formerly called Mr. Perch, the fish-and-chips stand located at the public dock offers delicious takeout.

Killarney Mountain Lodge 705-287-2242
The nightly menu of "comfort food" includes chicken, steak, grilled whitefish and a vegetarian entrée. Prime rib is served on Saturdays and roast turkey on Sundays. Don't miss the homemade breads and desserts.

The Sportsman's Inn. 705-287-2411
Triple-A steaks and North Atlantic seafood: oysters, barbequed whitefish and pickerel. Try the fresh Killarney blueberry pancakes and the eggs Benedict. The Georgian room features a view of the waterfront.

Sunset Grill. 705-287-2411
Enjoy burgers, steaks, pizza and fresh baked goods at this large outdoor patio on the waterfront at Sportsman's Inn.

More Info

Emergency 911
Killarney Health Centre 705-287-2300
Killarney Town Hall. 705-287-2424

and smoked fish at its on-the-water plant, as well as delicious fish and chips at its take-out stand, formerly known as **Mr. Perch**. It isn't actually perch anymore; since the early '90s, Herbert's has primarily pursued whitefish, considered by some to be just as tasty.

Finally, you might want to head across the channel to the **George Islanders' Marina**, a newish facility on the south side of the channel that accommodates as many as 20 boats up to about 80 feet long. At this popular spot for larger cruisers, power, water, pumpout, washrooms, showers and ice are all supplied. Because it's located on the other side of the channel, the marina has a quaint feel and a fabulous view. There are picnic tables, barbecues surrounded on all sides by trees and a garden to supply boaters with some fresh veggies.

In a pinch, you can try for a spot at **Pitfield's General Store**. Pitfield's docks are there primarily as a convenience for those who want to grocery shop without straying too far from the boat or who plan to stay at the nearby **Killarney Bay Inn** (800-265-9689). Its aisles feature a modest supply of fresh meat and produce, plus canned and dry goods. A laundromat is attached. For liquor or beer, you can pull right up to the government store next to Herbert's; it has its own dock. All of this makes Killarney a great place to stock up and attend to any mechanical quirks before striking out for

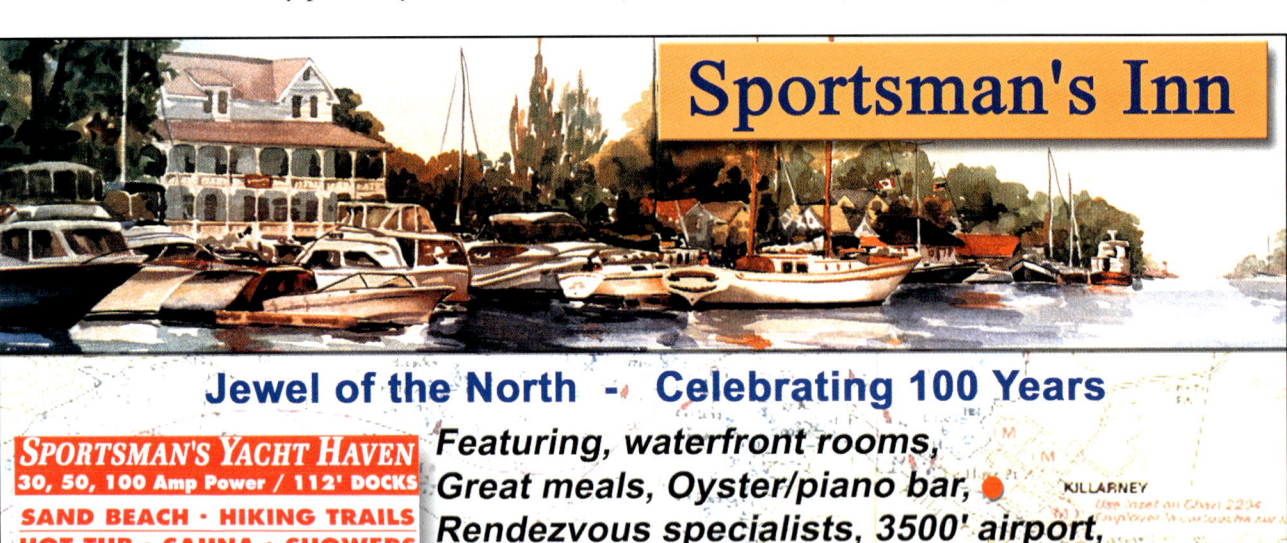

Sportsman's Inn

Jewel of the North - Celebrating 100 Years

SPORTSMAN'S YACHT HAVEN
30, 50, 100 Amp Power / 112' DOCKS
SAND BEACH · HIKING TRAILS
HOT TUB · SAUNA · SHOWERS
CABLE TV · WATER TAXI
Monitor VHF 68

Featuring, waterfront rooms,
Great meals, Oyster/piano bar,
Rendezvous specialists, 3500' airport,
Fresh meats and produce, Charts, Fuel
Reserved dockage & New Private Showers

HIGH SPEED INTERNET WIRELESS SERVICE

SUNSET GRILL PIER WEST

Visit the Sunset Grill on Pier West, Featuring Burgers,
Wraps, Steaks Plus take out Pizza & Bake Shop
Our New Outdoor Patio - On the Waterfront

West End, Killarney Harbour   www.sportsmansinn.ca   1-800-282-1913

Facilities information subject to change. We suggest you call ahead.

		Monitors VHF Channel	Transient Slips Available	ALTERNATE MOORING: Wall mooring / Rafting allowed	Maximum LOA	Minimum Depth at Dock	Power (amperage or volts)	HOOKUPS: Water / Cable TV	FUEL: Gas / Diesel / Pumpout	BASICS: Heads / Showers / Laundry	AMENITIES: Swimming pool / Whirlpool / Rec area / Grills / Picnic tables / Dog walk	Take Reservations	Take Credit Cards	Haulout (Capacity in tons or feet)	REPAIRS: Mechanical / Electronics / Fiberglass / Woodworking / Sails / Canvas	CONVENIENCE: Ship's store / Ice / Convenience foods & beverages
E	Gateway Marine & Storage 29 Channel St. 705-287-2333 Killarney, ON P0M 2A0	68	21	WR	55'	8'	30	W	P	HSL	PD	Y	Y	12		SIC
D	George Islanders' Marina Lot 22 705-287-2313 Killarney, ON P0M 2A0	68	20		78'	6'	15 30	W	P	HS	GPD	Y	N			I
C	Herbert Fisheries 21 Channel St. 705-287-2214 Killarney, ON P0M 2A0	68	6		100'	4'	200	W		H		N	N			IC
A	Killarney Mountain Lodge 3 Commissioner St. 800-461-1117 Killarney, ON P0M 2A0	68	30	WR	80'	4'	30 50	W	GDP	HS	SRGPD	Y	Y			IC
B	Pitfield's General Store 7 Channel St. 705-287-2653 Killarney, ON P0M 2A0	68		WR	125'	12'	60		GD	HSL	P	N	Y			SIC
H	Roque's Marina 65 Channel St. 705-287-9900 Killarney, ON P0M 2A0	68	10		40'	4'	30	W		HSL	PD	Y	Y		M	IC
G	The Sportsman's Inn 37 Channel St. 800-282-1913 Killarney, ON P0M 2A0	68	130	WR	175	18'	30, 50 100	WC	GDP	HSL	WRGPD	Y	Y			SIC

the North Channel.

But before you heave off, you might also want to spend a bit of time sampling the town's many charms. Aside from the dining opportunities afforded by Killarney's nicely maintained inns, the community also has a small **log cabin museum** and several **gift** and **craft shops**. And if you feel like walking off all the whitefish you've eaten, a hike to the **East Lighthouse** is highly recommended. A trail begins on Commissioner Street, just before the road entrance to Killarney Mountain Lodge. It's about a mile to the lighthouse, but it's a rewarding stroll. The lighthouse, which still has an operational foghorn, is perched among the pines on a pink granite outcropping. From here you can view the glittery expanse of Georgian Bay to the south or the long hump of Manitoulin Island to the west. If you want to explore the rugged, quartzite mountains to the north (the border of **Killarney Provincial Park**, a vast expanse of wilderness, is only about three miles away), ask around town for transportation—several opportunities are available.

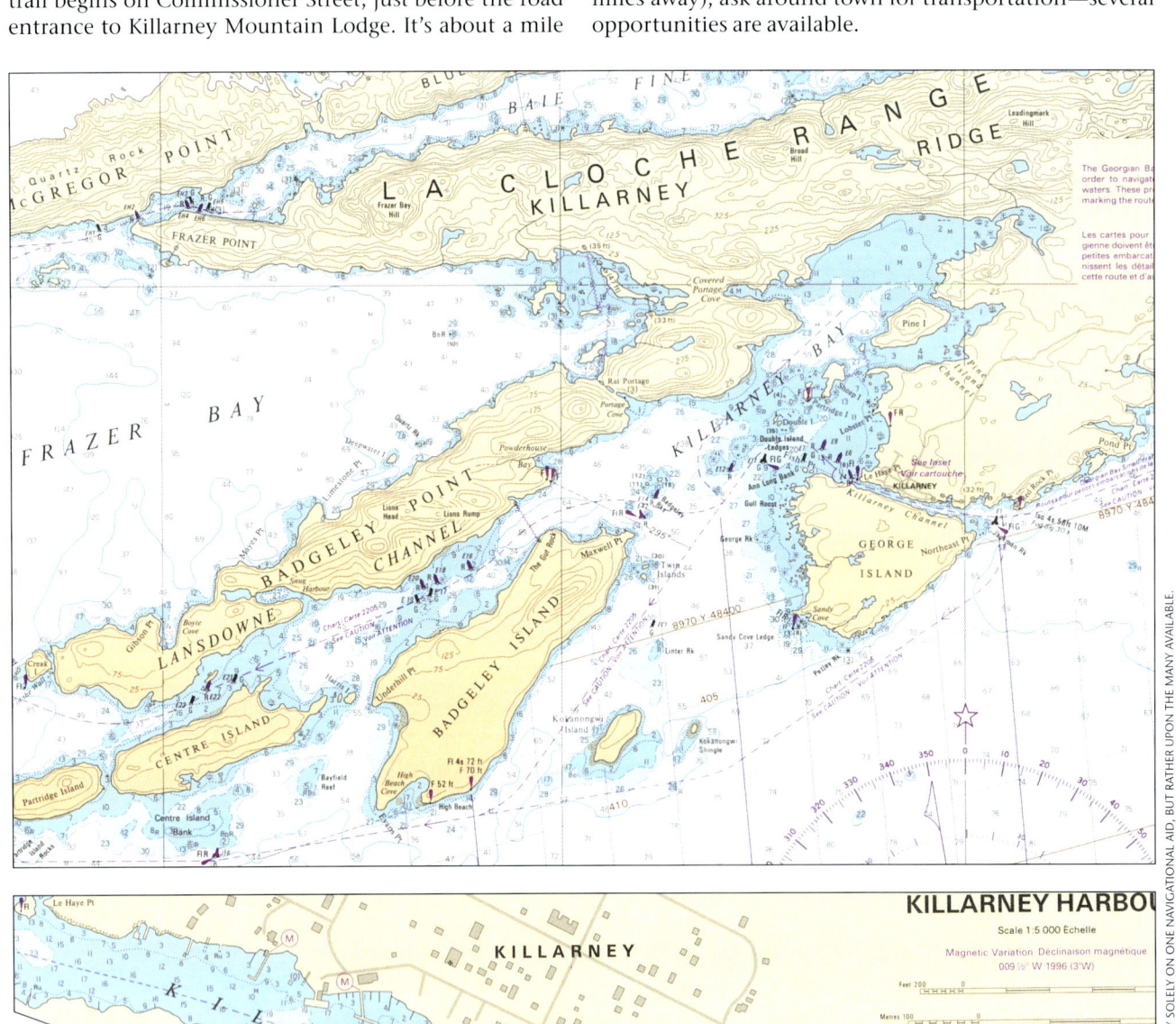

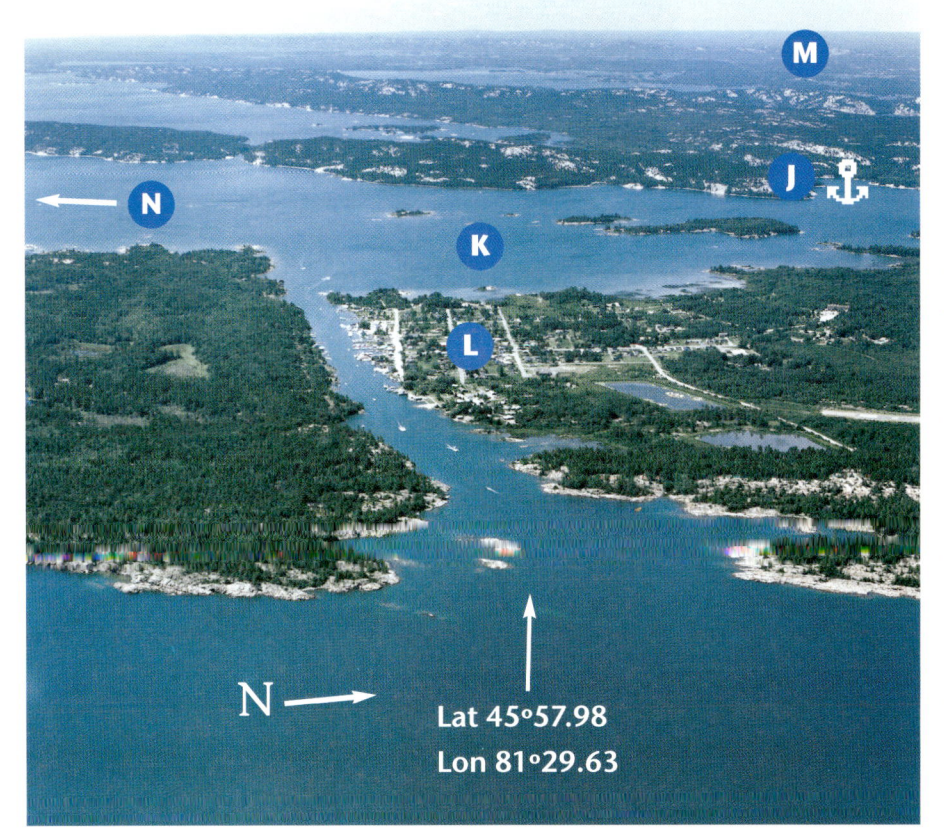

Brilliant white quartz mountains spotted with evergreens surround the turquoise waters of Killarney Bay **K**. There are at least five anchorages within three miles of the town of Killarney **L** and even more for adventurous types. One of the most popular anchorages in the whole of the North Channel and Georgian Bay is Covered Portage Cove **M** in the northwest corner of the bay (see below). Badgeley Point **N** defines the northern edge of the nine-mile-long Lansdowne Channel, the main thoroughfare between Killarney and Little Current.

Covered Portage Cove **O** is a spectacular and completely sheltered harbor on the west shore of Killarney Bay, slightly north and west of Sheep Island. There are sheer cliffs on the south side and plenty of hills to climb. As with everywhere else in these parts, when you get to the top, you will be rewarded with breathtaking vistas. Frazer Bay **P** extends to the west along the southern edge of Killarney Ridge **Q**.

Lansdowne Channel, Ontario

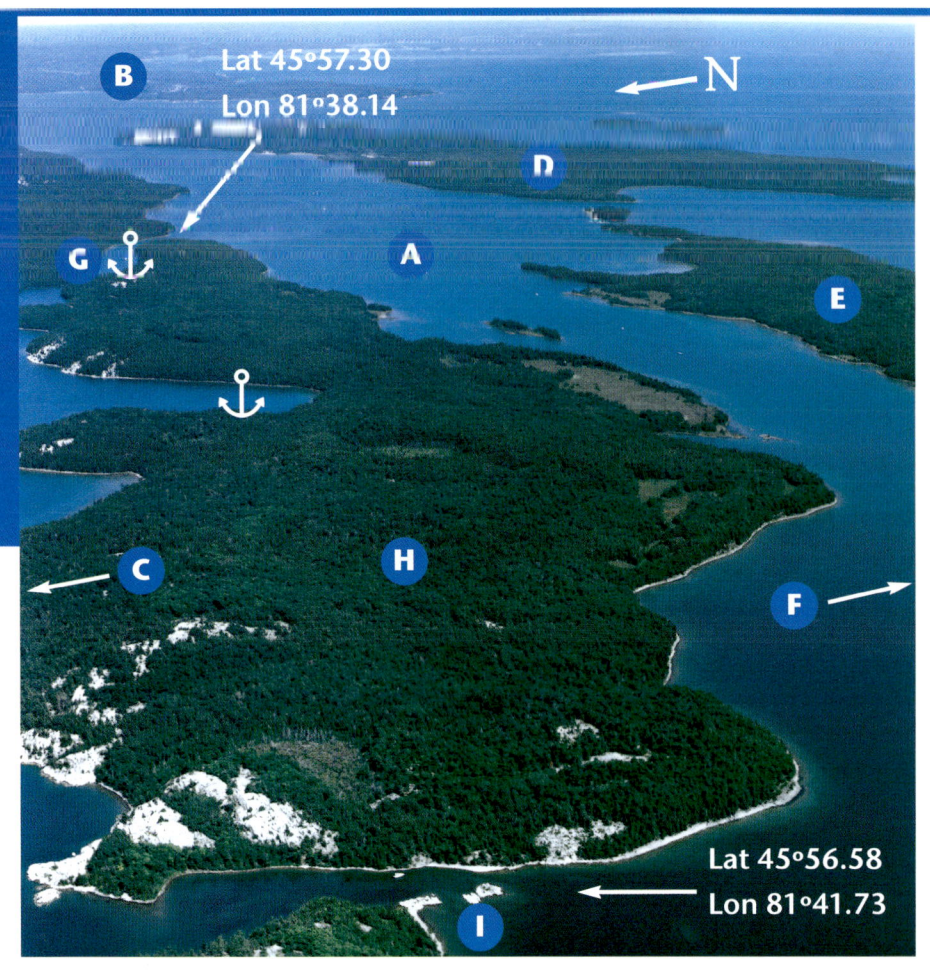

Lansdowne Channel **A** serves as a passage between Killarney **B** and Frazer Bay **C** to the west, sheltered by Badgeley **D**, Center **E** and Partridge **F** Islands. Snug Harbour **G** is a very secure anchorage on Badgeley Point **H**, a peninsula of the mainland that separates the Lansdowne from Frazer Bay. At the tip of Badgeley Point, you'll find Hole in the Wall **I**, a tight but picturesque short cut to Frazer Bay.

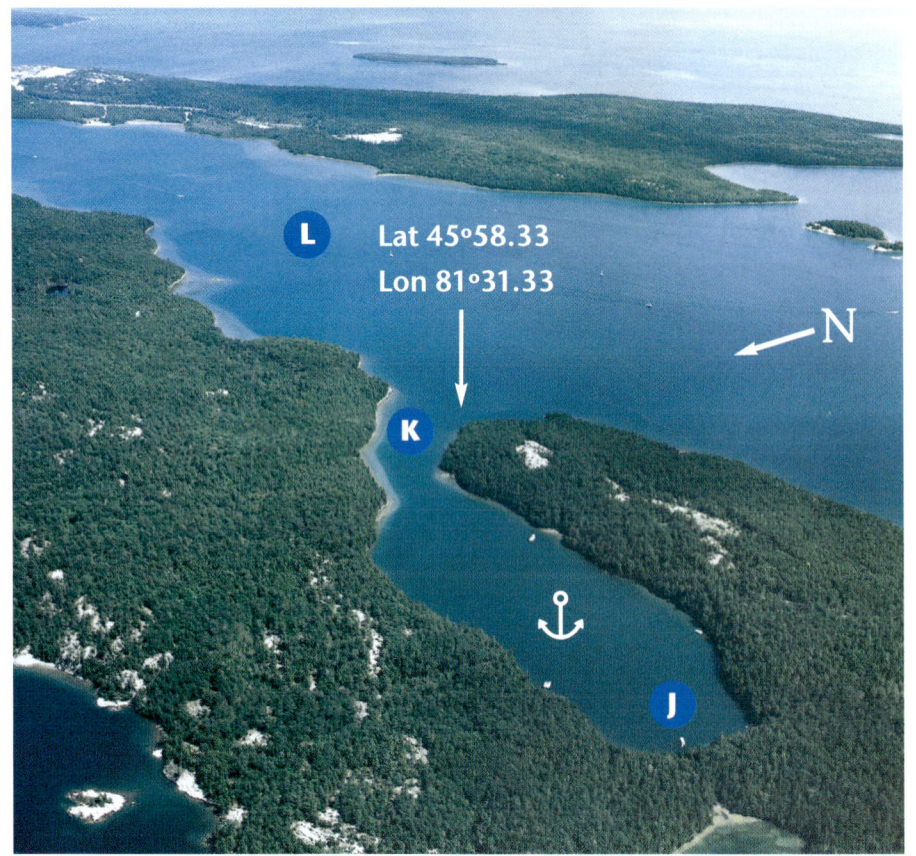

Snug Harbour **J** is so well-protected that a gale blowing outside it will hardly be noticed. Favor the western side of the entrance channel **K** to avoid shoals on the eastern side and you'll find deep water throughout the harbor. Lansdowne Channel **L** offers fairly wide-open cruising, with its shoals well marked.

LAKELAND BOATING · LAKE HURON CRUISING GUIDE

Baie Fine & McGregor Bay, Ontario

Lat 46°00.57
Lon 81°40.30

N

When entering Baie Fine, keep the red buoy **A** to starboard to avoid the shoal extending from Frazer Point **B**. There is no longer any transient dockage at Okeechobee Lodge **C**, which is now a private residence, but you can anchor across the bay. The marked course down the bay keeps to the southern shore until the first large island **D** before heading northeast to follow the northern shore the rest of the way in. Mary Ann Cove **E** on the southern shore about two miles in is a deservedly popular anchorage. Across the ridge of McGregor Point are the islands of McGregor Bay **F**.

Distances from Baie Fine (statute miles)
Killarney: 17 ESE
Little Current: 15 W

LAKELAND BOATING · LAKE HURON CRUISING GUIDE   **201**

THE NORTH CHANNEL

Baie Fine, Ontario
Don't pass this legendary anchorage by.

Generations of boaters have made Baie Fine their destination, drawn by its clear, protected waters, stunning quartz mountains and proximity to The Pool, the gorgeous anchorage at its far end. Pronounced "Bay Fin," the long cove and the gemlike Pool will enrapture anyone who cruises this way.

To get there, bear north across Frazer Bay from the western end of Lansdowne Channel toward McGregor Point. The southeastern end of Frazer Bay provides several beautiful, secluded anchorages, including one behind Blueberry Island and the horseshoe-shaped island to the north. Following the charted line easterly along the point's bluff, you'll clear the area's rocks and reach Frazer Point, the entrance to the 10-mile-long bay. As you enter from Frazer Bay, you'll see to starboard the fabled **Okeechobee Lodge**, known and loved by generations of boaters in good times and bad. The lodge was revived to a state of renewed glory by Gordon Blake, but after Gord's death in 1996, it was sold and is now a private residence—the beginning of a new era for one of cruising's top destinations.

Continuing east, there are numerous anchorages along the length of the bay. On the southern shore, two miles in, Mary Ann Cove is very lovely and is probably second in popularity only to The Pool at the far end. The Pool is one of those anchorages that cruisers talk about in reverential tones, sort of a yachtsmen's grail that everyone should seek out at least once in their cruising lives. The cottage at the entrance to The Pool is owned by Frances Langford, former USO singer and Hollywood star and widow of Ralph Evinrude. Her 110-foot yacht *Chanticleer* is a common sight at the dock, and cruisers are amazed that a ship that size can manage the tortuous route into The Pool.

Several trails lead from The Pool to picturesque lakes and offer unsurpassed views. Topaz Lake, about a half-mile hike uphill from the northeast corner of The Pool, is particularly splendid and well worth the climb.

Fishing is usually very productive in Baie Fine, and naturalists will enjoy scanning the shores for wildlife. Black bears are seen here, often with cubs, especially across from Okeechobee Lodge.

McGregor Bay

For the truly dedicated gunkholer, McGregor Bay, just

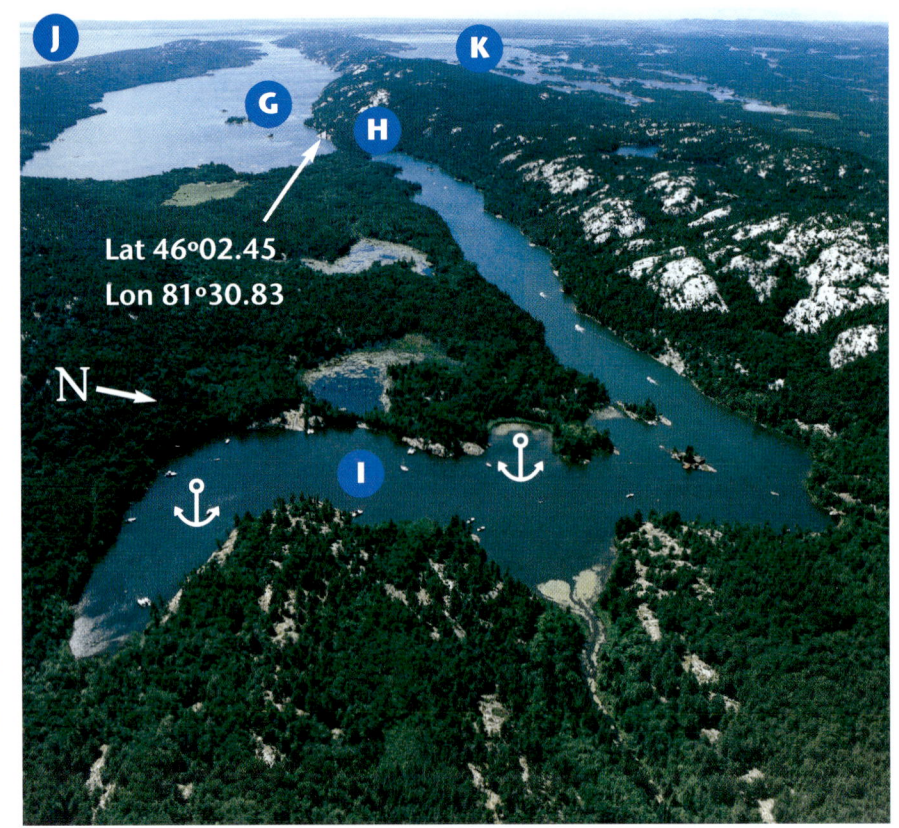

Lat 46°02.45
Lon 81°30.83

At the east end of Baie Fine you must keep between the Birthday Islands **G** and the northern shore before cautiously entering "the Narrows" **H** for the final approach to The Pool **I**. You can anchor just about anywhere. In the distance you can see Frazer Bay **J** and McGregor Bay **K**.

LAKELAND BOATING · LAKE HURON CRUISING GUIDE

Distances from McGregor Bay (statute miles)
Baie Fine: 4 S
Killarney: 14 SE
Little Current: 11 W

north of Baie Fine beyond McGregor Point, provides more than anyone could ask for. This bay's intricate channels and hundreds of islands make it perfect for shallow-draft, trailerable boats. The Great Lakes Cruising Club logbook will get you started toward exploring McGregor Bay's countless possibilities, and the new CHS chart 2206 makes cruising here much less scary than it used to be. Many cruisers with limited time bypass McGregor Bay and head to the vast cruising grounds to the west, which is truly a shame. Although not widely known, there is a boat passage from McGregor Bay to the Bay of Islands to the west that smaller craft use regularly to avoid the long haul to Little Current and around Great La Cloche Island. There is a nine-foot fixed bridge, but the Ontario Provincial Police boat, a 24-footer, uses the passage regularly.

More Info

Emergency	911
Manitoulin Health Centre	705-368-2300
Manitoulin Island Tourism	705-368-3021

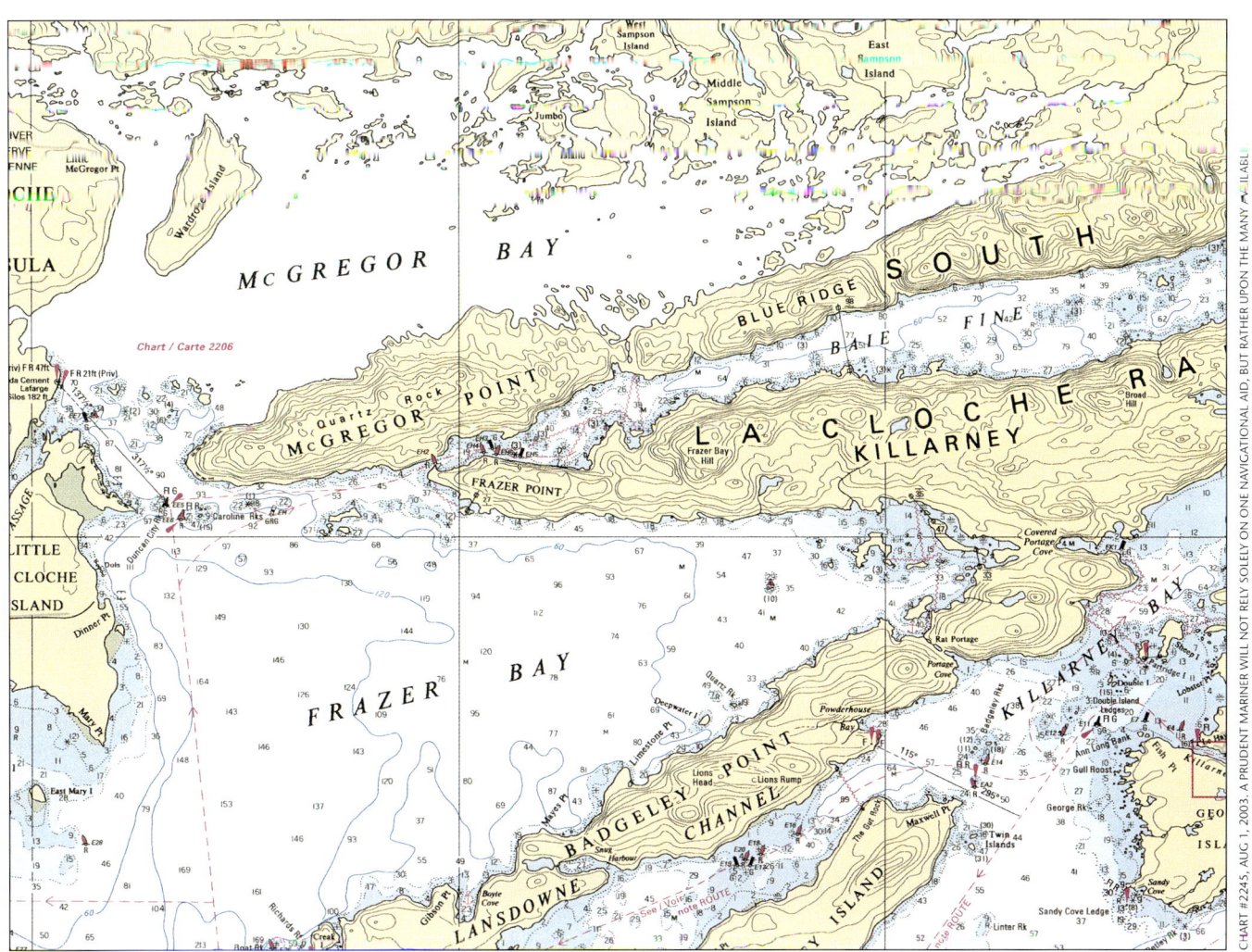

THE NORTH CHANNEL

Heywood Island, Ontario
Head to Browning Cove, one of the hidden gems of the North Channel.

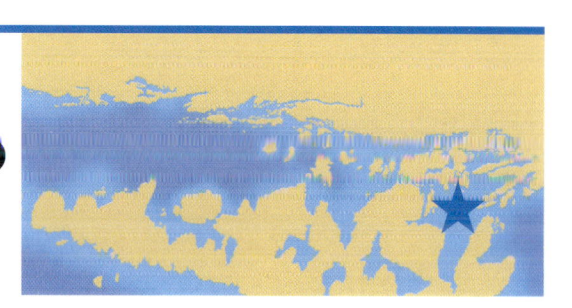

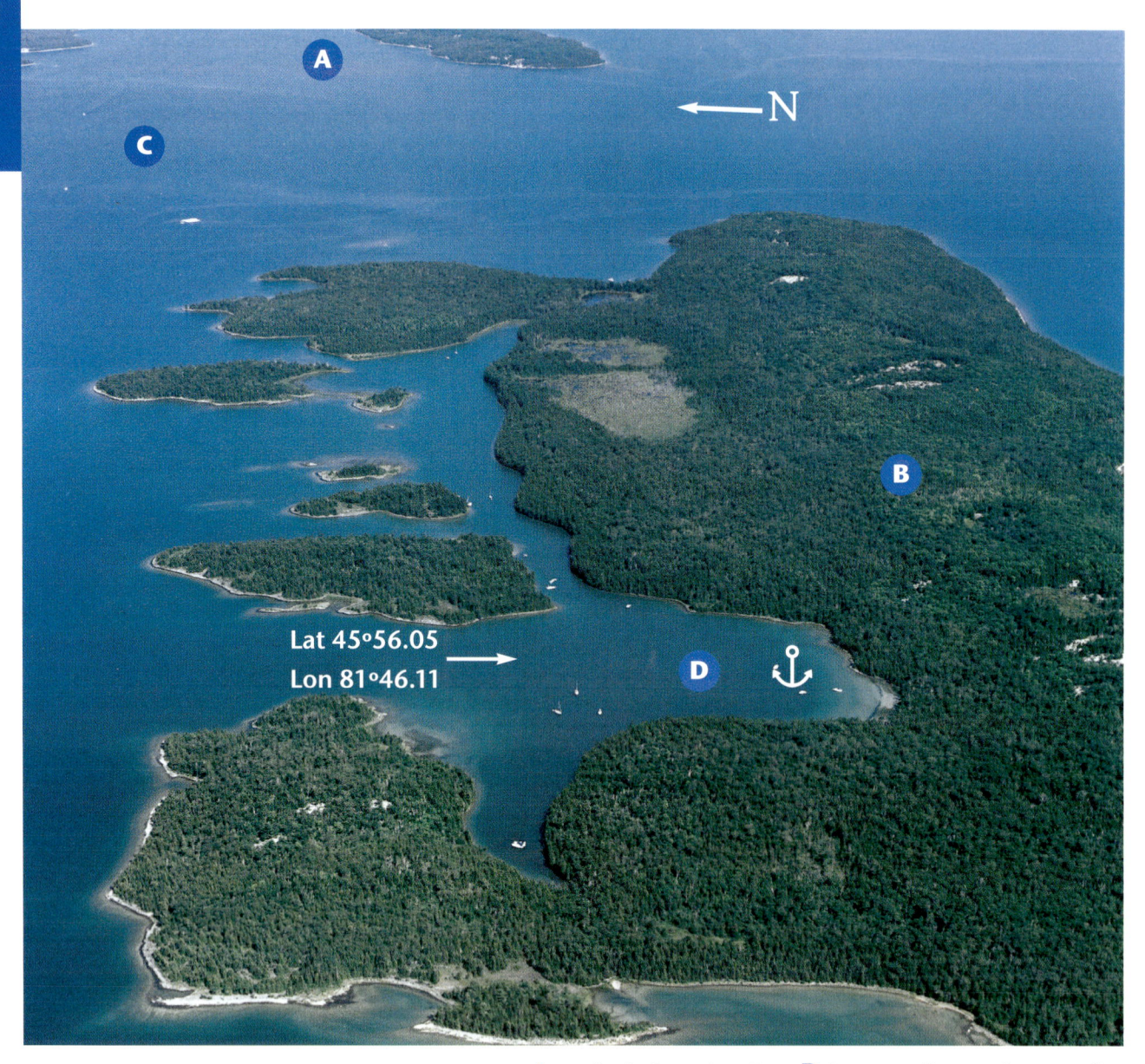

Lat 45°56.05
Lon 81°46.11

Continuing westward out of the Lansdowne Channel **A** toward Little Current, you pass Heywood Island **B** on your port side. Keep a sharp eye out for the black can **C** marking Split Rock and the more easily seen Heywood Rock. Browning Cove **D** is an excellent anchorage offering perfect protection and even a sand beach at low water. One of the little secrets of the North Channel, Browning Cove is wonderfully underused.

Distances from Heywood Island (statute miles)
Little Current: 8 W
Killarney: 16 E
Baie Fine: 7 NE

204 LAKELAND BOATING · LAKE HURON CRUISING GUIDE

THE NORTH CHANNEL

Bay of Islands & Whitefish Falls, Ontario
A delightful maze that's made for exploring.

Distances from the entrance to Bay of Islands (statute miles)
Little Current: 14 SW
Benjamin Islands: 21 W
Distances from Whitefish Falls
Little Current: 18 W and S
Benjamin Islands: 28 W

Red granite isles, thickly wooded with pine and birch, offer the intrepid boater innumerable anchorages in the Bay of Islands. But the shoaly, labyrinthine qualities that delight dedicated gunkholers and explorers also tend to scare off larger cruisers.

A buoyed channel runs the maze of the bay, and it's fairly easy to follow on a clear day with a pair of binoculars. The area is also well covered in the Great Lakes Cruising Club *Log Book*. At the eastern end of the bay—separated from McGregor Bay by the La Cloche Peninsula, which runs from Espanola to Little Current—lies the Whitefish River and the village of Whitefish Falls **A**. A **government dock** as well as several **resorts** operate here, providing ice, fuel, pumpout, water and heads for the transient boater. There is a launch ramp at the government dock. There is a small **general store** in town, along with a **tavern** that serves food. A mile or so up the river you'll find the falls that lend the town its name. The Bay of Islands has good fishing and the town is an excellent jumping-off point for trailerboaters, offering easy access to the Benjamins and McGregor Bay via La Cloche Channel.

About halfway along the buoyed channel you'll find **Island Lodge B** (800-461-1119), with dockage and a fine restaurant. Just before you swing north toward Island Lodge on your way to Whitefish Falls, a large bay opens to the south. This is Jumbo Bay; at its eastern end lies the La Cloche Channel, which leads to the boat passage into McGregor Bay.

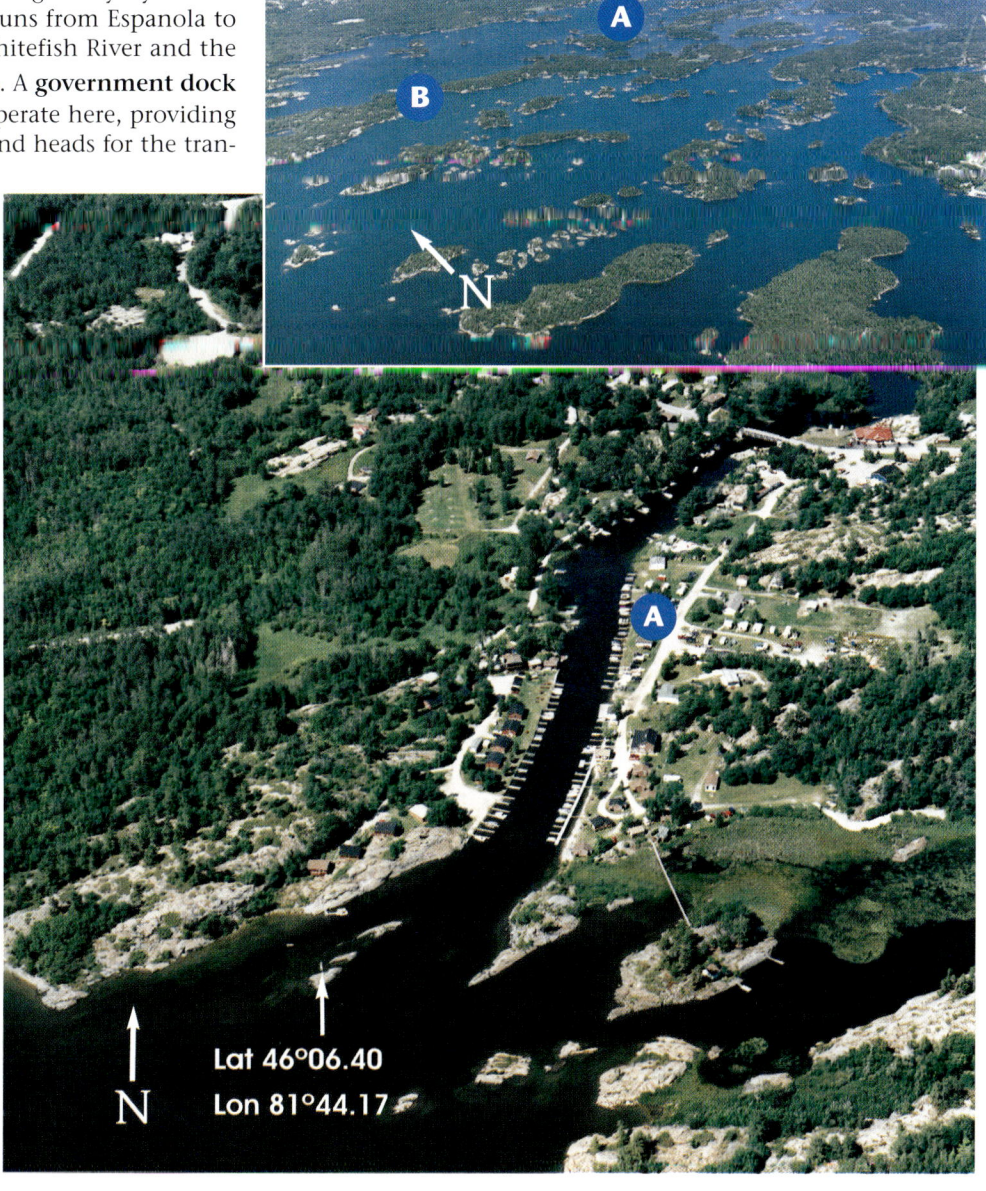

Lat 46°06.40
Lon 81°44.17

LAKELAND BOATING · LAKE HURON CRUISING GUIDE   **205**

Benjamin Islands Group, Ontario

Lat 46°05.15
Lon 82°15.06

Lat 46°05.02
Lon 82°15.53

The harbor between North Benjamin **A** and South Benjamin **B** islands is one of the North Channel's most popular, with good holding and tie-up spots in the bays of both islands. One entry is through the cut between the islands to the west of the harbor. Avoid Booth Rocks **C** and set a heading parallel to and passing the large rock on the south side of the cut **D** before making a slow turn into the harbor. The other entrance is between the Sow and Pigs Rocks **E**, just out of the photo, and Secretary Island over near Croker Island. Another anchorage on South Benjamin's southwest side **F** is easy to find, as are rocky outcroppings **G** to tie up to. Fox **H** and Eagle **I** islands are easy to see in the background.

Distances from Benjamin Islands (statute miles)
Little Current: 18 SE
Spanish: 14 NW
Gore Bay: 17 SW

206　LAKELAND BOATING · LAKE HURON CRUISING GUIDE

THE NORTH CHANNEL

Benjamin Islands Group, Ontario

Island-hopping, Canadian style.

For the modern-day explorer, arriving in the Benjamin Islands is like hitting the mother lode. Carved of beautiful pink granite, the Benjamins and the surrounding islands offer excellent anchorages within their bays and outcroppings and reward even short hikes with breathtaking vistas.

The ring of islands is situated 18 miles west-northwest of Little Current and 17 miles north-northeast of Gore Bay. Neighboring isles—Croker and Fox in particular—add to the charms of the Benjamins.

On a chart, the circular nature of the Benjamin Islands Group suggests a bull's-eye. Millions and millions of years ago, these islands were part of a bubble of molten rock that rose to the earth's surface, resulting in their concentric pattern. The glaciers then had their go at the area, carving the islands into their current shapes and generally leaving their footprints everywhere. A warning: one result of this igneous bubble is a strange magnetic anomaly that may affect your compass, requiring extra-diligent reckoning.

There are really only two recommended approaches into Benjamin Harbour, the main anchorage in the island group, located between North and South Benjamin. As you approach from the north and west, carefully avoiding Booth Rocks, look for the narrow cut between the two Benjamins. The clear channel is parallel to the large rock on the southern side of the cut. It's tight, and you should definitely post a bow watch, but there is 20 feet of water mid-cut and some really large yachts come this way. The easier approach, however, is from the south, staying east of the Sow and Pigs group of rocks and west of Secretary Island. There is pretty deep water here, though you'll want to watch for an awash rock just southwest of Secretary. Also, be sure you don't wander amongst the little "Piggies" and stub your bow. Study your chart closely.

North and South Benjamin Islands

The most popular anchorage on the North Channel is probably the natural harbor between the two Benjamins. During prime season, the area can hold as many as 50 boats. Though everyone should visit this beautiful spot at least once, many opt to avoid the crowds and head for alternate anchorages on subsequent visits. Most North Channel cruisers are fairly considerate and the multitudes only become a problem in northeast winds or when vying for favorite tie-up spots.

There are some really neat glacial grooves on both islands, and blueberries abound. The views from the high points are panoramic, particularly on North Benjamin. South Benjamin has a couple of nice spots on its southern shore, including a large bay that is protected in all but southerly winds, two sand beaches and rocky outcroppings that gather spiderweb collections of rafting boats. Go slow on approaches.

ROD MACLEOD

Benjamin Islands Group, Ontario

LAKELAND BOATING · LAKE HURON CRUISING GUIDE   **207**

Croker Island

Croker Island, defining the southeastern edge of the group, has a nice, large harbor, lots of impressive pink granite and a great selection of sand beaches. The approach is relatively easy, as long as you respect the Sow and Pigs group of rocks and the shoaly areas of Secretary Island as you split them before slowly turning into the harbor on Croker's southwest side. Protection is best to the south of the small island in the harbor, but a north wind can still produce an uncomfortable surge.

Fox Island

This island at the northern edge of the group is lower and flatter than the rest—which also means easy hiking—with several marshy lakes and great bird-watching. The main harbor is on Fox's southwest corner and requires an off-angle approach from the northwest before turning northeast into the deep, well-protected basin. An alternate anchorage on the island's east end, Gibson Cove, is easily accessed from McBean Channel.

McBean Channel

Directly north of the Benjamin Islands Group, above Fox Island, lies the McBean Channel, which runs 17 miles west to Little Detroit, the gateway to Whalesback Channel. McBean's most valuable secret lies above Hotham Island, through a barely noticeable opening on the north side of the channel between Hotham Island and Oak Point. Oak Bay is a whole cruising world of its own, with numerous well-protected anchorages, both on the mainland and in the bays on the north side of the island. McBean Harbour, found behind Beaudery Point at the channel's eastern end, is cottage-heavy, but is near a secluded anchorage north of Anchor Island. Mount McBean, at 620 feet, is not the highest of the La Cloche Mountains, which begin here, but it is one of the most readily identifiable hilltops on the North Channel.

More Info

Emergency 911

Mindemoya Hospital 705-377-5311

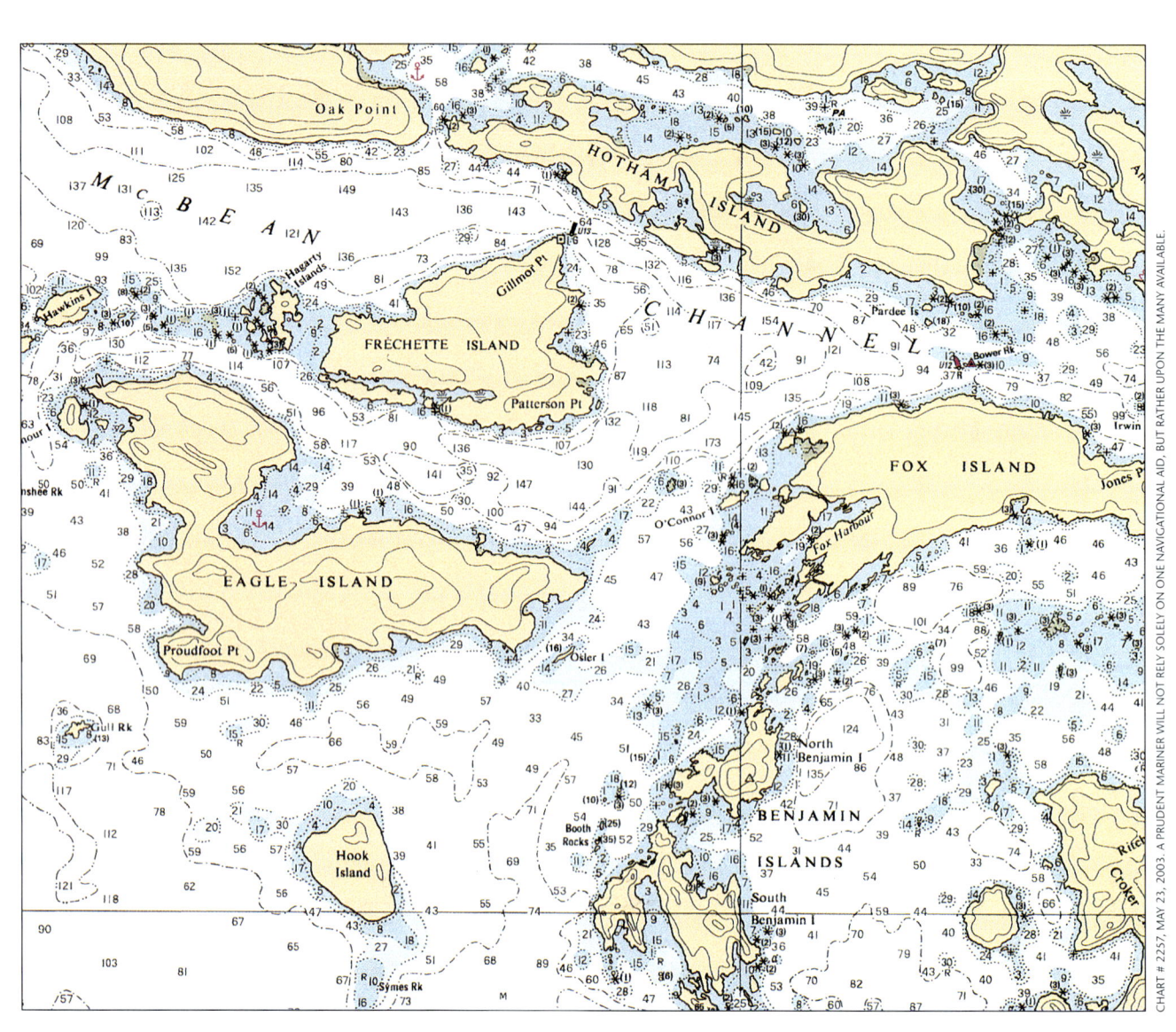

The approach into Croker Island's main harbor **J** stays east of Secretary Island **K** just outside the picture, before swinging slowly into the bay between Porcupine **L** and Croker **M** islands. The best anchorage is just to the south of the little island in the harbor.

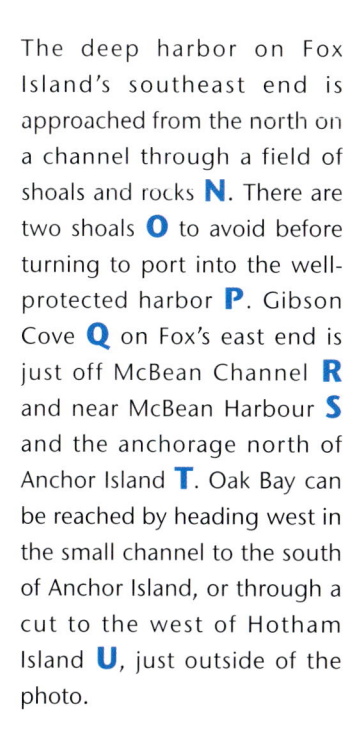

The deep harbor on Fox Island's southeast end is approached from the north on a channel through a field of shoals and rocks **N**. There are two shoals **O** to avoid before turning to port into the well-protected harbor **P**. Gibson Cove **Q** on Fox's east end is just off McBean Channel **R** and near McBean Harbour **S** and the anchorage north of Anchor Island **T**. Oak Bay can be reached by heading west in the small channel to the south of Anchor Island, or through a cut to the west of Hotham Island **U**, just outside of the photo.

LAKELAND BOATING · LAKE HURON CRUISING GUIDE **209**

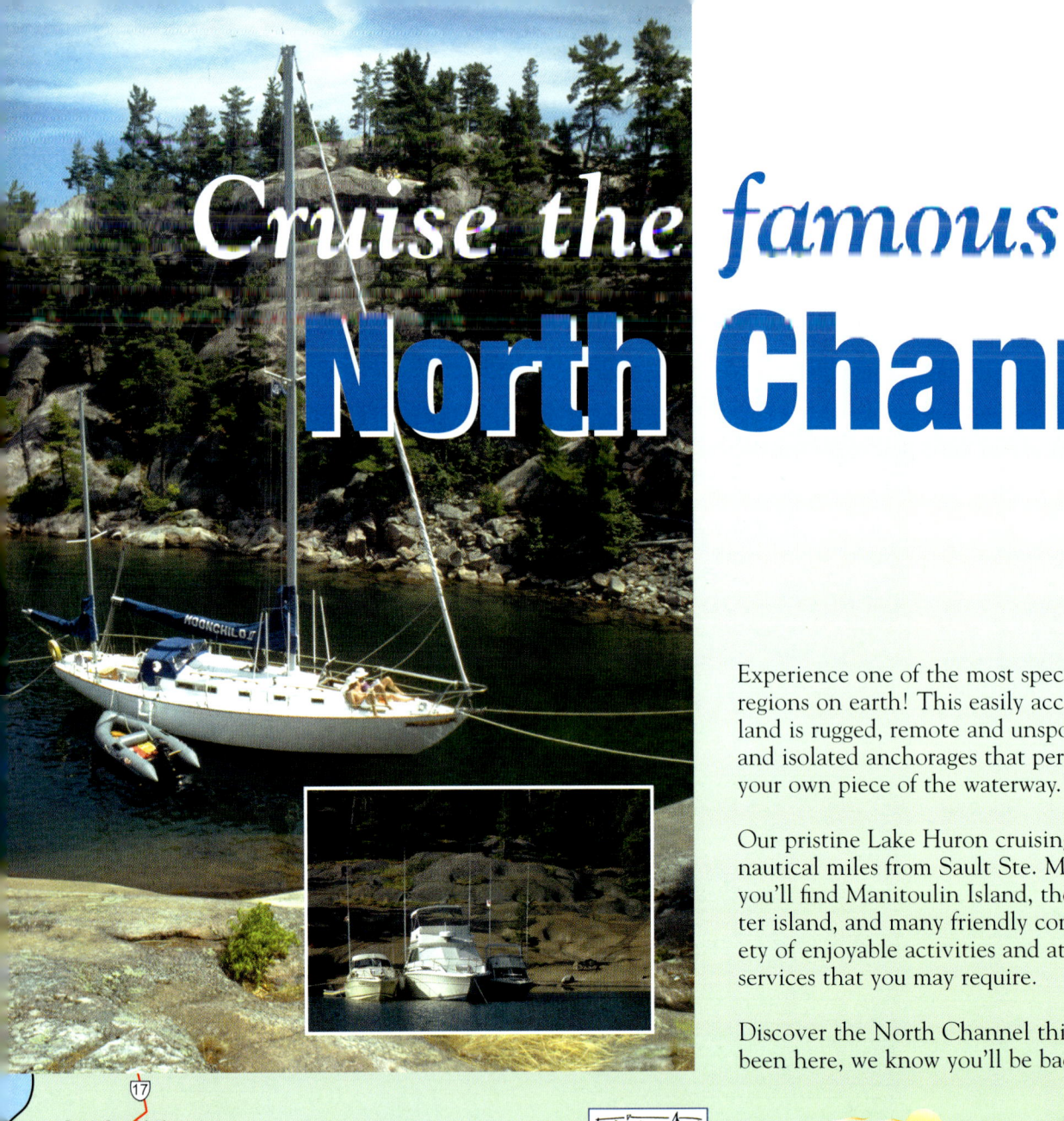

Cruise the *famous* North Channel

Experience one of the most spectacular freshwater sailing regions on earth! This easily accessible boating wonderland is rugged, remote and unspoiled with many hidden and isolated anchorages that permit private enjoyment of your own piece of the waterway.

Our pristine Lake Huron cruising waters stretch some 160 nautical miles from Sault Ste. Marie to Killarney. Here you'll find Manitoulin Island, the world's largest freshwater island, and many friendly communities offering a variety of enjoyable activities and attractions plus most of the services that you may require.

Discover the North Channel this season — once you've been here, we know you'll be back.

MANITOULIN–LaCLOCHE
Where Spirits Come to Play
www.manitoulinlacloche.com

Huron North
Legendary Waterways. Breathtaking Wilderness.
1-866-222-2261
www.huronnorth.on.ca

Canadian marinas monitor Channel 68

Ontario Northern Ontario Heritage Fund | Canada | FedNor

Blind River
Be sure not to miss this port! There are numerous restaurants and retail outlets for your enjoyment. Play a round of golf at Huron Pines Golf and Country Club. Visit Timber Village Museum and see the Northern Ontario Loggers Memorial, a huge bronzed and copper statue depicting river drivers working on a log jam.

Bruce Mines
Bruce Mines is a charming community rich in history. A visit to the Simpson Mine Shaft and the Bruce Mines Museum are "must do" activities. Don't forget to browse the local shops and dine at one of our friendly establishments.

Desbarats
The beauty and peacefulness of Desbarats has been attracting tourists from all over the United States and Canada since the late 1800's. The fishing is always exciting and local businesses are ready to provide warm, friendly service.

Drummond Island, Michigan USA
Drummond Island has a championship golf course "The Rock" and a small municipal golf course as well. There are several great restaurants, including the well known gourmet style "Bayside". If nature is of interest there are miles of trail and roads for biking or walking.

Gore Bay
A tree-lined downtown is home to a wide range of shopping, eating and banking establishments. Gore Bay's waterfront has proven popular with visitors to the Manitoulin LaCloche region, boasting a scenic boardwalk, tennis courts and marina facilities. The Gore Bay Summer Theatre and Gore Bay Museum are popular attractions.

Kagawong
Famous for its cruising waters, in the centre of Manitoulin Island on the North Channel, Kagawong borders on three of the largest lakes of Manitoulin Island – Lakes Kagawong, Mindemoya and Manitou. Visit Bridal Veil Falls, take a historic building walking tour and visit many quaint shops, some unique to the area. All amenities are convenient.

Killarney
Enjoy the beautiful scenery of Killarney while strolling along our quaint waterside streets, exploring our hiking trails, or having a picnic at the lighthouse overlooking the waters. Boat tours, canoe and kayak rentals are available. There are three direct access campsites and a medical centre. Visitors can relax and enjoy our famous fish & chips.

Little Current
Located at the gateway to Manitoulin Island, Little Current is the largest community on the island. All traffic to the island crosses the historic CPR swing bridge, which turns to allow passage of boat traffic sailing the North Channel's blue waters. The Town of Little Current features motels, restaurants, several bed & breakfasts, grocery stores, and provides supplies, fuel and repairs to boaters.

Manitowaning
Awaken your spirit on Highway 6, only twenty minutes from the ferry. Here, you will be hosted by a full-service community with a wide range of accommodation and dining experiences. There is a fully equipped marina and municipal airport for your convenience. The Norisle, one of Canada's few remaining steamships, is only steps away from the sandy public beach.

Meldrum Bay
Meldrum Bay, home of Canada's largest marine quarry, provides a look-out area to view the quarry operation. The community of Meldrum Bay offers a general store, accommodation, fishing charters and diving opportunities. The Net Shed Museum, located in a building once used by fisherman to repair and store their nets, is a tribute to the area's marine heritage.

Sault Ste. Marie
While in Sault Ste. Marie visitors can enjoy an excursion to scenic Agawa Canyon aboard the world famous Tour Train. Visit one of the many nearby attractions, including the Canadian Bushplane Heritage Centre, Ermatinger Clergue Heritage Site, Sault Canal National Historic Site of Canada, Museum Ship Norgoma, Sault Ste. Marie Museum, Art Gallery of Algoma and the Great Lakes Forest Research Centre. As well enjoy great shopping opportunities or visit the Sault Ste. Marie Charity Casino.

Spanish
Spanish is a small community with a big heart, offering a myriad of services including motels, restaurants and various entertainment. Visit the Spanish & Area Farmers Market open every Saturday from May until September. The many local businesses in the downtown core can provide all of the amenities needed.

Spragge
A cosy marina location for those seeking solitude. Nearby is the Wagoosh Hiking Trail that offers a view of the Murray Fault; of historical significance to the region as the primary source of all mineral extraction and industrial mining activity.

St. Joseph Island
When visiting either marina on St. Joseph Island be sure to bring along a bike so that you can explore all that the Island has to offer. Visit the Fort St. Joseph National Historic Site or play a round of golf at Island Springs Golf Resort. Take a pleasant stroll while exploring Adcocks' Woodland Garden. Visit the House of History and the St. Joseph Island Museum Complex to gather insight into the lives of the early settlers.

Thessalon
When strolling down Thessalon's main street be sure to stop in at Forestland, a beautiful gift shop you're sure to enjoy. Livingstone Creek Golf Course is located nearby and offers a challenge to golfers of all skill levels. Just north of the community is the Little Rapids Pioneer Museum which offers a glimpse into the history of the area.

Marina Information

Marina	Contact
Blind River	
Blind River Marine Park	(705) 356-7026 / brmarina@vianet.ca
Bruce Mines	
Bruce Mines Marina	(705) 785-3201 / brmines@algoma.net
Desbarats	
Holder Marine	(705) 782-6251
Drummond Island	
Drummond Island Yacht Haven Inc.	(906) 493-5232 / yachthaven@lighthouse.net
Gore Bay	
P. Purvis Marine Storage	(705) 282-2415 / purvisgb@vianet.on.ca
Canadian Yacht Charters	(705) 282-0185 / www.cycnorth.com
Gore Bay Marine	(705) 282-2906 / www.manitoulin.com/gorebay
Kagawong	
Kagawong Basin	(705) 282-2611
Northern Marina / Maple Ridge Yacht Charters	(705) 282-3330 / www.sailingcharterscanada.com
Killarney	
Killarney Mountain Lodge	(705) 287-2242
Killarney Public Dock	(705) 287-2214 / www.municipality.killarney.on.ca
Little Current	
Little Wally's Dock Service	(705) 368-2370
Harbor Vue Marina	(705) 368-3812 / harborvue@etown.net
Boyle's Marine	(705) 368-2239 / boylemarine@etown.net
Spider Bay Marina	(705) 386-3148
Manitowaning	
Manitowaning Marina	(705) 859-3700 / asigmtg@amtelecom.net
Manitowaning Lodge, Golf & Tennis Resort	(705) 859-3136 / www.manitoulin-island.com/manitowaninglodge
Meldrum Bay	
Meldrum Bay Marina	(705) 283-3252
Sault Ste. Marie	
Bellevue Marina	(705) 759-2838 / ssm.gardens@cityssm.on.ca
Roberta Bondar Marina	(705) 759-5430 / ssm.gardens@cityssm.on.ca
North Huron Charters Ltd.	(705) 253-9346 / sailing@soonet.ca
Spanish	
Municipal Marina	(705) 844-1077 / municipal@town.spanish.on.ca
Vance's Resort	(705) 844-2565 / www.ontariowalleye.ca
Vance's Marina	(705) 844-2442
Spragge	
North Channel Yacht Club	(705) 849-7446 / www.ncyachtclub.org
St. Joseph Island	
Hilton Beach Marina	(705) 246-2291 / www.hiltonbeach.com
Richards Landing Municipal Marina	(705) 246-0254 / www.stjosephisland.net/rl_marina
Thessalon	
Thessalon Municipal Marina	(705) 842-5188

Services available: Fuel, Repairs, Launch Ramp, Pumpout, Storage, Food, Restrooms, Showers, Laundromat, Ice, Hydro, Water, Accommodations

Spanish, Ontario

Located at the mouth of the Spanish River, a few miles from the intersection of Whalesback and McBean channels, Spanish serves as a natural pit stop for North Channel island-hoppers. The new Municipal Marina & Four Seasons Waterfront Complex **A** is located at the river mouth **B** and is recognizable by its white lighthouse replica, which used to be situated further upstream at the original Spanish government dock **C**. The new municipal marina is on the north bank of the river, just west of the old dock. Mitchells' Camp **D**, next to the old dock, has limited transient space. Vance's Marine, Ltd. & Resort **E** has a service dock and offers transient space at its nearby marina **F**. Whalesback Channel **G** extends west, separated from McBean Channel **H** by the Little Detroit narrows **I**.

Distances from Spanish (statute miles)
Little Current: 31 SSE
Gore Bay: 22 SSW
Meldrum Bay: 42 SW
Hilton Beach: 79 W

THE NORTH CHANNEL

Spanish, Ontario
Stock up and explore before setting off for some spectacular cruising.

How the town of Spanish gained its name may be an enigma—locals advance a number of theories, but the fact remains that no one here is of Spanish descent. Its popularity among boaters is no mystery, however. A convenient springboard to two popular anchoring spots, Spanish lies a mere 15 miles from the beautiful Benjamin Islands Group and closer still to the island-speckled waters of Whalesback and McBean channels. But the port is not just a place to simply fuel up and run. Visitors can enjoy the amenities of the downtown area, explore the rugged shoreline overlooking the river delta or head up the river itself, famous for its pickerel, pike and bass fishing.

On the way to Spanish from McBean Channel, prudent cruisers warn others of their approach before squeezing through Little Detroit Passage, a deep but narrow dogleg that links McBean to Whalesback Channel. A few miles north of the passage is the entrance to the Spanish River. When approaching from the west, respect the rocky tip of Gervase Island, keeping it to starboard, and stay between the buoys. As you enter the river mouth, be aware that the water becomes quite shallow; the soft silt bottom usually has a depth of 6 feet. The Spanish River delta is a habitat abundant with birds, fish and other wildlife.

The new **Municipal Marina & Four Seasons Complex** is located at the mouth of the river on the north side. A large wind turbine that supplies power to the facility stands on the bluff above the marina. Eighty percent of the 127 slips are designated for transient boaters. Gas, diesel, oil, pumpouts, maps, charts and sundries are available. All slips are serviced with water and shore power. End docks can accommodate vessels up to 65 feet. The large complimentary parking lot with 24-hour security, launch ramp and mast-stepper make Spanish an ideal place for trailer sailors. The Four Seasons Complex houses a laundry room, showers, a rec room, a sauna, a workout room, a tuck shop and a conference room. The building of a breakwall and dredging of the lagoon began in 1996. The two-story marina building echoes the lines of a boat. Directly behind the marina are two residential schools that were built in 1912. Here, in these two massive structures, native children were schooled until the early 1960s. Now boarded up, they are still worth checking out.

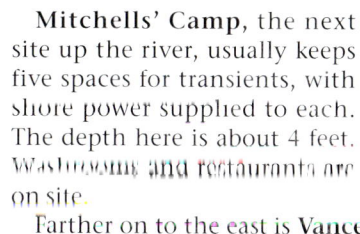

Mitchells' Camp, the next site up the river, usually keeps five spaces for transients, with shore power supplied to each. The depth here is about 4 feet. Washrooms and restaurants are on site.

Farther on to the east is **Vance's Resort**, which provides pumpout, bait, ice, fishing licenses, diesel, gasoline and oil at its service dock. A sauna, washrooms, showers and a screened-in waterfront gazebo with a barbecue are available for use. **Vance's Marine, Ltd.**, adjacent to the resort, has showers and washrooms and space for about 15 transient boats, with 20 amp power set aside for slips in the lagoon. Ice, bait and sundries are for sale at the marina office. The draft at the guest docks is roughly 6 feet. Vance's now offers haulout service for boats up to 35 feet long. Mechanics for both gas and diesel engines are available in town.

A half mile trek from Mitchells' and Vance's (and a mile from the new marina) are John and Front streets, where you'll find boat and auto repair at **Smith's Sales and Service**, along with a **bank**, a **post office** and a Saturday **farmers market**. While at the market, see Old Mustard Face (his real name is Jim) for a homemade sausage on a bun with fried onions and sauerkraut. Medical, dental and chiropractic services are also available in this neighborhood. Near the marina are a **bed & breakfast** and a **baseball diamond**.

The third weekend in August is the Spanish Fun Days celebration, with ballgames, a horseshoe tournament, food concessions, children's games and a dance. Lace up your hiking boots and hit the newly developed **Shoreline Discovery Trail**, which runs along the shore west of the municipal marina. The steep but short climb up to Baldy's Peak rewards hikers with a view of the reedy delta, which has been designated as a significant wetland, along with a portion of the Whalesback Channel to the west. Blueberries grow on Baldy's slopes. Another option is the rollerblading rink just a short walk from the marina. Finally, you may want to toss a line in the river. The best walleye fishing is in early summer. Bait is available at both the Municipal Marina and Vance's Marine.

After your sojourn in Spanish, fill your tank and take a deep breath. Some of the most spectacular cruising grounds in the world lie ahead.

LAKELAND BOATING · LAKE HURON CRUISING GUIDE

More Info

Emergency	911
St. Joseph's General Hospital	705-848-7181
Medical Clinic	705-844-2263
Dental Clinic	705-844-2886
Spanish Information Centre	800-844-1512
North Channel Marine Tourism Council	705-844-2522

Where to Eat

Lucky Snack Bar 705-844-2573
Takeout fries and burgers to eat at picnic tables on the rocks. Top off your meal with soft-serve ice cream.

Picnic Basket 705-844-2150
Home cooking and big breakfasts. Try the famous country-style chicken dinner. Call for delivery.

Vance's Motor Inn 705-844-2000
A family restaurant serving everything from burgers to seafood. The house specialty is the fresh pickerel. Open for breakfast, lunch and dinner.

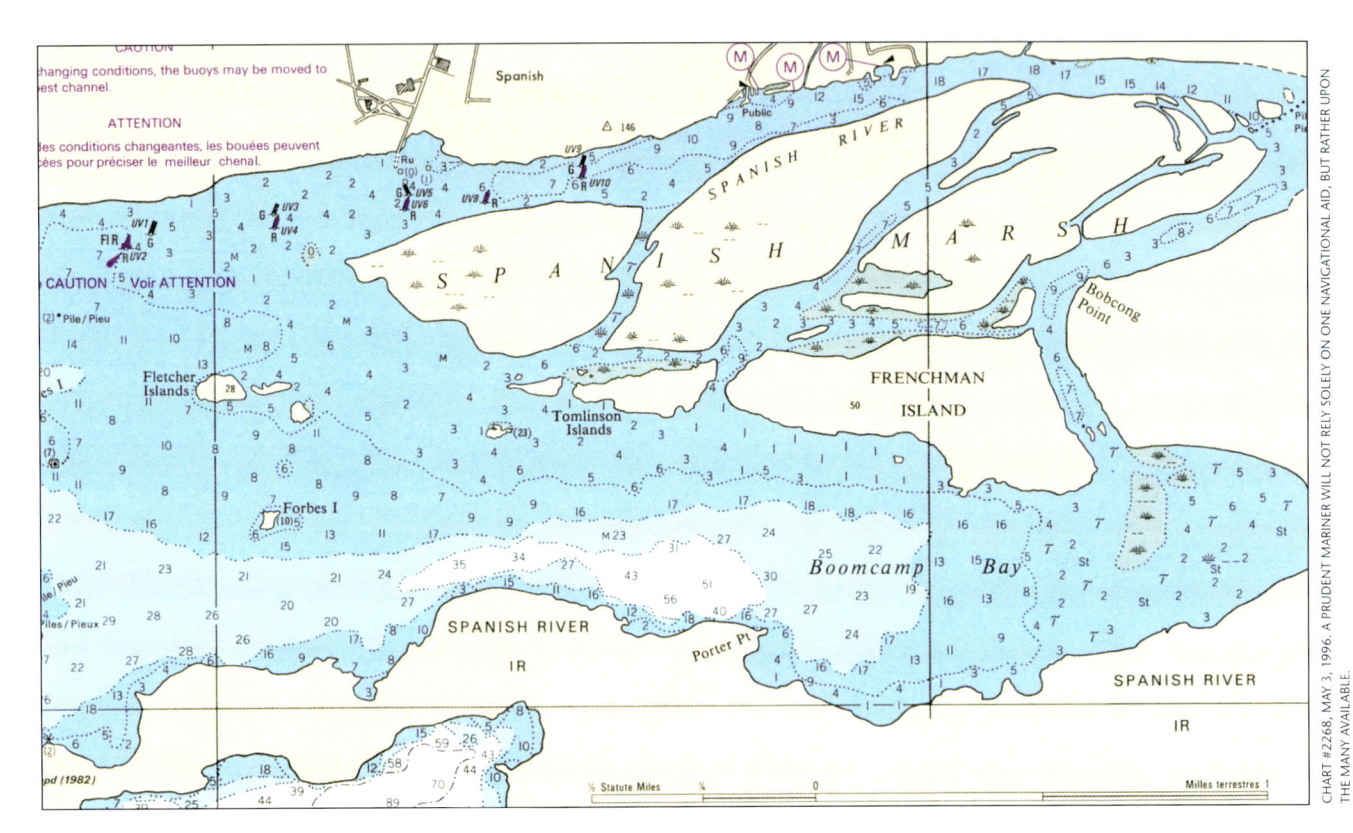

CHART #2268, MAY 3, 1996. A PRUDENT MARINER WILL NOT RELY SOLELY ON ONE NAVIGATIONAL AID, BUT RATHER UPON THE MANY AVAILABLE.

Facilities information subject to change. We suggest you call ahead.

		Monitors VHF Channel	Transient Slips Available	ALTERNATE MOORING: Wall mooring Rafting allowed	Maximum LOA	Minimum Depth at Dock	Power (amperage or volts)	HOOKUPS: Water Cable TV	FUEL: Gas Diesel Pumpout	BASICS: Heads Showers Laundry	AMENITIES: Swimming pool Whirlpool Rec area Grills Picnic tables Dog walk	Take Reservations	Take Credit Cards	Haulout (Capacity in tons or feet)	REPAIRS: Mechanical Electronics Fiberglass Woodworking Sails Canvas	CONVENIENCE: Ship's store Ice Convenience foods & beverages
D	Mitchells' Camp 69 Colonization Rd. 705-844-2202 Spanish, ON P0P 2A0		5	W	50'	4'	15			HS			GPD	Y	N	
A	Municipal Marina & Four Seasons Complex 705-844-1077 Spanish, ON P0P 2A0	68	90		65'		50	W	GDP	HSL			RPD	Y	Y	SIC
E	Vance's Marine, Ltd. & Resort Spanish Street 705-844-2442 Spanish, ON P0P 2A0	68	15		48'	6'	20	W	GDP	HSL			PD	Y	Y	35'	SIC

214 LAKELAND BOATING · LAKE HURON CRUISING GUIDE

Whalesback Channel, Ontario

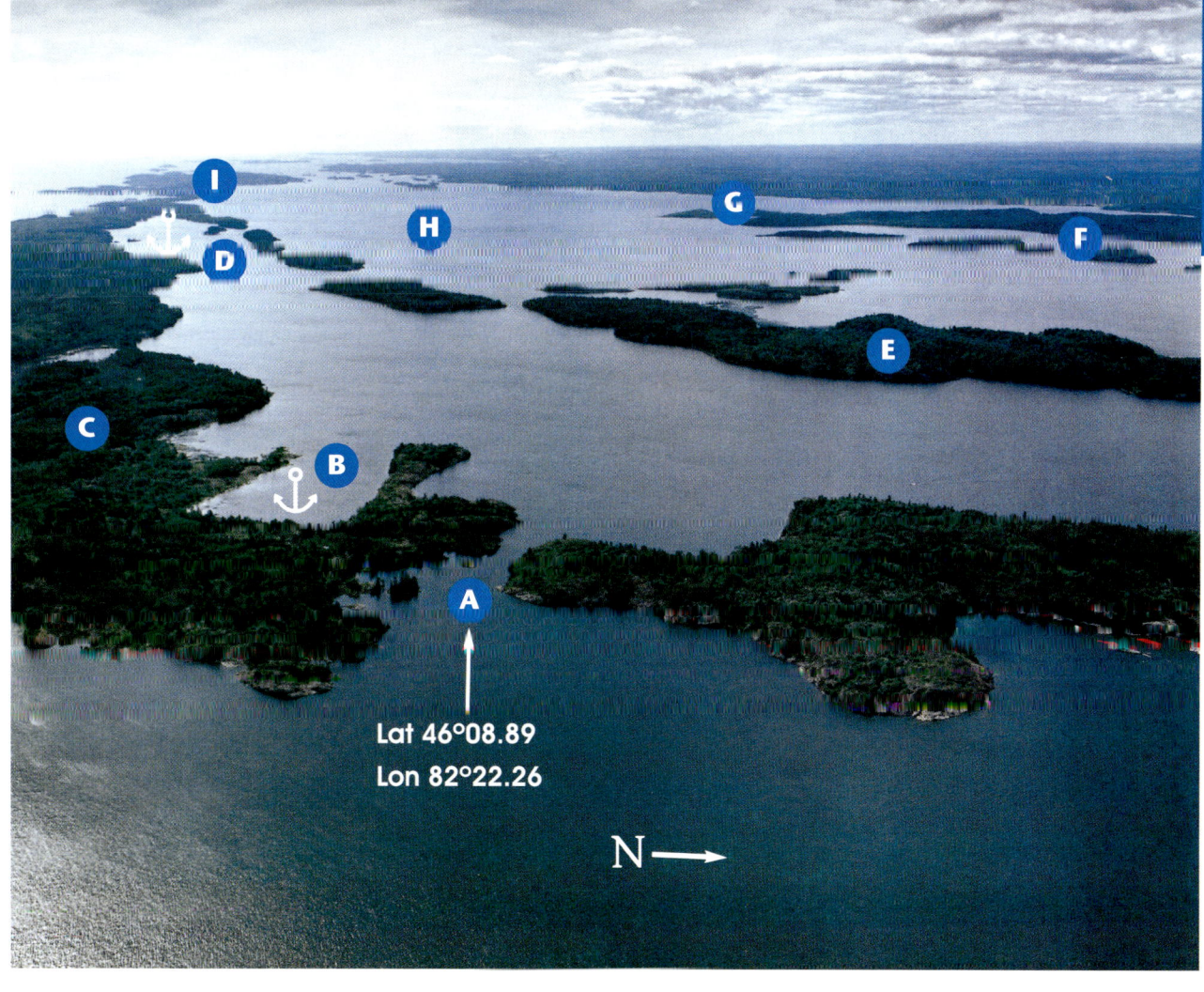

Lat 46°08.89
Lon 82°22.26

Range beacons mark the way through the dogleg passage at Little Detroit **A**. Heading straight west, you can anchor in the first cove to port **B** or farther along Aird Island **C** in the bay south of the Otter Islands **D**. Heading north and keeping Green Island **E** to port, you come to Spanish and Buswell bays (outside the photo), as well as Coursol Bay **F** and Aird Bay **G**. The main channel of the Whalesback **H** extends beyond, past John Island **I**.

Distances from Little Detroit (statute miles)
Little Current: 28 SE

Turnbull Island: 20 W

LAKELAND BOATING · LAKE HURON CRUISING GUIDE   **215**

THE NORTH CHANNEL

Whalesback Channel, Ontario

The storied passage that calls to boaters like the sirens of old.

One of the most fabled cruising areas in the North Channel, Whalesback Channel stretches westward for 20 miles from Little Detroit's tight passage out of McBean Channel to either Turnbull Island or La France Rock, depending upon whom you ask. Named for an evocatively shaped island in its western half, it is defined on the north by the mainland and on the south by a string of isles, including Aird and John islands. Granite slopes, deep blue water and panoramic views lure cruisers in the know back year after year. With well-marked exceptions, there is fairly open cruising here.

Little Detroit is the tightest passage of the North Channel—on the beaten track, that is! It is 75 feet wide after government blasting and 16 feet deep. Because of a blind spot at the turn, mariners usually give a *securité* call on channel 16 before entering. Range beacons clarify the course, but you'd be well advised to favor the mainland shore. Within the Whalesback's confines, you'll find numerous anchorages, from the deservedly popular to the sublimely isolated. The big names to seek out are Moiles Harbour (on the east end of John Island), John Harbour (on the island's west end), Cleary Cove (adjacent to John Harbour), Beardrop Harbour (on the north shore, north of John Island), Coursol, Birthday and Buswell bays (on the north shore above Aird Island) and the well-protected bay at the northwest end of Aird Island. The mouth of the Spanish River is located at the east-

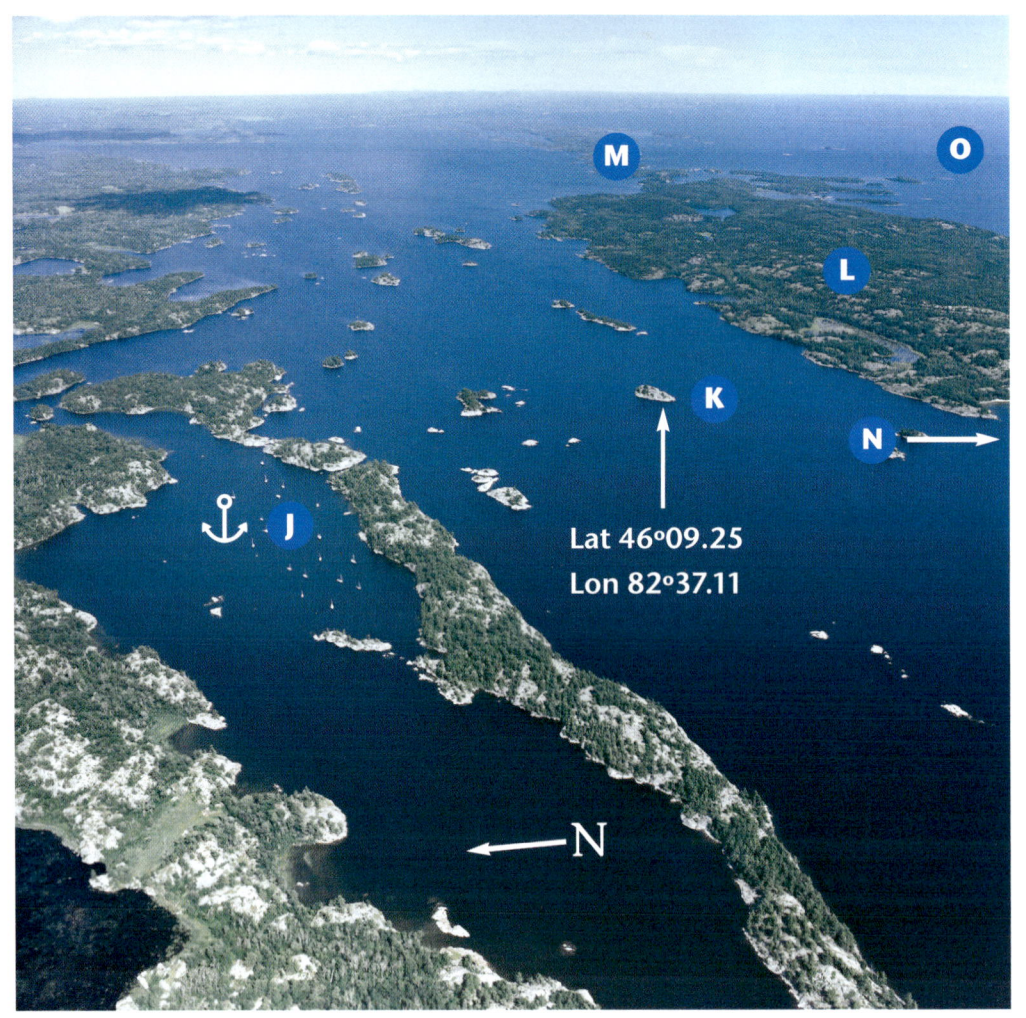

Lat 46°09.25
Lon 82°37.11

In this shot looking east, you get a close look at the popular Beardrop Harbour anchorage **J**, which is best approached from the west. The channel's namesake, Whalesback Island **K**, lies just to the north of John Island **L**. John Island has two fine anchorages, Moiles Harbour **M** at its east end and John Harbour **N**, just outside the photo. Cleary Cove (to starboard when entering John Harbour from the west) is a lovely spot. In the background you can see the North Channel **O**.

216   LAKELAND BOATING · LAKE HURON CRUISING GUIDE

ern end of the Whalesback Channel, and while relatively drab in this land of great vistas, it brings you to Spanish, where there are docking and provisioning facilities.

More Info

Emergency 911

St. Joseph's General Hospital 705-848-7181

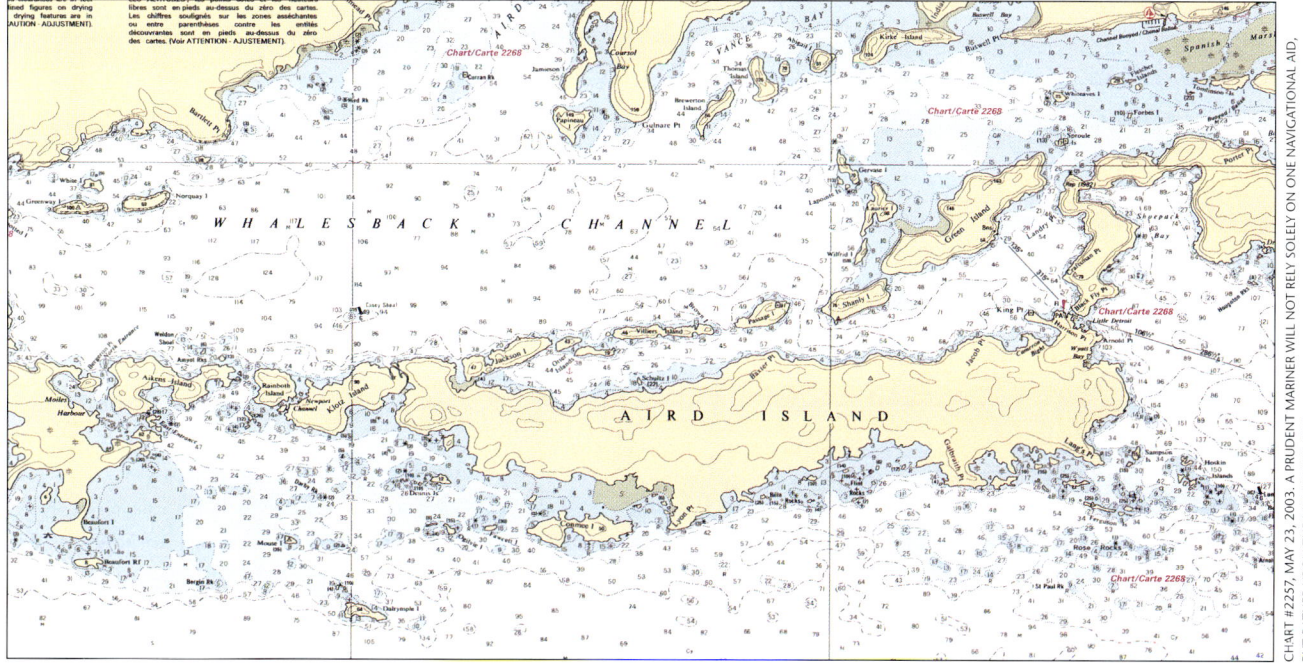

Spragge, Ontario

Lat 46°12.29
Lon 82°39.11

Nestled on the northern shore of Serpent Harbour **A**, just below the tiny community of Spragge, is the North Channel Yacht Club **B**—a picturesque, secluded facility that offers dockage and moorings to transient cruisers during the week (though usually not on weekends). Nobles Island **C** is visible in the distance. Just outside the harbor you can find a fine anchorage in Long Point Cove **D**, with entry through a tight cut.

Facilities information subject to change. We suggest you call ahead.

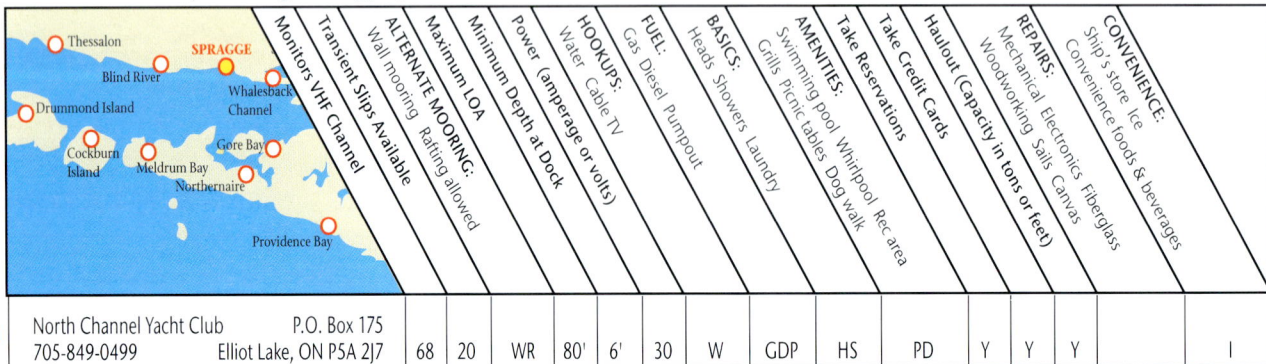

		Monitors VHF Channel	Transient Slips Available	ALTERNATE MOORING: Wall mooring, Rafting allowed	Maximum LOA	Minimum Depth at Dock	Power (amperage or volts)	HOOKUPS: Water, Cable TV	FUEL: Gas, Diesel, Pumpout	BASICS: Heads, Showers, Laundry	AMENITIES: Swimming pool, Whirlpool, Rec area, Grills, Picnic tables, Dog walk	Take Reservations	Take Credit Cards	Haulout (Capacity in tons or feet)	REPAIRS: Mechanical, Electronics, Fiberglass, Woodworking, Sails, Canvas	CONVENIENCE: Ship's store, Ice, Convenience foods & beverages
B	North Channel Yacht Club 705-849-0499 P.O. Box 175 Elliot Lake, ON P5A 2J7	68	20	WR	80'	6'	30	W	GDP	HS	PD	Y	Y	Y		I

THE NORTH CHANNEL

Spragge, Ontario
Discover the secret of Serpent Harbor.

Driving through Spragge, you'd never know there's a marine facility in the area. All you'd glimpse is a sprinkling of faded clapboard houses and a glint of water through the trees. But for the cruiser who has strayed northeast into Serpent Harbour, perhaps on a whim, the experience will be different. From this vantage, you will soon see a thicket of telltale masts. Before you know it—if you're ready to call it a day, that is, and the yacht club itself isn't too busy, which it usually isn't—you might just find yourself tied up for the night at one of the North Channel's best-kept secrets.

The **North Channel Yacht Club** is fairly easily to reach, whether you are heading east through South Passage, north following Turnbull Passage or west out of the Whalesback Channel toward Prendergast Island. Once in Serpent Harbour, the recommended route is to the north of Nobles Island, although the southern route is also manageable. Passing to the north, leave a generous distance between yourself and the submerged wreck off the island's northwest tip. The yacht club is just behind Hermann Point.

NCYC will accept transients on any weekday, while reserving its docks on weekends for members. The club has a total of 20 slips, which offer 0 feet of water. Moorings may also be rented. Power and water are supplied to the docks, while gas, diesel and pumpout can be secured at the fuel dock, where the draft is 12 feet. Ice, showers and washrooms are available. NCYC also has an extensive marine railway system, and though not itself a repair facility, it can haul boats out in emergency situations. There is no launch ramp. Marine engine and diesel mechanics can be arranged for through dockmaster Dave Morrissette, who is himself a small-engine mechanic.

Spragge itself is no more than a smattering of houses, and if you're after provisions, you're in the wrong place. The **KOA Campground**, located about a mile farther east (reachable by dinghy), does have a few groceries and also carries propane. But don't view this as a place to stock up. Think of the North Channel Yacht Club, with its peaceful atmosphere, friendly hosts and surrounding greenery, as a semi-rustic experience, something akin to a boaters' campground.

One of the North Channel's finest harbors is Long Point Cove, around the corner from Beardrop, just before the entrance to Serpent Harbour and Spragge. Protected from all winds, it's truly a delightful anchorage.

Where to Eat

Rocky's Restaurant 705-849-2780
Serving only home-cooked meals. Locals love the burgers, fries and all the other goodies. Accessible by water or road.

More Info

Emergency 911
St. Joseph's Health Centre 705-356-2265
North Channel Tourist Information 800-563-8719

Distance from Spragge (statute miles)
Blind River: 15 W

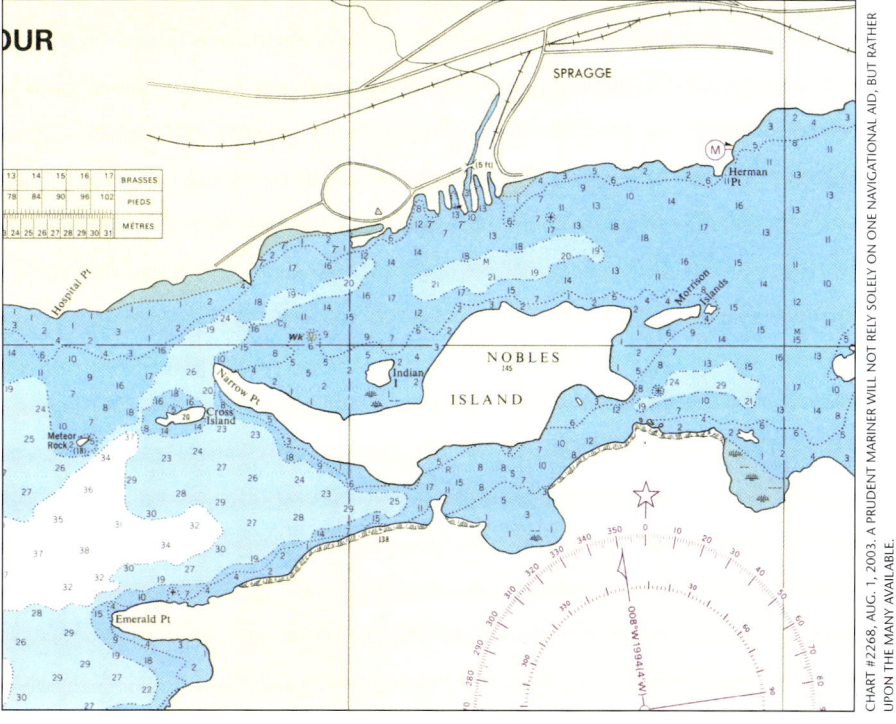

Blind River, Ontario

Lat 46°10.66
Lon 82°57.98

Blind River, a great place to stock up and stretch your legs, is situated about halfway up the North Channel on the mainland side. The only transient slips are at Blind River Marine Park **A**, a large, modern facility at the southwest end of town. There is a public boat launch **B** at the marina, and a popular sand beach **C** is only a five-minute walk away. The old federal wharf **D** is used solely for fishing tugs and is not available for pleasurecraft. The main street of Blind River **E** is about 10 minutes by foot from the marina. Enjoy the tennis and basketball courts next to the Community Centre **F**.

Distances from Blind River (statute miles)
Meldrum Bay: 20 SW
Thessalon: 30 W
Little Current: 56 SSE

THE NORTH CHANNEL

Blind River, Ontario
Take a beach break at the North Channel's halfway point.

With its sandy beaches and good mix of fun activities, Blind River—located halfway between Killarney and Sault Ste. Marie on the mainland side of the North Channel—is an excellent spot to stock up and chill out, whether you're midway through a lengthy cruise or just entering the North Channel from the west.

Once home to one of Canada's largest white pine mills, Blind River now employs only about 65 people in the manufacture of wood products. But with a population of about 4,000 people, the town still has a healthy, active feel, especially in the summer months, when boaters converge on the community from two sides. The town welcomes small boats coming down the Blind River from the north, as well as North Channel cruisers. Numerous sun-seekers flock to the beaches that scallop the town's shores.

If you're heading to Blind River from the south or southeast, steer west of Tenspot Shoal, which is marked with a red buoy. Otherwise, the approach is clear. Coming from the west or southwest, keep Mississagi Island, about five miles southwest of Blind River, to your northeast. As you approach the harbor, head for the Comb Point Light and follow the buoyed channel to the marina. Visible landmarks include the old water tank and incinerator of the veneer mill, located just northwest of the marina, and the large green water tank and communications towers a half-mile or so to the east.

Completed in 1992, **Blind River Marine Park** is a full-service facility that can accept up to 120 boats at its nicely protected docks. Draft at the guest slips is 10 to 12 feet, with electricity and water provided. Diesel, gas, propane and pumpouts are available at the service dock. The marina also boasts showers, restrooms, a coin-operated laundry, a shop selling charts, a boaters' lounge and a full-service restaurant. There are also two excellent launch ramps. **Blind River Boat Works,** located next to the marina, has a 35-ton Travelift and provides full hull and engine repairs and equipment sales.

The town of Blind River offers just about everything you could ask for—most of it within easy walking distance. The nearest grocery store, **ValuMart,** is located in a small mall about a 10-minute walk from the marina.

For eats, there's a wide variety of fast food, donut shops, fine dining and family eateries. The **Auberge Eldo Inn** is known across the North Shore for its whitefish and other fine dishes. If you're in the mood

LAKELAND BOATING · LAKE HURON CRUISING GUIDE 221

Where to Eat

Auberge Eldo Inn. **705-356-2255**
This restaurant, featuring non-greasy, home-style cooking, is a boater's favorite. Specialties include the grilled, fresh-cut whitefish dinner, barbequed chicken and the homemade raspberry pie. All non-smoking. Outdoor patio and shuttle from the marina is available.

Riverside Tavern. **705-356-7155**
A really, *really* old-fashioned Irish pub.

for a pleasant stroll, it's just under two miles east of the marina on Causley Street, but you can also have the courtesy van pick you up and bring you back. And if the notion of separate rooms for men and women strikes you as endearingly old-fashioned rather than annoyingly backward, check out the **Riverside Tavern** on Woodward. A traditional Irish pub, the sexes are indeed kept at a distance here. If a man is either accompanied or invited by a woman, he may stray into the women's quarters, but under no circumstances is a woman allowed in the men's area.

If you're curious about Blind River's lumbering past, pay a visit to the **Timber Village Museum**, situated about two miles from the marina on Causley Street. The marina offers guided tours and a gift shop. From there, it's only a short stroll over to the Northern Ontario Loggers Memorial, a huge monument and public art display that was constructed in 2001.

The town is also home to the fine 18-hole **Huron Pines Golf Club**. Located about two miles west of the marina, Huron Pines offers a championship layout, putting areas, a practice range and a fully licensed lounge. Visit them online at www.huronpines.com.

If you're looking for a lazier pastime, call the Eldo Inn and have them pick you up to spend some time in their outdoor hot tub. Or simply park yourself on a bench in the public park above the dam. You'll be back on the water soon enough, so you might want to just sit back and relax awhile.

More Info

Emergency	911
Blind River Health Centre	705-356-2265
Chamber of Commerce	705-356-2555
North Shore Travel Information Centre	800-563-8719
	www.blindriver.ca
Blind River Taxi	705-356-1651

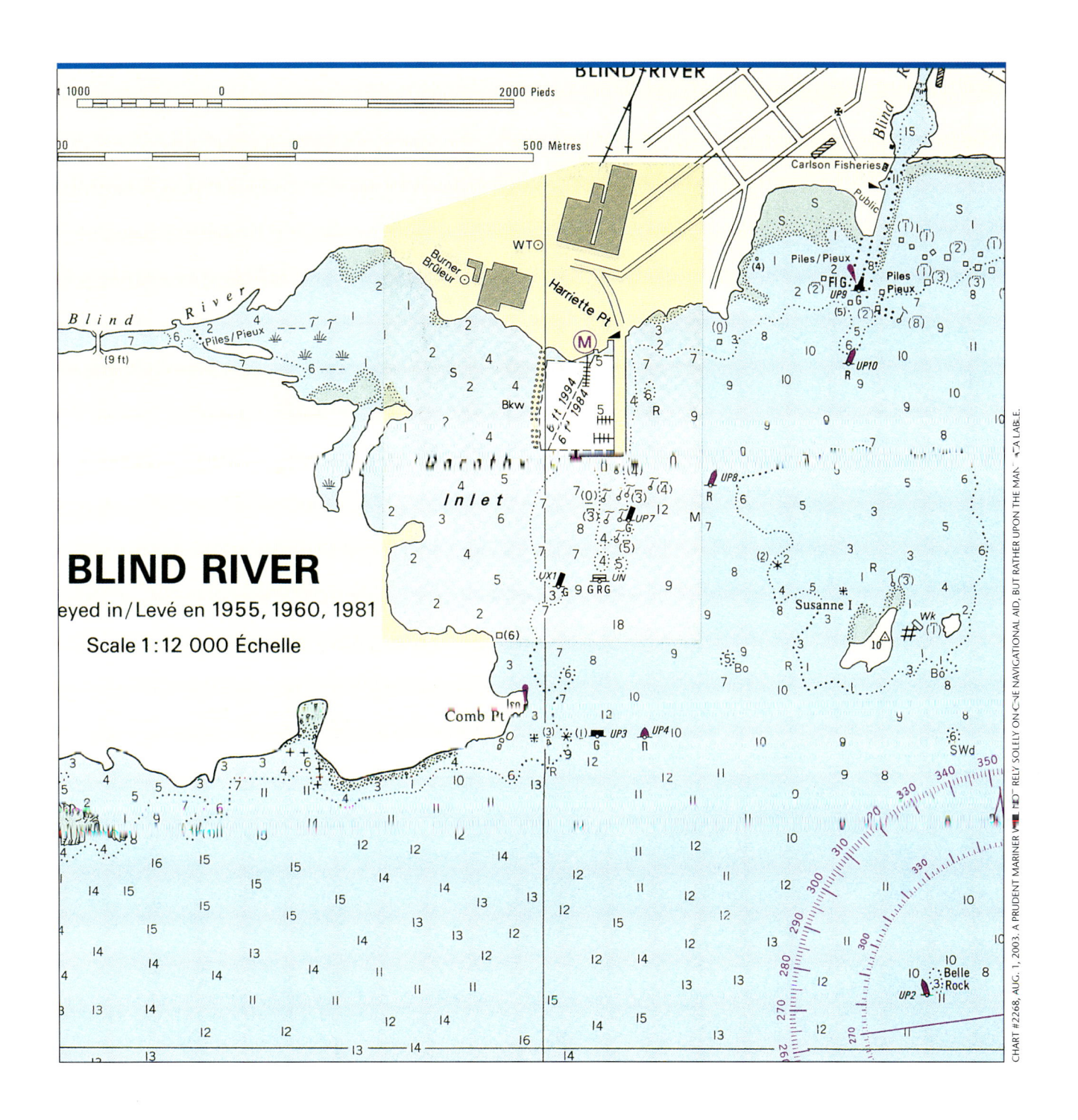

Facilities information subject to change. We suggest you call ahead.

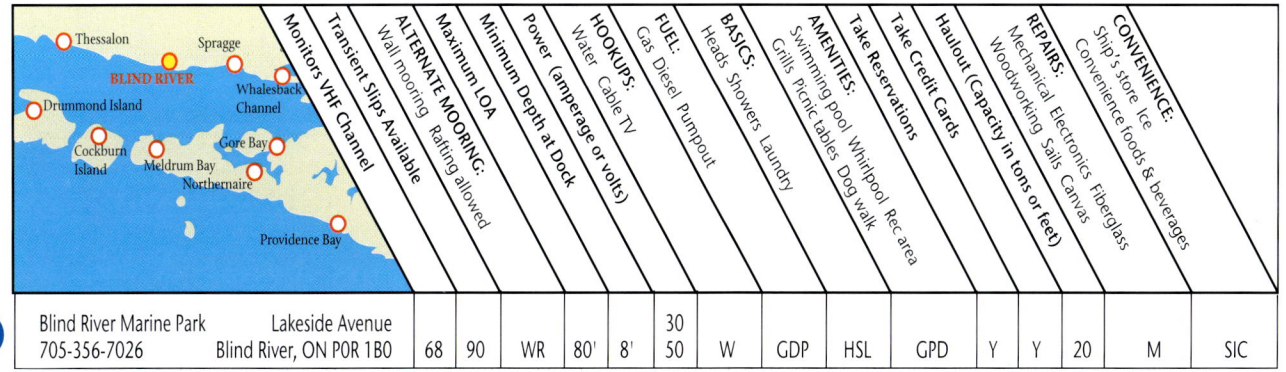

			Transient Slips Available / Monitors VHF Channel	ALTERNATE MOORING: Wall mooring / Rafting allowed	Maximum LOA	Minimum Depth at Dock	Power (amperage or volts)	HOOKUPS: Water / Cable TV	FUEL: Gas / Diesel / Pumpout	BASICS: Heads / Showers / Laundry	AMENITIES: Swimming pool / Whirlpool / Rec area / Grills / Picnic tables / Dog walk	Take Reservations	Take Credit Cards	Haulout (Capacity in tons or feet)	REPAIRS: Mechanical / Electronics / Fiberglass / Woodworking / Sails / Canvas	CONVENIENCE: Ship's store / Ice / Convenience foods & beverages
A	Blind River Marine Park 705-356-7026	Lakeside Avenue Blind River, ON P0R 1B0	68	90	WR	80'	8'	30 50	W		GDP	HSL	GPD	Y	Y	20	M	SIC

Lakeland Boating · Lake Huron Cruising Guide 223

Thessalon Ontario

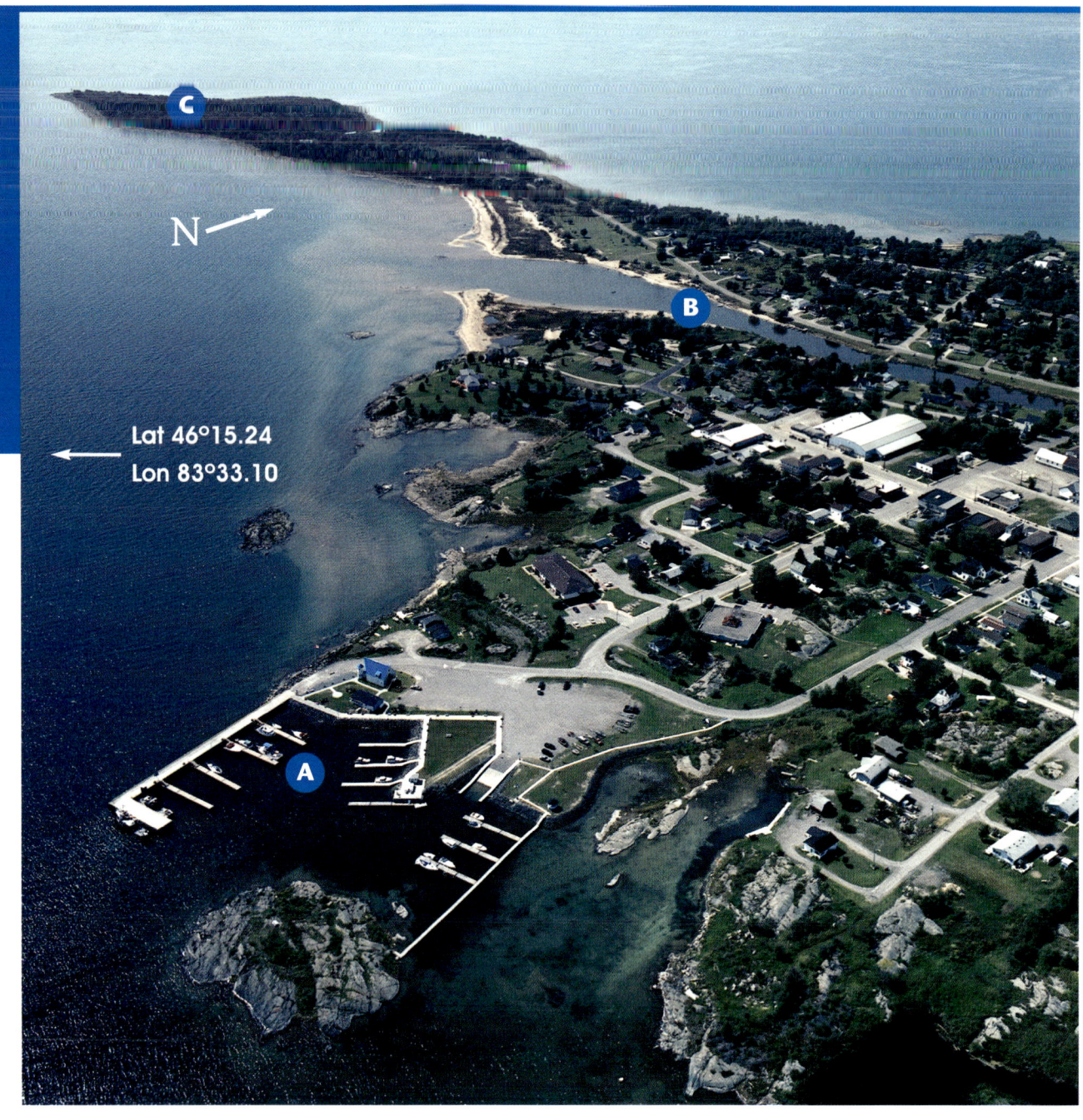

Lat 46°15.24
Lon 83°33.10

The harbor at Thessalon, located due north of Drummond Island on the north shore of Lake Huron, is home to the municipal Thessalon Marina **A**, built four years ago. New docks and a new building were added in 2003. The Thessalon River **B** runs south through the community, emptying into the bay at the base of Thessalon Point **C**, a mile-long spit that shields the harbor from a westerly blow.

Distances from Thessalon (statute miles)
Bruce Mines: 14 W
Blind River: 29 E

Meldrum Bay: 31 S
Little Current: 63 SE

224 LAKELAND BOATING · LAKE HURON CRUISING GUIDE

THE NORTH CHANNEL

Thessalon, Ontario
A bustling lumber town with heart.

Cruising down the pine- and spruce-lined banks of the Thessalon River, it's easy to forget that the lumber industry is still thriving in this port. While most of its neighbors along the north shore of Lake Huron have shut down their mills, Thessalon never stopped being a lumber town. Three mills still produce veneer, plywood and lumber. But the town feels much more like a wonderful place to relax and perhaps play a round of golf than a gritty industrial port.

The recently renovated, squeaky-clean **Thessalon Marina** has about 40 slips available for transients, each with 15- or 30-amp power and water hookups. Other services at the municipal facility include gasoline, diesel, pumpout, and a launch ramp. There's a generous draft of 12 to 14 feet at the inner docks, which now accommodate craft up to 96 feet long. Flower beds and a lamp-lit boardwalk add charm, while the newly completed marina building boasts washrooms, showers, laundry facilities and a lounge complete with stove, fridge, and counter space.

Reaching Thessalon is easy from any direction; the one exception might be if you are heading north along the east side of St. Joseph's Island. As you pass between Sulphur and Bigsby islands, leave a good buffer between your boat and Bigsby's outlying shoals. Using Thessalon's range—there is one light at the end of the marina dock, another at the bay head—make your way in past Thessalon Point, which is marked with a white light. Keep it, and the shoal that projects from its tip, to port. The bay itself is deep and the marina is protected from the south and east by two separate breakwaters.

Although the main street of Thessalon, about a five-minute walk from the docks, has a practical, no-frills feel

More Info
Emergency 911
Thessalon Hospital 705-842-2014
Thessalon Taxi Service 705-842-0633

Where to Eat

Carolyn Beach Restaurant 705-842-3330
Traditional family-oriented fare: fresh whitefish, roast beef and salads. Don't skip dessert—they're all homemade, and the pie is heavenly.

Shoreview Restaurant 705-842-3730
Enjoy homemade meals made with fresh ingredients while you enjoy the view of the water.

Sinton Tavern & Restaurant 705-842-2414
Close to the marina, this casual place serves up home cooking and pub fare.

to it, nearly everything you need can be found there. A **grocery store, hardware store, banks** with cash machines, a **liquor store** and **pharmacy** are all within walking distance. For dining, try the popular **Carolyn Beach Restaurant**. Most restaurants, including this one, will pick you up and return you to the docks. Other dining options are within walking distance. A taxi service is also available in town if you want to try out the nine-hole **golf course** just east of Thessalon. Or you can motor north to the **Kirkwood Forest**, an area that lays claim to a massive white pine, reportedly the largest tree east of the Rocky Mountains. If you feel like more walking, strike out along the banks of the Thessalon River in town or take the trail out to Thessalon Point. If you tackle the latter, budget a couple of hours for the trip there and back.

If you have mechanical problems while in Thessalon, don't panic: The marina keeps mechanics on call, as well as a diver—just in case you drop something overboard or get a line tangled in your prop. Haulout of boats up to about 40 tons and additional repairs are possible just across the bay at **Bill's Marine**, a dry-dock facility. Propane is available at **Outfit Supply**, located a couple of miles from the marina on Highway 17.

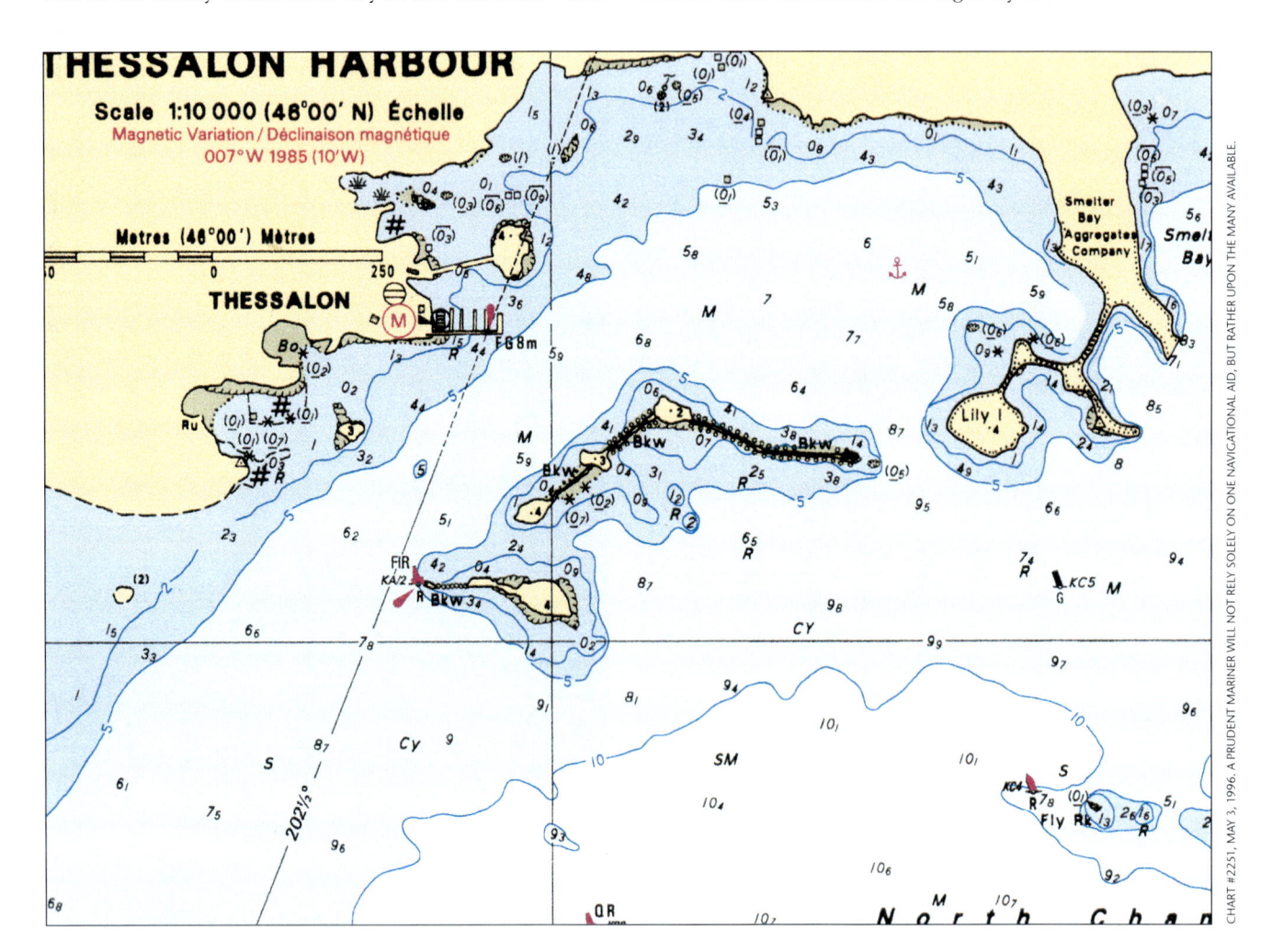

Facilities information subject to change. We suggest you call ahead.

			Monitors VHF Channel	Transient Slips Available	ALTERNATE MOORING: Wall mooring Rafting allowed	Maximum LOA	Minimum Depth at Dock	Power (amperage or volts)	HOOKUPS: Water Cable TV	FUEL: Gas Diesel Pumpout	BASICS: Heads Showers Laundry	AMENITIES: Swimming pool Whirlpool Rec area Grills Picnic tables Dog walk	Take Reservations	Take Credit Cards	Haulout (Capacity in tons or feet)	REPAIRS: Mechanical Electronics Fiberglass Woodworking Sails Canvas	CONVENIENCE: Ship's store Ice Convenience foods & beverages
A	Thessalon Marina 705-842-5188	56 Algoma St. E. Thessalon, ON P0R 1L0	68	40	WR	96'	12'	15 30	W		GDP	HSL	RPD	Y	Y	33	ME	IC

Fish Dishes

Sweet Soy Garlic Broil

- 8- to 10-pound trout or salmon, skinned and filleted
- 1 tablespoon chopped garlic
- 1 tablespoon brown sugar
- Soy sauce

Whisk garlic and brown sugar in enough soy sauce to cover fillets. Marinate in refrigerator for half an hour. Broil the fillets, turning once, about four minutes per side.

Blackened Fish Fillets

- six 8- to 10-ounce salmon, whitefish or walleye fillets
- 3 tablespoons Paul Prudhomme's Blackened Redfish Magic seasoning
- ½ stick unsalted butter, melted

Heat a large cast-iron skillet over a high flame until it's extremely hot. With a spoon, spread a little of the butter on one side of the fillets. Sprinkle with Blackened Redfish Magic and place fillets in skillet, seasoned side down. Spread more butter on the side facing up and sprinkle with seasoning. Turn frequently until the fish starts to flake, about four minutes. Repeat with remaining fillets, wiping out pan between batches. Serve piping hot.

This will create lots of smoke, so make sure your kitchen is well-ventilated or else prepare this dish outside. Chef Paul Prudhomme's seasonings are available at www.paulprudhomme.com.

–Marty Kovarik

Heavenly Battered Walleye

- six 8- to 10-ounce walleye fillets
- 2 cups flour (or 1 cup each flour and cornmeal)
- 2 tablespoons salt
- 1 rounded tablespoon garlic powder
- 1 rounded tablespoon onion powder
- 1 rounded tablespoon baking powder
- 1 heaping tablespoon powdered sugar
- 1 heaping tablespoon powdered buttermilk
- 1 heaping tablespoon Hungarian hot paprika, cayenne or black pepper
- 2 cans beer or 7-Up
- 1 egg

Mix together dry ingredients. In another bowl, whisk beer or 7-Up and egg. Dredge fillets first in plain flour, then dip in the egg mixture. Coat thoroughly with seasoned flour mixture and deep fry at 375 degrees until golden.

–Dave Mull

Bruce Mines Ontario

Lat 46°17.82
Lon 83°47.41

Bruce Bay, a shallow but protected harbor, lies in the northwestern corner of the North Channel. Bruce Mines Marina **A** offers the area's only transient dockage. The shopping area of Bruce Mines **B** is just a short walk from the marina.

Facilities information subject to change. We suggest you call ahead.

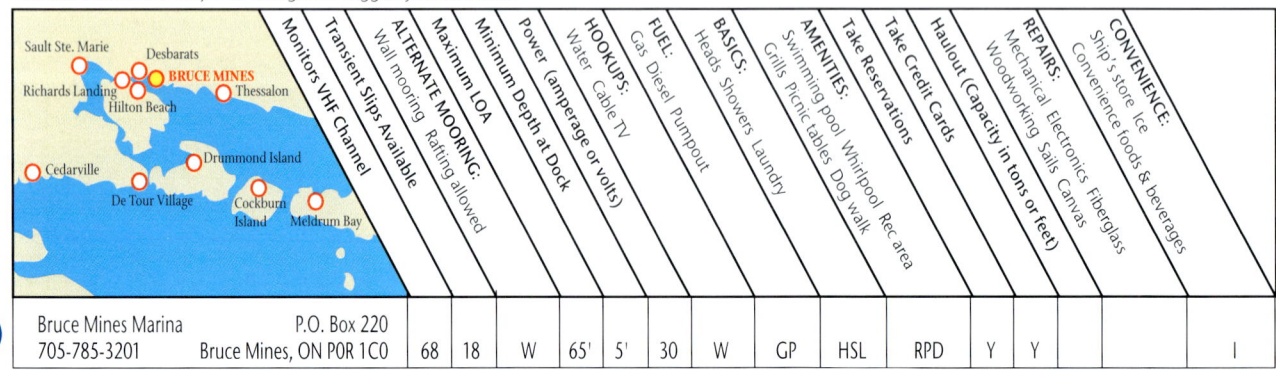

		Monitors VHF Channel	Transient Slips Available	ALTERNATE MOORING: Wall mooring / Rafting allowed	Maximum LOA	Minimum Depth at Dock	Power (amperage or volts)	HOOKUPS: Water / Cable TV	FUEL: Gas / Diesel / Pumpout	BASICS: Heads / Showers / Laundry	AMENITIES: Swimming pool / Whirlpool / Rec area / Grills / Picnic tables / Dog walk	Take Reservations	Take Credit Cards	Haulout (Capacity in tons or feet)	REPAIRS: Mechanical / Electronics / Fiberglass / Woodworking / Sails / Canvas	CONVENIENCE: Ship's store / Ice / Convenience foods & beverages
A	Bruce Mines Marina 705-785-3201 P.O. Box 220 Bruce Mines, ON P0R 1C0	68	18	W	65'	5'	30	W		GP	HSL	RPD	Y	Y			I

228   LAKELAND BOATING · LAKE HURON CRUISING GUIDE

THE NORTH CHANNEL

Bruce Mines, Ontario
Stop in this friendly port for convenient provisioning.

ETHAN MELEG

Settled by miners from Cornwall, England, Bruce Mines was the home of Canada's first copper mine. With just 600 residents, its days as a sprawling boomtown are over, but the community has remodeled itself over the years into a gracious host to marine travelers, offering both a modern docking facility and some of the most accessible shopping in the North Channel.

Heading toward the port from the west, south or east, steer a straight course between McKay Reef and Prout Rock, both marked with buoys. Keep an eye out for the lighthouse on McKay Island, which signals the entrance to Bruce Bay. Once past the lighthouse, the channel to the marina is marked with a string of green spar buoys. A red light is mounted above the fuel dock.

The **Bruce Mines Marina**, a federal wharf managed by the town, has about 50 slips, 18 of which are reserved for transients. The depth at the long finger docks and the fuel dock is approximately 5 feet. Elsewhere, to the east and west of the marina, the bay grows remarkably shallow, with as little as 3 feet of water in some places. To the east, there are also old dock and railway pilings, not to mention old mine tailings, so don't think of Bruce Bay as prime anchoring territory.

Rather, pull into the docks and find yourself a space. Thirty-amp power, water, pumpout, gasoline, garbage disposal, ice, charts, a gin pole, launch ramp, showers and washrooms are among the many amenities offered here.

Some repair work can be performed by a local garage mechanic in an emergency. However, it is recommended that if you are at all seaworthy, you should seek mechanical help elsewhere.

The one clear advantage to Bruce Mines as a port of call is

Distances from Bruce Mines (statute miles)
Thessalon: 13 E
Hilton Beach: 7 W
Meldrum Bay: 43 SE

More Info

Emergency 911
Thessalon Hospital 705-842-2014
Bruce Mines Tourist Information 705-785-3370

LAKELAND BOATING · LAKE HURON CRUISING GUIDE

Where to Eat

Bavarian Inn 705-785-3447
A German/Canadian restaurant that prides itself on its schnitzel. Also try the Friday all you can eat fish fry, featuring pickerel, perch and whitefish.

Bobber's Restaurant 705-785-3485
They have a little bit of everything, but most people order the tasty fish and chips.

Sharon's Cafe 705-785-3522
Located near the marina, this cafe offers sandwiches, salads, soups and a variety of specials.

---

that everything is close at hand. You can plot your shopping route by studying the map posted on the wall of the marina—though even this seems somewhat beside the point, since you'll be seeing it all firsthand within minutes anyway. Strung along Highway 17 in Bruce Mines are a **bank** with a cash machine, a **laundromat**, a **post office**, a **liquor store** and **Foster's Freshmart** grocery store.

You'll also find two historical attractions at the heart of Bruce Mines: the **Bruce Mines Museum Welcome Centre**, housed in a white turreted building; and the **Simpson Mine Shaft**, a re-creation of an 1848 mining operation, which can be toured throughout the summer. The tour begins just east of the liquor store building. The **waterfront park** near the marina features live music in the summer and plays host to a heritage festival in August.

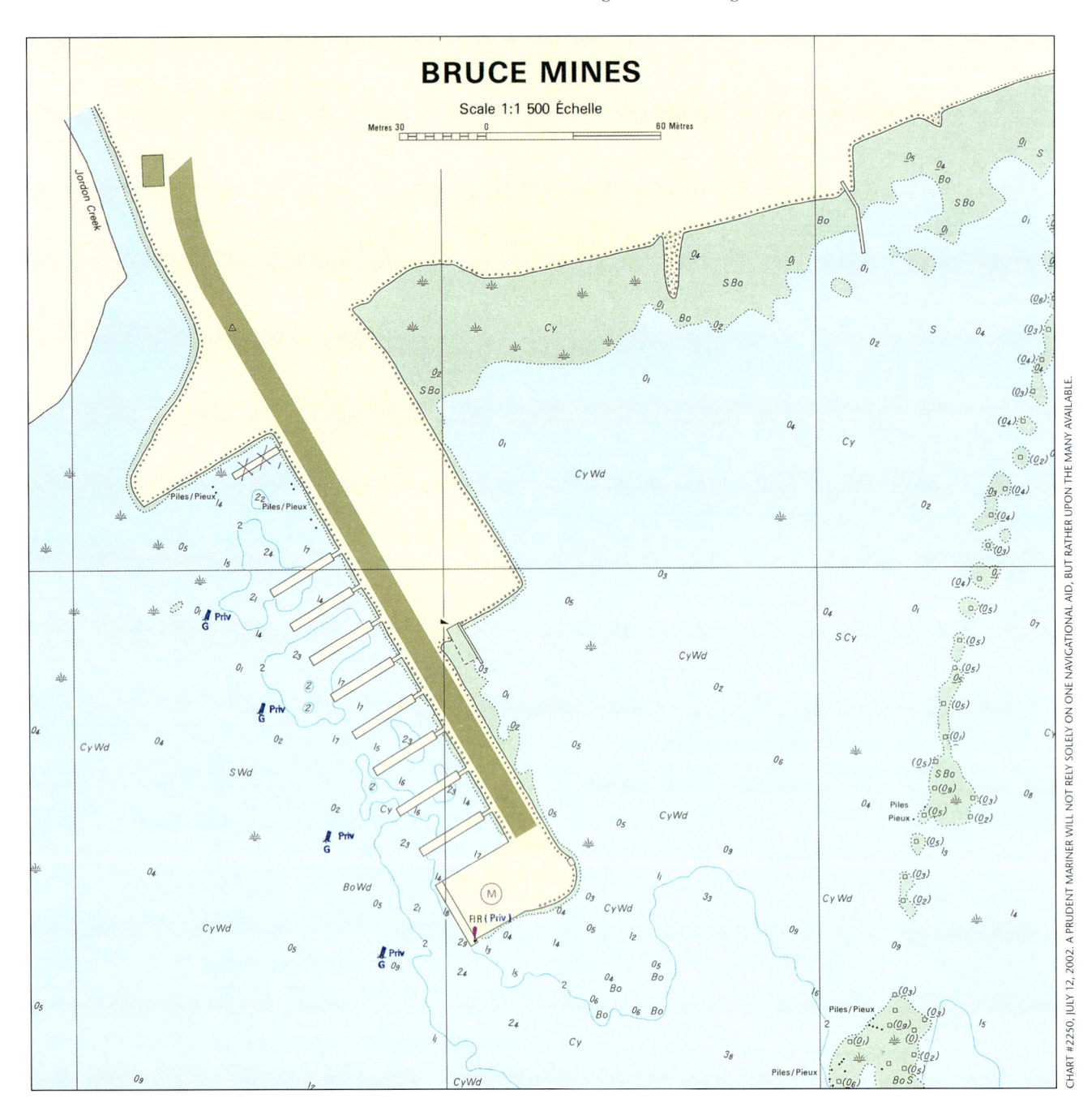

FISHING TIPS

The Angler's 10 Commandments

Obey these tips and you'll be blessed with a heavenly catch.

1. Love thy line. Start with a premium-grade brand name. Neglect it not: change or swap it end-to-end at least once during the season. Check it regularly all the day long. If it be nicked, retie it—especially after doing battle with the Big One.

2. Use thy owner's manual and maintain thy reels according to the laws written therein. Lubricate to prevent corrosion and promote smooth operation.

3. Ensure that thy drag system works smoothly. Reject all dirt, sand and corrosion lest it corrupt the mechanism, causing it to stick. Anoint it with oil and adjust the drag so that the hook can set solidly, while still allowing the Big One to take line. One method is to secure the line and pull on it until the rod strains. Then ease off on the drag until line grudgingly slips off of thy reel.

4. Seek thou the fishing hotspots. Change thy fishing location from week to week. Go thou forth among the bridge pilings, riprap, docks, deep creek bends, brush piles and incoming streams. In waters both shallow and deep shalt thou abide.

5. Thou shalt not be a one fish angler. Offer up a lightweight spinning rig to the bluegills and crappies. Use thy jigs and smaller crankbaits to bring unto thy vessel all species of fish. Teach thy children well the joys of fishing by focusing on the easiest of catches.

6. Thou shalt try more than one or two lures. Look upon the different colors, sizes, shapes and patterns. Try jigging a spinnerbait or swimming a jig. Experiment and be merry.

7. Thou shalt not use dull hooks. Keep always a hook hone, quality-made sharpener or file in thy tacklebox.

8. Watch for signs upon the waters. When the baitfish jump and scatter, rejoice, for the Big Ones are feeding. If a lunker be caught in 11 feet of water on the edge of a rock bar, know thou that this is the spot. If a fish spitteth out a 4-inch perch minnow in thy livewell, 'tis a sign unto thee to switch to a similar plug.

9. Be ever-vigilant. Keep thy concentration, even unto the sixth hour, lest the Big One get away.

10. Thou shalt keep a positive mental attitude. Concentrate and follow these commandments. Be confident of thy methods and verily, thy net and thy stomach will be full.

Hilton Beach, Ontario

Hilton Beach Marina **A**, in Fisher Bay on St. Joseph Island's northeast shore, is a spacious, full-service facility. The village of Hilton Beach **B** is a few steps away. Take a pleasant hike along the shore to the beach **C** that gave this town its name. The North Channel **D** extends into the distance beyond Gravel Point **E**.

Facilities information subject to change. We suggest you call ahead.

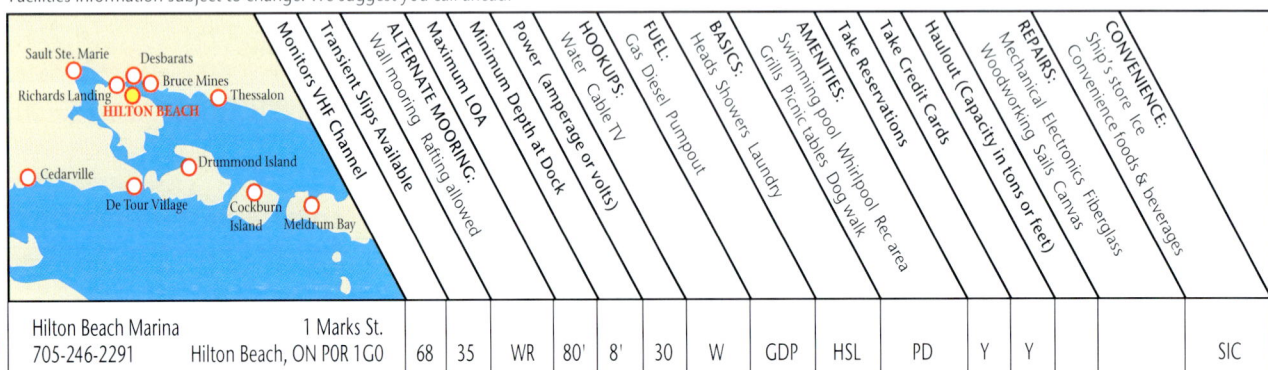

		Monitors VHF Channel	Transient Slips Available	ALTERNATE MOORING: Wall mooring Rafting allowed	Maximum LOA	Minimum Depth at Dock	Power (amperage or volts)	HOOKUPS: Water, Cable TV	FUEL: Gas, Diesel, Pumpout	BASICS: Heads, Showers, Laundry	AMENITIES: Swimming pool, Whirlpool, Rec area Grills, Picnic tables, Dog walk	Take Reservations	Take Credit Cards	Haulout (Capacity in tons or feet)	REPAIRS: Mechanical, Electronics, Fiberglass Woodworking, Sails, Canvas	CONVENIENCE: Ship's store, Ice Convenience foods & beverages
A	Hilton Beach Marina 705-246-2291 1 Marks St. Hilton Beach, ON P0R 1G0	68	35	WR	80'	8'	30	W		GDP	HSL	PD	Y	Y			SIC

THE NORTH CHANNEL

Hilton Beach, Ontario
A true beauty on St. Joseph Island.

You won't find casinos or supermarkets in Hilton Beach, a quiet port that has retained its charm even while growing to become the second-largest community on St. Joseph Island. The village, which was linked to the mainland by bridge in 1972, is a pleasant collection of restaurants and shops, surrounded on all sides by maple trees, wildflowers and country gardens.

Where fishing tugs and the odd passenger ferry once tied up, the long, curving breakwall of the **Hilton Beach Marina** now provides shelter to North Channel cruisers. Fisher Bay offers clear, deep water to cruisers coming up the east coast of St. Joseph Island from DeTour or Drummond Island or approaching through St. Joseph Channel from the North Channel or the Sault. Once you've reached the bay, the marina entrance is clearly marked with a starboard-side light and red/green range lights and daymarks. Fisher Bay, though deep, does not offer many protected anchorages, with the exception of one small cove inside Gravel Point, on the harbor's southeast shore. This will suffice if the wind is strictly easterly.

The marina keeps at least 30 of its slips available for guests. The minimum depth at the overnight docks is 8 feet, with 15 feet of water at the fuel dock. Slips are supplied with 20- and 30-amp power and water. Boat length should not be a problem, because the marina puts larger boats at the ends of each of the piers. The smallest dock is also being reconfigured to accommodate larger vessels, and 60 new slips are being added for the 2004 season. Gasoline, diesel and pumpout are available at the fuel dock, while the office building houses showers, washrooms and laundry facilities. Ice is available, as are Canadian and American charts. The marina has an excellent launch ramp and a gin pole for those sailors who want to cruise the St. Joseph Channel without worrying about clearing the bridge.

For propane, go right next door to the **Hilton Beach Tourist Park**, which also carries some fishing supplies. A **hardware store** is about a mile away on the highway. The rest of Hilton Beach, however, is right at your feet. You'll find rooms and a hot tub at the new **Hilton Harbour Resort** overlooking the marina. Next door, at the **Hilton Beach Restaurant**, located in the **Hilton Beach Hotel**, you can enjoy dinner and cocktails on the wide

Distances from Hilton Beach (statute miles)
Thessalon: 18 E
Sault Ste. Marie: 37 WNW
Meldrum Bay: 49 SSE
Little Current: 82 E

LAKELAND BOATING · LAKE HURON CRUISING GUIDE   **233**

Where to Eat

Gophers Café 705-457-1296
Choose from the full-course menu or snack on pizza or wings. Homestyle cooking, including homemade bread and pastries.

Hilton Beach Restaurant 705-246-2204
There's something for everyone here, from burgers and steaks to specialty salads and fish. A licensed eatery in the Hilton Beach Hotel.

More Info

Emergency 911
Matthows Momorial Hospital 705-246-2670
Village of Hilton Beach 705-246-2242
www.hiltonbeach.com

waterfront veranda. **Gophers**, a café-style restaurant that features home-baked goods, is located immediately to the west.

For groceries, visit the **Hilton Beach General Store**, which carries alcohol and a good selection of meat, including homemade sausage. The general store also makes deliveries to boats. For fresh produce, catch the open-air **farmers market**, held every Saturday in the summer on Mark Street. You'll find unique crafts at the several small shops scattered throughout the village. Arts at the Dock, a juried show, features local artists and artisans displaying their work outside at the marina and indoors at the **Community Hall**.

Visit the **Hilton Web Connection** in the **Waterfront Centre** if you want to check your email or need to use a computer. This community access site has online PCs and even a scanner. The village's website includes a list of activities, including Community Night and the Coffee House.

One drawback to Hilton Beach is that there isn't a bank, although several businesses do have debit machines. The attraction of the small community is its nicely maintained marina, its quiet atmosphere and its nearby hardwood forests, which are bursting with white-tailed deer in the summer months—and of course, its sand beach, located about a mile from the marina and connected by a waterfront boardwalk.

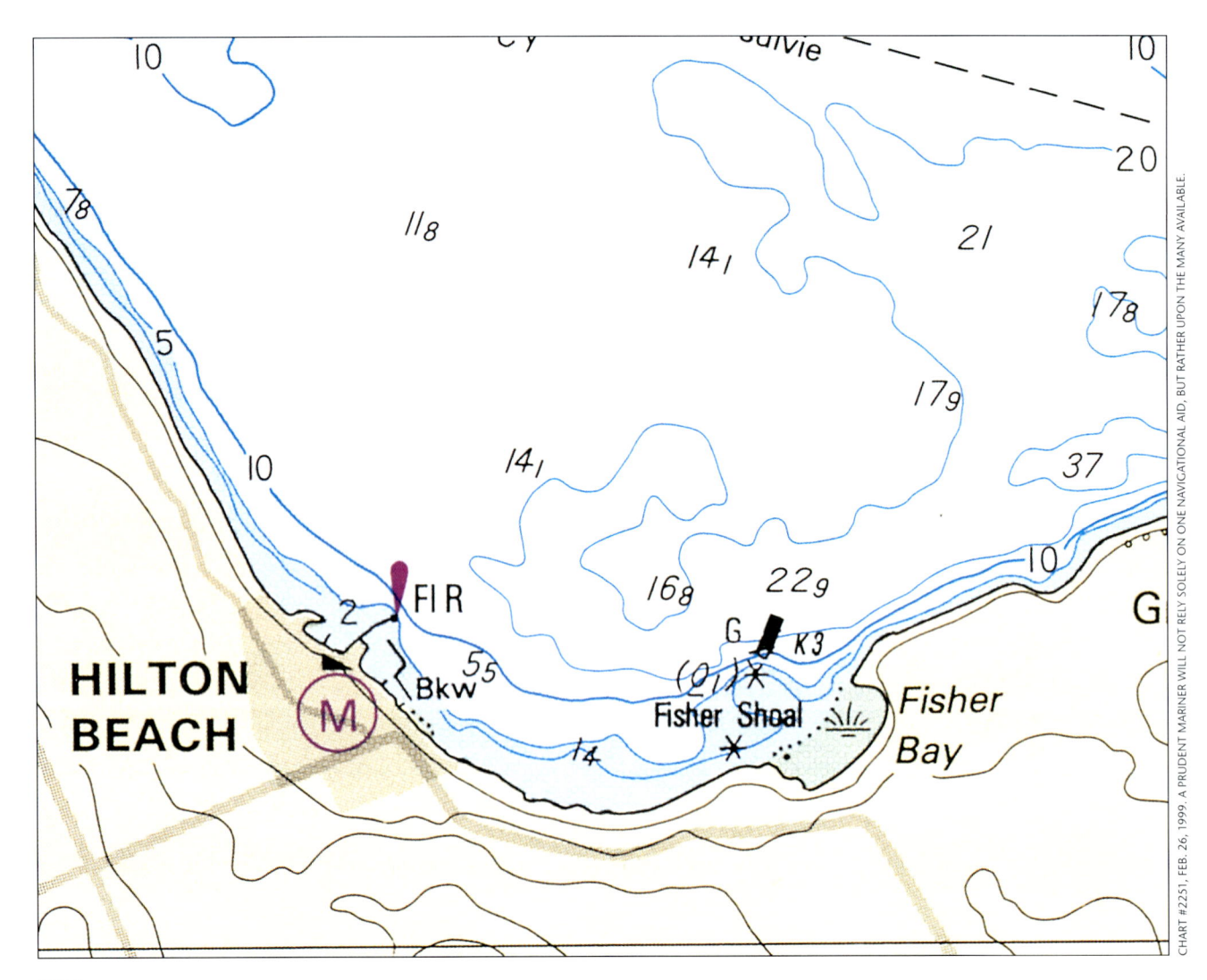

234   LAKELAND BOATING · LAKE HURON CRUISING GUIDE

Maritime Museums Around Lake Huron

MICHIGAN

Colonial Michilimackinac.......... 231-436-5563
Mackinaw City
May through October

Fort Mackinac.................... 231-436-5563
Mackinac Island
May through October

Historic Mill Creek............... 231-436-5563
Mackinaw City
May through October

Huron Lightship Museum.......... 810-982-0891
Port Huron
Open year-round

Port Huron Museum............... 810-982-0891
Port Huron
Open year-round

Museum Ship Valley Camp......... 906-632-3658
Sault Ste. Marie
May through October

New Presque Isle Lighthouse......... 989-595-9917
Presque Isle
May through October

**Old Presque Isle Lighthouse
& Museum**..................... 989-595-2787
Presque Isle
May through October

ONTARIO

Collingwood Museum............ 705-445-4811
Collingwood
June through September

Discovery Harbour.............. 705-549-8064
Penetanguishene
May through September

**Fathom Five National
Marine Park**.................... 519-596-2233
Bruce Peninsula
June through September

Huron County Museum.......... 519-524-2686
Goderich
Open daily; closed Saturdays in winter

Marine & Rail Museum........... 519-371-3333
Owen Sound
Open daily with reduced winter hours

**Mississagi Lighthouse
Heritage Park**.................. 705-283-1084
Meldrum Bay
May through September

Museum Ship Norgoma.......... 705-256-7447
Sault Ste. Marie
Mid-June through Canadian Labour Day

**Nancy Island
National Historic Site**............ 705-429-2728
Wasaga Beach
Weekends in May, then daily June through September

Ste. Marie-Among-the-Hurons....... 705-526-7838
Midland
May through October

**West Parry Sound
District Museum**................ 705-746-5365
Parry Sound
Open daily with reduced winter hours

LAKELAND BOATING · LAKE HURON CRUISING GUIDE   **235**

Desbarats, Ontario

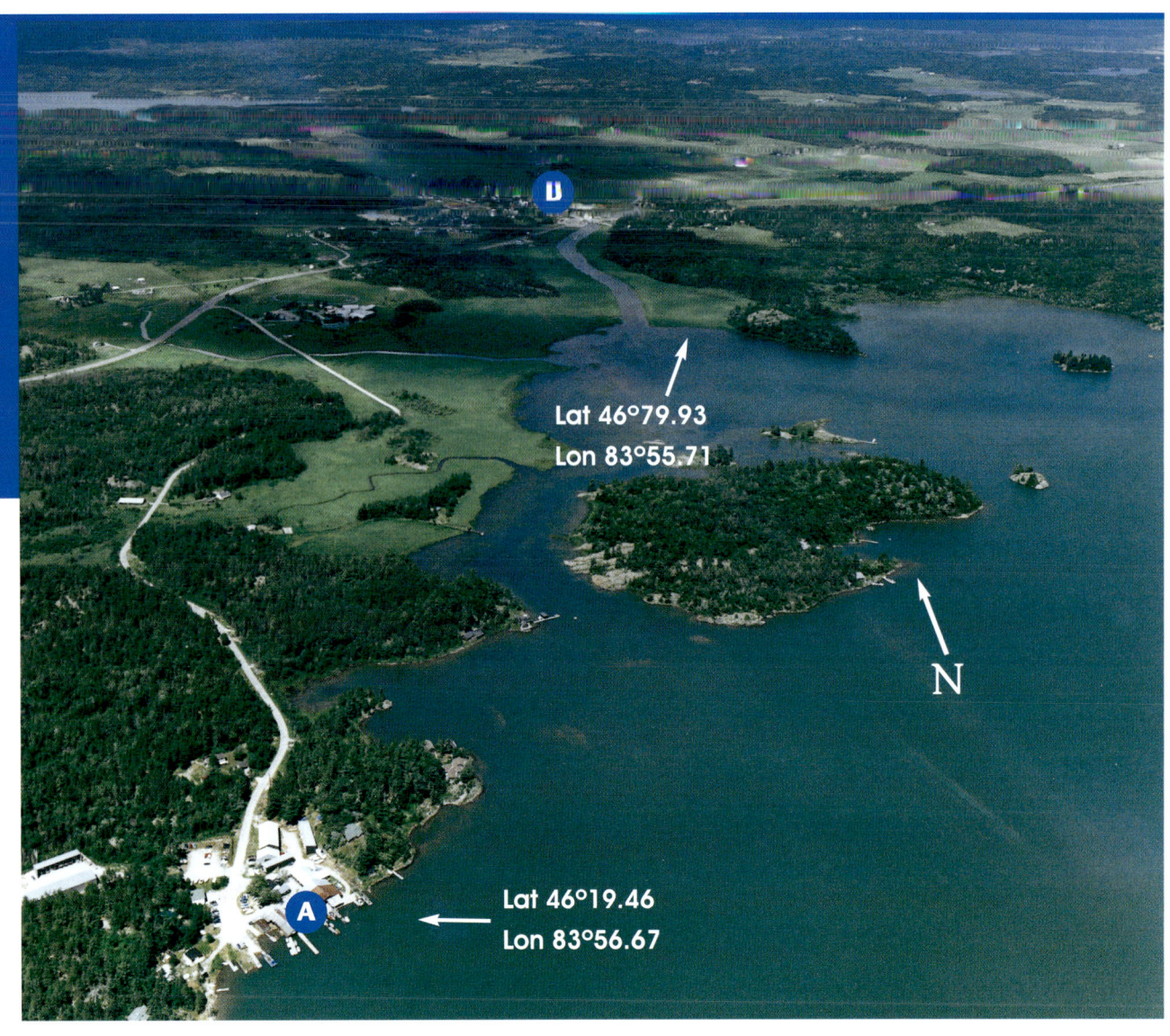

Lat 46°79.93
Lon 83°55.71

Lat 46°19.46
Lon 83°56.67

Just east of the Gilbertson Bridge (which links St. Joseph Island to the mainland) on the north shore of St. Joseph Channel are two facilities that should be of interest to the cruiser, though neither officially accepts overnight guests. Holder Marine **A**, located at Kensington Point, offers gasoline and performs engine and hull repairs. The Desbarats Municipal Docks **B**, about two miles east of Kensington Point on the Desbarats River, welcome visitors during the day and are close to a useful cluster of stores.

Facilities information subject to change. We suggest you call ahead.

			Monitors VHF Channel	Transient Slips Available	ALTERNATE MOORING: Wall mooring, Rafting allowed	Maximum LOA	Minimum Depth at Dock	Power (amperage or volts)	HOOKUPS: Water, Cable TV	FUEL: Gas, Diesel, Pumpout	BASICS: Heads, Showers, Laundry	AMENITIES: Swimming pool, Whirlpool, Grills, picnic tables, Dog walk, Rec area	Take Reservations	Take Credit Cards	Haulout (Capacity in tons or feet)	REPAIRS: Mechanical, Electronics, Fiberglass, Woodworking, Sails, Canvas	CONVENIENCE: Ship's store, Ice, Convenience foods & beverages
B	Desbarats Municipal Docks No phone	Highway 17 Desbarats, ON P0R 1E0	12			20'	6'										
A	Holder Marine 705-782-6251	39 Kensington Rd. Desbarats, ON P0R 1E0				35'	5'			G		D	Y			MEFWSC	SC

236   LAKELAND BOATING · LAKE HURON CRUISING GUIDE

THE NORTH CHANNEL

Desbarats, Ontario
Two handy spots for smaller craft.

For cruisers who have rounded St. Joseph Island from either the east or west, or who are striking out for the North Channel from Sault Ste. Marie, St. Joseph Channel not only provides many interesting nooks and crannies, it also has two facilities on its mainland shore that could prove quite handy. However, because the Gilbertson Bridge has a clearance of only 37 feet, taller boats—sailing vessels in particular—are stymied at this point, and the channel remains unexplored by many boaters.

Holder Marine's main clientele are the numerous cottagers who summer in the area, but the facility offers fuel for the cruiser. It's also one of the few places for repair work between Sault Ste. Marie and Thessalon. Holder services both inboard and outboard engines and offers competent fiberglass hull repairs. The marina has no Travelift but can hoist boats up to 30 feet long with its marine railway system. The team at Holder also specializes in restoration of antique wooden boats, which might prove interesting if there is a work in progress when you happen to stop in. The facility carries propane on site and is situated next to a government launch ramp.

Desbarats (pronounced "Debra") is a small community situated about two miles east of Holder Marine on the Desbarats River. The river is relatively shallow—between 6 and 8 feet—and is hemmed in on both sides by reedy banks. But mid-sized and small boats should be able to motor the short distance upriver from the channel to the **Desbarats Municipal Docks**. Here, about 12 spaces are available, with a draft of 6 to 7 feet. Visitors are welcome to tie up, but there are neither services nor supervision. The tiny hamlet offers a handful of nearby stores, which makes Desbarats a good pit stop if you're running low on supplies. Across Highway 17 and a few paces north you'll find **Desbarats Market**, which sells liquor and beer; **McLelland's Hardware**, which has fishing tackle and supplies; and a **post office**. Closer to the docks are a **gas station** and **convenience store**.

If you're poking around in the numerous bays of St. Joseph Channel searching for one of many anchorages, Desbarats offers two marina facilities and a number of handy services.

More Info
Emergency 911
Matthews Memorial Hospital 705-246-2570

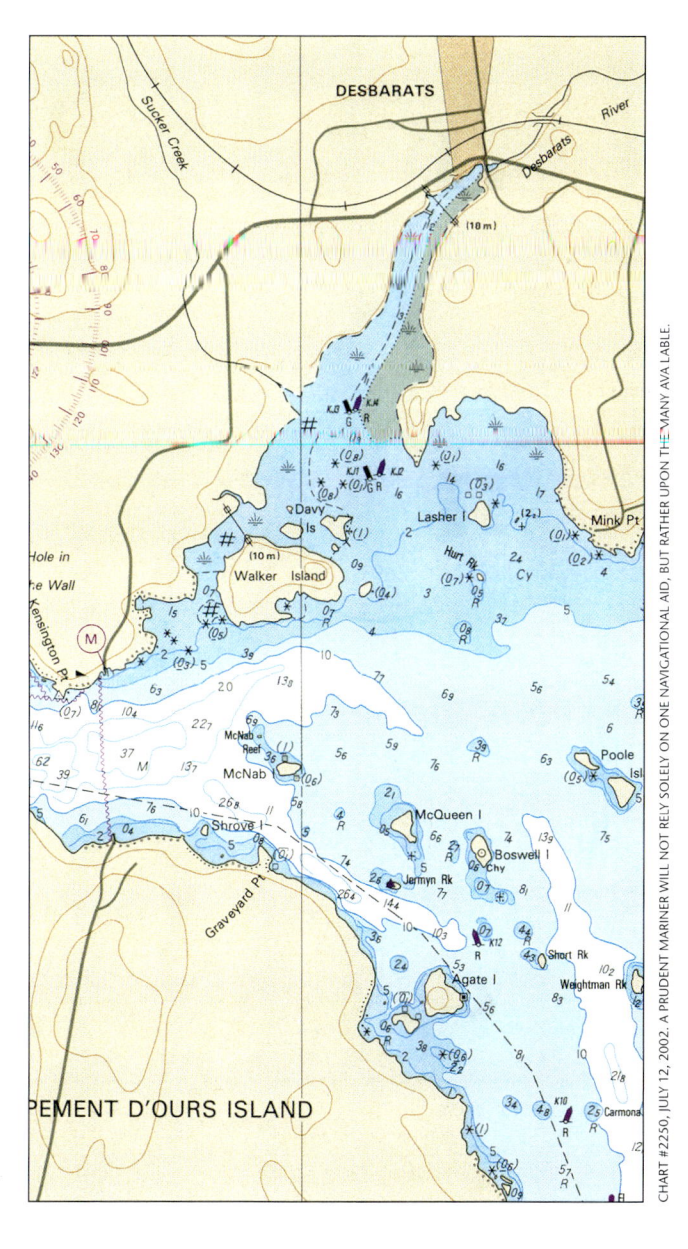

Distances from Desbarats (statute miles)
Richards Landing: 7 SW
Hilton Beach: 7 S

LAKELAND BOATING · LAKE HURON CRUISING GUIDE **237**

Richards Landing, Ontario

Lat 46°17.70
Lon 84°02.20

Located on a short spit extending from the town of Richards Landing on the north shore of St. Joseph Island, the newly expanded Richards Landing Municipal Marina **A** is within sight of both the Gilbertson Bridge to the east and the St. Marys River to the northwest. The western basin **B** has been expanded to accommodate more cruisers. The village **C**—a good place to stock up or begin an exploration of the island—commences right at the foot of the marina.

Facilities information subject to change. We suggest you call ahead.

Marina	Monitors VHF Channel	Transient Slips Available	ALTERNATE MOORING: Wall mooring Rafting allowed	Maximum LOA	Minimum Depth at Dock	HOOKUPS: Power (amperage or volts) Water Cable TV	FUEL: Gas Diesel Pumpout	BASICS: Heads Showers Laundry	AMENITIES: Swimming pool Whirlpool Rec area Grills Picnic tables Dog walk	Take Reservations	Take Credit Cards	Haulout (Capacity in tons or feet)	REPAIRS: Mechanical Electronics Fiberglass Woodworking Sails Canvas	CONVENIENCE: Ship's store Ice Convenience foods & beverages
Clark's Cove Marina Highway 548 705-246-1118 Richards Landing, ON P0R 1J0		10	WR	40'	6'					PD	Y	N	15	
Richards Landing Municipal Marina Rural Route 1 705-246-0254 Richards Landing, ON P0R 1J0	68	30	WR	100'	5'	15 30, 50	W	GDP	HSL	RGPD	Y	Y		IC

THE NORTH CHANNEL

Richards Landing, Ontario

Step ashore and explore St. Joseph Island.

Cruisers navigating the St. Joseph Channel en route to either Sault Ste. Marie or the North Channel will find the charming town of Richards Landing a convenient place to provision as well as an excellent base for exploring St. Joseph Island. The port boasts the island's best shopping, easy access to a number of natural and cultural attractions and an expanded and improved municipal marina.

Coming along St. Joseph Channel toward Richards Landing, stay in the marked channel. When passing under the bridge, don't assume you can cruise through just anywhere. You might have a great deal of clearance above you (the bridge accommodates 37-foot-high boats) but there are rocks underneath, as well as a frequently brisk current. Stick to the recommended passage.

Distances from Richards Landing (statute miles)
Hilton Beach: 12 E
Sault Ste. Marie: 25 W

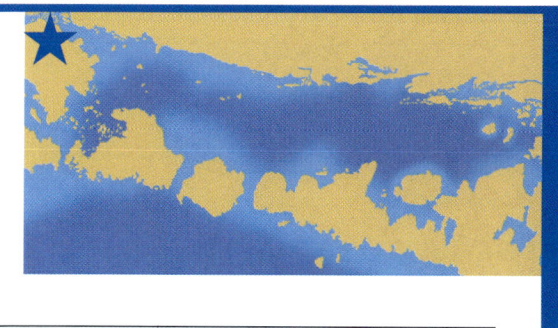

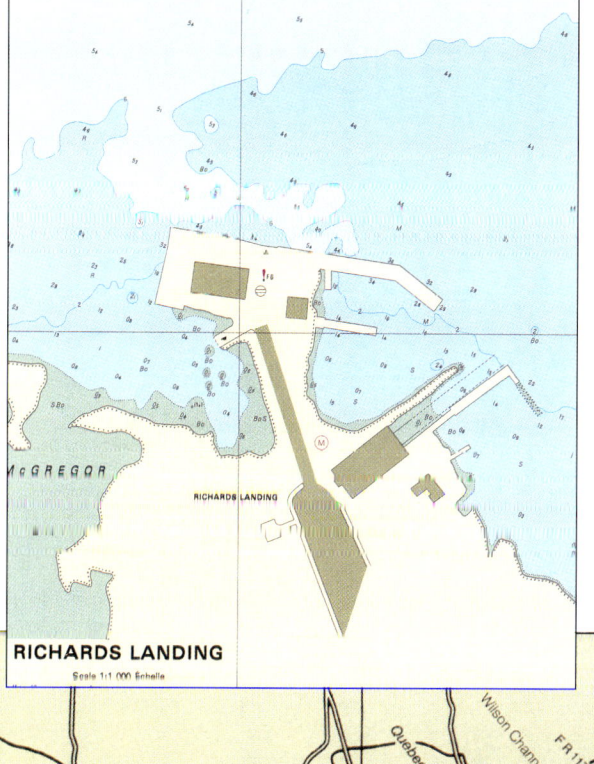

LAKELAND BOATING · LAKE HURON CRUISING GUIDE   239

More Info

Emergency 911
Matthews Memorial Hospital 705-246-2570
Township of St. Joseph 705-246-2625

Where to Eat

Anne's Café 705-246-3122
A local favorite for pizza, this establishment serves tasty fries, too. Take-out available.

Coming from the west, you will pass the Shoal Island Light, better known to locals as the B-Line Lighthouse, on the shore of St. Joseph Island at the junction of the St. Marys River and St. Joseph Channel. There is a green light at the marina.

Richards Landing Municipal Marina, which was taken over by the town in 1998, was expanded in 2003 to provide dockage for 75 boats. The new slips are situated within two new breakwalls on the marina's west flank and have water hookup. The draft at the original dock ranges from 5 to 10 feet. Boats up to 55 feet long can be accommodated inside the new west basin and larger craft on the outside of the main wharf. Fifty-, 30- and 15-amp power is available, with gasoline, diesel and pumpout provided at the service dock. The new marina complex also features a 45-seat restaurant and waterfront deck, a boaters' lounge, restrooms and laundry facilities, and an excellent boat launch in the new western basin. The marina is also a **Canada Customs** check-in point.

Clark's Cove Marina, just west of the town docks, usually has about 10 no-frills spaces for transients. The slips, which can accommodate boats up to 40 feet, have no water or power and the marina has few amenities. It can, however, haul out boats up to 15 tons.

Anne's Café, a pizzeria famed for its pies and homemade breads, is less than a five-minute walk from the marina. Along the short walk up Richards Street, you'll be able to take care of most of your other needs: There is a **grocery store**, a **bank** with a cash machine, a **liquor store**, a **folk art gallery**, a **children's library** and a **post office**.

For enjoyment, trek over to the town **park** and **beach**, which houses a tennis court. St. Joseph Island's country roads were made for cycling, and boaters who bring their bikes will find a number of interesting sights within comfortable cycling range, including artists' studios. **Stribling Point Park** and the **Sailors Encampment** are both nice spots to watch international ships traveling up and down the St. Marys River.

2004 MARINAS & DESTINATION GUIDE

Boating Ontario

A friend you will not leave at the dock...

Boating Ontario: Marinas & Destination Guide is a friend indeed, providing listings of facilities and services of marinas throughout the province as well as extensive destination articles. Ontario provides the best freshwater cruising in the world with over a quarter million lakes, countless rivers and bays.

For a free copy of the Guide visit your local marina, Ontario Tourism Information Centres, or www.marinasontario.com.

www.marinasontario.com

Lat 46°30.19
Lon 84°20.32

The Sault Ste. Maries, Ontario and Michigan, are two bustling, history-steeped ports that sit at the junction of lakes Huron and Superior and form a link between Canada and the United States. Transient cruisers motoring west along the St. Marys River will probably want to pass the Algoma Sailing Club and the Bellevue Marina on the Canadian shore and head directly for the Roberta Bondar Transient Marina **A**, a well-equipped option at the heart of Sault Ste. Marie, Ontario. Nearby attractions include the Roberta Bondar Pavilion **B**, site of concerts and farmers markets, and the Civic Center **C**. On the American side, the George Kemp Downtown Marina, a $3 million, 61-slip facility, was opened in 1998. This municipal marina offers full-service capabilities and lies close to downtown, beside the Museum Ship *S.S. Valley Camp* **D**, just outside the picture. If you are approaching "the Soo" from Lake Superior, you can use either the U.S. locks **E** or the Canadian locks **F**, located beyond the rapids **G**.

Distances from Sault Ste. Marie (statute miles)
Grand Marais: 88 NW
DeTour Village: 48 S
Mackinac Island: 88 SW
Richards Landing: 29 E
Hilton Beach: 37 E
Blind River: 86 E
Meldrum Bay: 86 SE

LAKELAND BOATING · LAKE HURON CRUISING GUIDE   **241**

THE NORTH CHANNEL

Sault Ste. Marie, Michigan & Ontario
Double your fun in this international port.

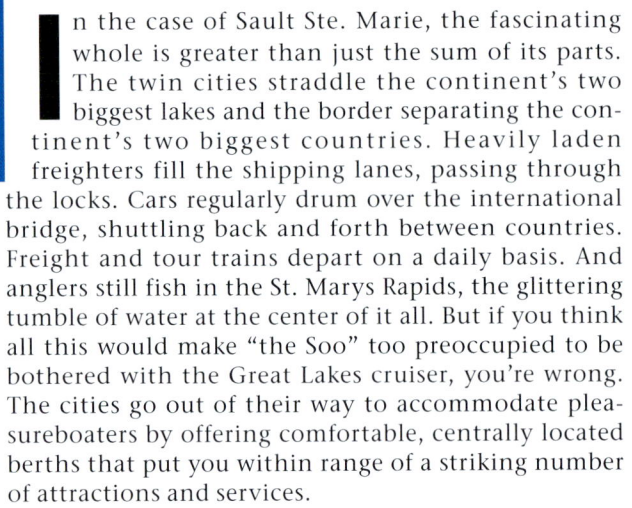

In the case of Sault Ste. Marie, the fascinating whole is greater than just the sum of its parts. The twin cities straddle the continent's two biggest lakes and the border separating the continent's two biggest countries. Heavily laden freighters fill the shipping lanes, passing through the locks. Cars regularly drum over the international bridge, shuttling back and forth between countries. Freight and tour trains depart on a daily basis. And anglers still fish in the St. Marys Rapids, the glittering tumble of water at the center of it all. But if you think all this would make "the Soo" too preoccupied to be bothered with the Great Lakes cruiser, you're wrong. The cities go out of their way to accommodate pleasureboaters by offering comfortable, centrally located berths that put you within range of a striking number of attractions and services.

If you are approaching Sault Ste. Marie from downriver, follow the shipping lanes up the east side of Neebish Island. The downbound channel passes to the west of the island. The inside, or east, channel of Sugar Island is not recommended, because the deepest part is 10 feet or less, with as little as 5-foot depths elsewhere. The shipping lanes boast at least 27 feet of water; however, the river is considerably shallower in some places outside the channel, so stick between the buoys. As you near Sault Ste. Marie, you will encounter several marine facilities on both the U.S. and Canadian sides before you reach the area's main transient options.

Built in 1998, the $3 million, 61-slip **George Kemp Downtown Marina** on the Michigan side offers full service and well-secured docking just five minutes from town. This is the only option for transients in the Michigan Soo now that the **Charles T. Harvey Marina**, located in a side channel near Frechette Point, no longer takes transients.

The **Algoma Sailing Club of Sault Ste. Marie**, the most easterly of the Canadian facilities, occasionally welcomes sailboats but shouldn't be considered a primary transient option. The **Bellevue Marina**, just west of the sailing club, has more than 100 slips, but

242 LAKELAND BOATING · LAKE HURON CRUISING GUIDE

Welcome to the George Kemp Downtown Marina

Where Great Lakes Meet

- ☆ Soft drinks, bag & block ice
- ☆ Gift shop next door at Museum Ship
- ☆ Great cruising & fishing
- ☆ Watch lakers entering & departing Soo locks
- ☆ Located right downtown in the heart of the tourist district
- ☆ Open 15 May through 15 October
- ☆ Open 24 hours during peak season
- ☆ Accommodate vessels up to 75'
- ☆ Average mean depth of basin 9'
- ☆ Dog run

- ☆ 30-amp or 50-amp service
- ☆ Dock lighting & fresh water at each slip
- ☆ Midgrade gasoline, Premium diesel, Pumpout
- ☆ ADA Accessible well-maintained restrooms
- ☆ Private showers
- ☆ Coin-operated laundry facility
- ☆ Courtesy bicycles
- ☆ Courtesy Weber gas grills public telephone
- ☆ Internet modem connection FAX available Complimentary coffee in lounge each morning

GEORGE KEMP MARINA
Sault Ste. Marie Michigan

George Kemp Downtown Marina
TEL 906-635-7670 FAX 906-635-7690 VHF Channel 9
www.kempmarina.com • E-mail harbormaster@kempmarina.com
admin@kempmarina.com • 185 Water St., Sault Ste. Marie, Michigan 49783

---

most are reserved for local boaters. Continue upriver to the **Roberta Bondar Transient Marina**, a white building with a red roof located just west of the large glass **Civic Centre** building and just east of the big tent pavilion. Named after Canada's first female astronaut, the Roberta Bondar Marina offers 57 transient slips with 30-amp power, water and a generous draft. The marina also provides gasoline, diesel, pumpout, showers, laundry facilities, charts and other products. The *M.S. Norgoma*, the last passenger ferry construct-

More Info

Ontario
Emergency	911
Tourism Information	705-945-6941
	www.ontariotravel.net
Chamber of Commerce	705-949-7152
	www.ssmcoc.com
7500 Taxi	705-945-7500

Michigan
Emergency	911
War Memorial Hospital	906-635-4460
Chamber of Commerce	906-632-3301
	www.saultstemarie.org
Independent Taxi	906-632-4664

Where to Eat

Arturo Ristorante 705-253-0002
Seafood, steaks, veal, rack of lamb, live lobster and more.

Cesira's Italian Cuisine 705-949-0600
Licensed white-tablecloth dining with Italian specialties and seafood.

Freighters Restaurant 906-632-4211
Full menu with daily specials, such as prime rib sandwiches and pasta primavera. Located in Michigan's Ojibway Hotel at the edge of the locks.

Lockview Restaurant 906-632-2772
The menu offers hamburgers, chicken and steak, but most people come for the seafood. On the U.S. side.

Swiss Chalet and Roost Bar 705-256-2677
Excellent dining to please your taste buds.

Time Out Steakhouse & Bar 705-949-0127
More than 20 surf and turf combos to choose from in a relaxed atmosphere.

Vincenzo's 705-256-8241
Homemade Canadian and Italian cuisine.

ed on the Great Lakes, is tied up in the same harbor. Although not fully restored yet, it is open to the public.

Part of a new waterfront development, the marina is close to a host of other interesting points. Right next door, the **Roberta Bondar Pavilion** features local festivals, music, exhibitions and farmers markets (on Wednesdays and Saturdays). A **snack bar** sells ice cream and fast food. A few steps down the boardwalk, which stretches almost a mile in total, is the office of **Lock Tours Canada Boat Cruises**. The tour boat leaves four times daily throughout the summer, taking passengers on a two-hour cruise through the American and Canadian locks. After closing temporarily due to a crack in the structural wall, the **Sault Canal Historic Site** is now fully restored and open to pleasurecraft. The lock is 250 feet long and 50 feet wide, with a draft of 9.8 feet. The site has much to offer, including guided tours of historic buildings and engineering works, indoor and outdoor exhibits, an observation deck, a 2.2-kilometer trail to the famous St. Marys Rapids, excellent birdwatching, world-class fly fishing and picnic areas. The site also boasts the only remaining emergency swing bridge dam in the world. For lockage information, call 705-941-6262 or visit parkscanada.pch.gc.ca/sault on the Internet.

Before the **Soo Canals** were opened in 1855, boatmen laboriously hauled boats from Lake Huron to Lake Superior, a distance of about 63 miles. The schooner *Algonquin* made the trip from December 1839 to April 1840, passing through the main streets of Sault Ste. Marie en route.

The area has a number of marine repair services, though none are located right at the waterfront. **Purvis Marine, Ltd.** can haul your boat out at the government dock if necessary, with crane and slings. **Superior Marine** provides a mobile mechanic, as do **H&S Outdoor Equipment** and **ProMotion Power Sports**. **Eagle Engine and Machine Co.** offers mobile

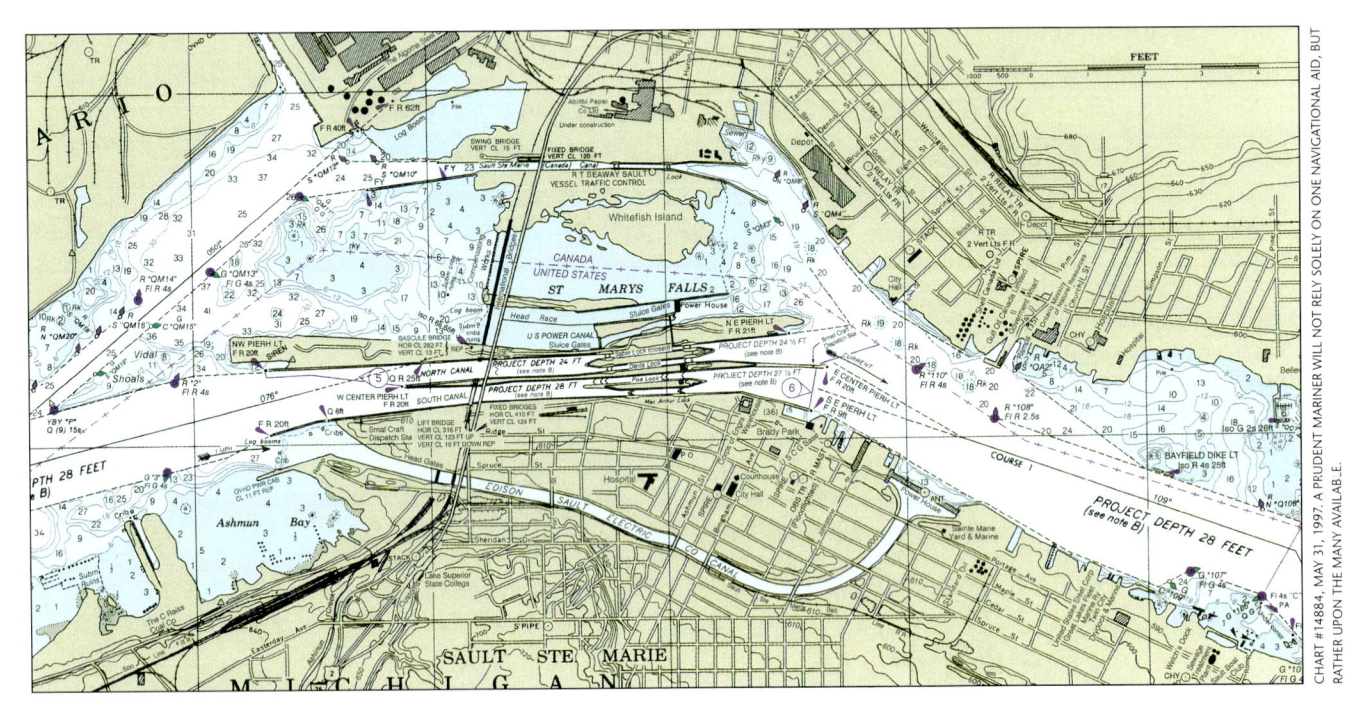

Facilities information subject to change. We suggest you call ahead.

		Monitors VHF Channel	Transient Slips Available	ALTERNATE MOORING: Wall mooring, Rafting allowed	Maximum LOA	Minimum Depth at Dock	Power (amperage or volts)	HOOKUPS: Water Cable TV	FUEL: Gas Diesel Pumpout	BASICS: Heads Showers Laundry	AMENITIES: Swimming pool Whirlpool Grills Picnic tables Dog walk Rec area	Take Reservations	Take Credit Cards	Haulout (Capacity in tons or feet)	REPAIRS: Mechanical Electronics Woodworking Sails Canvas Fiberglass	CONVENIENCE: Ship's store Ice Convenience foods & beverages
D	George Kemp Downtown Marina 485 E. Water St. 906-635-7670 Sault Ste. Marie, MI 497839	45			110'	10'	50	30 W	GDP	HSL		RGPD	Y	Y		IC
A	Roberta Bondar Transient Marina 99 Foster Dr. 705-759-5430 Sault Ste. Marie, ON P6A 5N1	68	57	W	80'	10'	30	W	GDP	HSL		PD	Y	Y		SIC

244 LAKELAND BOATING · LAKE HURON CRUISING GUIDE

service for diesel engines. There are also two electronic repair businesses in town, which can send someone to your boat if you have radio or radar problems.

Good launch ramps are located at the George Kemp Downtown Marina on the Michigan side and at Bellevue Marina at the foot of Pine Street on the Ontario side.

For provisioning, dining or entertainment, nothing is too far away from Bondar Marina. The **Station Mall**, located just west of the pavilion, boasts 130 stores and services including a **postal outlet**, a **laundromat**, a **liquor store**, **banks**, a **pharmacy**, a **grocery store**, a multi-screen **cinema** and restaurants. Queen Street, the main drag of Sault Ste. Marie, Ontario, is just two blocks north of the marina.

If you have a whole day to spare, the **Agawa Canyon Train Tour** offers a stunning wilderness trip by rail. It departs from the Station Mall at 8 a.m. and returns by 5 p.m. En route to the canyon you will skirt numerous rivers and lakes, cross immense trestles and experience the same rugged beauty that inspired Canada's Group of Seven painters to put brush to canvas years ago.

For history buffs, there are a number of spots, all within walking distance of each other. The **Sault Ste. Marie Museum**, housed in a former post office, is at the corner of Queen and East streets. **Ermatinger's Old Stone House**, said to be northwestern Ontario's oldest residence, is a couple of blocks away. And the **Canadian Bushplane Heritage Centre** has a wide variety of float planes on display in its waterfront hangar, a 15-minute hike from the docks at Bay and Pim streets. There are also two **art galleries**: one right downtown, the other on the waterfront.

Across the river lies Sault Ste. Marie, Michigan, and the popular **Vegas Kewadin Casino**, as well as the **Museum Ship *S.S. Valley Camp***. The Michigan-based **Dial-A-Ride** (906-632-6882) departs from the Canadian bus terminal six days a week for a buck.

But who needs a lift to a casino when there's one located right on the waterfront? Boaters can simply stroll down the boardwalk to the west end of the Station Mall to find themselves on the doorstep of the **Sault Ste. Marie Casino**, with 450 slot machines and 60 tables. For information, call 800-828-8926.

For golf enthusiasts, the Soo boasts the fantastic 18-hole **Sault Ste. Marie Golf Club** on Queen Street East. Another option is **Tanglewood Marsh Golf Course**. Those seeking a shorter excursion should head to the nine-hole executive course at **Queensgate Greens**.

For tourist information, the **Chamber of Commerce** operates a kiosk in the mall seven days a week. Friendly travel counselors make recommendations. The website, www.ssmcoc.com, provides up-to-date listings of special events, attractions, dining, tours—even a live weather camera.

By the time you say goodbye to "the Soo," you are guaranteed to be well-stocked and satiated, even if you haven't managed to sample all of the city's many offerings.

Cruise the heart of the Great Lakes

SAULT STE. MARIE
Ontario Canada
naturally gifted

Roberta Bondar Marina is situated at the heart of two of North America's greatest cruising ranges — the scenic splendour of the North Channel of Lake Huron and the rugged beauty of the North Shore of Lake Superior. The full-service marina is located adjacent to the downtown of Sault Ste. Marie, just a short walking distance to shopping, restaurants, entertainment and area attractions including the Agawa Canyon Tour Train, Casino Sault Ste. Marie, the Sault Ste. Marie Canal and Recreational Lock, Canadian Bushplane Heritage Museum, Ermatinger/Clerque Heritage Site, Sault Ste. Marie Museum and Art Gallery of Algoma.

FREE SAULT STE. MARIE VISITORS GUIDE
1-800-461-6020 or www.sault-canada.com

- 38 serviced slips up to 45'
- Over 500' of serviced dock wall
- Minimum depth of 10'
- Fuel, showers, pump-out, laundry, ice, charts, radio watch on Channel 68
- Canada Customs check-in point
- On St. Marys River boardwalk at Roberta Bondar Park & Pavilion

Roberta Bondar Marina
Your Full-Service Port-of-Call
INFORMATION/MARINA RESERVATIONS
1-800-361-1522 or (705) 759-5430
ssm.gardens@cityssm.on.ca

Section Three

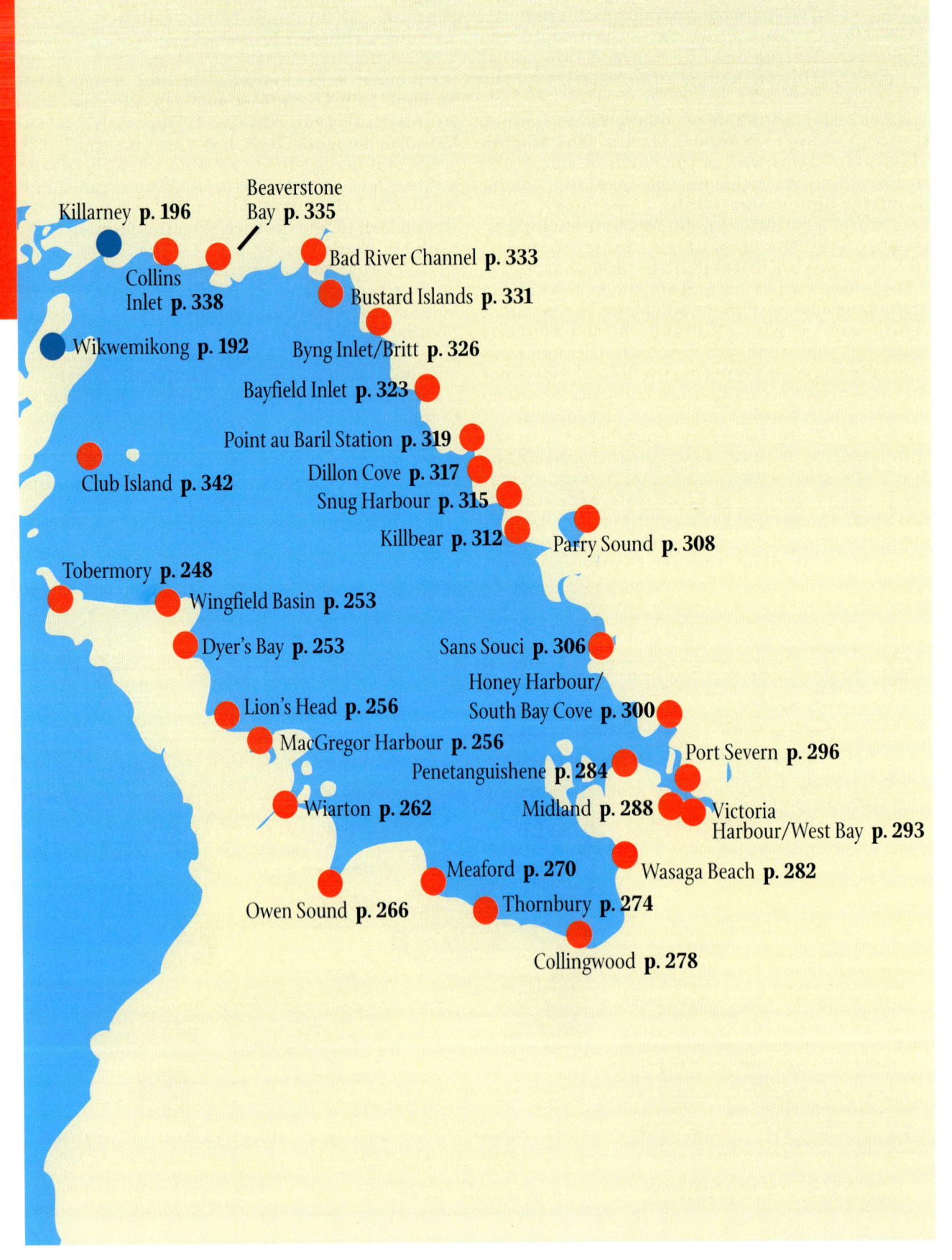

- Killarney **p. 196**
- Beaverstone Bay **p. 335**
- Bad River Channel **p. 333**
- Collins Inlet **p. 338**
- Bustard Islands **p. 331**
- Wikwemikong **p. 192**
- Byng Inlet/Britt **p. 326**
- Bayfield Inlet **p. 323**
- Point au Baril Station **p. 319**
- Dillon Cove **p. 317**
- Club Island **p. 342**
- Snug Harbour **p. 315**
- Killbear **p. 312**
- Parry Sound **p. 308**
- Tobermory **p. 248**
- Wingfield Basin **p. 253**
- Dyer's Bay **p. 253**
- Sans Souci **p. 306**
- Honey Harbour/South Bay Cove **p. 300**
- Lion's Head **p. 256**
- MacGregor Harbour **p. 256**
- Port Severn **p. 296**
- Penetanguishene **p. 284**
- Wiarton **p. 262**
- Midland **p. 288**
- Victoria Harbour/West Bay **p. 293**
- Meaford **p. 270**
- Wasaga Beach **p. 282**
- Owen Sound **p. 266**
- Thornbury **p. 274**
- Collingwood **p. 278**

246 LAKELAND BOATING · LAKE HURON CRUISING GUIDE

Georgian Bay: The Sixth Great Lake
Islands and coves call to sweetwater mariners.

Although technically part of Lake Huron, Georgian Bay is separated by a series of islands stretching from Manitoulin (the world's largest freshwater island) southward to the tip of the Bruce Peninsula.

The archipelago guards Georgian Bay's 6,000 square miles of water and over 90,000 islands.

French explorer Samuel de Champlain came here in 1615 and called Georgian Bay *la mer douce*—the sweetwater sea. He traveled up the Ottawa River to the French River and into the bay at its northeastern corner. From there he went south through the thousands of islands to present-day Penetanguishene and Midland. Boaters following in his footsteps will find extraordinary beauty, fascinating history and dozens of unique ports.

Our *Lake Huron Ports O' Call* offers a smorgasbord of cruising in the lake proper, Georgian Bay and the North Channel. We begin our travels on the Michigan side of Lake Huron and continue up the Canadian side to Cape Hurd and Tobermory—the southwestern entrance to Georgian Bay and the North Channel.

Leaving Tobermory, we head east to Wingfield Basin at Cabot Head, following the headlands of the Niagara Escarpment, which runs from Niagara Falls up the Bruce Peninsula. There it slips underwater (except for its many island outcroppings) before rising up as Manitoulin Island, and from there continues westward, with outcroppings all the way to Wisconsin's Door Peninsula on Lake Michigan.

The eastern shore of the Bruce Peninsula, called the "Saugeen Shore," is made up of towering gray limestone bluffs—all part of the Niagara Escarpment—topped by twisted pines several hundred feet tall.

Protected from the prevailing westerlies, this shore provides excellent cruising. While only Lion's Head, Wiarton and Owen Sound provide dockage, food and supplies, there are a number of protected anchorages along the route.

After you pass Meaford, Thornbury and Collingwood along the bottom of Nottawasaga Bay, you'll find mostly beach communities with no dockage and few facilities along the eastern shore from Wasaga Beach north (until you reach Albert's Marina at Gidley Point, which itself only provides gas and emergency dockage).

Rounding Cedar Point on the mainland and going inside Christian Island past its lighthouse, you'll see islands—30,000 of them (actually, 90,000 have been charted so far!).

Continuing on to Breubeof Light, turn south to Penetanguishene and Midland.

Tobermory Ontario

From the southwest, the most direct route is through the well-charted Cape Hurd Channel **A**. You can round Big Tub light **B** on Lighthouse Point and enter Big Tub Harbour **C** to reach Big Tub Resort **D**, under new ownership since May 2003. The resort offers gas, diesel and pumpout. Tobermory Marina **E**, which plans to have gas and diesel installed by the 2004 season, also has an excellent launch ramp. The ferry terminal on Middle Point **F** is the embarkation point for travelers taking the *Chi-Cheemaun* ferry **G** to South Baymouth on Manitoulin Island. Gas should soon be available at the old Coast Guard dock **H**.

Distances from Tobermory (statute miles)

Wingfield Basin: 21 E
Killarney: 53 NNE
Little Current: 66 NNW
Parry Sound: 83 E
Midland: 101 SE
Sarnia: 176 S
Goderich: 134 S
Presque Isle, Michigan: 84 W
Harbor Beach, Michigan: 107 SSW

248 LAKELAND BOATING · LAKE HURON CRUISING GUIDE

GEORGIAN BAY

Tobermory, Ontario
Stop and smell the orchids.

Perched at the tip of the beautiful Bruce Peninsula, Tobermory wows visitors with scenic stone ledges, towering cliffs and the world-famous Bruce Hiking Trail. The surrounding waters rival the Caribbean in color and clarity and are filled with wrecks—some of which can be explored by snorkel from the shore. Adding to the area's charms is its unique flora—including 41 species of orchid and a carnivorous plant or two.

Once a booming fishing and lumbering capital, Tobermory is now a sleepy little town of 400 residents—and boatloads of tourists—that is surrounded by the vast **Bruce Peninsula National Park** and **Fathom Five National Marine Park**. It is the last layover before tackling the wilderness to the north and east.

Called Collins Harbour by Admiral Bayfield, who surveyed much of the Bruce, the name was changed by one of the early Scottish settlers. Prior to the arrival of the English, the entire area was occupied by members of the Huron Nation, and in the early 17th century, French explorers headed by Samuel de Champlain arrived. However, the French quest for a passage to the Orient carried most of those early voyageurs through the North Channel to western points. Tobermory thrived as long as the fish and lumber lasted, then fell into a sleepy period, revived in this century by the arrival of tourism. This is an area of great natural beauty, both on the land and on the water. A cruiser would do well to spend at least a few days exploring before moving on.

Tobermory Harbour is divided into two sections: Big Tub Harbour to the southwest is home to Big Tub Resort, while Little Tub Harbour to the southeast is where the yacht basin is located, along with restaurants and shops. The entrance is marked to the west by Big Tub Light on Lighthouse Point, which shows a flashing red from a white hexagonal tower 43 feet high. To the northeast is the North Point light, with a flashing green from a square skeleton tower 34 feet high.

The area has a fascinating geologic history. The thick limestone bed known as the Niagara Escarpment runs from Niagara Falls along the entire Bruce Peninsula to Tobermory before diving under water, resurfacing as Manitoulin Island and, even farther on, as Wisconsin's Door County. This slab, which is 1,500 feet thick in places, is responsible for the cliffs and ragged islands that were carved and cut by the last glacier. The straight sides of Little and Big Tug harbors were also cut by this glacier and not by human hands. This fossilized coral was formed during the Silurian Period some 420 million years ago. While the Bruce Peninsula is 45 degrees north of the equator—or halfway between the equator and the North Pole—the coral reef was located 10 degrees south of the equator at the time it was formed.

The glacier-ravaged remnants of the escarpment off the tip of the Bruce combine with the possibility of fog to require sharp navigational skills when approaching from any direction except east, which is straightforward.

From the west, the safest route is to round Cove Island on the north side, keeping the O'Brien Patch to port until you can round the lighthouse. Then carry a southerly course between North Otter and Echo islands to the harbor mouth.

From the south, there are four channels to choose from that pass between Cove Island and the mainland: North Channel has a group of shoals obstructing its southeast

ETHAN MELEG

LAKELAND BOATING · LAKE HURON CRUISING GUIDE

Where to Eat

Bootlegger's Cove Pub 519-596-2191
Located at the Big Tub Resort, it offers tasty pub food, beer, spirits and a view of Big Tub Harbour.

Craigie's Harbourview Restaurant 519-596-2867
This casual eatery at the head of Little Tub Harbour has been serving its speciality, fresh fish and homemade french fries, since 1932.

Crowsnest 519-596-2575
On the edge of Little Tub Harbour, this place features a cafe/deli downstairs and a pub upstairs.

The Grandview Restaurant 519-596-2220
At the eastern end of the harbor, this charming restaurant has knockout views and great food.

The Lighthouse Restaurant & Tavern 519-596-2281
Located a block from Little Tub, char-broiled steaks and fresh Georgian Bay fish are the house specialities.

Little Tub Bakery 519-596-8399
Delicious snacks near the harbor.

More Info

Emergency 911
Tobermory Health Clinic 519-596-2305
Chamber of Commerce   519-596-2452
                                                         www.tobermory.org
Ransbury Pro Hardware 519-596-2202

end, making it unsuitable for passage for those without local knowledge; MacGregor Channel is not recommended in the Canadian Sailing Direction because it is unmarked and has strong currents; Devil Island Channel is marked by a light range and buoys and can be used when visibility allows; and Cape Hurd Channel is well-marked and the fastest approach to Tobermory. This is the route taken by 99 percent of cruisers. Currents as strong as 6 knots have been reported in the Devil Island and Cape Hurd channels following changes in wind direction.

The street that rings the yacht basin is lined with shops and stores catering to tourists and boaters. The **Blue Heron Co.'s** chart shop on the north side of Little Tub has one of the biggest collections of area charts available anywhere. It also offers a large selection of books, clothing and nautical gear. **Peacock's Meat & Grocery**, with fresh meat and an in-store laundry facility, is on the north side of Little Tub. Next to it is the **Little Tub Laundromat**. Must-stops for art lovers are **Kent Wilkens' Golden Gallery** and the **Circle Arts Gallery**, which has been showcasing some of Canada's finest artists and artisans since 1969. A two-mile-long shoreline walkway follows the south side of the harbor and affords vistas of breathtaking sunsets. The wooden boardwalk ends at the **Tugs**, where four wreck sites are accessible from shore. Numerous glass-bottom and dive boats are available for more ambitious underwater exploring.

Significant Waypoints
1 mile north of Tobermory harbor:
45°16.67N/81°40.20W
Entrance to Tobermory harbor:
45°15.67N/81°40.20W

Local events

May
Whitefish Fry
Bruce Peninsula Orchid Festival

June
Chi-Cheemaun Festival

July
Canada Day Celebration

July & August
Chicken Bar-B-Que

September
Celtic Ceilidh

Legendary Flowerpot Island

According to legend, Shining Rainbow, an Ottawa princess, and Bounding Deer, a prince of the Chippewa, fell in love long ago, though their tribes were at war. One night, the two lovers decided to escape by canoe. Pursued by Shining Rainbow's tribesmen, the pair bravely decided to try to take refuge on what is now Flowerpot Island, from which no one had ever returned. As Shining Rainbow and Bounding Deer neared and prepared to jump onto shore, Little Spirit, the powerful island deity, decided to prove his strength. As the canoe touched ground, he turned the lovers into stone.

Flowerpot Island's two natural stone structures are said to have more closely resembled human figures before the wind and water eroded them. However, if you look closely at the larger "flowerpot," you just might see what appears to be a face. The island is the only one in the marine park with public facilities. And Little Spirit hasn't turned anyone into stone for quite some time.

Facilities information subject to change. We suggest you call ahead.

			Monitors VHF Channel	Transient Slips Available	ALTERNATE MOORING: Wall mooring, Rafting allowed	Maximum LOA	Minimum Depth at Dock	Power (amperage or volts)	HOOKUPS: Water, Cable TV	FUEL: Gas, Diesel, Pumpout	BASICS: Heads, Showers, Laundry	AMENITIES: Swimming pool, Whirlpool, Rec area, Grills, Picnic tables, Dog walk	Take Reservations	Take Credit Cards	Haulout (Capacity in tons or feet)	REPAIRS: Mechanical, Electronics, Fiberglass, Woodworking, Sails, Canvas	CONVENIENCE: Ship's store, Ice, Convenience foods & beverages
D	Big Tub Resort 519-596-2191	236 Big Tub Rd. Tobermory, ON N0H 2R0	68	10	WR	50'	30'	15	W	GDP	HS	PD	Y	Y			IC
E	Tobermory Marina 519-596-2731	15 Bay St. South Tobermory, ON N0H 2R0	68	38		60'	8'	15 30	W	GDP	HSL	PD	N	Y		M	I

The Chi-Cheemaun Ferry

Hop aboard the *Chi-Cheemaun* ferry if you want to explore the wilderness of Manitoulin Island while you're in Tobermory. The largest passenger/vehicle ship operating on the Great Lakes, it makes trips from early May to mid-October.

While its name literally means "the big canoe" in Ojibway, you'll find this ship quite a bit more comfortable. A venerable local icon, the ferry runs from Tobermory to South Baymouth and back. During the busy cruising months, the *Chi-Cheemaun* makes four trips a day. One-way fare for an adult is $12; $10.70 for seniors; $6 for kids 5 to 11; and children less than 5 years old ride free. Vehicles cost an additional $26.15 and up, depending on weight. If you don't have a car, you can get the Family Plan: two adults and all dependent children can ride for $40.35. There are reduced rates for same-day-return walk-ons.

There is ample seating, a cafeteria and a play area to amuse the kids. The $12 million ferry can hold 143 vehicles and 638 passengers in its belly. Stroll along the mezzanine decks and watch as the boat nears the island. It's the next leg of your journey—and adventure awaits.

For more information, call 800-265-3163 or check out www.chicheemaun.com.

Fathom Five National Marine Park

While you're in Tobermory, you can check out some 22 historical shipwrecks in the amazingly clear waters of the Fathom Five National Marine Park. The park encompasses 45 square miles of water and an archipelago of 19 islands and features numerous large caves and dense forests submerged beneath the surface of Georgian Bay. As you swim by the massive limestone cliffs, look for the remains of ancient corals—evidence that these were once tropical waters.

Among the tragic group of wrecks is the *Forest City*, which was lost in a heavy fog in 1904. Its forward bow lodged into the cliff on the northeast shore of Bears Rump Island. The crew was rescued, but the ship eventually filled with water and slipped beneath the waves. As you swim around the wreckage, you'll notice that the bow plating is still jammed into the rock.

A more recent shipwreck site to poke around is that of the *W.L. Wetmore*, a steam-engine-powered vessel. It sank as a result of the November Gales, an infamous fall weather condition in these parts. The *Wetmore* lies 30 feet below the surface, near Russel Island. Examine the boiler of this 213-foot-long ship and follow the chain that leads to the anchor.

Dive shops, sightseeing and charter vessels and the National Park Diver Registration & Information Centre (519-596-2233, www.parkscanada.gc.ca), are all located within walking distance of Little Tub Harbour in Tobermory.

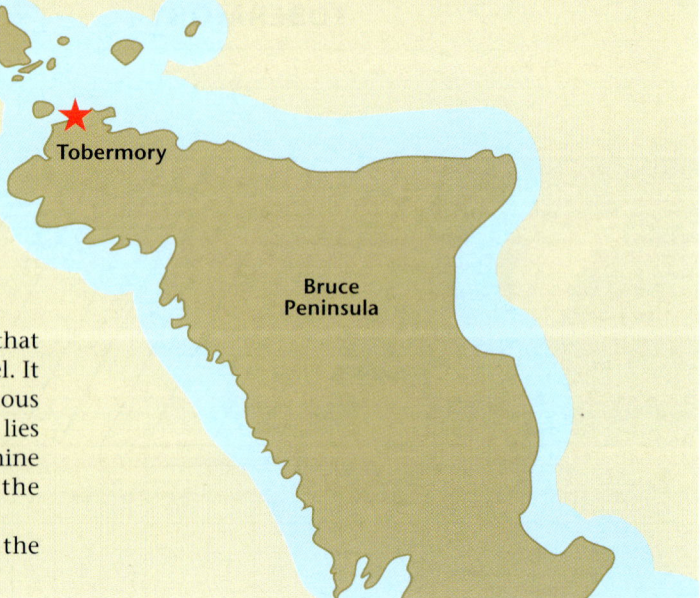

Wingfield Basin & Dyer's Bay, Ontario

Wingfield Basin is easy to spot from the water, thanks to Cabot Head's stunning limestone cliffs **A** and the Cabot Head Light **B** just east of the basin. Fixed red range lights **C** at the head of the bay will lead you on a course of 167.5°T into the buoyed channel. The western half of the basin offers better anchoring and a closer view of the partly submerged wreck of the steam tug *Gargantua* **D**. The lat/lon given is 350 yards north of Wingfield Basin on range 167½.

Distances from Wingfield Basin (statute miles)
Tobermory: 25 W
Club Harbour: 25 N
Killarney: 53 N
Lion's Head: 18 S

LAKELAND BOATING · LAKE HURON CRUISING GUIDE   **253**

GEORGIAN BAY

Wingfield Basin & Dyer's Bay, Ontario

Glaciers, *Gargantua* and a legendary gale.

Wingfield Basin is impossible to miss, thanks to the three limestone cliffs that identify Cabot Head. The basin was created during the last ice age by a giant glacier, which also excavated the ports of Little Tub and Big Tub with such amazing precision. Today Wingfield Basin is a well-protected harbor surrounded on all sides, except for the 350-foot entrance facing due north. Drop the hook in 12 to 21 feet of water and enjoy the isolation. Tobermory is 18 miles west and Dyer's Bay is eight miles south.

Entrance to the basin is marked with flashing red range lights and white daymarks with orange stripes on the south shore. Come in on a course of 167.5°T. The rocky bar at the mouth has been dredged to a depth of 15 feet and a width of 21 feet. There is a small sand beach on the south shore. The Cabot Head Lighthouse is being operated as a museum and visitor center. A dinghy ride to the landing just east of the harbor entrance will set you a short walk to the lighthouse, which offers an excellent view. There you'll find picnic and barbecue facilities and a road to Dyer's Bay. The wreck of the steam tug *Gargantua* lies in the northwest corner of the harbor, partially exposed.

The basin was the scene of a legendary feat of seamanship that ended with the loss of two ships. In October 1886, a storm from the east developed into an unprecedented, now mythic, gale. The sailing barge *Bentley* had just set off from Parry Sound when the first winds struck. With the gale gusting past 65 mph, a violent roll sent the deck cargo overboard. Barely regaining control, the captain set a desperate course for the cliffs of Cabot Head and the protection of Wingfield Basin, knowing that if he missed the cut, they'd be dashed on the cliffs. Amazingly, he made it safely into the basin, only to have a wind change wreck the *Bentley* on the beach. The owner sent a steam barge with a schooner in tow to salvage what was left of the cargo. While en route to their home port, the two ships were hit by another gale. The tow line snapped and the schooner foundered on the rocky shore. This autumn storm became known as the Bentley Gale. The story should serve as a cautionary reminder that this rocky coast of the Bruce is no place to be caught out in a storm.

About four miles to the southwest is Dyer's Bay, an open bight between Cabot Head and Cape Chin. Anchoring is possible on a sandy ledge that extends out from heavily wooded limestone cliffs, but the bay is exposed to all winds except west and northwest, providing little or no protection from Georgian Bay seas. There is a deteriorating public dock with a sign reading "Unsafe: Use at own risk."

More Info

Emergency	911
Grey Bruce Health Services	519-793-3424
Bruce County Tourism	800-268-3838
	www.naturalretreat.com

Where to Eat

Rocky Raccoon Café 519-795-7586
Situated over the dock at Dyer's Bay, this local favorite features global cuisine, including some vegetarian dishes. On calm days, boaters can tie up at the dock in Dyer's Bay, or they can call from Wingfield Basin and the café will send a car to pick them up. Open daily throughout the summer, with limited hours in the fall.

LAKELAND BOATING · LAKE HURON CRUISING GUIDE

Just south of Dyer's Bay on the imposing cliffs of the Niagara Escarpment is Devil's Monument, a land-based flowerpot created by erosion just after the retreat of the glaciers, when the levels of Georgian Bay were much higher. If the day is clear and the waters still, a shoreline cruise from Wingfield Basin to Dyer's Bay to Cape Chin is worthwhile. There are no obstructions and you can cruise several hundred yards from shore, viewing the escarpment, the Bruce Trail and its lookouts. You can even drop the hook for a cool swim.

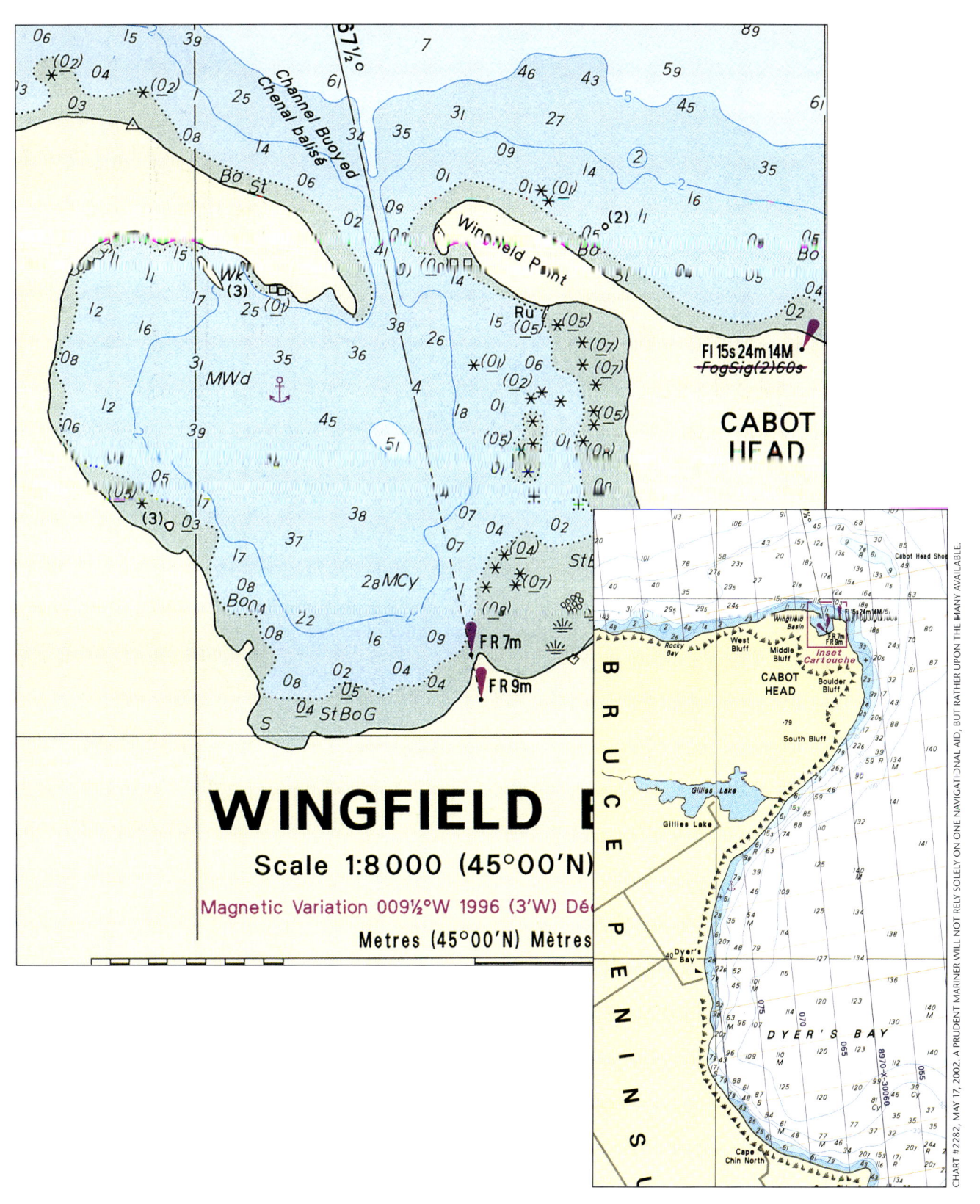

LAKELAND BOATING · LAKE HURON CRUISING GUIDE   **255**

Lion's Head & MacGregor Harbour, Ontario

Lat 44°59.60
Lon 81°14.80

N

At the head of Isthmus Bay, Lion's Head Harbour is an easy approach. Watch for Jackson Shoal, two miles northeast of Lion's Head Point, then make your way into the bay looking for the lighthouse replica **A** or the flashing red light on the end of the rubble pier **B**. Lion's Head Marina's transient docks **C** are immediately to starboard, next to the gas dock **D** and the locals' dock **E**. There is a launch ramp here, too. The village is a short walk away.

Distances from Lion's Head (statute miles)
Wiarton: 36 SE

Midland: 77 ESE

GEORGIAN BAY

Lion's Head & MacGregor Harbour, Ontario
Dramatic cliffs lead the way to quiet bays.

Now a popular and picturesque summer resort with about 500 inhabitants, Lion's Head was once an important commercial port. Logs churned from the mill in Dyer's Bay were brought here for export, and the area was also home to a large fishing fleet. Lion's Head, named for the profile of a lion that can be seen on a nearby cliff overlooking Georgian Bay, lies on the 45th parallel, halfway between the equator and the North Pole.

The port lies at the very southern head of Isthmus Bay. The coast itself consists of high gray cliffs of the Niagara Escarpment falling straight to a limestone beach. It is as dramatic a vista as can be found anywhere, stretching for 50 miles from Cabot Head to Owen Sound. North American Indians and the early voyageurs avoided this coastline, preferring instead to portage across the Bruce. Approaching from Wingfield Basin or Dyer's Bay to the south, there are no dangers, but Jackson Shoal a mile and a half off Lion's Head to the east has red and green spar buoys and a depth of 4 feet or less. If you do approach closely, bear in mind that the green marks the northern edge of the shoal and the red the southern edge. Do not split them!

The harbor is protected on its north side by a long breakwater with a flashing red light at a 27-foot elevation and a red and white rectangular daymark. At night, the light may be difficult to distinguish from the taillights of cars. **Lion's Head Marina**, in the northern bight of the harbor, has most services. Anchoring is not recommended. There is an excellent launch ramp at the north end of the town docks. On the other side of the bay is **Peninsula Marine** (519-793-3122), which doesn't take transients but does all kinds of repairs and has haulout for boats up to 40 feet. It also operates a ship's store.

The tidy village is just a short walk away. It has a **hardware store**, a **bank**, Hellyer's **Food Market** and a government **liquor store**. Lion's Head also hosts a show of local professional artists the first week of August. The Bruce Trail comes right through town along the shore.

Lat 44°55.63
Lon 81°01.91

Tucked into the bight of Cape Croker, MacGregor Harbour **F** is a well-protected anchorage between Barrier Island and Benjamin's Point. You'll need to keep an eye out for Lamorandiere Bank off of Pine Tree Point. In addition to MacGregor Harbour, Melville Sound offers beautiful anchorages in Sydney Bay **G** and Hope Bay **H** with Jackson Cove at its mouth (not shown).

LAKELAND BOATING · LAKE HURON CRUISING GUIDE   **257**

Local events

June
Annual Yard Sale Trail

Canada Day Parade

July
Family Fishing Weekend

August
Bruce Peninsula Art Show

Sandcastle Contest

Bruce Peninsula Fish Derby

Where to Eat

Colonel Clark's Restaurant & Tavern 519-592-5865
Located in a one-of-a-kind building, featuring roadhouse-style food and weekend entertainment.

The Lion's Head Inn 519-793-4601
Convenient to the harbor, the inn serves lunch, afternoon tea and dinner.

The Village Inn 519-599-6100
On Main Street, this casual restaurant is the place for family dining, as well as takeout.

A rustic alternative to Lion's Head is anchoring out in MacGregor Harbour, one of the three bays in Melville Sound. About 10 miles to the southeast of Lion's Head, MacGregor is the most sheltered anchorage in Melville Sound, protected from winds in all directions. Anchor in three fathoms of water on a mud bottom that provides excellent holding. The settlement here is part of the **Cape Croker Indian Reserve**, where visitors are allowed ashore.

More Info

Emergency 911
Lion's Head Hospital 519-793-3424
Bruce County Tourism 800-268-3838

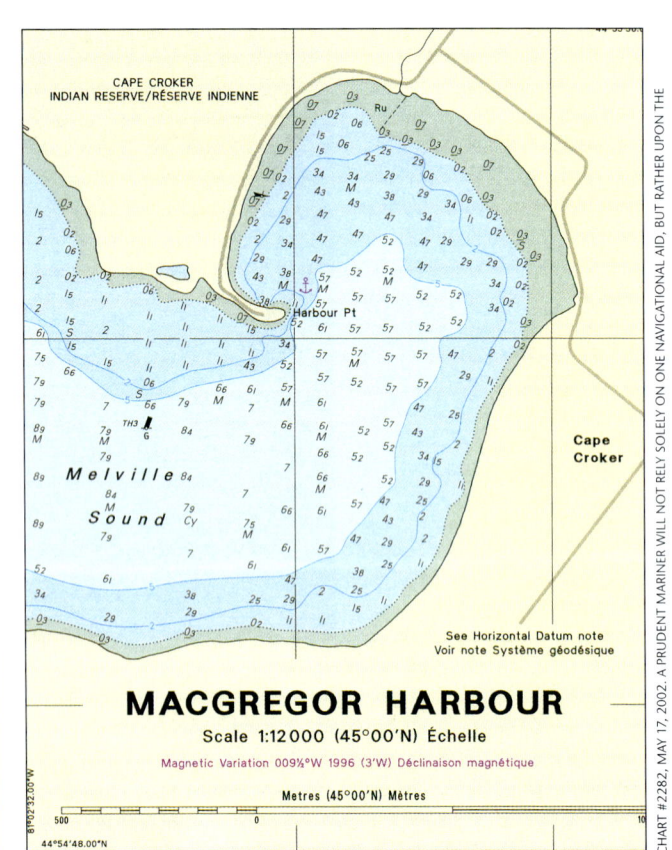

MACGREGOR HARBOUR
Scale 1:12 000 (45°00'N) Échelle
Magnetic Variation 009½°W 1996 (3'W) Déclinaison magnétique

258 LAKELAND BOATING · LAKE HURON CRUISING GUIDE

LION'S HEAD HARBOUR

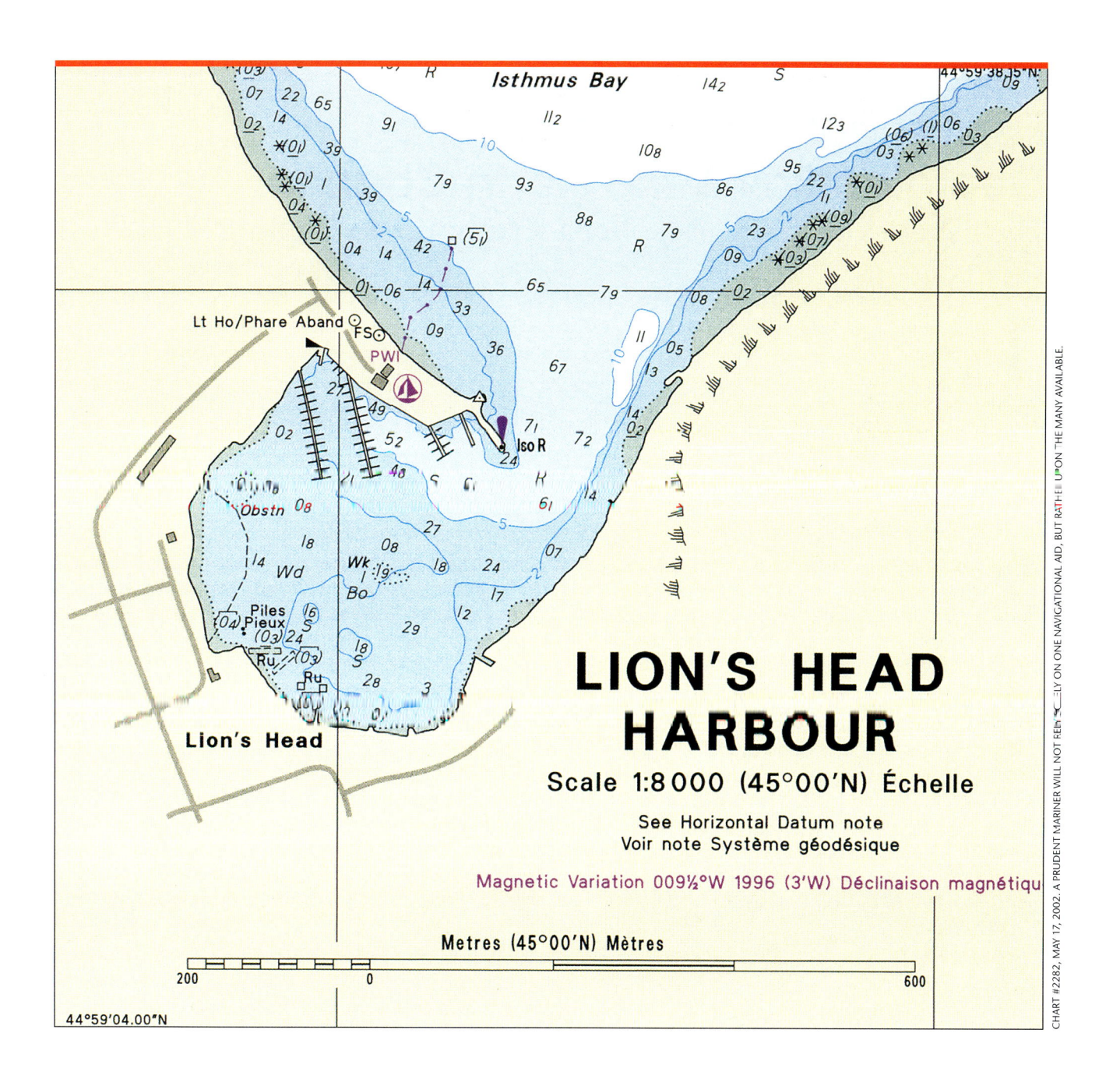

Scale 1:8 000 (45°00'N) Échelle

See Horizontal Datum note
Voir note Système géodésique

Magnetic Variation 009½°W 1996 (3'W) Déclinaison magnétique

Metres (45°00'N) Mètres

Facilities information subject to change. We suggest you call ahead.

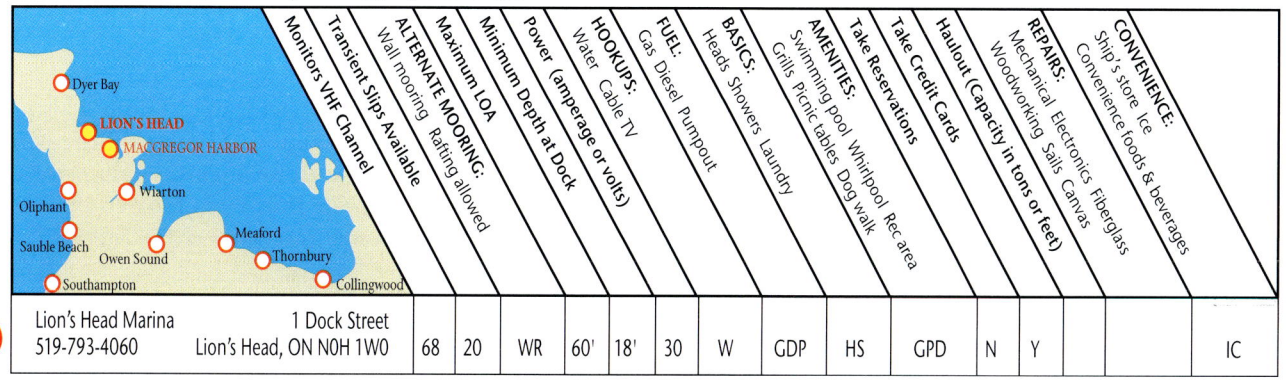

		Monitors VHF Channel	Transient Slips Available	ALTERNATE MOORING: Wall mooring / Rafting allowed	Maximum LOA	Minimum Depth at Dock	Power (amperage or volts)	HOOKUPS: Water / Cable TV	FUEL: Gas / Diesel / Pumpout	BASICS: Heads / Showers / Laundry	AMENITIES: Swimming pool / Whirlpool / Rec area / Grills / Picnic tables / Dog walk	Take Reservations	Take Credit Cards	Haulout (Capacity in tons or feet)	REPAIRS: Mechanical / Electronics / Fiberglass / Woodworking / Sails / Canvas	CONVENIENCE: Ship's store / Ice / Convenience foods & beverages
C	Lion's Head Marina 519-793-4060	1 Dock Street Lion's Head, ON N0H 1W0	68	20	WR	60'	18'	30	W	GDP	HS	GPD	N	Y			IC

FISHING FRENZY

Fishing Hotspots of Lake Huron, Georgian Bay & the North Channel

The fisheries of this region offer a scrumptious sampling of species.

1 Munuscong and Potagannissing Bays: These bays at the Lake Huron end of the St. Marys River offer stellar year-round fishing within 15 miles of each other. Dozens of small islands, numerous points and acres of shoals provide an ideal habitat for smallmouth bass, northern pike and walleye. In fact, some call Munuscong's grassy flats and rocky shore the best walleye fishin' hole in the Michigan. Troll the break lines or cast over the rocky points to take walleye, yellow perch, crappie, large muskellunge and up-to-20-pound northern pike. Smallmouth fishermen will find plenty of great action here as well in early summer. In late summer and early fall, salmon move into the outer channels of Potagannissing's southwest rim along the northern edge of Drummond Island. The 4th of July weekend marks the peak of one of the most unique fisheries in the state, as schools of herring arrive. *Call Ron Papin at Papin's Resort on Drummond Island at 906-493-5234 or Fisherman's Point Resort at 906-287-6671.*

2 Les Cheneaux Islands: Once the premier yellow perch fishery in the state, this area is quietly becoming known as quality smallmouth bass water. The cluster of islands, rocky structure and continuous influx of cooler, forage-rich Lake Huron water keeps the bass fat and feisty. Inshore areas provide good early action before the water warms, but don't forget the barrier islands and the steep dropoffs after midsummer, when inshore waters heat to uncomfortable levels. *Call Linda at the Les Cheneaux Tourist Information Center in Cedarville at 888-364-7526.*

3 Rockport: To many northern Lake Huron trollers, Rockport is at the heart of the best salmon and trout action in the entire lake. Lake trout and steelhead—mixed with a sprinkling of brown trout—start the season soon after ice-out. The parade of big kings lasts well past Labor Day weekend. Fisheries biologists describe this area as the most chinook-rich habitat in Michigan and introduce nearly 1 million salmon fry each spring in Swan Bay and other natal streams. Deep water with plenty of good structure offers even novice trollers the opportunity to catch chinook without much prospecting. Fishing grounds include the wreck of the freighter Nordmere, Middle Island and the rocky structures off the lighthouses of Presque Isle. From Rogers City, try the 100-foot depths off the 40 Mile Point Lighthouse and Adams Point. *Call guide Bob Willick in Alpena at 517-595-2272 or Terry's Bait & Tackle in Rogers City at 517-734-4612.*

4 Oscoda: Calling cards at the Au Sable River mouth include large numbers of lake trout and salmon, as well as steelhead, rainbow, brown trout, walleye, catfish and perch. The river itself yields these species, plus northerns and smallies. In spring, the river's runoff attracts smelt, which in turn draw hungry trout and salmon. Beginning in June and lasting through September, trollers will find ample salmonid success working the 22-mile run from Au Sable Point north to Harrisville. By mid-August, salmon are poised to run the Au Sable, and the sport takes on a carnival atmosphere. Summer evenings find the Oscoda pier packed with anglers in lawn chairs enjoying one of the Great Lakes' best catfish bites. *Call guide Terry Walsh at 517-876-8318.*

5 Charity Islands: These two small islands in northern Saginaw Bay are, perhaps, the premier walleye hotspots on the lake. Primarily a troll fishery with Hot-N-Tots, the rocky structure near the islands attracts and holds many walleye feeding on the area's abundant forage base. Occasionally, large northern pike and trophy-class smallmouth bass can be taken. Fish deep and stock up on baits because the uneven, rocky structure quickly exhausts a tacklebox. *Call Frank's Great Outdoors in Linwood at 517-697-5341.*

6 Grindstone City/Port Austin: Traditionally known for its perch fishery, the tip of Michigan's Thumb is also a strong option for browns, kings and lakers. By mid-May, most trollers chasing chinook run toward the Port Austin Lighthouse, where water warms more quickly, attracting smelt and alewives. By August, salmon crowd the shoreline (trollers must be reef vigilant!). If you are chasing yellow bellies, be aware that zebra mussels have so cleared these waters to make finesse a virtue. The best perch bite occurs in fall. *Call Capt. Fred Davis at 517-738-5271 or the Greater Port Austin Chamber of Commerce at 517-738-7600.*

By Mike Modrzynski

7 **Port Huron:** Fishermen wait impatiently for ice-out, heralding the arrival of the parade of predators moving in to feed on smelt near shore. Steelhead and brown trout are joined by walleye picked up in southern Lake Huron. Drop into the St. Clair River up to a mile and start trolling patterns that cover shoals and deep water breaks. Diving crankbaits and minnow-body plugs are favorite baits for all species. The deep shoulders of the shipping channel are favorite haunts of summer and early fall anglers. *Call the Greater Port Huron Chamber of Commerce at 313-985-7101.*

8 **Owen Sound:** The Sydenham River is the bull's-eye of a great chinook fishery that is entirely supported by fishing club stocking and natural reproduction. The gin-clear water prompts ice-out trollers to start at 100 feet and then work deeper. Other hotspots are off the Beaver and Bighead Rivers in Nottawasaga Bay. The Sydenham also supports a stellar rainbow fishery. *Call Going Fishing Television host Darryl Choronzey at 519-371-1916.*

9 **French River:** This remote river mouth, located in a quiet corner of Georgian Bay, offers boaters the chance to get lost. Hundreds of small humpback granite islands provide a buffer for the actual delta, attracting nearly every species of gamefish that swims in the lake. The islands are the perfect structure for smallmouth; the protected bays and flats are ideal for trophy-class northern pike and muskellunge; and the narrows and chutes between the larger islands washed by the river's spilling current attract good numbers of walleye. *Call Ian McMillan at the Algoma Kinniwabi Travel Association in Sault Ste. Marie, Ontario at 705-254-4293.*

10 **Manitoulin Island:** The waters around this island offer a smorgasbord of season-long possibilities, starting in the spring. Offshore trolling is available for lake trout, steelhead and the occasional salmon; nearshore action is available for smallmouth bass. Once the waters begin to warm, large northern pike, muskellunge, walleye and yellow perch appear. By mid-summer, salmon add to the excitement, and deep water is just minutes from the dock. If your boat is not rigged for offshore action, there's plenty to keep you occupied along the rocky shoreline. This area is not as heavily fished as some southern ports, so you could be fishing alone. *Call Ian McMillan in Sault Ste. Marie, Ontario at 705-254-4293.*

Wiarton, Ontario

Lat 44°44.53
Lon 81°08.08

As you enter Colpoy's Bay you pass the popular anchorage of Kidd Bay on White Cloud Island **A**, which sits in the center of the entrance to the bay. There's a public wharf with no services on the island. Transients can dock at full-service Wiarton Marina, Ltd. **B** or at the bare-bones Government Dock **C** at Bluewater Park. The Barley Bin **D**, an excellent restaurant, is located near the waterfront, as is a large supermarket **E** and government liquor store **F**.

Distances from Wiarton (statute miles)
Owen Sound: 30 S   Midland: 74 E

GEORGIAN BAY

Wiarton, Ontario
Cruise down Colpoy's Bay on Groundhog Day.

At the head of scenic Colpoy's Bay, Wiarton combines a quaint, old-fashioned atmosphere with up-to-date shopping and services. Protected from seas in all directions but the northeast, Colpoy's Bay is ringed by high limestone cliffs and is considered one of the best sailing grounds in Georgian Bay. You can anchor in Kidd Bay on White Cloud Island, which lies at the entrance to Colpoy's Bay. This is a popular weekend anchorage, with deep water to the shore, about 10 miles from Wiarton's harbor. Its neighbor, Griffith Island, is privately owned.

The town sits at the foot of the beautiful Bruce Peninsula, a rocky finger of land pointing north between Lake Huron and Georgian Bay. With more than 500 miles of majestic shoreline, the peninsula offers sweet air, clear waters and unique scenery and is a paradise for photographers, birders and amateur botanists. The fishing, the views along the renowned **Bruce Hiking Trail**, and the scuba diving over the area's many shipwrecks are all unsurpassed.

Many vacationers arriving by plane use **Wiarton Airport**, the largest such facility between Toronto and the Sault. It will accommodate 727's and executive jets.

Entrance to the harbor is straightforward. A dredged basin lies behind a long breakwater that extends from the western shore. **Wiarton Marina**, behind the breakwater, provides all services. The **Government Dock** at the southwest end of the harbor has no hookups but features a fine launch ramp. A sewage outfall marked by a buoy is east of the Government Dock.

Soak up the sun on the white sand shoreline of the bay. **Bluewater Park**, which runs alongside the harbor, has a large community pool, a baseball diamond, the **Pirate Ship Play Park** and a sheltered picnic area. The centerpiece of the park is the restored **Wiarton Railway Station**, built in 1904, which now serves as the **Wiarton Information Centre**. The station was formerly an important rail terminus for shipping lumber, fish, cattle and furniture to U.S. and Canadian cities. The park hosts con-

More Info

Emergency	911
Wiarton Hospital	519-534-1260
Bruce County Tourism	800-268-3838
	www.naturalretreat.com
Information Centre	519-534-2592
Bluewater Taxi	519-534-1086
Grab-A-Cab	519-534-2474
Deacon's Home Hardware	519-534-0500
McKenzie's Pharmacy	519-534-0230

Where to Eat

Barley Bin................519-534-0195
In a turn-of-the-century flour mill just two blocks from the marina. Homemade soups, pies and whitefish.

Coalshed Willie's............519-534-2727
Enjoy a local whitefish dinner or have a burger—they've got it all. And take in one of the best views on the peninsula: beautiful Bluewater Park.

The Green Door.............519-534-3278
A popular lunch spot for boaters.

Wiarton Inn................519-534-3400
Enjoy a great meal in a great atmosphere within walking distance of the marina. The English-style pub and restaurant offers a wide selection of entrees and features a patio and elegant dining room.

LAKELAND BOATING · LAKE HURON CRUISING GUIDE

certs on Sundays throughout the summer.

Another of the town's claims to fame is an albino groundhog, **Wiarton Willie**. When Willie emerges from his burrow and doesn't see his shadow, he predicts an early spring. Take a stroll through Bluewater Park and check out the statue of Willie, sculpted by Dave Robinson out of a piece of limestone from the Adair Quarry at Hope Bay. Well-loved Willie even has a website, www.wiartonwillie.com.

The area around Wiarton is a nature-lover's delight. **Colpoy's Bluff** is a nature preserve on the north shore of the bay. Another favorite spot is **Spirit Rock Conservation Area**, which has a sign marking the spot where the apparition of a North American Indian maiden who leapt to her death is said to appear. The wave-cut caves of the **Bruce Caves Conservation Area** are one of the geological wonders of the area. Rare ferns also abound throughout the region.

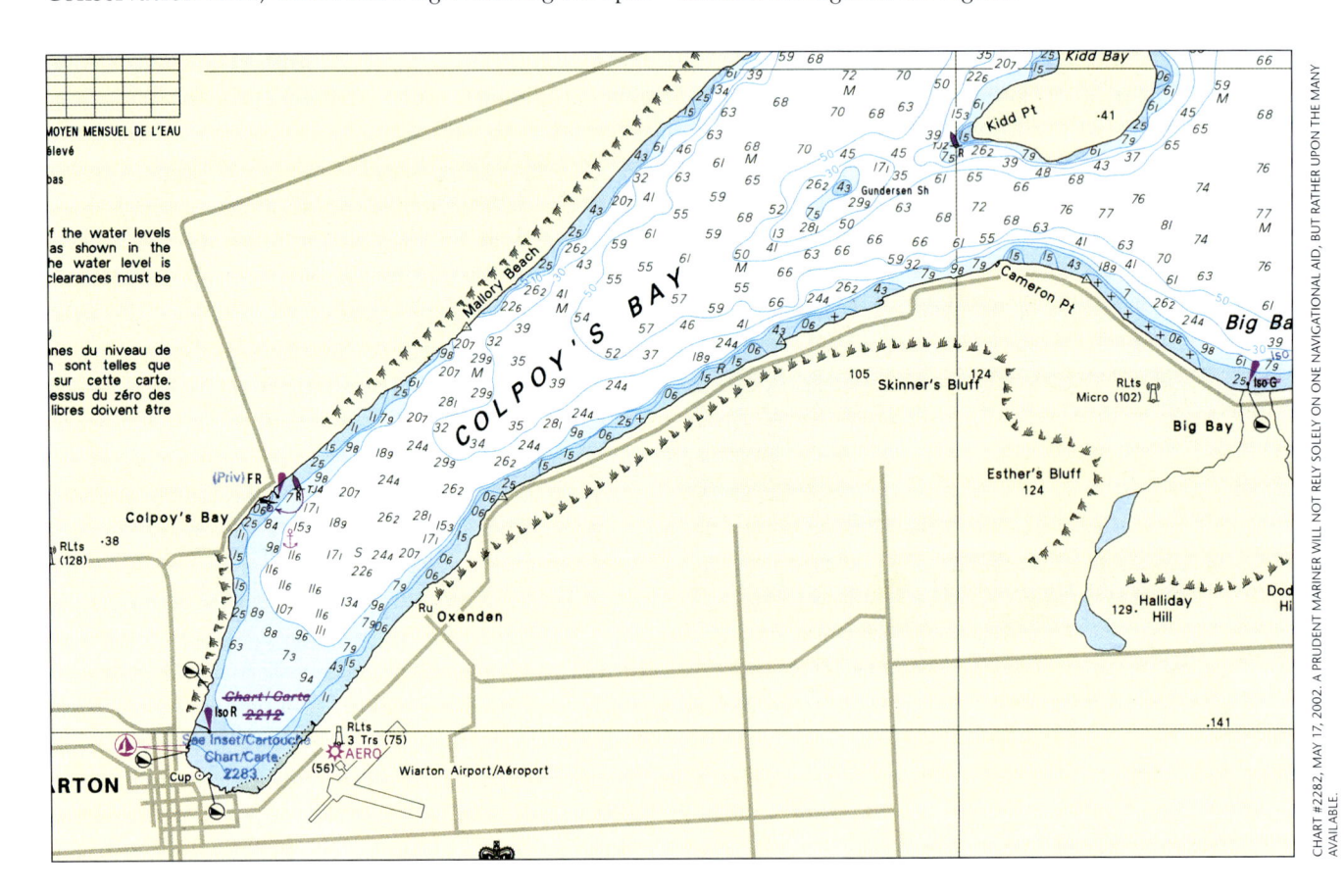

Significant Waypoints

Mouth of Colpoy's Bay, off White Cloud Island:
44°50.25N/81°00.00W

½ mile east of Wiarton Marina in Colpoy's Bay:
44°45.00N/81°07.70W

			Monitors VHF Channel	Transient Slips Available	ALTERNATE MOORING: Wall mooring Rafting allowed	Maximum LOA	Minimum Depth at Dock	Power (amperage or volts)	HOOKUPS: Water Cable TV	FUEL: Gas Diesel Pumpout	BASICS: Heads Showers Laundry	AMENITIES: Swimming pool Whirlpool Rec area Grills Picnic tables Dog walk	Take Reservations	Take Credit Cards	Haulout (Capacity in tons or feet)	REPAIRS: Mechanical Electronics Fiberglass Woodworking Sails Canvas	CONVENIENCE: Ship's store Ice Convenience foods & beverages
C	Government Dock 519-534-2592	400 William Wiarton, ON N0H 2T0		5	W	35'	4'				H		SRGPD	N	N		IC
B	Wiarton Marina, Ltd. 519-534-1301	827 Bay St. Wiarton, ON N0H 2T0	68	25	WR	70'	8'	30	W	GDP	HS		RPD	N	Y	25	MFW	SI

264 LAKELAND BOATING · LAKE HURON CRUISING GUIDE

TRAILERBOATING

Head for the Border

Towing your boat offers a new way to explore Georgian Bay and the North Channel.

Cruising Georgian Bay or the North Channel is the point of "going north" on Lake Huron. Getting there by water will mean at least two days of hard running from Port Huron, Michigan. From Chicago, the run can take three or four days—if the weather remains favorable. Double those days for the return trip home, then subtract the total from the cruising time available. Suddenly, there's not much time left for blueberry pancakes in the Benjamins.

A trailerboat can make the same Port Huron-to-Tobermory run in an easy day's drive. Two days will do the same for the trip from Chicago, Indianapolis or Pittsburgh. The obvious result is that the trailerboat owner spends more time cruising the northern wilds. The Small-Craft Route in Georgian Bay and most of the North Channel offer ample protection from the big lake for smaller boats. Indeed, there are many prime cruising areas—Bay of Islands, MacGregor Bay and much of the 30,000 Islands—that are more accessible to trailerable boats than to larger cruisers. And, of course, launch ramps are abundant.

Bringing trailerable boats into Canada poses no problems. Be sure to have your registration papers for the boat and proof of auto insurance for your car.

While you don't need one of those pesky I-68 forms from U.S. Customs to bring your boat back into this country, you may be asked to prove that you did not purchase expensive electronic equipment (a GPS receiver or a digital depthsounder) while in Canada. Sales receipts or serial numbers on an insurance policy should suffice.

Canadian laws on marine toilets are more restrictive than U.S. laws. All heads must be permanently installed, and they can only be emptied through a deck pumpout. Portable toilets that are removed from the boat for emptying don't meet these requirements.

If you have a VHF marine radiotelephone, you may be required to have both a station license for the transmitter and a restricted operator permit for yourself. Neither of these are required if you operate in U.S. waters. Both can be obtained from the Federal Communications Commission. Most marine electronics stores have the necessary forms.

It is illegal to bring a handgun into Canada or to possess one there without a special, extremely restricted permit. The penalties are severe!

Driving in Canada is pretty much the same as in the United States. The main difference is the use of kilometers per hour instead of statute miles per hour in measuring speed limits. Nearly all late-model U.S. cars have kilometers per hour on their speedometers, so this should not be a problem.

U.S. money is accepted almost everywhere in Canada. However, you'll get a much friendlier reception if you use the local currency. There are money-changing stations at every major border crossing. The exchange rate is usually about $1.50 Canadian to $1 U.S.

Owen Sound, Ontario

Lat 44°35.07
Lon 80°56.37

The approach to Owen Sound is simple. Most of the facilities of interest to boaters are lined up along the western shore. The Georgian Bay Yacht Club **A** doesn't cater to transients, but the Owen Sound Marina **B** does. Enter from the east through the opening in the breakwall **C**, which is marked by a flashing red light. Larger boats that don't require any services can tie up at the Owen Sound Harbour **D**, located to port inside the mouth of the Sydenham River. The city's Visitor Information Centre **E** can be found on the west side of the harbor, in the old Canadian National rail station, along with the Owen Sound Marine and Rail Museum. Beautiful Kelso Beach Park **F** is a short stroll from the marina, with a sign-marked trail connecting to the harbor walkway.

Distances from Owen Sound (statute miles)
Meaford: 32 E

Midland: 68 NE

GEORGIAN BAY

Owen Sound, Ontario
Where nature, art and history meet.

European settlers were lured to Owen Sound by its deep harbor, lush valleys and breathtaking beauty. The sound, which is surrounded by a dramatic curve of the Niagara Escarpment, has long flourished as a shipping center because of its protected location, and today the port is still bustling. The town is situated at the end of what is known as the Bluff Coast, and from here, the lakeshore is low-lying, with sandy beaches and many populous summer resort towns.

Formerly home to a number of manufacturers, including shipbuilders, cement-makers and furniture factories, Owen Sound is more of a refitting port than a resort stopover for the cruising yachter, though it offers many of the services of a small city.

Samuel de Champlain and Father LeCaron were, in 1616, most likely the first Europeans to discover Owen Sound. At that time it was a large cedar swamp inhabited by the Petun Indians. This band was driven out by the Iroquois, who were in turn ousted by the Ojibway tribes that lived north of Lake Superior.

It wasn't until 1788 that the next recorded Caucasian visitor came to the area. That was Capt. Gother Mann, who named the harbor Thunder Bay because of the inclement weather. Capt. William Owen and Lt. Henry Bayfield arrived by ship in 1815, surveyed the harbor and surrounding area and named it Owen Sound in honor of the captain's elder brother. The original town was laid out in 1837 and was known as Sydenham, but became Owen Sound in 1851. It was settled largely by Scots, Irish and English. As the

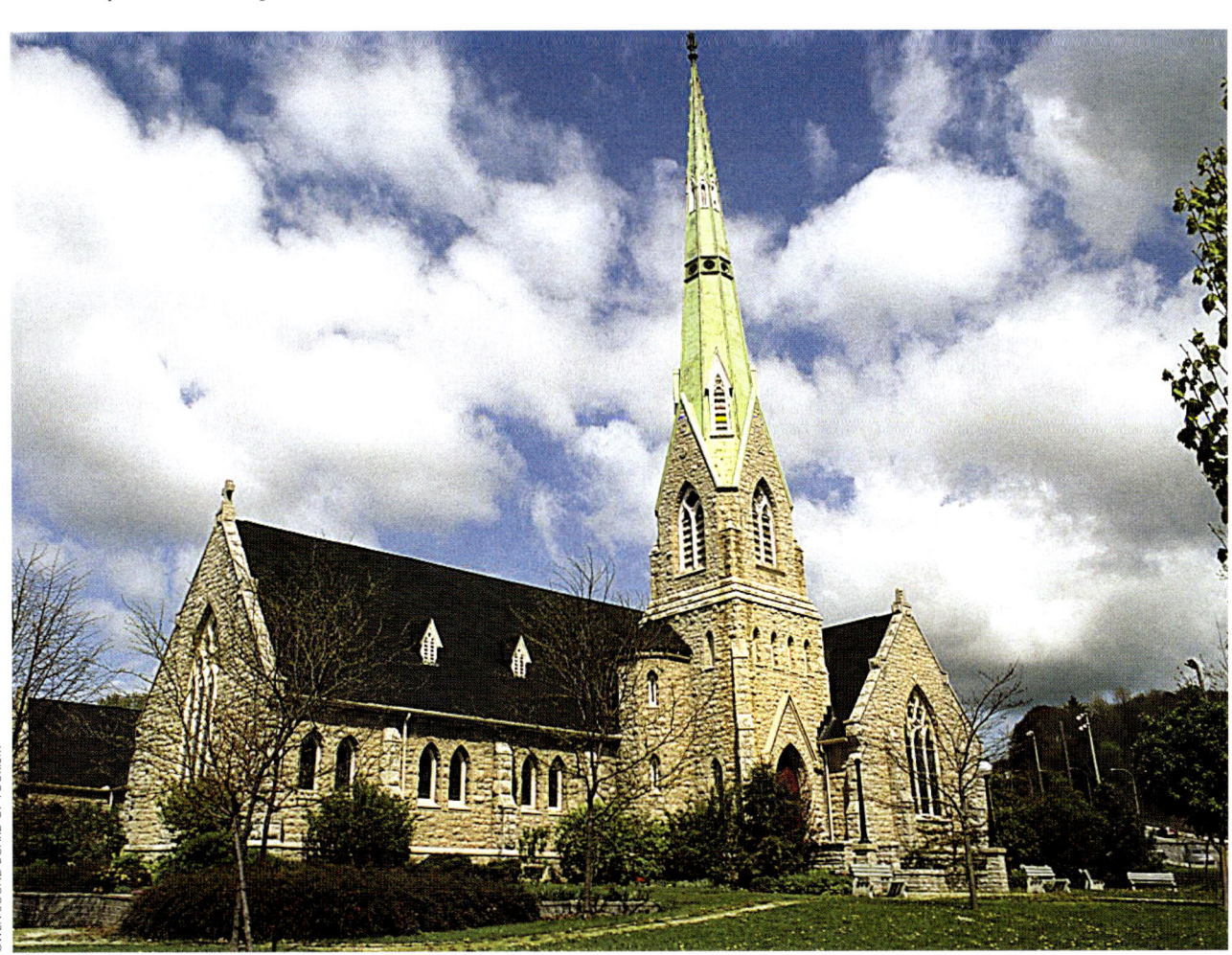

LAKELAND BOATING · LAKE HURON CRUISING GUIDE   **267**

northern terminus of the Underground Railway, many escaped slaves settled here. With the lumber boom, the town thrived as a shipping port. A rail line arrived in 1873, ensuring its commercial future.

The approach to Owen Sound is a no-brainer. The yacht facilities are located on the west bank. The first one you'll reach, the **Georgian Yacht Club,** does not cater to transients but may honor other clubs' memberships. The entrance to the **Owen Sound Marina** is marked by a flashing red light on the north side of the entrance, which opens to the east.

Just to the south is **Kelso Beach Park**. The inner harbor is marked with range lights that lead in on a course of 15.5°T.

The harbor is industrial, but boats can tie up along the dock, known locally as **Owen Sound Harbour**, for a couple of nights free of charge. The area is not staffed and there are no amenities available. However, the downtown district includes a large supermarket, conveniently located at the tip of the harbor.

There are three major conservation areas within close proximity that feature dramatic waterfalls and a variety of wildlife. The **Hibou Conservation Area**, overlooking Paynter's Bay on Owen Sound just half a kilometer northeast of the city, has two natural sand beaches.

The town is also home to numerous museums and galleries. The **Tom Thomson Memorial Art Gallery** features the work of Thomson and the Group of Seven. The **Owen Sound Artists' Co-Op, Inc.** downtown showcases fine collectibles from area artists.

The **Owen Sound Marine and Rail Museum** has a fascinating collection of ship and railway displays that depict the days when the town was known as the "Liverpool of the North." Grab a map at the **Visitor Information**

Significant Waypoints
Mouth of Owen Sound:
44°44.75N/80°50.00W
Flashing RT12 in Owen Sound Harbour on range 195½ just east of Georgian Yacht Club:
44°35.20N/80°56.32W

Where to Eat

Harrison Park Inn 519-376-5151
Great food in a beautiful setting overlooking the Sydenham River. After dinner, treat yourself to what has been called the best old-fashioned ice cream cone in Georgian Bay.

The Rusty Gull 519-371-8198
Located in the marina proper, this is a large, friendly restaurant overlooking the water, featuring steaks and seafood. Downstairs there is a separate pub.

More Info

Emergency 911
Grey/Bruce Regional Health Centre 519-376-2121
Visitor Information Centre 888-675-5555
                                                    www.owensound.ca
Bayshore Taxi 519-371-5555
Pro Hardware 705-282-2433
Robertson's Drug Store 705-282-2147

Centre, located on the west side of the inner harbor in the old Canadian National rail station along with the museum, and take the self-guided tour of the area's historic buildings.

For a night out, there is the **Roxy Theatre**, home to the **Owen Sound Little Theatre Company**.

OWEN SOUND MARINA

"If you're Gonna do it...do it right!"

- 450 protected slips
- washroom, shower and laundry facilities • double launch ramp
- complete sales and service
- water and hydro on 300 slips
- fully stocked chandlery
- gas, diesel, pump-out
- mobile repair service
- restaurant and convention centre

**195 24th Street West
Owen Sound, Ontario N4K 6H6
519-371-3999
Toll Free 1-888-565-2628
www.owensoundmarina.com**

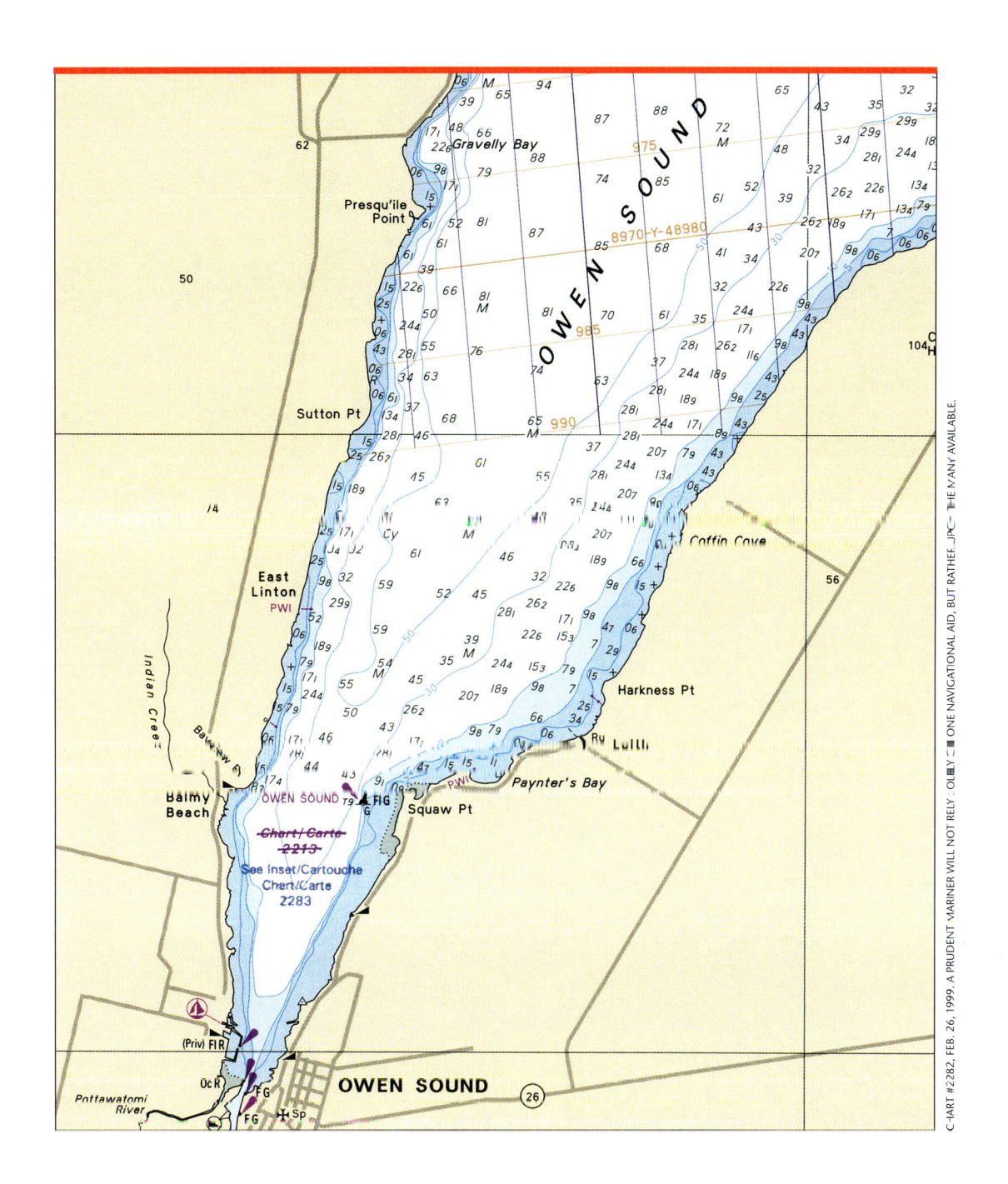

Facilities information subject to change. We suggest you call ahead.

			Monitors VHF Channel	Transient Slips Available	ALTERNATE MOORING: Wall mooring / Rafting allowed	Maximum LOA	Minimum Depth at Dock	Power (amperage or volts)	HOOKUPS: Water / Cable TV	FUEL: Gas / Diesel / Pumpout	BASICS: Heads / Showers / Laundry	AMENITIES: Swimming pool / Whirlpool / Grills / Picnic tables / Dog walk / Rec area	Take Reservations	Take Credit Cards	Haulout (Capacity in tons or feet)	REPAIRS: Mechanical / Electronics / Woodworking / Sails / Canvas / Fiberglass	CONVENIENCE: Ship's store / Ice / Convenience foods & beverages
D	Owen Sound Harbour 519-376-4081	R.R. 2 Owen Sound, ON N4K 5N4			WR	120'+	15'								N	N	
B	Owen Sound Marina 519-371-3999	195 24th St. W. Owen Sound, ON N4K 4L7	68	40		65'	8'	30	WC	GDP	HSL	RGPD	Y	Y	30	MEFWSC	SIC

Meaford, Ontario

Lat 44°36.72
Lon 80°35.33

The approach to Meaford is easy—just stay five miles offshore of the firing range on the point that separates this town from Owen Sound to the west. Facilities are located at the mouth of the Bighead River. There are two basins: The entrance to the east basin **A** is marked by a flashing red; the entrance to the west basin **B** is marked by a flashing green. The east basin is now the primary one due to silting along the eastern shore of the west basin from the Bighead River. Access to Cliff Richardson Boats, Ltd. **C**, which has dockage and some services in the west basin, is not a problem. The Municipality of Meaford Harbour **D** offers dockage, pumpout and restrooms. A Coast Guard station **E** is also located in the east basin.

Distances from Meaford (statute miles)
Thornbury: 8 SE
Midland: 47 NW

Significant Waypoint
200 yards north of harbor between the two entrances: 44°36.80N/80°35.20W

GEORGIAN BAY

Meaford, Ontario
The Big Apple of the bay.

In the heart of apple country, Meaford blossoms in spring and retains its beauty year-round. Founded in 1837 on the banks of the Bighead River, today the town has a population of about 4,000. The area has numerous trails, beaches and parks to explore.

With its well preserved brick buildings, Meaford is a convenient and attractive layover, offering everything a cruising yachtsman might need. Years ago, Meaford functioned as a rail center and transfer point for western grain, and the town's brick mansions, quaint tree-lined streets and impressive city hall are a testament to its prosperity during that era.

Approaching from the west, observe the restricted tank firing range that extends five miles offshore from the point that separates Meaford from Owen Sound. A green water tower inscribed with the name Meaford is located in the southwest part of town. The upper section of the tower is floodlit and marked with red lights.

The harbor consists of two basins extending east and west on either side of the central breakwater, which juts out 700 feet offshore from the mouth of the Bighead River. The north breakwater extends west and east from the outer end of the central breakwater. The flashing green Meaford Breakwater Light is situated near the west end of the north breakwater at an elevation of 30 feet. Buoys lead into the old, western harbor, the location of **Cliff Richardson Boats, Ltd.**, which has gas, dockage and marine services. Despite the narrow and shallow entrance to the basin, access to Cliff Richardson is good. The boatyard built the *Spume*, one of the last wooden search-and-rescue vessels in service and today a tourist attraction located on shore in the new basin. Transients should check in with the harbormaster in the west harbor to get their dock assignments at the **Meaford Municipal Harbour**.

Entrance to the east basin—now the main one due to silting in the west basin—is marked by the Meaford Harbour Light, an oscillating red light at the east end of the north breakwater. It is shown at an elevation of 42 feet from a white circular tower with a red upper portion. Washrooms and showers are a short walk from this basin. There is also an excellent launch ramp at the municipal marina.

Ashore, Meaford has a wide range of unique speciality shops and sightseeing attractions. The **Meaford Museum**, next to the harbor, is located in the original pumphouse, built in 1895. The **Meaford Opera House** is an imposing building that first opened to the public in 1909. It still offers a full schedule of professional theater and music. The **Fire Hall**, which was built in 1887, is another striking architectural landmark. And don't miss **Beautiful Joe Park,** dedicated to the dog in Margaret Marshall Saunders' children's book *Beautiful Joe*.

Local events

May
Apple Blossom Orchard Tours

June-September
Tuesday Cruise Nights at the Harbour

June
Annual Garden Tour

Take a Kid Fishing Bass Derby

July
Canada Day Giant Fireworks Display

Fine Arts Fest

July-August
Georgian Bay Sailing Regatta

August
50/50 Salmon Derby

Lion's Club Car Show

Meaford & St. Vincent Agricultural Society Fall Fair

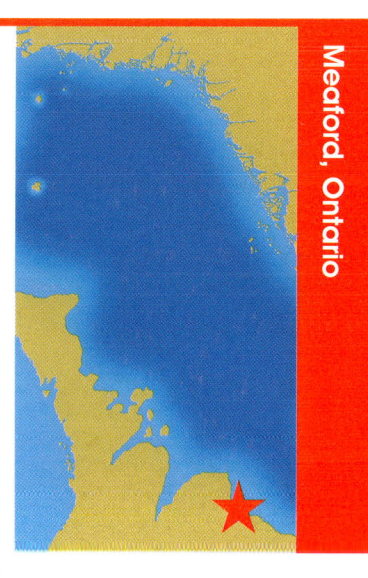

More Info

Emergency 911
Meaford General Hospital 519-538-1311
Chamber of Commerce 519-538-1640
www.meaford.com
Dooley's Taxi 519-538-4411

TESS SCARROW

LAKELAND BOATING · LAKE HURON CRUISING GUIDE

Where to Eat

Fisherman's Wharf Restaurant.......... 519-538-1390
Across the street from the harbor, offering mussels steamed in white wine and fresh local fish.

The Harbour Moose................. 519-538-0014
Also situated across the street from the harbor, the Moose offers a wide variety of foods.

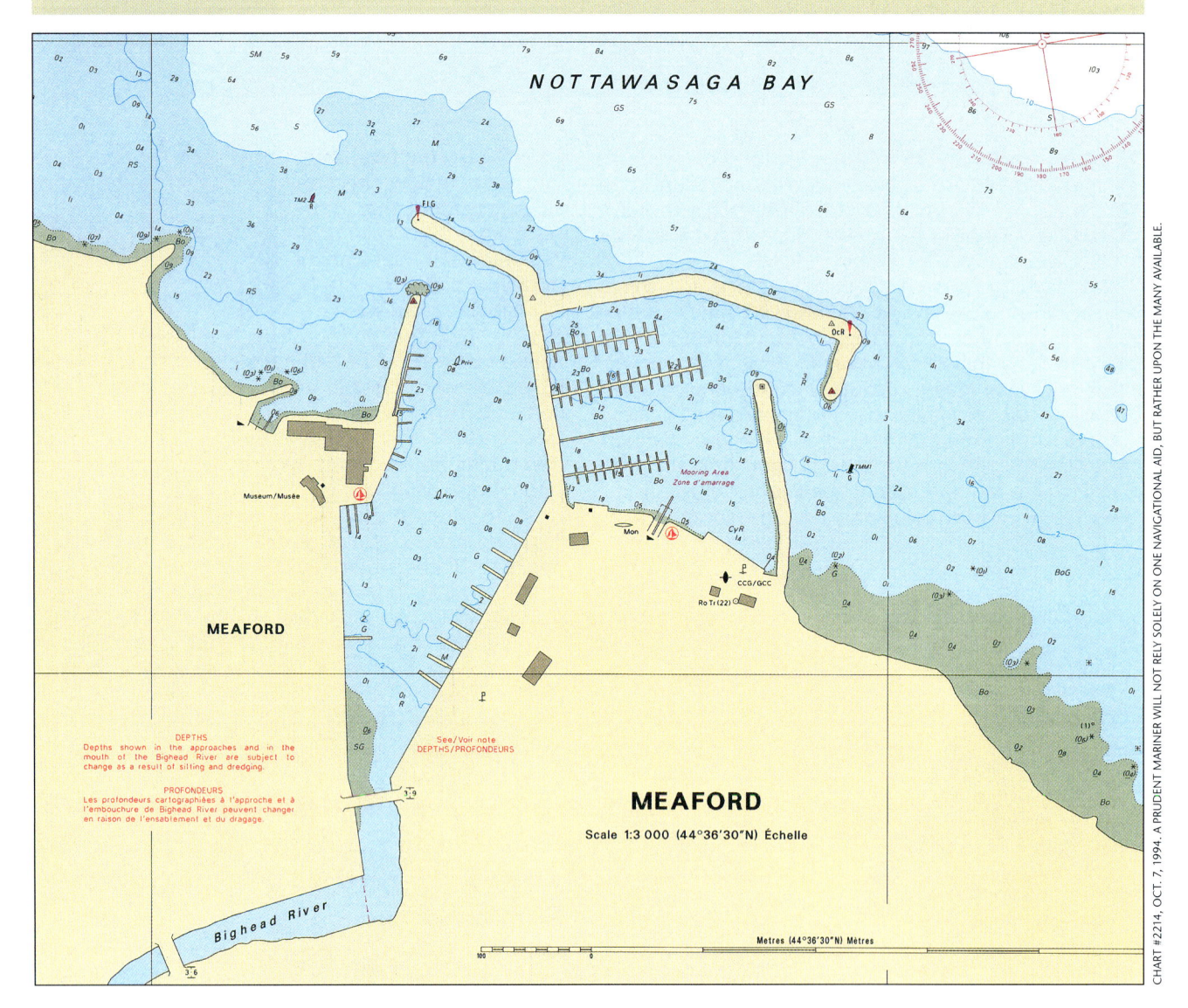

Facilities information subject to change. We suggest you call ahead.

		Monitors VHF Channel	Transient Slips Available	ALTERNATE MOORING: Wall mooring Rafting allowed	Maximum LOA	Minimum Depth at Dock	Power (amperage or volts)	HOOKUPS: Water Cable TV	FUEL: Gas Diesel Pumpout	BASICS: Heads Showers Laundry	AMENITIES: Swimming pool Whirlpool Grills Picnic tables Dog walk Rec area	Take Reservations	Take Credit Cards	Haulout (Capacity in tons or feet)	REPAIRS: Mechanical Electronics Fiberglass Woodworking Sails Canvas	CONVENIENCE: Ship's store Ice Convenience foods & beverages
C	Cliff Richardson Boats, Ltd. 103 Bayfield St. 519-538-1940 Meaford, ON N4L 1N4	68	5		55'	6'	30 50	W	GDP	HSL	RPD	Y	Y	35	MFW	SIC
D	Municipality of Meaford Harbour 3 St. Vincent St. 519-538-5975 Meaford, ON N4L 1A1	68	12	WR	45'	5'5"	30		P	HSL	RPD	Y	Y			I

272 LAKELAND BOATING · LAKE HURON CRUISING GUIDE

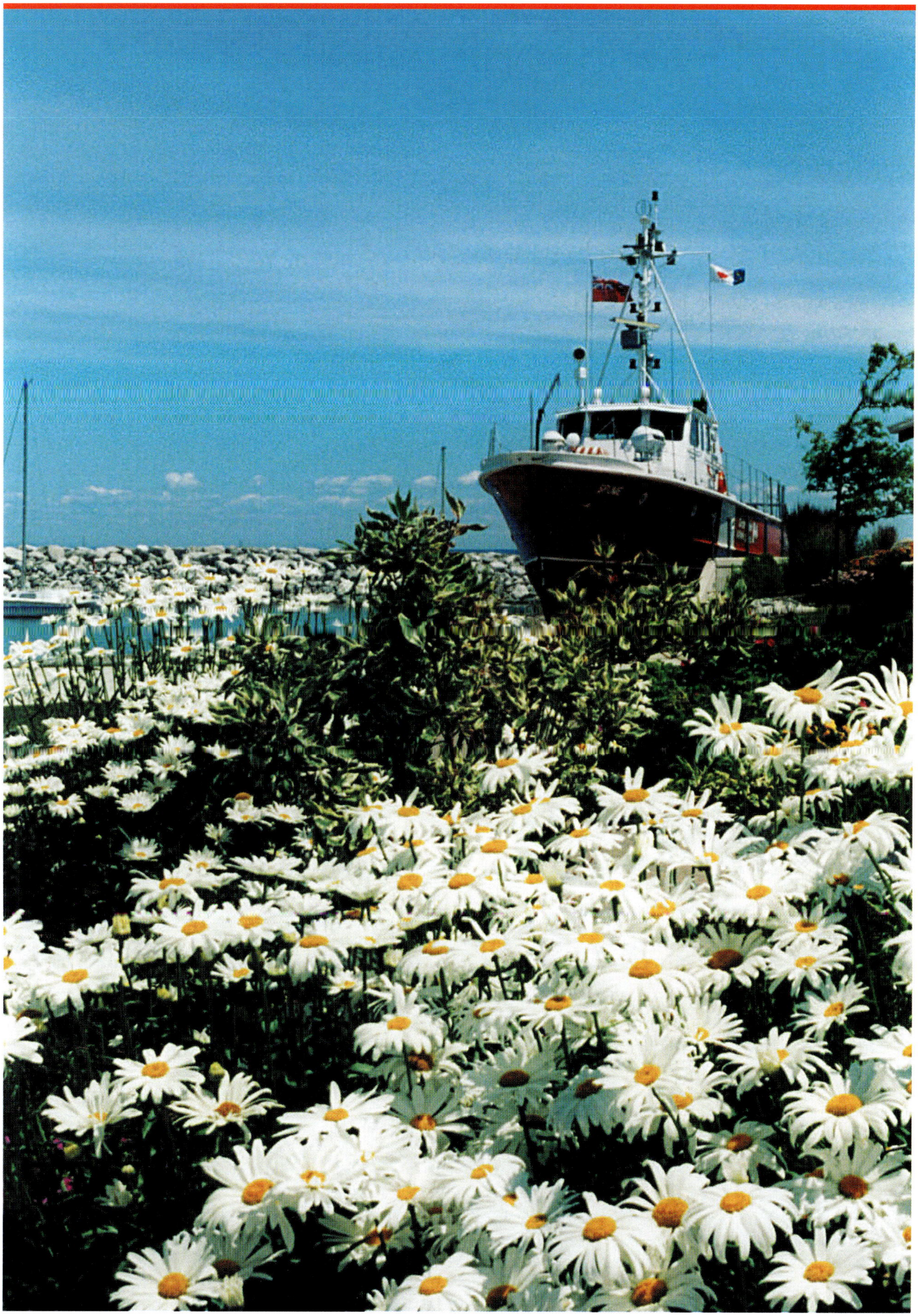

GILLIAN BIGL

Thornbury, Ontario

Lat 44°34.25
Lon 80°26.81

A

B

N

The approach to Thornbury is straightforward. Avoid shoals east of here by entering on a range of 212°T. The lat/lon given above is for a point off the photograph, about 600 yards north of the entrance to the harbor. A new flashing white sector light replaces the former fixed forward and rear red lights marking the entrance to the harbor **A**. Thornbury Municipal Marina **B** has ample transient space and sells fuel.

Distances from Thornbury (statute miles)
Collingwood: 15 SE
Midland: 44 NE

Significant Waypoint
Approximately five miles north of entrance to harbor:
44°38.95N/80°26.80W

274 LAKELAND BOATING · LAKE HURON CRUISING GUIDE

GEORGIAN BAY

Thornbury, Ontario
A scenic harbor nestled under the Blue Mountains.

Thornbury has a reputation as a boutique village and a shopping and dining destination, drawing visitors year round. For transient boaters, the area also offers a well-protected harbor, ample shoreside services and scenic caves to explore.

The village of Thornbury, population 1,500, is located where the Beaver River and the Beaver Valley meet the shoreline, about seven miles east of Meaford and 13 miles west of Collingwood. The village remains a boating-friendly community within the larger municipality known as the Town of the Blue Mountains. The harbor lies in the lee of the scenic Blue Mountains, for which the town is named.

Approaches from the west are straightforward, but from the east mariners must avoid the dangerous Mary Ward Ledges (see the Collingwood sailing directions). A white water tower that reads "Thornbury" is located in the southwestern part of the community and is visible from the water. The manmade harbor lies at the mouth of the Beaver River, where a world-record chinook salmon weighing 43 pounds, 7 ounces was caught. Boats are protected from any surge, thanks to overlapping breakwaters. Range lights in line bear 212°T. A flashing white sector light has replaced the flashing red light formerly located at the end of the breakwater and the rear light at the shore's end. Dock at **Thornbury Municipal Harbour**.

There is a shingle and sand swimming beach close to the west pier, and miles of bike trails can be accessed directly from the harbor. An excellent launch ramp is located next to the harbor office.

The village center is a short walk up the hill from the harbor. It has a **bank**, **liquor** and **beer stores**, **laundry facilities**, shops and restaurants catering to the four-season tourist trade. The **Thornbury Bakery** will make you hefty sandwiches to go, as well as provide fresh bread for the galley. **Green's Valu-Mart** is the local grocery store. **Gyles Sails & Service**, which has marine supplies and charts, is located in the **Harbour Mews** on Bruce Street. Serious antique collectors should head for the **Art Glass & Antiques Studio**. Local attractions include the new dam and fish ladder on the Beaver River below the picturesque Mill Pond.

From the last week in July through the first week in August, Thornbury plays host to a sailing regatta that draws a fleet from all over Georgian Bay. And every

Local events

May
Apple Blossom Ball

June
Bazaar on the Bruce
Easter Seal Regatta

July
The Blue Mountains Chili Cookoff
Historical Society Pot-luck Picnic

July-August
Georgian Bay Sailing Regatta

August
Blues at Blue Festival
Summer Antiques Show
Summer Sundown Celebration

September
Beaver Valley Fall Fair

BILL BRENNAN

LAKELAND BOATING · LAKE HURON CRUISING GUIDE   **275**

Where to Eat

Fitzgerald's on the Pond 519-599-2217
This charming inn has a fine dining room and a more casual pub. Located on Thornbury Mill Pond off of Highway 26.

Harbour Café 519-599-3553
Overlooking the harbor, this restaurant has a wide menu, a deck for dining and an ice cream pavilion.

The Mill Café 519-599-7866
Perched on the banks of the Beaver River, this restaurant combines charm with tasty cuisine.

Sisi on Main 519-599-7769
A 90-seat restaurant with a martini and mussel bar, steaks and seafood. Dinner only; closed Tuesdays.

Sterios 519-599-5319
An intimate eatery that's singled out by residents for the quality of its steak and seafood.

White House on the Hill 519-599-6261
A charming restaurant at the top of Thornbury's main street, offering continental cuisine. Open Thursday to Sunday for dinner only.

More Info

Emergency 911
Meaford General Hospital 519-538-1311
Chamber of Commerce 519-599-3223
www.greycounty.on.ca

Georgian Triangle
Tourist Association 705-445-7722
town.thebluemountains.on.ca

Meaford Taxi 519-538-3139

July, the village is the site of the Blue Mountains Chili Cookoff, which sends its top three winners to the Terlingua International Cookoff in Texas. There are dozens of teams, great entertainment and the chance for visitors to cast ballots for the People's Choice Awards. Also check out the **Bruce and Georgian trails**, which offer the opportunity for some brisk exercise in a beautiful setting.

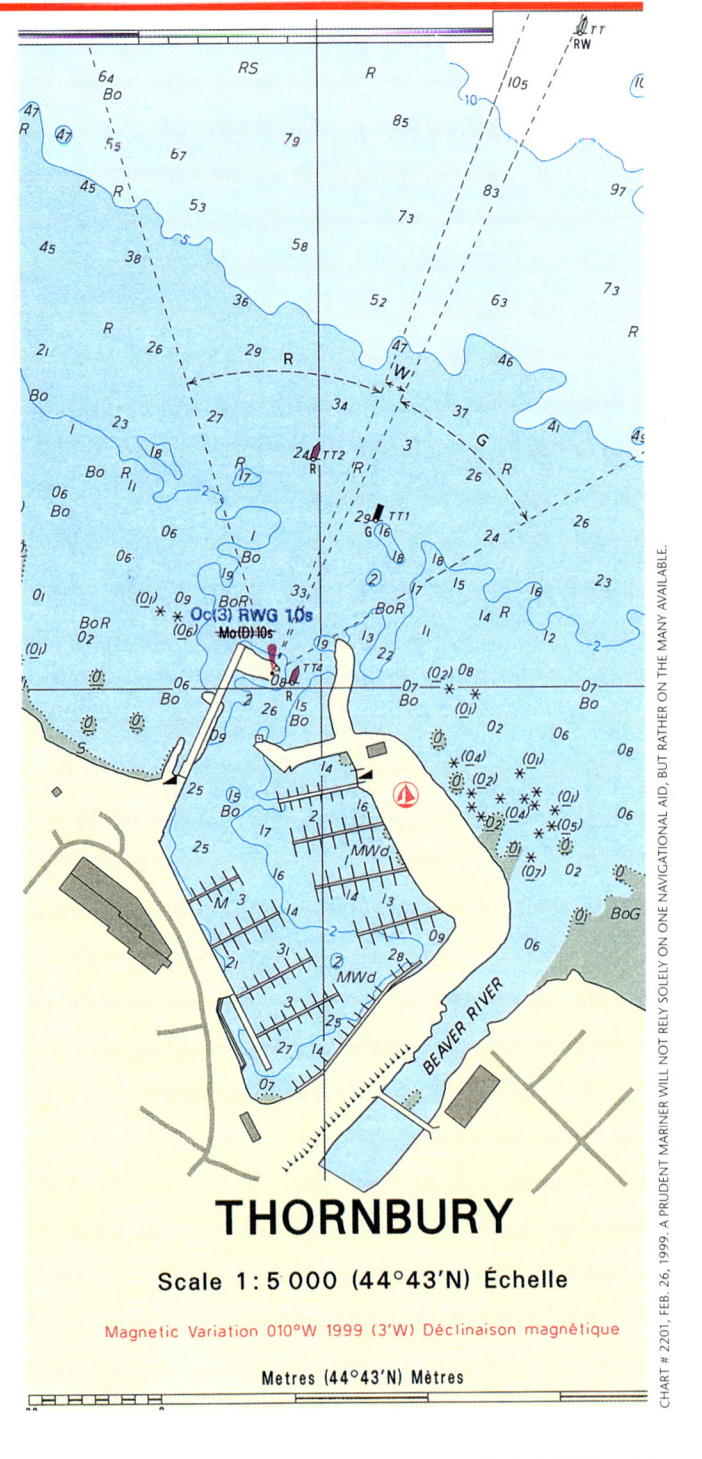

THORNBURY
Scale 1:5 000 (44°43'N) Échelle

Magnetic Variation 010°W 1999 (3'W) Déclinaison magnétique

Metres (44°43'N) Mètres

Facilities information subject to change. We suggest you call ahead.

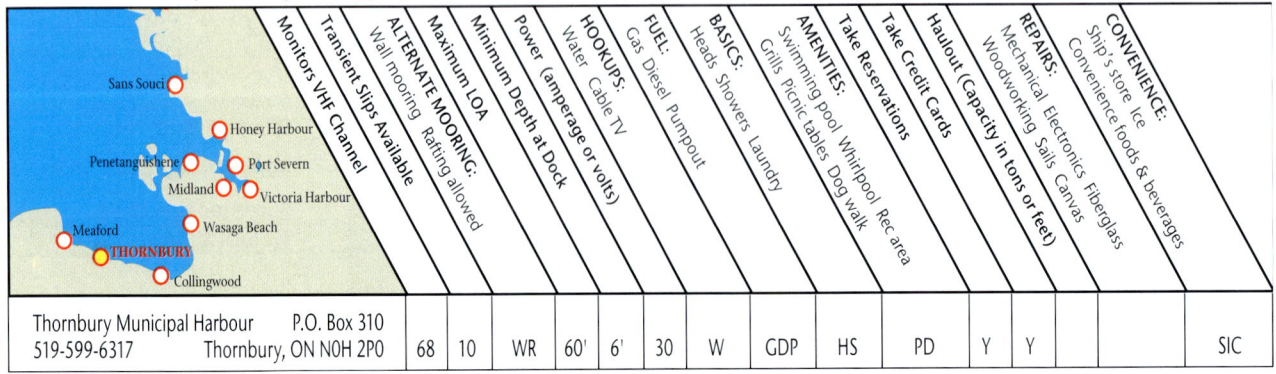

	Monitors VHF Channel	Transient Slips Available	ALTERNATE MOORING: Wall mooring / Rafting allowed	Maximum LOA	Minimum Depth at Dock	Power (amperage or volts)	HOOKUPS: Water / Cable TV	FUEL: Gas / Diesel / Pumpout	BASICS: Heads / Showers / Laundry	AMENITIES: Swimming pool / Whirlpool / Rec area / Grills / Picnic tables / Dog walk	Take Reservations	Take Credit Cards	Haulout (Capacity in tons or feet)	REPAIRS: Mechanical / Electronics / Fiberglass / Woodworking / Sails / Canvas	CONVENIENCE: Ship's store / Ice / Convenience foods & beverages
B Thornbury Municipal Harbour P.O. Box 310 519-599-6317 Thornbury, ON N0H 2P0	68	10	WR	60'	6'	30	W		GDP	HS		PD	Y	Y	SIC

276 LAKELAND BOATING · LAKE HURON CRUISING GUIDE

VHF Radio Frequencies
Frequency (MHz)

Channel Desig.	Ship Transmit	Ship Receive	Intended Use
06	156.300	156.300	Intership Safety
07A	156.350	156.350	Commercial
08	156.400	156.400	Commercial (intership only)
09	156.450	156.450	Commercial & Non-Commercial
10	156.500	156.500	Commercial
11	156.550	156.550	Commercial
12	156.600	156.600	Port Operations
13	156.650	156.650	Navigational (bridge-to-bridge)
14	156.700	156.700	Port Operations
15		156.750	Environmental (receive only)
16	156.800	156.800	International Distress, Safety & Calling
17	156.850	156.850	State Control
18A	156.900	156.900	Commercial
19A	156.950	156.950	Commercial
20	157.000	161.600	Port Operations (duplex)
20A	157.000	157.000	Port Operations
21A	157.050	157.050	US Coast Guard Only
22A	157.100	157.100	US Coast Guard Liaison
23A	157.150	157.150	US Coast Guard Only
24	157.200	161.800	Public Correspondence
25	157.250	161.850	Public Correspondence
26	157.300	161.900	Public Correspondence
27	157.350	161.950	Public Correspondence
28	157.400	162.000	Public Correspondence
65A	156.275	156.275	Port Operations
66A	156.325	156.325	Port Operations

Channel Desig.	Ship Transmit	Ship Receive	Intended Use
68	156.425	156.425	Non-Commercial
69	156.475	156.475	Non-Commercial
70	156.525	156.525	Data-Computer Only
71	156.575	156.575	Non-Commercial
72	156.625	156.625	Non-Commercial (intership only)
73	156.675	156.675	Port Operations
74	156.725	156.725	Port Operations
77	156.875	156.875	Port Operations (intership only)
78A	156.925	156.925	Non-Commercial
79A	156.975	156.975	Commercial. Non-Commercial in Great Lakes Only
80A	157.025	157.025	Commercial. Non-Commercial in Great Lakes Only
81A	157.075	157.075	US Government Only—Environmental Protection Operations
82A	157.125	157.125	US Government Only
83A	157.175	157.175	US Coast Guard Only
84	157.225	161.825	Public Correspondence
85	157.275	161.875	Public Correspondence
86	157.325	161.925	Public Correspondence
87	157.375	161.975	Auto ID System duplex repeater
WX1		162.550	Weather (receive only)
WX2		162.400	Weather (receive only)
WX3		162.475	Weather (receive only)
WX4		162.425	Weather (receive only)
WX5		162.450	Weather (receive only)
WX6		162.500	Weather (receive only)

Note: The addition of the letter "A" to the channel number indicates that the ship receive channel used in the United States is different from the one used by vessels and coast stations of other countries. Vessels equipped for U.S. operations only will experience difficulty communicating with foreign ships and coast stations on these channels.

Collingwood, Ontario

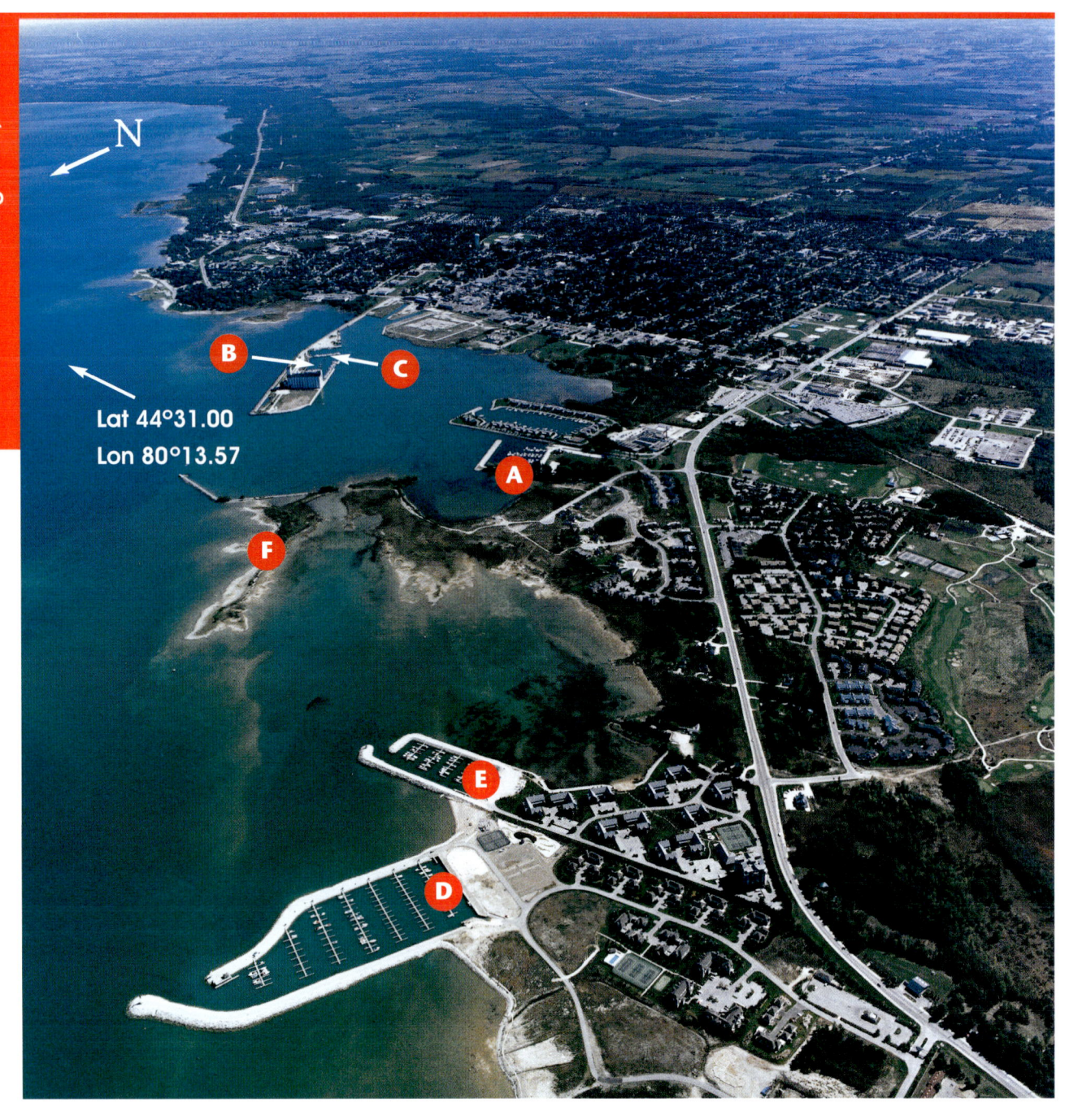

Lat 44°31.00
Lon 80°13.57

When approaching Collingwood, stay five miles offshore until you reach the flashing red buoy light north of New Bank. Line up the sector lights and enter the harbor on a course of 180°T. The lat/lon given here is for 500 feet off the entrance on range. The best bet for transient dockage is at Cranberry Marina **A**, in the northwest part of the harbor. Slips are occasionally available at Collingwood Yacht Club **B**, behind the grain elevator on the southwest side of the eastern breakwater. The whole eastern breakwater area is currently being recreated as "Harborland Park." In a pinch, the Public Dock **C** has tie-up space for those who don't require any services. Lighthouse Point **D** and Rupert's Landing **E** are private facilities closed to transients just beyond Hen and Chicken Island **F**.

Distances from Collingwood (statute miles)
Wasaga Beach: 9.5 E
Penetanguishene: 41 NNE
Midland: 43 NNE

GEORGIAN BAY

Collingwood, Ontario
Shake, rattle and roll on in to this thriving port.

Collingwood, named in honor of one of Admiral Nelson's commanders at the Battle of Trafalgar, has also been known as Hurontario and as "the Hen and Chickens," because of the peculiar shape of the island and the formations off of its coast. But whatever you call it, this port is a thriving, picturesque tourist town.

Settled in the 1830s, Collingwood was originally an agricultural community, and the area still produces some of the best apples in Canada. Nestled in the southwest pocket of Nottawasaga Bay, the town is located on a low-lying plain, guarded by the Blue Mountains and part of the Niagara Escarpment. In the past, Collingwood functioned as a busy commercial port and shipbuilding center, though the grain elevators and shipyards are now closed.

Approaches to Collingwood from the west must be plotted with care to avoid the off lying and dangerous Mary Ward Ledges. The reefs and rocks extend for six miles northwest of Nottawasaga Island and four miles north of Delphi Point. Nottawasaga Island is wooded and low-lying, with a light situated near the northwest end atop a 94-foot-high white circular tower. Simcoe Bank, Lockerbie Rock and Lafferty's Home all lie in the approaches to Collingwood.

Fisherman Point Range Lights, in line bearing 149.5°T, lead boaters clear of the offshore hazards. The flashing green front light is located on Sunset Point and is shown from a tripod tower 38 feet high. Proceed in on the range heading, all the while looking for the Collingwood Shore Range Light, a flashing red that sits on a crib in the harbor. The rear range light is shown from open park land. The lights intersect the Fisherman Point Range and lead in on a course of 180°T to the well-buoyed harbor channel.

A tall blue water tank is visible from the bay. The **Public Dock** and the small white building of the **Collingwood Yacht Club** are both located next to the clearly visible grain elevator on the southwest side of the eastern breakwater that forms the harbor. The public dock has no facilities aside from an excellent launch ramp and is mostly taken up by local boaters, while the yacht club's dockage is limited and usually is available only to members. **Kaufman Marina** is private. The more scenic **Cranberry Marina** in the northwest part of the harbor is your best bet and also has a nice launch ramp. There are shore facilities in nearby Cranberry Village.

Collingwood boasts many excellent restaurants. **Christopher's** on Pine Street and the **Spike and Spoon** on Hurontario are both housed in historic homes. Hockey fans will want to head to Canadian hockey announcer Don Cherry's Sports Grill on First Street.

Elvis fans will want to schedule their arrival for late July, when Collingwood hosts the **Canadian National Elvis Tribute and Convention**. It features shows, dinners, street dances, a film festival and of course competitions at dozens of venues around town.

The **Collingwood Museum** is located opposite the harbor. Of particular interest is the municipal office building, which is crowned with a clock tower. There is a large **Loblaws Supermarket** within walking distance of the public dock, as well as banks and shops of all varieties on Hurontario Street, which runs inland from the water. Collingwood is also home to the **Blue Mountain Pottery Factory Outlet**, as well as the showrooms of the **Kaufman Furniture Factory**. It is recommended that you rent a car to see all that is offered in the area—the scenic caves of the Niagara Escarpment are a must.

Significant Waypoint
Buoy TN12, a quarter mile north of entrance:
44°31.74N/81°13.68W

Where to Eat

Spike and Spoon 705-446-1629
A classic local establishment reminiscent of an English country house. Fresh, natural ingredients are the staples here.

Terra Cotta Restaurant 705-445-2623
This is located in a magnificently restored brick building. The decor is striking, with batik tablecloths and original paintings. Specialities include chicken chausseur.

More Info

Emergency	911
Collingwood General Hospital	705-445-2550
Georgian Bay Triangle Association	705-445-7722
Chamber of Commerce	705-445-0221
Ace Cab	705-445-3300

LAKELAND BOATING · LAKE HURON CRUISING GUIDE

COLLINGWOOD

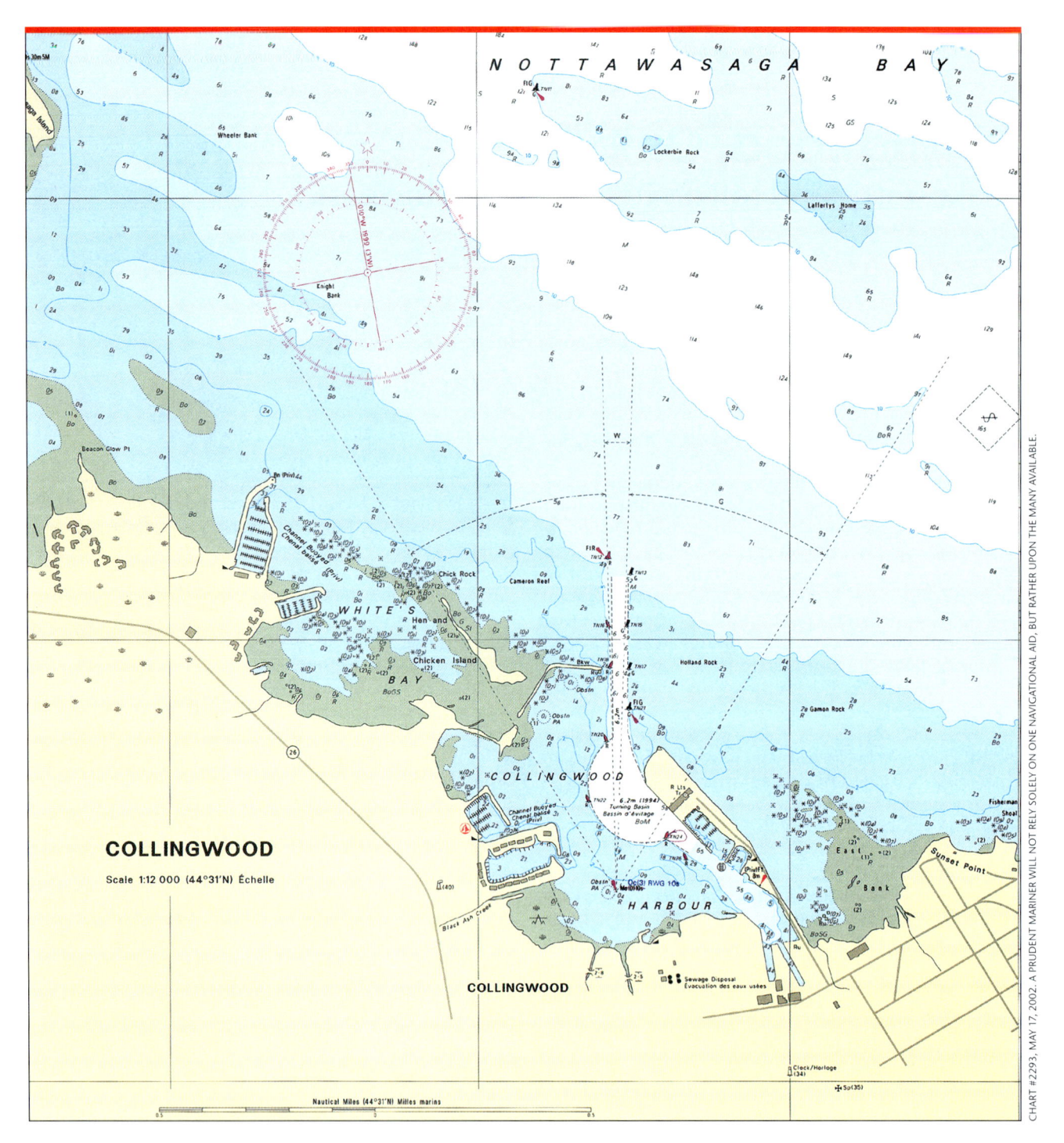

Scale 1:12 000 (44°31'N) Échelle

Facilities information subject to change. We suggest you call ahead.

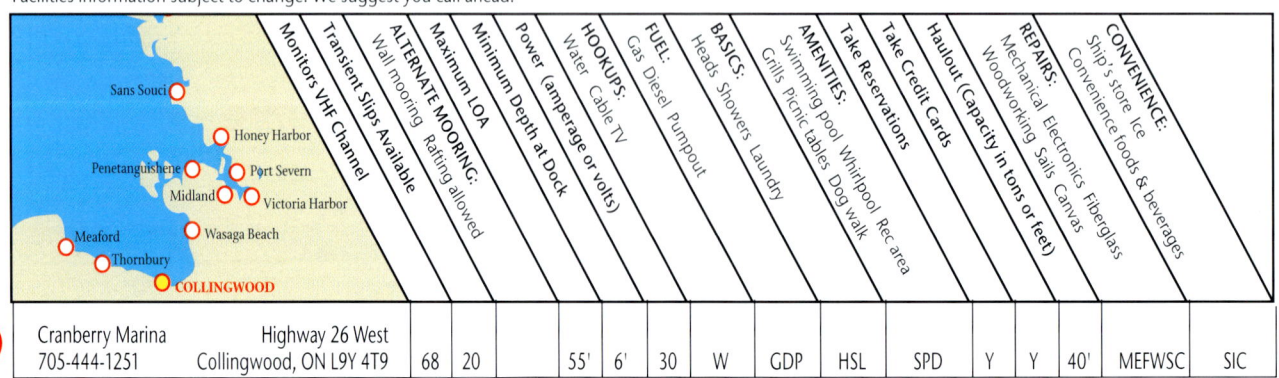

		Monitors VHF Channel	Transient Slips Available	ALTERNATE MOORING: Wall mooring, Rafting allowed	Maximum LOA	Minimum Depth at Dock	Power (amperage or volts)	HOOKUPS: Water, Cable TV	FUEL: Gas, Diesel, Pumpout	BASICS: Heads, Showers, Laundry	AMENITIES: Swimming pool, Whirlpool, Rec area, Grills, Picnic tables, Dog walk	Take Reservations	Take Credit Cards	Haulout (Capacity in tons or feet)	REPAIRS: Mechanical, Electronics, Fiberglass, Woodworking, Sails, Canvas	CONVENIENCE: Ship's store, Ice, Convenience foods & beverages
A	Cranberry Marina 705-444-1251	Highway 26 West Collingwood, ON L9Y 4T9	68	20		55'	6'	30	W	GDP	HSL	SPD	Y	Y	40'	MEFWSC	SIC

280 LAKELAND BOATING · LAKE HURON CRUISING GUIDE

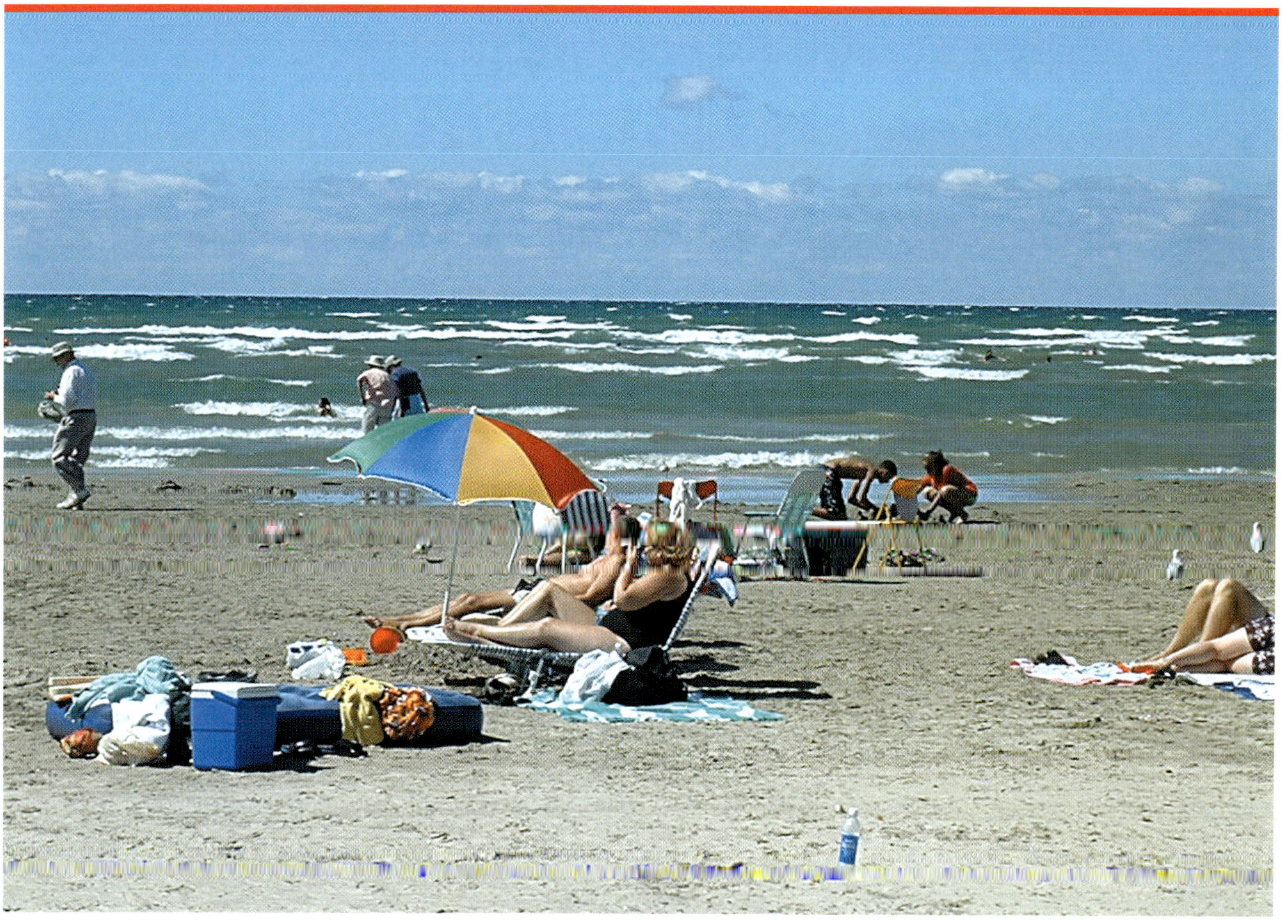

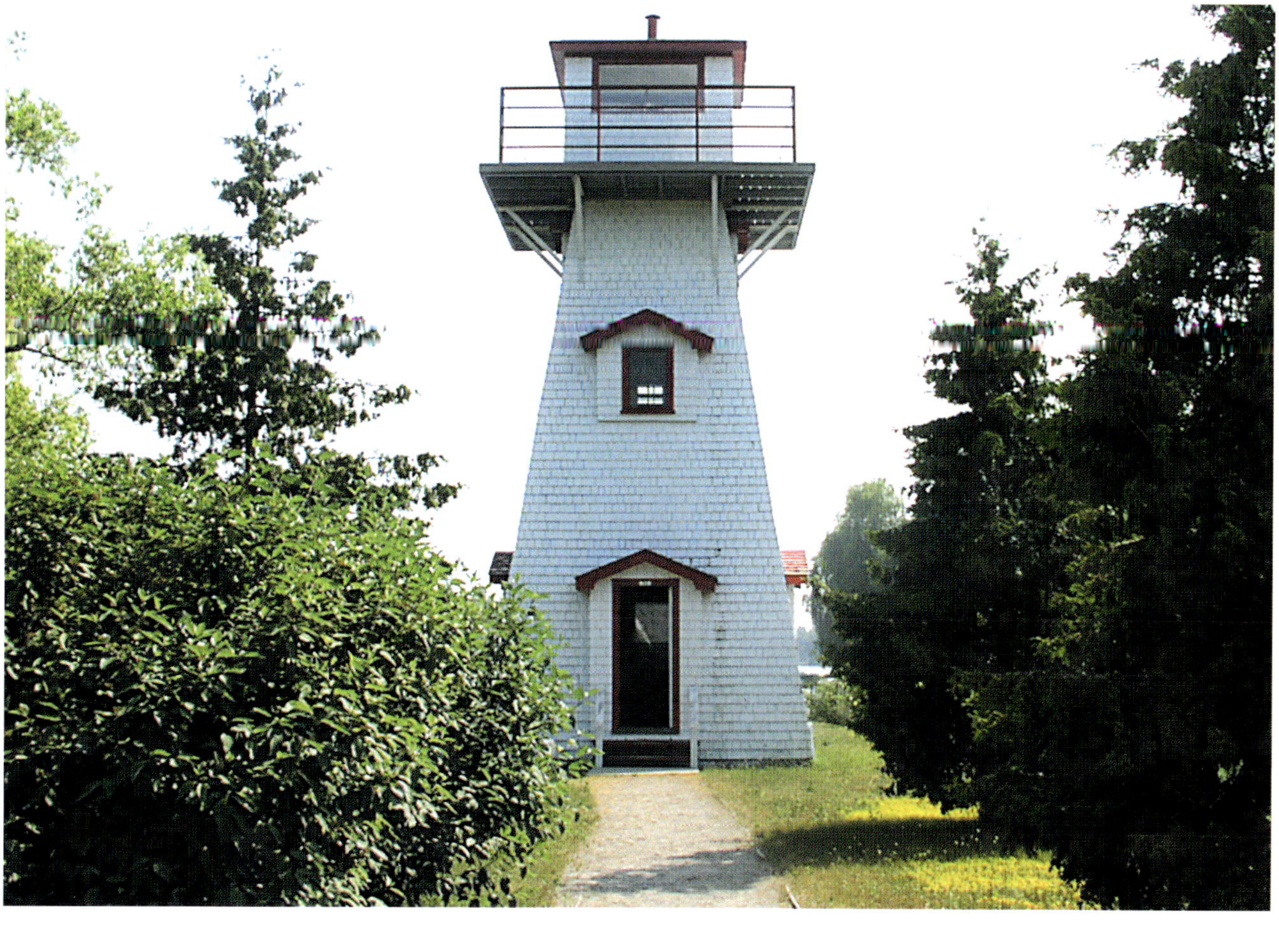

WASAGA BEACH CHAMBER OF COMMERCE

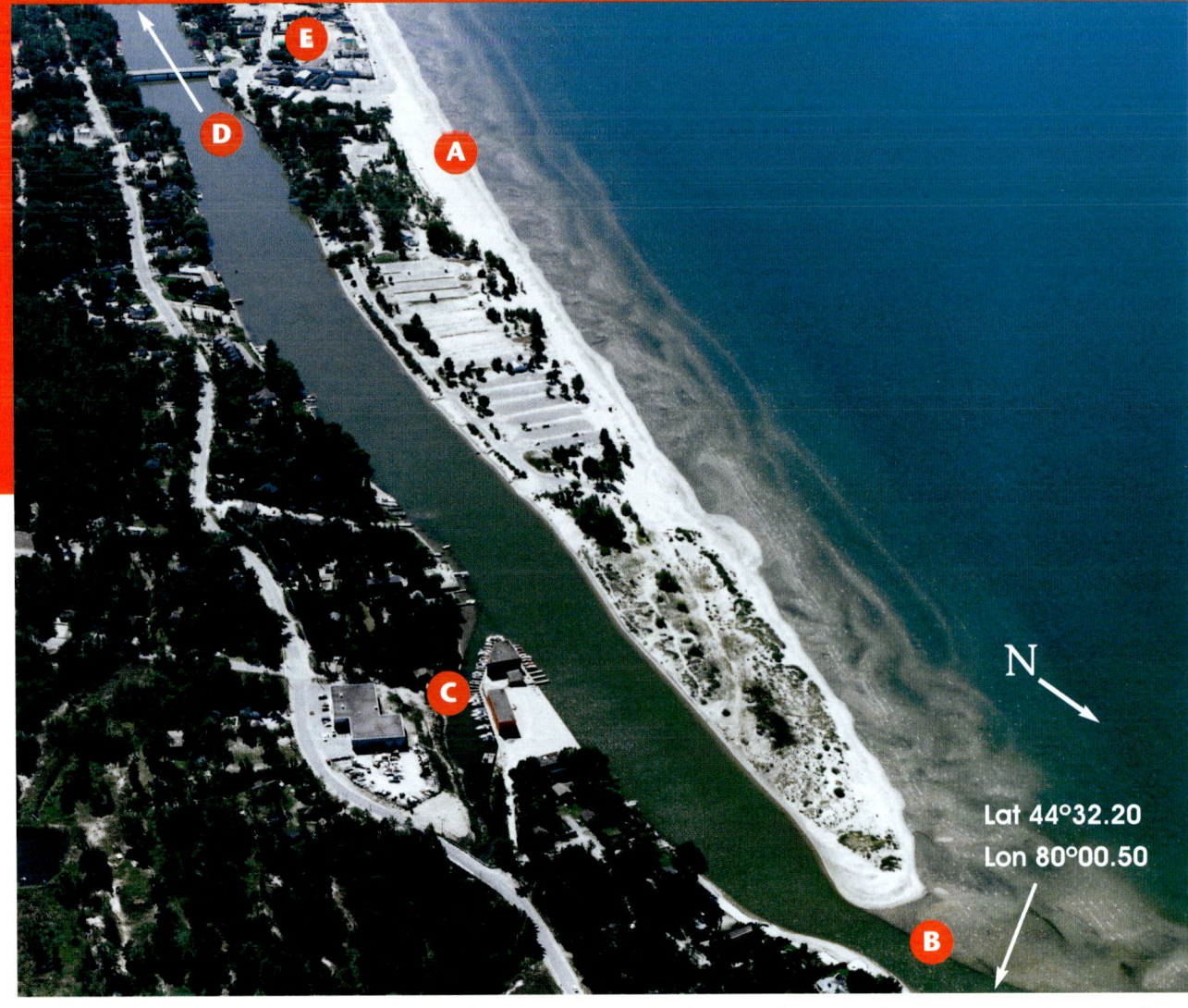

Wasaga Beach, Ontario

Lat 44°32.20
Lon 80°00.50

The town of Wasaga Beach is a popular holiday destination, thanks to its fantastic eight-mile-long beach **A**. The approach and entrance to the Nottawasaga River **B** is exposed to Georgian Bay, as there are no breakwalls or piers. Entry can be a problem in extreme conditions due to the narrow channel and sandbars. Look for the Nottawasaga River Light and follow the buoys in. The town lies on the strip that separates the river from Georgian Bay. Transients can dock at Sturgeon Point Marina, Ltd. **C**, to port, or continue two miles upriver to Wasaga Marine **D**, to starboard just beneath the highway bridge. Depending on water levels, you'll have approximately 12 to 13 feet of clearance. An amusement park **E** is located nearby.

Facilities information subject to change. We suggest you call ahead.

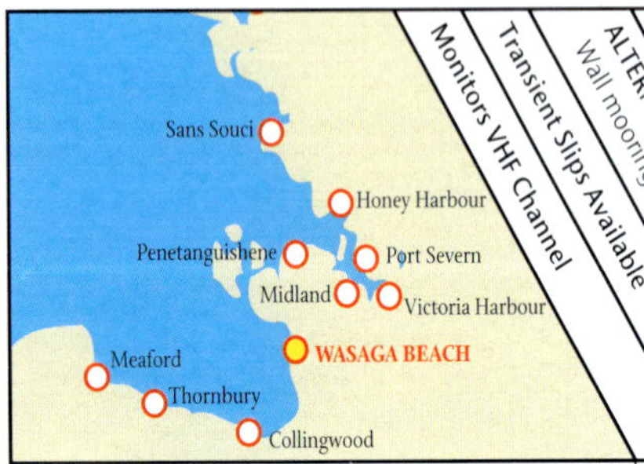

		Monitors VHF Channel	Transient Slips Available	ALTERNATE MOORING: Wall mooring / Rafting allowed	Maximum LOA	Minimum Depth at Dock	Power (amperage or volts)	HOOKUPS: Water / Cable TV	FUEL: Gas / Diesel / Pumpout	BASICS: Heads / Showers / Laundry	AMENITIES: Swimming pool / Whirlpool / Rec area / Grills / Picnic tables / Dog walk	Take Reservations	Take Credit Cards	Haulout (Capacity in tons or feet)	REPAIRS: Mechanical / Electronics / Fiberglass / Woodworking / Sails / Canvas	CONVENIENCE: Ship's store / Ice / Convenience foods & beverages
C	Sturgeon Point Marina Ltd. 350 River Rd. E. 705-429-2934 Wasaga Beach, ON L0L 2P0		10	WR	35'	3'5"	15		G	HSL	PD		Y	4	MEFWSC	SIC
D	Wasaga Marine 1237 Mosley St. 705-429-2100 Wasaga Beach, ON L0L 2P0		20		32'	4'	110v		GP	H	PD	Y	Y	5	MEFC	SI

282   LAKELAND BOATING · LAKE HURON CRUISING GUIDE

GEORGIAN BAY

Wasaga Beach, Ontario
Something for everyone along this stretch of sand.

With eight miles of sand, Wasaga Beach is a mecca for those looking for fun in the sun. In summer, the town swells by 100,000 as daytrippers flock to the waterfront. Within the town itself, Wasaga Beach Provincial Park features 350 acres of beach, picnic areas, and hiking and biking trails. There's also a waterside amusement park. The Nottawasaga River runs behind the beach for almost four miles and the shore is dotted with vacation cottages. The town center, located on the strip between the river and the bay, has most services for mariners.

The approach to the river is straightforward from any direction. The shore from Sunset Point southeast to what was once called Brock's Beach has boulders and rocks extending offshore for a tenth of a mile. The Nottawasaga River entrance light is shown at an elevation of 41 feet, from a white mast with a red and white rectangular daymark. Privately maintained buoys lead into the river. Normal depth over the bar is reported by local marinas to be 6 feet, but it can be as little as 3, depending on wind and other factors. **Sturgeon Point Marina** is located to port near the river entrance. **Wasaga Marine** is located about two miles upriver on the starboard bank, just below the highway bridge. Docking is limited to boats under 35 feet. There is a launch ramp at Wasaga Marine.

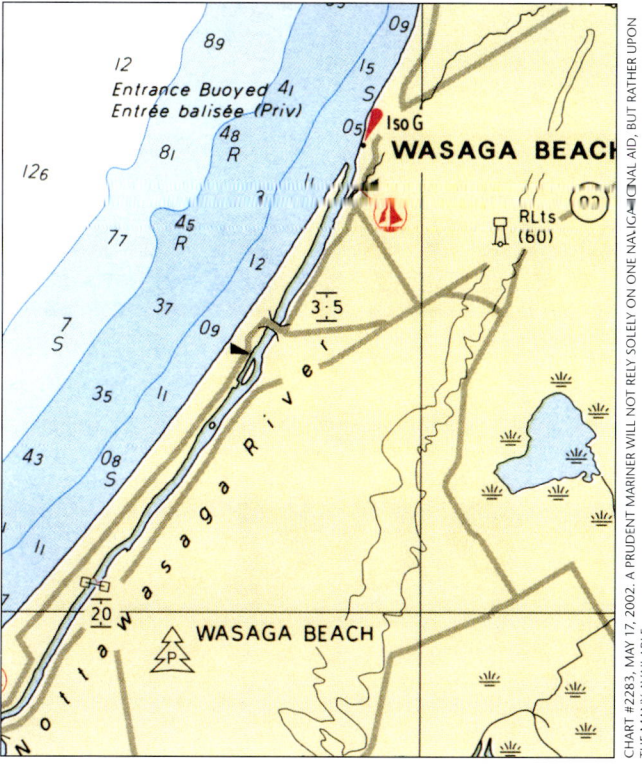

Distances from Wasaga Beach (statute miles)
Penetanguishene: 41 NNE
Midland: 43 NNE

More Info
Emergency	911
River Road Health Centre	705-429-8270
Visitor Information Centre	866-292-7242
Wasaga Taxi	705-429-5611

Where to Eat
Little Marina Restaurant.........705-429-2626
A popular eatery with pasta specialties that is also open for breakfast.

The Sizzling Shrimp.............705-429-8193
Fresh lobster, chicken, crab and pasta dishes. Patio dining is available.

LAKELAND BOATING · LAKE HURON CRUISING GUIDE

Penetanguishene, Ontario

Lat 44°50.22
Lon 79°53.39

Just after you enter the channel, the two-mile-long bay takes a 90-degree turn to port. The lat/lon given here is for Pinery Point Light at the mouth of the harbor. You'll know you're getting close to town when you pass Magazine Island **A** on your port side. On the same side, you'll pass Dutchman's Cove Marina **B** and Bay Moorings Yacht Club **C**. To starboard, you'll see Northwest Basin Marina, Ltd. **D**. Farther along to port is the town's public wharf and dock complex, the Historic Port of Penetanguishene **E**. Follow that shoreline around and you'll find the other marinas: Beacon Bay Marina **F**, Hindson Marine, Ltd. **G** and Harbour West Marina **H**.

Distances from Penetanguishene (statute miles)
Midland: 10 E
Tobermory: 101 W
Honey Harbour: 10 ENE
Victoria Harbour: 12 E
Port Severn: 15 E
Parry Sound: 55 N

Georgian Bay

Penetanguishene, Ontario
Bienvenue à this French-English port.

Penetang—as many locals call it—is considered the oldest town in Ontario and the second-oldest in Canada. After visiting in 1793, Upper Canada governor John Graves decided the deep waters of the narrow bay were perfect for a naval base. Building began after the War of 1812, and the garrison grew fast after the British Navy selected the town as its primary post in the upper Great Lakes.

French Canadians and Métis also relocated here, settling on the western shores of Penetanguishene Bay. Farmers from Quebec settled in Penetang in the 1840s and established the town as a dual-language oasis, today home to one of Ontario's few bilingual school systems.

You will see **Discovery Harbour**, a recreated naval and military base, behind Magazine Island on your starboard side as you cruise along the well-marked channel down this two-mile-long bay on the east side of the Penetang Peninsula. The bay is ringed with marinas.

Northwest Basin Marina, Ltd. lies to starboard opposite Magazine Island. After passing this, you will see **Bay Moorings Yacht Club** on the south shore. The town wharf, the **Historic Port of Penetanguishene**, is at the bottom of the bay; there is an excellent launch ramp and parking just south of the municipal wharf at the foot of Main Street. Following the shoreline south from here, you'll find other possibilities for transient dockage, such as **Beacon Bay Marina, Hindson Marine, Ltd.** and **Harbour West Marina**.

Raise the wharf master's office on VHF channel 68 as you approach. Officials there will direct you to overnight moorings that offer services such as water taxis, fuel, pumpout, chart shops, scuba diving and bait for fishing the area's perch, pike, bass, and rainbow and lake trout.

The town center is close to the public wharf area. The bay's eastern shore boasts delightful shady parks and a small beach, and a variety of shops and restaurants line the main street. The **Penetanguishene Centennial Museum**, housed in a period logging mill and store, is well worth a visit. The town also hosts the Georgian Bay Poker Run on the last weekend of July.

Gidley Point & Thunder Beach

Fourteen miles north of Wasaga Beach on Nottawasaga Bay lies Gidley Point, just south of Christian Island. Here you'll find **A.C. Marina**, the only refuge between Wasaga Beach and Thunder Beach. The former Albert's Cove Marina has a range and a buoyed entrance. The lat/lon is 44°44.48N/80°07.00W. They have transient slips, perform all repairs and sell gas and nautical supplies.

Thunder Beach, a residential cottage and tourist area, occupies the top of the Penetang Peninsula but is still miles away from town. A **government dock** is located halfway down the bay on the west shore at Thunder Beach; it offers no overnight transient dockage or marina services but does sell gas. Harbor manager Louis Delisa lives across the road from the dock; call 705-533-2730 to reach him. A friendly corner store and restaurant featuring Greek food, great burgers and ice cream is just 200 yards from the beach.

Thunder Beach is a beautiful big bay with a sandy bottom and beaches all along its shores. It is not surprising

PENETANGUISHENE HARBOUR

Where to Eat

Bay Moorings Inn 705-549-5237
Excellent steak, veal, pasta and seafood at reasonable prices. At Bay Moorings Yacht Club.

Bullwinkle's Lodge 705-549-1472
A casual Canadian restaurant serving burgers, steaks and pasta.

Captain Ken's 705-549-8691
This casual, friendly place is a local favorite for fish.

Dock Lunch 705-549-8111
Sit on the patio and enjoy the famous burger. Also offers fish, chicken and onion rings.

Olympia Gardens 705-549-4802
This Mediterranean restaurant serves primarily Greek cuisine. Known for its roast lamb.

Pan Mai 705-549-0526
Chinese-Canadian cuisine.

More Info

Emergency 705-526-5466
Penetang General Hospital 705-549-7431
Chamber of Commerce 705-549-2232
       www.southerngeorgianbay.on.ca
Union Taxi 705-549-7666

Tomb—guard the bay's entrance. You can find anchorage and sand beaches on the eastern shore of Christian Island and on the eastern and western shores of Beckwith Island. You can also drop anchor near the golden beaches on the south shore of Hope Island, the southeast side of Giants Tomb Island and Methodist Bay on the mainland. Both Giants Tomb and Methodist Bay are a part of **Awenda Provincial Park**.

> **Significant Waypoint**
> One-tenth mile north of Asylum Point light:
> 44°48.67N/79°56.00W

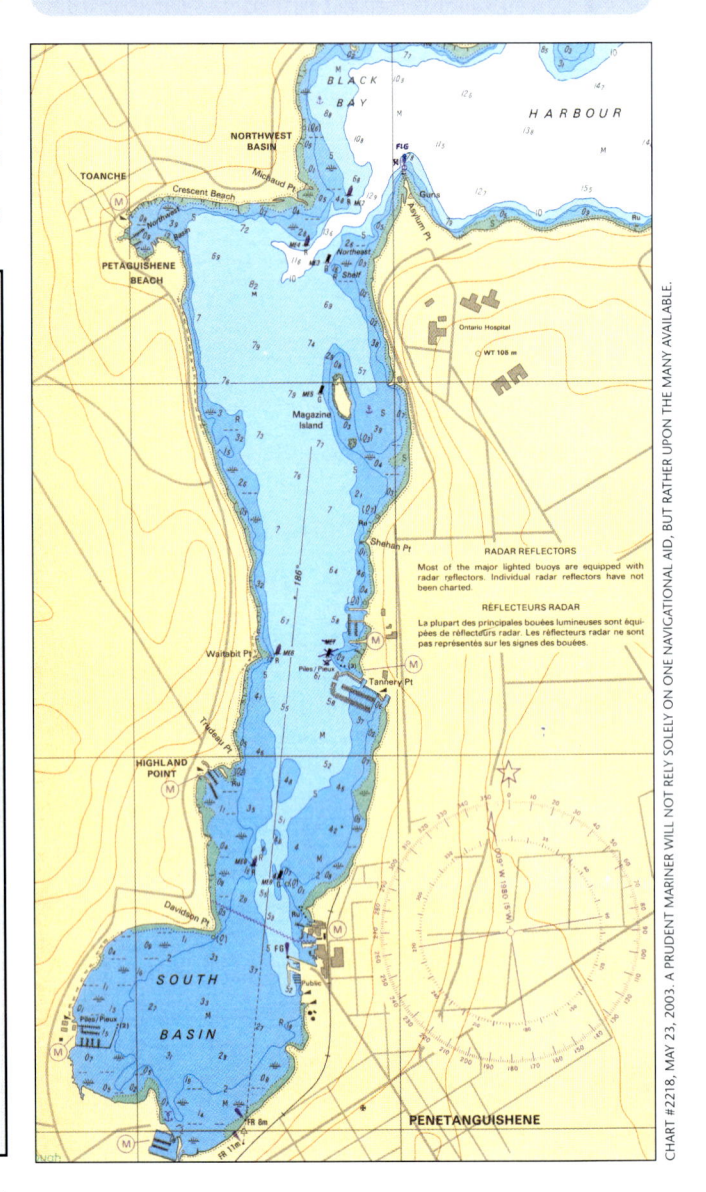

that a large number of Great Lakes cruisers pull in here. The area looks rather like a tropical bay you might find in Antigua or Grenada. It is also an ideal, sheltered spot for both motor- and keel-boat cruisers to slip into and hide when Georgian Bay weather and waves get a little rambunctious.

Four islands—Beckwith, Hope, Christian and Giants

Port of Historic Penetanguishene

Services Available:

- Slips to accommodate up to 45 feet •12 foot draft
- 20-30 Amp Service available •Free showers with slip fees • Reservations accepted • Pump-out services
- Launching ramp • Tourist information office • Fully trained marina staff •Charts available •Close to marine service, supplies and fuel •Within walking distance of shopping, entertainment, dining and tourist attractions •Competitive rates available

VHF Monitoring Channel 68
Located on Strip Chart 2202 (Sheet 1 of 5)
Latitude 44°46.5N Longitude 79°56.5W

FOR AREA INFORMATION & EVENTS
Please contact the Penetanguishene-Tiny Chamber of Commerce at **1-800-263-7745** or the Town of Penetanguishene at **1-705-549-7452**.

"Your Destination...Our Heritage"

**10 Robert Street West
Penetanguishene, Ontario L9M 2G2**
www.town.penetanguishene.on.ca

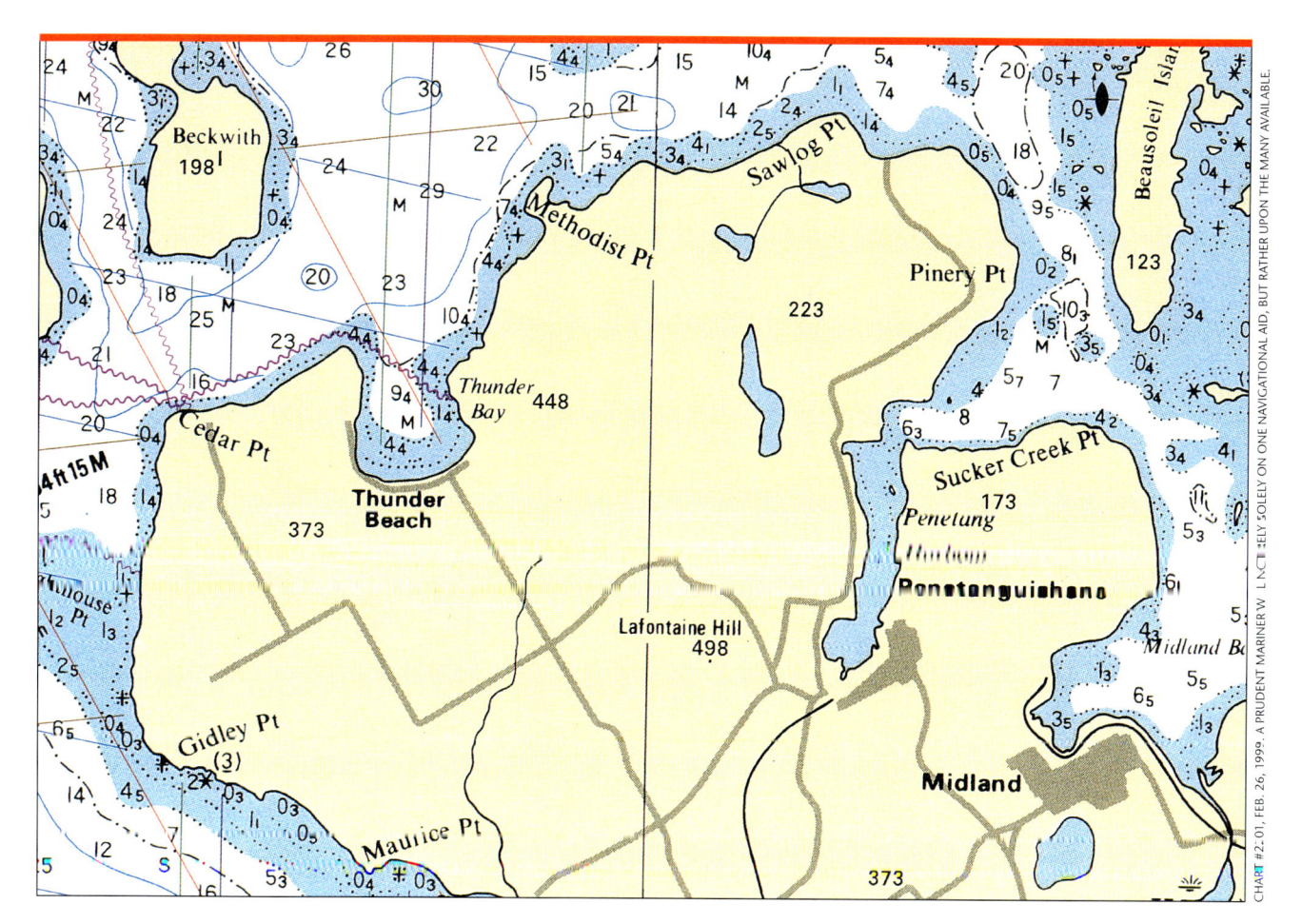

Facilities information subject to change. We suggest you call ahead.

Marina	Address	Monitors HF Channel	Transient Slips Available	ALTERNATE MOORING: Wall mooring / Rafting allowed	Maximum LOA	Minimum Depth at Dock	Power (amperage or volts)	HOOKUPS: Water/Cable TV	FUEL: Gas Diesel Pumpout	BASICS: Heads Showers Laundry	AMENITIES: Swimming pool Whirlpool Rec area / Grills Picnic tables Dog walk	Take Reservations	Take Credit Cards	Haulout (Capacity in tons or feet)	REPAIRS: Mechanical Electronics Fiberglass / Woodworking Sails Canvas	CONVENIENCE: Ship's store Ice / Convenience foods & beverages
A.C. Marina 705-533-2024	13 Sunrise Ct. Penetanguishene, ON L9M 1R3	16 68				5'	30	W	GP	H	PD	N	Y	18	MC	SIC
C Bay Moorings Yacht Club 705-549-6958	1-200 Fox St. Penetanguishene, ON L9M 1E7	68	10		60'		30	W	GDP	HSL	SRGPD	Y	Y	50	MEFW	SIC
F Beacon Bay Marina 705-549-2075	1-37 Champlain Rd. Penetanguishene, ON L9M 1S1	68	20		60'	10'	30 50	W	GDP	HSL	SRPD	Y	Y	40	MEFWSC	SIC
B Dutchman's Cove Marina 705-549-2641	222 Fox St. Penetanguishene, ON L0K 1P0	68	1		45'	6'	15 30	W	P	HS	RPD	Y	N	30	MEFWC	I
H Harbour West Marina, Ltd. 705-549-9378	3-319 Champlain Rd. Penetanguishene, ON L9M 1S3	68	20		40'	4'	30 60	W	P	HS	PD	Y	Y	22	MEFC	SIC
G Hindson Marine, Ltd. 705-549-2991	67 Champlain Rd. Penetanguishene, ON L9M 2G2	68	20	W	60'	7'	30	W	GDP	HSL	SGPD	Y	Y	35	MEFWC	SIC
E **Historic Port of Penetanguishene** **705-549-7777**	**2 Main St. Penetanguishene, ON L9M 1T1**	68	30	W	100'	14'	30	W	P	HSL	PD	Y	Y	20		
D Northwest Basin Marina, Ltd. 705-549-2655	579 Champlain Rd. Penetanguishene, ON L9M 1R2	68	15		30'	5'	30	W	P	HS	PD	Y	N	30	M	IC

LAKELAND BOATING · LAKE HURON CRUISING GUIDE **287**

Midland, Ontario

Midland Bay, between Midland Point and Flat Point **A**, has two smaller bays within it. The town of Midland is located in the southwest bay. Just before coming to this bay, you'll see Doral Marine Resort **B** in the southeast bay, known as Tiffin Basin **C**. To starboard in the southwest bay is the red Midland Bay shoal light buoy, marked M20. You'll know you're close to Midland when you see the huge grain elevators at the bottom of the bay. Keep your bow on the A.D.M. Milling Co. Grain Elevators **D**. As you enter the bay you'll see Midland Marine **E** and beside it the Midland Harbour Town Docks **F**. On the west shore is Bay Port Marina, Ltd. **G**. The adjacent Midland Sailing Club **H** is private and does not cater to transients. There is an excellent launch ramp next to the club at the city-owned Pete Peterson Park **I**.

Lat 44°45.58
Lon 79°53.87

Distances from Midland (statute miles)
Port Severn: 11 E
Penetanguishene: 10 W
Parry Sound: 56 N

Tobermory: 101 NNW
Victoria Harbour: 7 E
Honey Harbour: 8 NE

GEORGIAN BAY

Midland, Ontario
A haven for history buffs.

Home to museums, reconstructed historic sites and even a shrine to martyred missionaries, Penetanguishene's sister city is a must-visit port for history lovers. It also just happens to be located in one of the Great Lakes' premier cruising grounds.

Along with an array of antique shops, delis, cafes and supermarkets, Midland offers plenty of marina options. The **Midland Harbour Town Docks** have a roofed, sheltered picnic area. The town's main street is close to the docks, so everything you might need is handily located in a nearby marina or within strolling distance.

The massive **Doral Marine Resort** offers 100 transient spaces, along with all services, including fuel, laundromat, pool, haulout and all repairs.

Midland Marine has several slips for transients. They sell gas and diesel and do some repairs, too.

The vast yet green **Bay Port Marina, Ltd.** has 40 transient slips, a beautiful pool and a ship's store. They also provide gas, diesel, pumpout and full repairs. The scenic **Rotary Waterfront Trail** joins the marina to the downtown, taking you past historical sites to restaurants and shops.

For fun, take the walking tour of more than 30 historic murals downtown. Dockside, you'll find **Scully's Crab Shack** near the harbormaster's office. The *Miss Midland*, a 300-passenger cruise ship, will take you on a two-and-a-half-hour sightseeing tour through the 30,000 Islands of Georgian Bay.

Don't leave Midland without heading to the **Martyrs' Shrine**, designated "a holy place" by Pope John Paul II. It's on the outskirts of town, just a few minutes away by cab. Or you can dinghy up the buoyed Wye River to the shrine, entering just beyond the entrance to the Doral Marine Resort in Tiffin Basin. There you'll see a twin-spired stone church dedicated to eight Jesuit martyrs who were killed in this region between 1642 and 1649 and subsequently canonized.

Just across Highway 12 is **Sainte-Marie Among the Hurons**, a reconstruction of a mission village that flourished here about the same time the martyrs were killed. Check out the website: www.saintemarieamongthehurons.on.ca.

Nearby is the **Wye Marsh Wildlife Centre**, a wildlife area complete with nature tours on boardwalks over wetlands, canoe trips, informational talks and a theater. They'll arrange transportation to and from your berth; call 705-526-7809.

Huronia Museum & Huron Indian Village is less than a mile from the docks. Located in **Little Lake Park**, the reconstructed Wendat village illustrates how the Huron people lived before the arrival of the first Europeans. Visitors can play Huron games and listen to shaman chants. The collection includes artifacts

More Info

Emergency 911
Huronia District Hospital 705-526-3751
Chamber of Commerce 705-526-7884
www.southerngeorgianbay.on.ca
Deluxe Taxi 705-527-4444

LAKELAND BOATING · LAKE HURON CRUISING GUIDE

Local events

May
Bed & Breakfast Tour & Tea Party of Southern Georgian Bay

June
Doors Open Huronia
National Aboriginal Day

July
Canada Day
Huronia Open Old-Time Fiddle & Step Dance Contest
Festival du Loup de Lafontaine
Georgian Bay Poker Run
Waterfest

that date back more than 10,000 years. New exhibits open every couple of months. After taking in some history, check out the souvenirs at **Mundy's Bay Store**.

Showcasing historic uniforms and medals, **Royal Canadian Legion Museum Branch 80** is also an excellent source for reference materials about World Wars I and II.

Hit the links at **Balm Beach Golf Club** (nine holes), **Brooklea Golf & Country Club** (27 holes) or **Midland Golf & Country Club** (18 holes). Or hit a bucket at **Chelsea Target Range**.

Midland is truly a boater's haven. The nearly 100,000 visitors who come off the water to spend a few days each year in this beautiful, historic spot can't be wrong.

Where to Eat

Arch 705-526-7313
Popular with locals and cottagers, this place serves excellent steaks, fish and Greek dishes.

Casey's Bar & Grill 705-527-1122
A grill house offering Canadian cuisine, located in the Huronia Mall.

The Cellarman's Ale House 705-526-8223
Behind the Roxy Theatre. Choose from the pub-style menu. Breakfast is also served, and patio dining is available. Some nights are hopping, with live music and dancing.

Henry's South 705-528-1919
With the same owners as Henry's in Sans Souci, this lakeside restaurant is renowned among boaters. It serves a complete range of local fish, with pickerel a mainstay.

The Library 705-528-0100
Five-star nouvelle cuisine in a restored, red-brick Victorian building. The ambience is delightful, and the food is well worth the price.

Scully's Crab Shack 705-526-2125
Good prices and a prime waterfront location. A favorite with locals and boaters alike.

Cruise Georgian Bay and visit the
Port of Midland

- 200 stores and restaurants within walking distance
- Tour vessels
- Public park with green space, waterfall and paved trails adjacent to docking area
- Numerous tourist attractions

705-526-4275

Town of Midland, 575 Dominion Avenue
Midland, Ontario L4R 1R2
705-526-9971 fax www.town.midland.on.ca

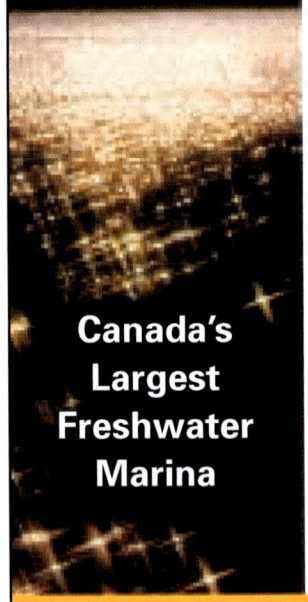

On Beautiful Georgian Bay Midland, Ontario

Canada's Largest Freshwater Marina

75 minutes north of Toronto and minutes from the most incredible boating in the world - Georgian Bay's 30,000 Islands

Dockage, Marine Supplies, Sales and Service are all located here at the "King of Marinas®"

DORAL
Doral Marine Resort

What boating is all about!

3282 Ogden's Beach Road, P.O. Box 40, Midland, ON L4R 4K6
Tel: 705-526-0155
866-253-6725
Email: info@doralmarineresort.com
www.doralmarineresort.com

Discover Boating
Discover Georgian Bay
Discover Bay Port

Bay Port Yachting Centre is located on the doorstep of the best fresh water boating in North America. Only 90 minutes north of Toronto, we offer everything you need for a great summer on the water. And, if you choose Bay Port year-round, your boat will get top quality care and maintenance throughout the winter. In the spring, it will be ready to go when you are.

This summer, come to Georgian Bay and make Bay Port your Home Port!

- Heated pool and spa
- Children's playground
- Private washrooms with showers
- Complete mechanical/structural repair for sail & power
- Close to Midland's historic sites, shopping and great cuisine

BAY PORT
MIDLAND

165 Marina Park Avenue
P.O. Box 644
Midland, Ontario
CANADA L4R 4P4

Tel: 705 527-7678
Fax: 705 527-4190
Toll Free: 1 888 229-7678
e-mail: bayport@bayport.on.ca

MERCURY
The Water Calls

ZODIAC

Significant Waypoint

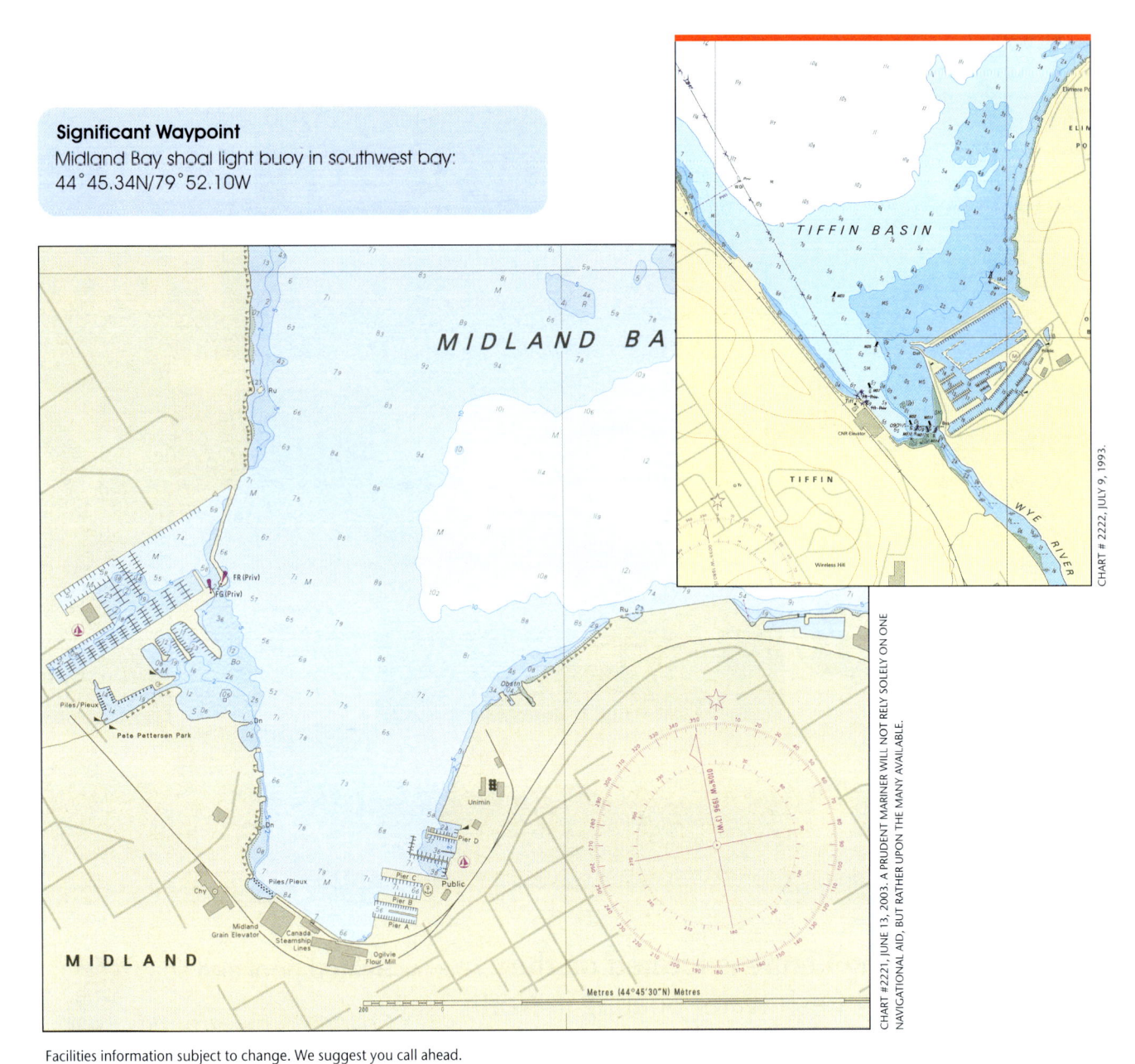

Midland Bay shoal light buoy in southwest bay:
44°45.34N/79°52.10W

Facilities information subject to change. We suggest you call ahead.

		Monitors VHF Channel	Transient Slips Available	ALTERNATE MOORING: Wall mooring, Rafting allowed	Maximum LOA	Minimum Depth at Dock	Power (amperage or volts)	HOOKUPS: Water, Cable TV	FUEL: Gas, Diesel, Pumpout	BASICS: Heads, Showers, Laundry	AMENITIES: Swimming pool, Whirlpool, Grills, picnic tables, Dog walk, Rec area	Take Reservations	Take Credit Cards	Haulout (Capacity in tons or feet)	REPAIRS: Mechanical, Electronics, Woodworking, Sails, Canvas, Fiberglass	CONVENIENCE: Ship's store, Ice, Convenience foods & beverages
B	Doral Marine Resort 3282 Ogden's Beach Road 705-526-0155 Midland, ON L4R 4K6	68	100		80'	6'	30	W	GDP	HSL	SPD	Y	Y	50	MEFWC	SIC
G	**Bay Port Marina, Ltd. 165 Marina Park Ave.** **888-BAYPORT Midland, ON L4R 4P4**	68	40	W	100'	20'	30 50	WC	GDP	HSL	SWPD	Y	Y	35	MEFWSC	SIC
F	Midland Harbour Town Docks 575 Dominion Ave. 705-526-4610 Midland, ON L4R 3S3	68	55	W	200'+	18'	15 30	W		HSL	PD	Y	Y		MEFW	IC
E	Midland Marine 171 King St. 705-526-4433 Midland, ON L4R 4L1	68	8		50'	18'	15 30	W	GP	HS	GPD	Y	Y	15	MEFW	SI

292   LAKELAND BOATING · LAKE HURON CRUISING GUIDE

Victoria Harbour, Ontario

Lat 44°45.25
Lon 79°46.75

The best bet for dockage is at the luxurious Queen's Cove Marina **A**, a private facility that sells fuel and accepts transients. Other options are the Federal Dock **B** and the Town Dock **C**, where depths are considerably shallower. Neither of these public facilities has services.

Facilities information subject to change. We suggest you call ahead.

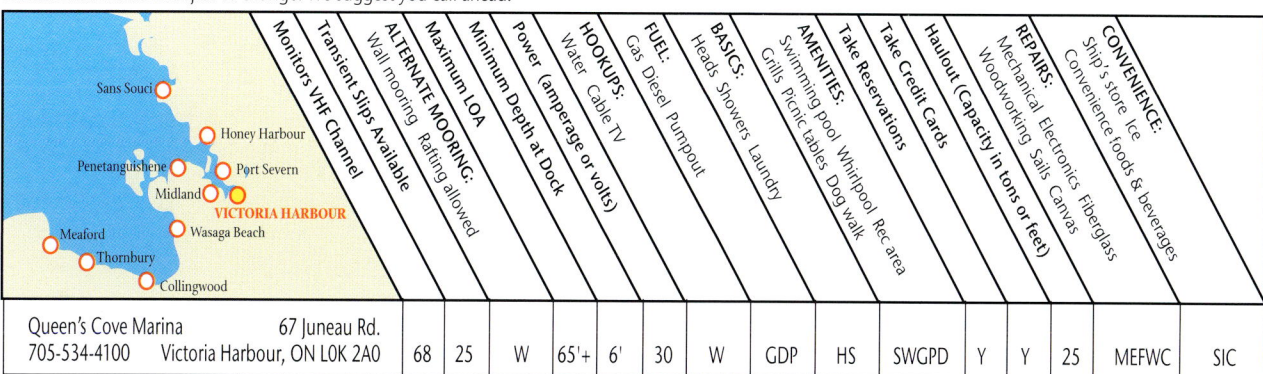

		Monitors VHF Channel	Transient Slips Available	ALTERNATE MOORING: Wall mooring / Rafting allowed	Maximum LOA	Minimum Depth at Dock	Power (amperage or volts)	HOOKUPS: Water / Cable TV	FUEL: Gas / Diesel / Pumpout	BASICS: Heads / Showers / Laundry	AMENITIES: Swimming pool / Whirlpool / Rec area / Grills / Picnic tables / Dog walk	Take Reservations	Take Credit Cards	Haulout (Capacity in tons or feet)	REPAIRS: Mechanical / Electronics / Fiberglass / Woodworking / Sails / Canvas	CONVENIENCE: Ship's store / Ice / Convenience foods & beverages
A	Queen's Cove Marina 705-534-4100 / 67 Juneau Rd. Victoria Harbour, ON L0K 2A0	68	25	W	65'+	6'	30	W	GDP	HS	SWGPD	Y	Y	25	MEFWC	SIC

GEORGIAN BAY

Victoria Harbour, Ontario
Get the royal treatment at this beautiful port.

When lumber baron John Waldie bought the mills in Victoria Harbour in 1885, he built company housing for his workers and gave each family in town a turkey every Christmas. You'll feel similarly well taken care of here, because **Queen's Cove Marina** is one of the best-managed and most luxurious marinas in these waters. It features a tennis court, a pool and spa and 285 slips, as well as more than 1,700 feet of covered walkways, affording protection should a summer shower catch you by surprise. The marina also boasts a fine launch ramp.

There's additional no-frills dockage at the **Federal Dock**, located east of Queen's Cove Marina, and at the **Town Dock** to the southeast.

Nestled in Hog Bay, just west off of the larger Sturgeon Bay, Victoria Harbour rests in the far southeastern toe of Georgian Bay, right across from Port McNicoll.

The center of town is within easy strolling distance. Here you'll notice Waldie's other contribution to the port—the bright yellow hue of some of the buildings. When Waldie built up the town, he painted all the housing as well as the company store canary yellow, leading people to dub the port "Canary Towne"—a moniker still in use by some.

Within blocks, you'll find a post office, a **liquor store**, a **library**, a **coin-operated laundry** and an **IGA supermarket**. The **Corner Café** at Albert and William streets serves breakfast, lunch and dinner. **MacKenzie Park** has a public beach located near the marina.

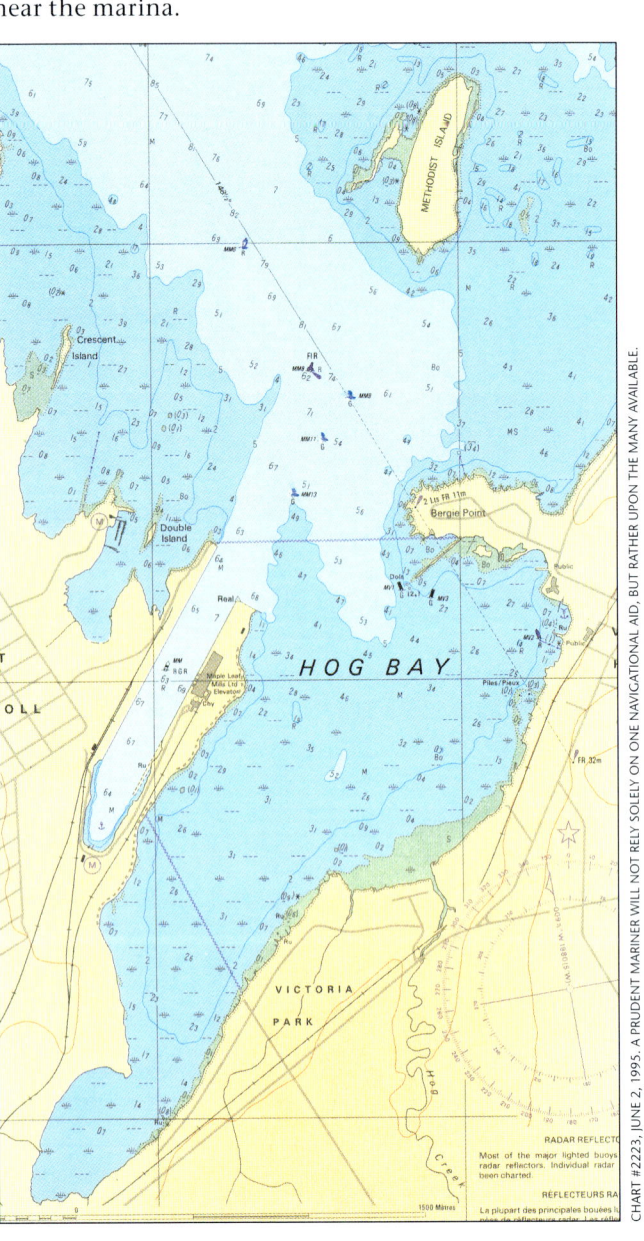

Distances from Victoria Harbour (statute miles)
Midland: 7 W
Port Severn: 10 E
Honey Harbour: 10 N
Parry Sound: 54 N

More Info
Emergency 888-310-1122
Huronia District Hospital 705-526-3751
Chamber of Commerce 705-526-7884
www.southerngeorgianbay.on.ca
A-1 Bats Taxi 705-534-1122

Where to Eat
LC's Restaurant 705-534-0323
Stop by for some great home cooking.

294   LAKELAND BOATING · LAKE HURON CRUISING GUIDE

Waubaushene, Ontario

This bustling port offers several options for transients.

At the eastern end of Severn Sound, Waubaushene has three marina facilities, a **public dock** on the south shore carrying 12 feet alongside and a growing shopping district, which includes a **grocery store** around the corner from the marinas. All have facilities for repair as well as dockage. There are launch ramps at each marina as well as at **Waubaushene Marine Industries** (705-538-2266), which boasts an excellent ramp and a large parking area. The ramp at Pier 69 is outside the bridges.

Significant Waypoint
Directly west of harbor entrance:
44°47.68N/79°48.45W

Distances from Waubaushene (statute miles)
Port Severn: 4 NE   Victoria Harbour: 6 W

Facilities information subject to change. We suggest you call ahead.

Marina	Address	Transient Slips	VHF	Mooring	LOA	Depth	Hookups	Fuel	Amenities	Reservations	Credit Cards	Haulout	Repairs	Convenience
Marsh's Waubaushene Marina 705-538-2285	5 Duck Bay Rd. Waubaushene, ON L0K 2C0	20			24'	6'			H	D	Y	Y	M	SIC
Pier 69 705-538-2867	10 Duck Bay Rd. Waubaushene, ON L0K 2C0	68	8	WR	48'	4.5'	15	GDP	HS	RGPD	Y	Y	18	MESC	SIC
Twin Bridge Marina 705-538-2295	35 Duck Bay Rd. Waubaushene, ON L0K 2C0	3		W	25'	4'		G		PD	Y	Y	25	ME	SC

Behind our welcome mat is the best marina on Georgian Bay

Royal treatment for you and your boat. On the threshold of the Thirty Thousand Islands

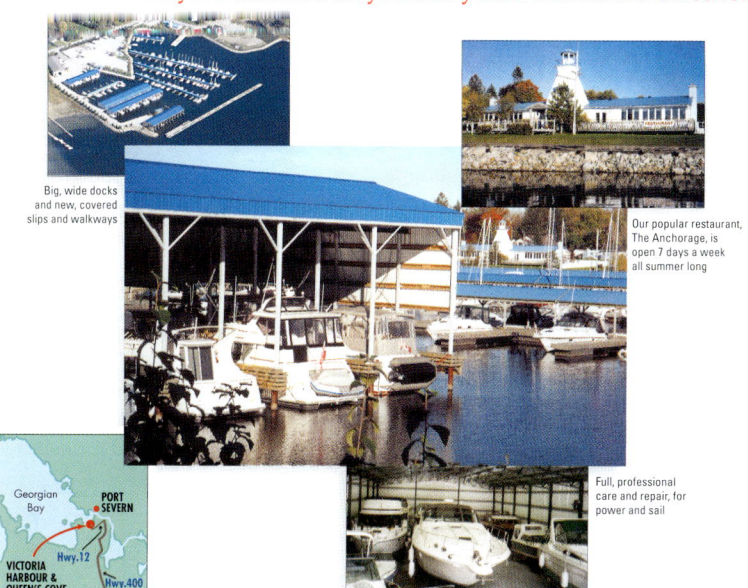

Whether you're planning to visit overnight or to moor with us for the season, you simply won't find a friendlier spot than Queen's Cove Marina. We'll treat you royally from the moment you arrive.

Start with the extra-wide berths and big, full-service docks with covered walkways. Enjoy the BBQ pit, with picnic tables and Muskoka chairs; head for the tennis court; entertain under our covered gazebos; or relax in the heated pool and spa.

We'll treat your boat royally, too. Our first-class facilities are open year round, with complete hull and mechanical repair services, custom woodworking, ship's chandlery, and indoor (heated) and outdoor storage.

If you're looking for a great cruising base, you can't beat Queen's Cove Marina, either. Located in Victoria Harbour, we're the first full-service marina on Georgian Bay out of Lock 45, only a 45-minute trip from the Trent-Severn. By car, we're 1 1/2 hours from Toronto — just 10 minutes off Hwy. 400.

And, best of all, we're right on the threshold of the Thirty Thousand Islands. So you don't waste your time getting to the best cruising in Georgian Bay – at Queen's Cove Marina, you're already there. And welcome.

Big, wide docks and new, covered slips and walkways

Our popular restaurant, The Anchorage, is open 7 days a week all summer long

Full, professional care and repair, for power and sail

QUEEN'S COVE MARINA

Tel: (705) 534-4100 or Toll Free: 1-800-461-BOAT • Fax: (705) 534-3118 Box 333, Victoria Harbour, Ont. L0K 2A0

Port Severn, Ontario

Lat 44°48.21
Lon 79°43.22

Lock 45 is the western terminus of the Trent-Severn Waterway, which stretches from the Bay of Quinte at Trenton on the St. Lawrence to Georgian Bay. Approaching from Georgian Bay **A**, keep the red buoys to port and green to starboard to avoid the shallows extending into Tugs Channel from Shields Island **B** and Sawdust Island, immediately south. As you pass under the bridge **C**, if you don't plan to lock through and enter the waterway right away, dockage is available at Jim Earle Marine **D** just below the lock **E**. Above the lock are the Driftwood Cove Resort Marina **F**, Severn Marina, Ltd. **G**, Bush's Marine **H** and Severn Boat Haven, Ltd. **I**.

Distances from Port Severn (statute miles)
Midland: 11 W
Honey Harbour: 11 W
Parry Sound: 55 NW
Killarney: 178 WNW
Tobermory: 110 WNW

296   LAKELAND BOATING · LAKE HURON CRUISING GUIDE

GEORGIAN BAY

Port Severn, Ontario
Explore the historic waterways of the Trent-Severn.

Lock 45 is the western terminus of the Trent-Severn Waterway and Mile 0 of the Small-Craft Route in Georgian Bay, and while it is the smallest lock on the Trent-Severn, it is also one of the busiest. Boaters from around the world visit Port Severn to go through the lock, which is still manually operated. Like the old veteran square-rigged sailors weighing anchor by walking around the windlass, the lockmaster walks around a mechanical device called a sweep, which closes and opens the lock doors. The deck hands then stand by to hold their boat away from the walls as it is dropped 14 feet to the Georgian Bay level.

About 8,000 recreational boats passing through each year. Some are entering Georgian Bay, while others are heading for the waterway, then over to Lake Ontario and through the Erie Canal to end up in the Atlantic. From there they often cruise to the Gulf of Mexico, up the Mississippi and then to Chicago. The lock system has become a tourist draw with a steady stream of picnickers in the small public park beside the lock.

Boaters and tourists who want to hang around and enjoy the atmosphere can stay at **The Inn at Christie's Mill**, situated on the water's edge just above the lock. It offers accommodations, dockage, a fine dining room, tennis courts and a small beach. Within walking distance of the dock are **grocery stores** and a **post office**.

Rawley Lodge, which overlooks the locks, also offers fine food and accommodations. The lodge, which was once a boarding house for a mill, is an ideal place to relax and watch the boats pass by.

Transient options include **Bush's Marine**, **Severn Boat Haven, Ltd.**, **Severn Marina, Ltd.** and **Jim Earle Marine**. All of these facilities offer full services, including gas, repairs, haulout and ship's stores. **Big Chute Marina, Ltd.**, another transient option, has gas, pumpout, new docks, a grocery store and restaurant and sells hunting and fishing licenses. **Driftwood Cove** is a luxury marine resort west of the lock.

Where to Eat

Driftwood Cove Café. 705-538-0334
Open seasonally. Located at Driftwood Cove Resort Marina.

The Inn at Christie's Mill. 705-538-2354
Open for breakfast, lunch and dinner, serving fish, steak and pasta. Try the ribs—touted as "the tenderest around"—and the fresh pickerel.

Muskoka Joe's. 705-538-0986
Great fare at reasonable prices. The establishment has reduced hours during the colder months, so call ahead.

Rawley Lodge 705-538-2272
Good home cooking, Canadian-style. The menu ranges from steaks to homemade dessert.

More Info

Emergency. 911
Huronia District Hospital. 705-526-3751
Muskoka Tourism. 800-267-9700
www.discovermuskoka.ca
Port Severn Lock Office 705-538-2586

LAKELAND BOATING · LAKE HURON CRUISING GUIDE **297**

The Trent-Severn Waterway

Welcome to the 240-plus miles of twisting, narrow canals, rivers and lakes that make up the Trent-Severn Waterway. While the waters have long been enjoyed by Canadian cruisers, their neighbors to the south haven't been as quick to catch on.

Open from May to October, the waterway cuts a scenic path through marshes, gentle farmlands and bustling town centers, leading to the pristine granite fields in the north. You'll glide past gardens of lily pads floating under drooping willows, and cattails and silver birches pointing heavenward. You'll spot ducks and kingfishers flying over vacation cottages and small fishing boats reeling in sunfish, smallmouth bass and rockfish.

Pleasure traffic is the only sort that navigates the 43 locks found here. Most of the gates are still manually operated, manipulated by large turnstiles similar to the capstans on old sailing vessels. The water can be shallow, sometimes only 5 feet deep. You'll rarely need to glance at your charts, although it does take some practice to spot the slender Canadian buoys that guide the way.

In cities near the locks, there will often be a municipal marina to accommodate you. Boaters can also tie up along the lock walls for a fee.

ramps at Driftwood Cove and Severn Marina.

There are no facilities in Port Severn for stepping masts for sailboaters entering or leaving the locks; instead try **Queen's Cove Marina** at Victoria Harbour (7 miles) or Midland's **Doral Marine Resort** (9 miles) or Bay Port Marina (11 miles).

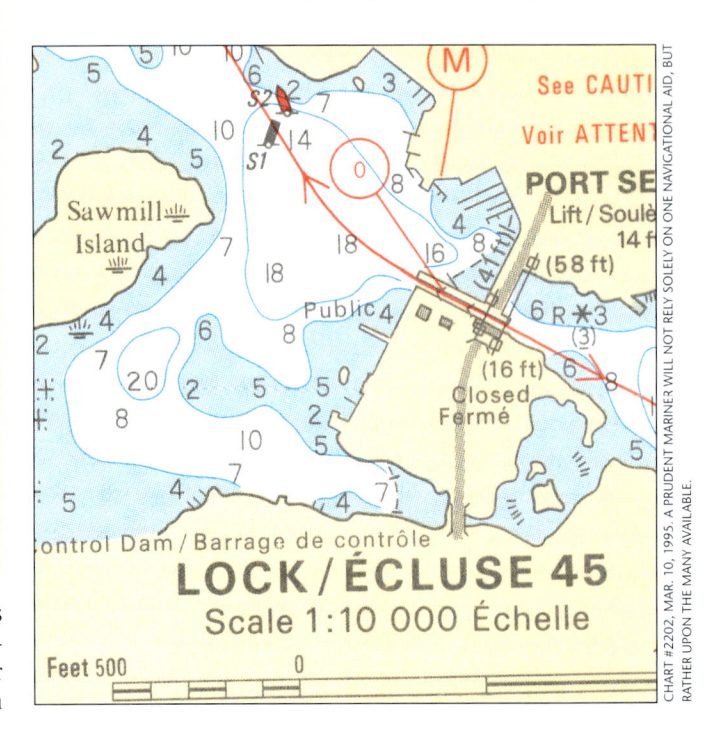

Local events

July
Wooden Boat Drive-By

Classic Wooden Boat Show

Reservations are recommended for docking or for one of its cottages. The resort also boasts a hot tub and café with laundromat. You'll find an LCBO store that sells liquor and beer within walking distance of the marina, and there are launch

Facilities information subject to change. We suggest you call ahead.

Marina	Address	Monitors VHF Channel	Transient Slips Available	ALTERNATE MOORING: Wall mooring / Rafting allowed	Maximum LOA	Minimum Depth at Dock	Power (amperage or volts)	HOOKUPS: Water/Cable TV	FUEL: Gas/Diesel/Pumpout	BASICS: Heads/Showers/Laundry	AMENITIES: Swimming pool/Whirlpool/Rec area/Grills/Picnic tables/Dog walk	Take Reservations	Take Credit Cards	Haulout (Capacity in tons or feet)	REPAIRS: Mechanical/Electronics/Fiberglass/Woodworking/Sails/Canvas	CONVENIENCE: Ship's store/Ice/Convenience foods & beverages
Big Chute Marina, Ltd. 705-756-2641	Rural Route 1 Coldwater, ON L0K 1E0	68	14		60'	15'	30	W	GP	HS	RPD	Y	Y	42	MEFC	SIC
H Bush's Marine 705-538-2378	Port Severn Road Port Severn, ON L0K 1S0		12	R	48'	8'	30		G	HS	RGPD	Y	Y	6	MEFW	SIC
F Driftwood Cove Resort Marina 705-538-2502	Port Severn Road Port Severn, ON L0K 1S0	68	var.	W	50'	4'	30	WC	P	HSL	WRGPD	Y	Y			SIC
D Jim Earle Marine 866-JIM-EARLE	2754 Marine Dr. Port Severn, ON L0K 1S0		10		55'	5'	30	W	G	H	D	Y	Y	35	MEFWC	SIC
I Severn Boat Haven, Ltd. 705-538-2975	Kelly's Road Port Severn, ON L0K 1S0	68	10		65'	6'	30	W	GDP	HS	RPD	Y	Y	25	MEFWC	SIC
G Severn Marina, Ltd. 705-538-2571	P.O. Box 95 Port Severn, ON L0K 1S0	68		W	40'	5'	15	W	GP	S	GPD	Y	Y	45'	MEFWSC	SIC

298 LAKELAND BOATING · LAKE HURON CRUISING GUIDE

Georgian Bay

The 30,000 Islands, Ontario
One hundred miles of jaw-dropping beauty.

Stretching more than 100 miles from Port Severn to the French River, the 30,000 Islands are among the world's most beautiful—and best-charted—cruising grounds. Many consider this spectacular region to be comparable world wide only to the Greek Isles.

The channels twist, revealing magnificent new panoramas with each turn. In some places, the rocks seem to squeeze you in, making speeding a dangerous idea. So take your time and enjoy the beauty. The Georgian shoals and granite are part of the Canadian shield, carved by glaciers millions of years ago, and are very unforgiving!

These islands, which provide protection from Georgian Bay, demand a watchful eye because of their density and hardness. But the 30,000 Islands (actually 83,000 at current count) are also among the best-charted cruising grounds in the world, and a yachtsman's vigilance will be rewarded with breathtaking vistas and serene anchorages.

Leaving Port Severn, boaters can head north either by open water or by taking the "Inside Passage," also called the "Small-Craft Route." But do not be misled by this name. Fifty-footers and larger ply these waters regularly. The Canadian government states the limiting depth at 6 feet. Since this is not always the case, boaters should follow their charts to bypass problem areas if necessary. The Canadian Hydrographic Service book, CHS *CEN 306 (Sailing Directions: Georgian Bay)*, along with strip charts 2202, 2203, 2204 and 2205, will guide boaters safely from Port Severn to Little Current on the North Channel, while also pointing out numerous anchorages.

Steamers have plied these protected waters for more than a century, and their passage route is shown by dotted red lines on the charts. Since this route has deeper water than the solid red line of the Small-Craft Route, many larger cruisers opt for parts of this route to avoid the Small-Craft Route's shallower and tighter sections.

The 30,000 Islands offer hundreds of anchorages, coves and bays—far too many to list. The best resources for visiting these waters are the CHS *CEN 306* and the logbook of the Great Lakes Cruising Club. Members of the GLCC have access to detailed information on the Small-Craft Route and the harbors and anchorages found along the way. Check out the club's website at www.glcclub.com.

Though anchorages abound, there are only a few true ports of call. Bear in mind that this is cruising country. Most of the ports we cover in this section are primarily basic supply and fuel stops with occasional laundry facilities and dockage for restocking between anchorages. These are not entertainment centers!

But then, that is precisely the beauty of the 30,000 Islands and one of the reasons so many cruisers are drawn to this area. After a visit, you're sure to agree.

Honey Harbour, Ontario

Honey Harbour is located just off the Small-Craft Route, which winds its way among the area's many islands. Watch your charts carefully—shallow water abounds here. For transient boaters, Honey Harbour itself offers several well-equipped marinas. Honey Harbour Boat Club **A** and Admiral's Marina, Inc. **B** are located on an inlet to port that leads to Village Marina **C**. Continue along this route to reach South Bay Cove. Delawana Inn **D** lies just around the bend. Stay on the Small-Craft Route to find Paragon Marina, Inc. **E**, Picnic Island Resort **F**, which is a gas and grocery stop, and Bayview Marine Resort **G**.

Distances from Honey Harbour (statute miles)
Midland: 10 SW
Tobermory: 100 W
Parry Sound: 44 N
Sans Souci: 28 N

South Bay Cove **H** is a stone's throw from Honey Harbour **I**, just off of the Small-Craft Route. The friendly South Bay Cove Marina **J** is your only transient option—and is probably the finest docking and dining facility on all of Georgian Bay. Head west, passing Jack Rock **K** and Lownie Island **L** to port to reach Honey Harbour or to rejoin the Small-Craft Route **M**.

LAKELAND BOATING · LAKE HURON CRUISING GUIDE   **301**

GEORGIAN BAY

Honey Harbour &
South Bay Cove, Ontario

Honey Harbour & South Bay Cove, Ontario
A sweet spot for island cruising.

Honey Harbour might be named for the sweet stuff made by bees, which early settlers here produced, but the hamlet could just as well have been given this moniker by boaters, who find the area's thousands of passages, channels, sheltered bays and coves, well, sweet.

Situated at the southeast toe of Georgian Bay, Honey Harbour makes an ideal jumping-off point for island-hopping. Stop in at the **National Parks office** in town

PETER MARREK

302   LAKELAND BOATING · LAKE HURON CRUISING GUIDE

and pick up a visitors guide, which provides information on cruising through the islands and mooring off them. Dinghy ashore at the special picnic spots and use the shore facilities. Some of these spots have docks for a day's tie-up.

Beausoleil Island, the largest of the Canada Parks islands, is just about one mile offshore. From there, some 59 park islands stretch northward for 80 miles up the Small-Craft Route heading toward the north shore of Georgian Bay.

Honey Harbour itself is one of those small hamlets that has a warm, welcoming atmosphere. In town, you'll find the **post office** and **library** on the waterfront and a **liquor store** within strolling distance. The **Honey Harbour Towne Centre** is a store that sells groceries, hardware and marine supplies. If you need some boating information, stop in at **Our Lady of Mercy Catholic Church** beside the store and talk to Father Bill Fellion. The church operates a boat launch ramp and has a parking lot. There are also excellent ramps at **Honey Harbour Boat Club** and **Admiral's Marina**.

Other area marinas and service shops include the **Village Marina, Ltd., Gerry's Marine & Welding Service** and **Jim Langley Marine**. If you wish to stay ashore, the luxurious **Delawana Inn** offers 150 rooms on 48 acres, a pool with sauna and a dining room with fine cuisine. More rustic is **The Elk's Hide-Away**, an eight-room bed and breakfast, which contains a former chapel that was operated by nuns. The tempting smell of fresh-baked bread fills the air here, because the Hide-Away is also a bakery.

South Bay, a small, sheltered cove just a few miles from Honey Harbour, is home to one of the most luxurious marinas in the area. The modern **South Bay Cove Marina** was opened to serve the growing number of boaters who frequent this popular boating route. With 120 slips, there is plenty of room for transient visitors. Free newspapers and weather reports are delivered to guests' boats each morning on weekends, making the marina a popular spot for both cruisers and long-term visitors.

The marina boasts a fine dining restaurant with a terrific chef. It also has a sauna, laundry facilities and a patio bar; a golf course is nearby. The marina can handle fixed-keel sailors, with 8-foot depths around the docks. Visitors can motor or sail a short distance offshore and quickly find a sheltered inlet or bay. From here, boats readily find the nearby Small-Craft Route.

Local events

July
JulyFest BBQ & Auction

August
Annual Regatta

Moonlight Madness Dance

More Info

Emergency 911
Huronia District Hospital 705-526-1300
Muskoka Tourism 705-689-0660
www.discovermuskoka.ca

Where to Eat

Amicci's 705-756-4619
A bistro roadhouse serving everything from burgers and pizza to steak and salads. Try the weekend specials.

Delawana Inn 705-756-2424
The Thursdays dinner buffet features steak and barbequed ribs. On Sundays, the buffet includes shrimp, salads and crab. It's a good place for a large gathering—the dining room seats up to 450. Also open for breakfast.

The Elk's Hide-Away 705-756-2993
Famous for their pies, this bakery and pizzeria serves up delicious breads, donuts and deli sandwiches. Takeout only.

Papa Joe's 705-756-0822
This Italian place at Admiral's Marina also serves great burgers.

Top of the Cove Restaurant 705-756-3399
Fine French dining in a beautifully designed and decorated building with a spectacular view.

LAKELAND BOATING · LAKE HURON CRUISING GUIDE

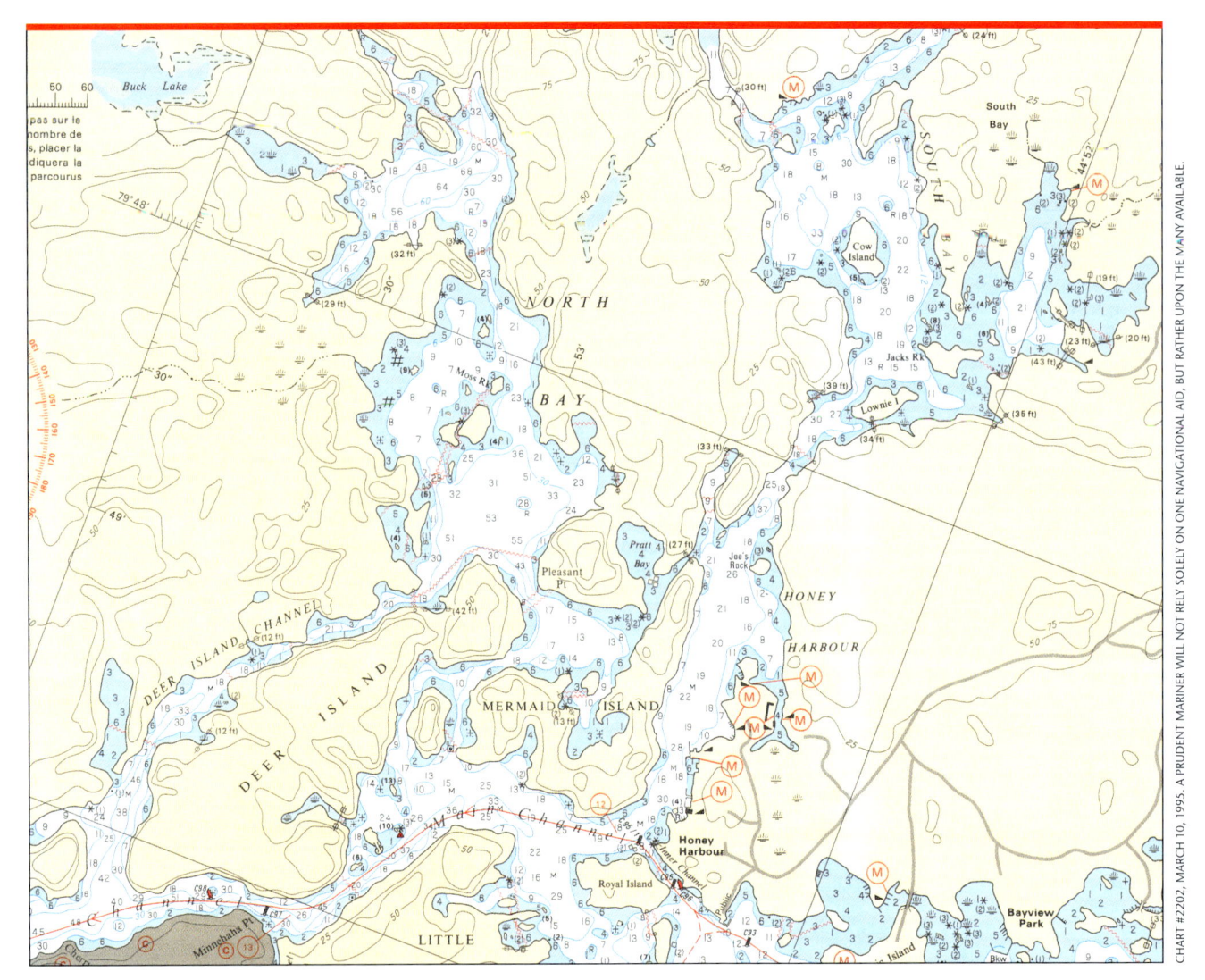

Facilities information subject to change. We suggest you call ahead.

		Monitors VHF Channel	Transient Slips Available	ALTERNATE MOORING: Wall mooring, Rafting allowed	Maximum LOA	Minimum Depth at Dock	Power (amperage or volts)	HOOKUPS: Water, Cable TV	FUEL: Gas, Diesel, Pumpout	BASICS: Heads, Showers, Laundry	AMENITIES: Swimming pool, Whirlpool, Rec area, Grills, Picnic tables, Dog walk	Take Reservations	Take Credit Cards	Haulout (Capacity in tons or feet)	REPAIRS: Mechanical, Electronics, Fiberglass, Woodworking, Sails, Canvas	CONVENIENCE: Ship's store, Ice, Convenience foods & beverages
B	Admiral's Marina, Inc. Honey Harbour Road 705-756-2432 Honey Harbour, ON P0E 1E0		6		40'	10'	30	W	GDP	HS	PD	Y	Y	12.5	MEFWSC	SIC
G	Bayview Marine Resort 387 Baxter Loop 705-756-2482 Honey Harbour, ON P0E 1E0	16 68	6	W	35'	5'	20	W	G	HS	PD	Y	Y	12	MEFWC	SIC
D	Delawana Inn P.O. Box 279 705-756-2424 Honey Harbour, ON P0E 1E0		12	W	50'	10'	15			H	S	Y	Y			SIC
E	Paragon Marina, Inc. P.O. Box 98 705-756-2402 Honey Harbour, ON P0E 1E0	68	10		48'	6'	30	W	GP	HS	D	N	Y	30	MEFWC	SIC
J	South Bay Cove Marina South Bay Road 705-756-3333 Port Severn, ON L0K 1S0	68	30		100'	7'	30	WC	GDP	HSL	RGPD	Y	Y		MEFS	SIC
C	Village Marina, Ltd. P.O. Box 160 705-756-2706 Honey Harbour, ON P0E 1E0		6	W	25'	8'			G	H		Y	Y	10	MEFWC	SIC

A Port to Call Home.

There are no limits to great hospitality.
At South Bay Cove Marina, we cater to meet your every need.
We want you to sit back, relax and enjoy our services,
fine dining and superb facilities.

Located in Honey Harbour, on the southern tip of the beautiful
Georgian Bay, South Bay Cove Marina has a reputation of
being one of the finest marina's in North America.
Come and see for yourself, you will not be disappointed!

Facilities Include: Accommodation for vessels up to 100' • Depth at dock 8ft
• 30 or 50 amp service • Fresh water, pumpout and cable at each slip
• Fuel facilities, gas, diesel and propane • Assistance in and out of slip
• Modern, clean restrooms and private showers • Gift and Marina Shop
• Fine dining and catering to vessel • Sandy beach and deck • Laundry facilities
• Barbeque and picnic areas • Club House with fireplace for morning coffee
• Newspaper delivery daily • Wireless high speed satellite internet
• Golf courses nearby • Shuttle service available

South Bay Cove Marina
375 South Bay Road, RR#1
Port Severn, Ontario
L0K 1S0
Tel.: (705) 756-3333
Fax: (705) 756-3223
Restaurant: (705) 756-3399
www.southbaycove.com

Marina monitors
Channel 68

SOUTH BAY COVE MARINA

Sans Souci, Ontario

Continue about 28 miles northwest of Honey Harbour along the Small-Craft Route and you'll find the cozy hamlet of Sans Souci, where the route threads between Fryingpan **A** and Sans Souci **B** islands. The light on Barrel Point **C** is a handy landmark. LeBlanc's Marina **D** does repairs and sells gas, but does not offer transient dockage. The docks **E** owned by the Sans Souci Association do not cater to transients. Henry's Fish Restaurant **F** is the best bet for overnight dockage, gas and diesel—and the best freshwater fish on the bay.

Lat 45°10.07
Lon 80°07.35

Facilities information subject to change. We suggest you call ahead.

		Monitors VHF Channel	Transient Slips Available	ALTERNATE MOORING: Wall mooring / Rating allowed	Maximum LOA	Minimum Depth at Dock	Power (amperage or volts)	HOOKUPS: Water / Cable TV	FUEL: Gas Diesel Pumpout	BASICS: Heads Showers Laundry	AMENITIES: Swimming pool Whirlpool Grills Picnic tables Dog walk Rec area	Take Reservations	Take Credit Cards	Haulout (Capacity in tons or feet)	REPAIRS: Mechanical Electronics Fiberglass Woodworking Sails Canvas	CONVENIENCE: Ship's store Ice Convenience foods & beverages
F	Henry's Fish Restaurant 705-746-9040 Fryingpan Island Sans Souci, ON P0C 1H0	68	24		50'	15	30		G	HS	PD	N	Y			I
D	LeBlanc's Marina 705-746-5598 Fryingpan Island Sans Souci, ON P0G 1L0								GDP	H					MF	SIC

306 LAKELAND BOATING · LAKE HURON CRUISING GUIDE

GEORGIAN BAY

Sans Souci, Ontario
Leave your cares—but not your swimsuit—behind.

Fryingpan Island may not look much like a frying pan, but with all the fish just waiting to be caught, it wouldn't hurt to have a frying pan on hand to prepare them.

The island has a peaceful air to it, which perhaps led to the cozy hamlet located here being dubbed Sans Souci, which is French for "without a care." Located on the east side of the island, Sans Souci offers sheltered waters and, for a small settlement, most amenities. It's a natural travel route and an excellent base for exploring the area between Parry Sound and Honey Harbour, which has hundreds of excellent anchorages—including Go Home Bay, Monument Channel, Indian Harbour, Twelve Mile Bay, Moon Bay and Echo Bay, just to name a few.

If you're approaching from the south, come through O'Donnell Channel to the mouth of Twelve Mile Bay. You'll be in the clear, for the most part, when you get to McCurry Rocks, until you reach Fryingpan Island.

From the north, head between the black beacon on Ajax Rock and the red triangular beacon on Wildgoose Island. Keep the green spar buoy to starboard. As you pass the red triangular beacon on Bull Rock, turn slightly to starboard.

One of the only places on the eastern shoreline of Georgian Bay where you can eat fish caught that very morning is at the world-renowned **Henry's Fish Restaurant**. A gas dock, along with a marina that caters to transients, is located on the premises. The water is deep here, with a minimum depth of about 20 feet.

LeBlanc's Sans Souci Marina performs engine repairs and has gas, diesel and pumpout.

Enjoy this rustic stopover, which offers myriad spots to fish or swim. Visit the stone cross monument to explorer Samuel de Champlain's passage in 1615. A very long trek by dinghy will take you to **Moon River Falls**, where you can picnic by the 30-foot-high cascade.

Distances from Sans Souci (statute miles)
Parry Sound: 16 NE
Midland: 38 SSW

More Info

Emergency 800-267-7270
Parry Sound General Hospital 705-746-9321
Parry Sound
Chamber of Commerce 800-461-4261

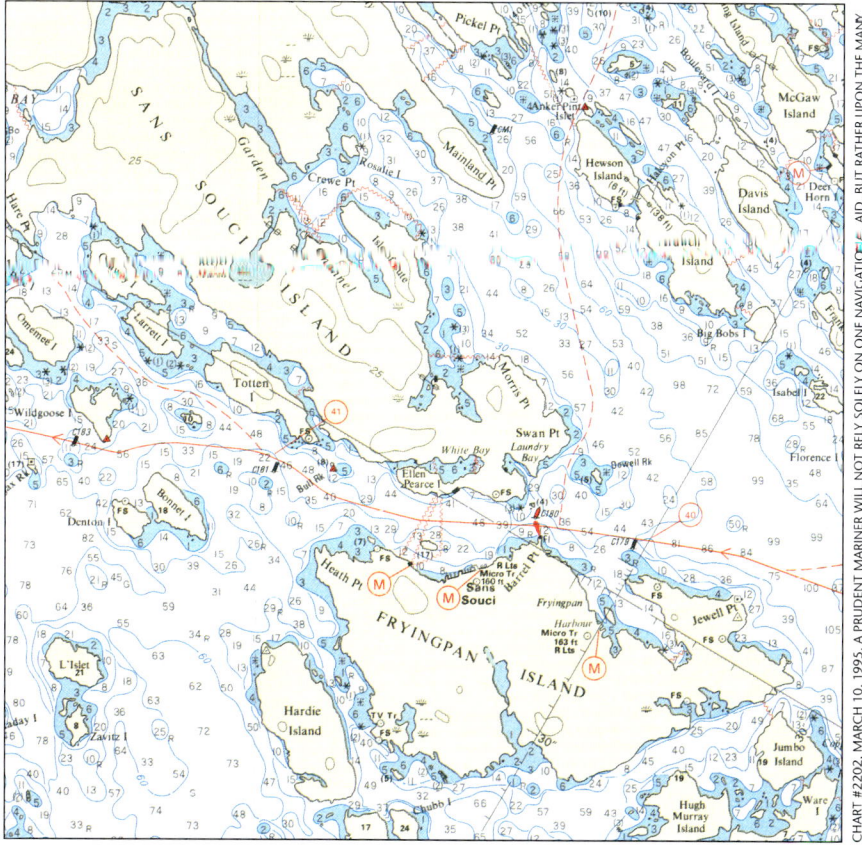

Where to Eat

Henry's Fish Restaurant 705-746-9040
Serving freshwater fish, including trout and their famous pickerel. Don't miss the award-winning butter tarts.

LAKELAND BOATING · LAKE HURON CRUISING GUIDE 307

Parry Sound, Ontario

Approaching Parry Sound from the bay along the Small-Craft Route, pass under the swing bridge **A** to enter Parry Sound Harbour **B**. To starboard south of the bridge, Glen Burney Marina **C** does not cater to transients; neither does Rose Point Marina **D**. Overnight options include Georgian Bay Marina **E**, Point Pleasant Marina **F** and Holiday Cove Marina **G**. Sound Boat Works, Ltd. **H** offers dockage, fuel and repairs. The Parry Sound Town Dock **I** offers transient dockage. Big Sound Marina **J** offers the most transient slips. Parry Sound Marine **K** does repairs and can occasionally accommodate a few transients. Parry Sound **L** is visible in the distance.

Distances from Parry Sound (statute miles)

Sans Souci: 16 SW
Midland: 56 SSW
Tobermory: 83 W

Killarney: 123 NNW
Byng Inlet/Britt: 63 N
Pointe Au Baril: 37 N

Lat 45°19.31
Lon 80°02.50

GEORGIAN BAY

Parry Sound, Ontario
Head in from the wilderness to this restful port.

After exploring the unique channels of Georgian Bay's 30,000 islands, head to the area's capital, Parry Sound. You'll find just about everything you need right at the water's edge, along with lots of options for rest and relaxation.

This is a place of history—of lumber barons, Ojibway hunters and fishermen, explorers, railway magnates and British naval officers. Today, it's also known as the birthplace of Boston Bruins great Bobby Orr, who has his own hall of fame. And it's the home port of the *Island Queen,* the largest sightseeing cruise ship in Canada. Culture buffs will enjoy the newly opened **Charles W. Stockey Centre for the Performing Arts.**

As you approach the harbor, keep the CP railway bridge in front of you, then turn to port and tie up at **Big Sound Marina** or the **Parry Sound Town Dock,** transient facilities operated by the chamber of commerce. Staff there will help you find mooring, fuel, accommodations and food.

Parry Sound Marine offers launch facilities, a hoist, repairs, and boat and motor sales and service. It's also a good place to buy fuel (gas, propane and diesel), marine supplies and tackle. The marina is an authorized dealer for Canadian Hydrographic Service nautical charts. Stretching southward are more facilities: **Sound Boat Works, Holiday Cove Marina, Point Pleasant Marina** and **Georgian Bay Marina.** You'll find launch ramps at the salt docks near the Coast Guard Station, beyond Bob's Point in the sound, and in the harbor on the south shore between Parry Sound Marina and Sound Boat Works.

Everything in Parry Sound is in or close to the harbor area. If you would like to explore the islands while someone else does the driving, take a morning or afternoon trip aboard the *Island Queen* or catch the smaller *M.V. Chippewa* to nearby Fryingpan Island. For either lunch or dinner, walk over to Bay Street across from the public wharf to the Bay Street Café, which has a heated patio overlooking the bay.

Just a five-minute walk from the waterfront, the downtown offers restaurants and cafés, boating and hardware supplies, unique shops with hand-

Local events

May
Home, Cottage & Sportsman Show

Spring Fling

Salomon Adventure Challenge

Raid the North Adventure Race Series

June
Dragon Boat Festival

Canada Day Celebrations

Sportbike Festival

July
Art in the Park

July through August
Festival of the Sound

August
Native Awareness Days

Civic Holiday Festival

Provincial Canoe Championships

Multi-Sport Canada Triathlon

Tugfest

LAKELAND BOATING · LAKE HURON CRUISING GUIDE

Where to Eat

Sound Grill House 705-746-3712
Fresh seafood, spinach dip and hearty salads.

Bay Street Cafe 705-746-2882
Specializing in seven kinds of delicious fish.

More Info

Emergency 911
West Parry Sound Health Centre 705-746-9321
Parry Sound Area
Chamber of Commerce 800-461-4261
Parry Sound Taxi 705-746-1221

crafted Canadiana, and kayak and canoe outfitters.

If you want to stretch your legs, a recreational trail starts at the public wharf, threads its way through town and joins up with the larger **Seguin Trail** just outside of Parry Sound on Highway 69, which runs south to Toronto. That trail links up with other Ontario walking trails. The **Rainbow Theatre** puts on live performances five days a week during the summer. In July and August, Parry Sound hosts the Festival of the Sound, Canada's foremost chamber music festival.

And if all that isn't enough to keep you busy while in this charming port, you can canoe, fish for lake trout or take a charter seaplane for an aerial tour of the islands.

PARRY SOUND HARBOUR

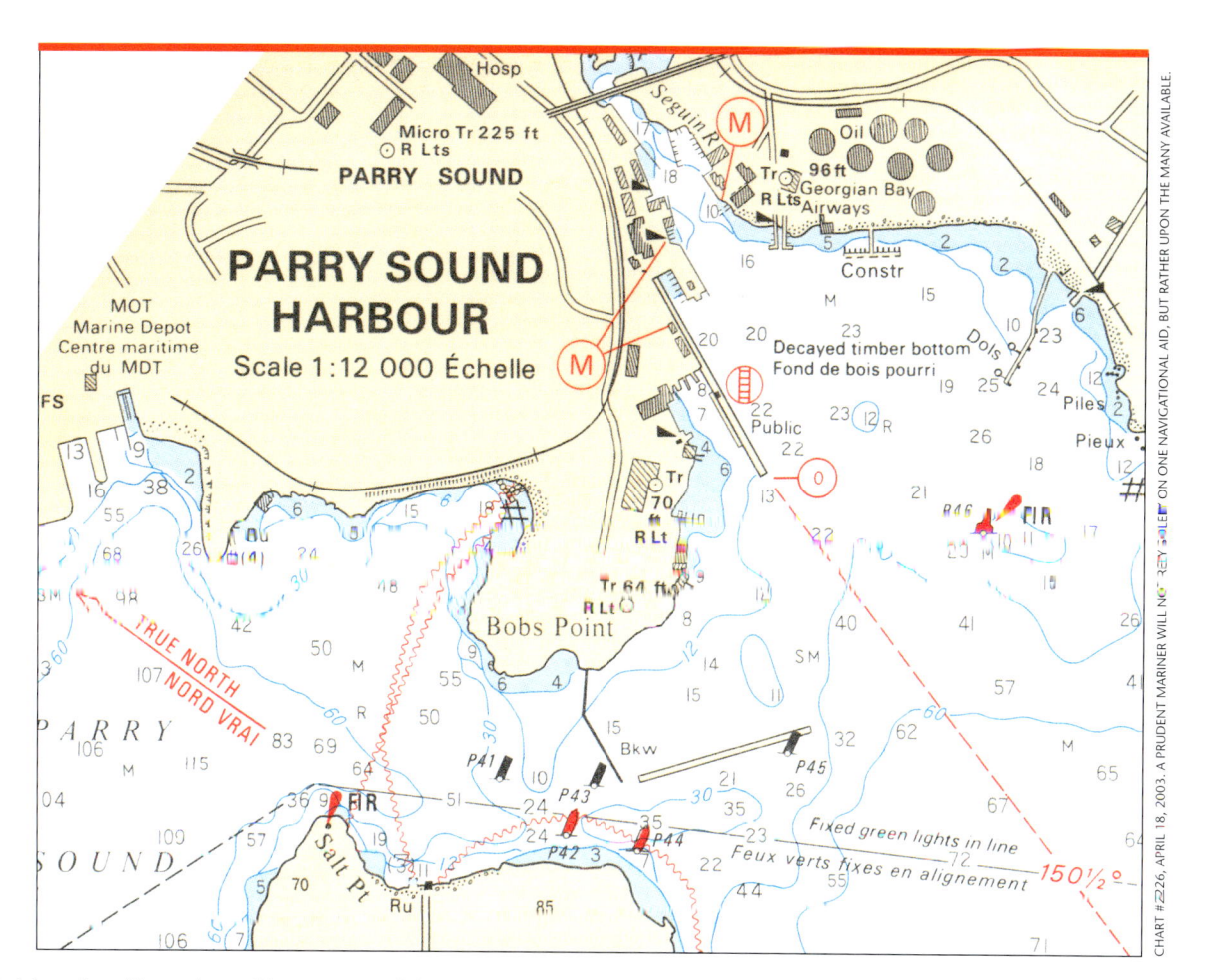

Scale 1:12 000 Échelle

Facilities information subject to change. We suggest you call ahead.

	Marina	Address	Monitors VHF Channel	Transient Slips Available	ALTERNATE MOORING: Wall mooring, Rafting allowed	Maximum LOA	Minimum Depth at Dock	Power (amperage or volts)	HOOKUPS: Water, Cable TV	FUEL: Gas, Diesel, Pumpout	BASICS: Heads, Showers, Laundry	AMENITIES: Swimming pool, Whirlpool, Rec area, Grills, Picnic tables, Dog walk	Take Reservations	Take Credit Cards	Haulout (Capacity in tons or feet)	REPAIRS: Mechanical, Electronics, Fiberglass, Woodworking, Sails, Canvas	CONVENIENCE: Ship's store, Ice, Convenience foods & beverages
J	Big Sound Marina 705-746-7642	9 Bay Street Parry Sound, ON P2A 1S4	68	136		60'	18'	30	W		HSL	PD	Y	Y			IC
E	Georgian Bay Marina 705-746-9559	99 Rose Point Rd. Parry Sound, ON P2A 2X5	68	5		26'	6'	15	W	G	H	D	Y	Y	25'	MEFWSC	SIC
G	Holiday Cove Marina 705-746-2250	Rose Point Rd. Parry Sound, ON P2A 2X3		10		24'	12'			G			Y	Y			IC
	Wasauksing Marina 705-746-7212	P.O. Box 250 Parry Sound, ON P2A 2X5		30		40'	4'	15 30		G	HS	GPD	Y	Y		M	IC
K	Parry Sound Marine 705-746-5848	51 Great North Rd. Parry Sound, ON P2A 2N9	68	2		36'	5'	30	W	GDP	H		Y	Y	8	M	SIC
I	Parry Sound Town Dock 705-746-7373	11 Bay St. Parry Sound, ON P2A 2X3	68	15	W	150'	25'	30	W	P	HS	PD	Y	Y			IC
F	Point Pleasant Marina 705-746-9671	Point Pleasant Road Parry Sound, ON P2A 2X2	68	5		40'	25'	30	W	GP	HS	PD	N	Y	12	ME	SIC
H	Sound Boat Works, Ltd. 705-746-2411	Box 190 Parry Sound, ON P2A 2X3	68	10		40'	10'	30		GDP	HS	GPD	Y	Y	30	MEFWC	I

LAKELAND BOATING · LAKE HURON CRUISING GUIDE 311

Killbear, Ontario

Lat 45°21.30
Lon 80°14.30

There's little here but Killbear Marina **A**—but what more could you ask for? This is one of the best service facilities in these parts. There's even a great dining room, not to mention an incomparable view.

Distances from Killbear (statute miles)
Parry Sound: 12 SE
Snug Harbour: 5 N
Dillon Cove: 10 N
Pointe au Baril: 24 N

312 LAKELAND BOATING · LAKE HURON CRUISING GUIDE

GEORGIAN BAY

Killbear, Ontario
Sandy beaches and stunning sunsets.

There might be only one marina in Killbear—but it's probably the best service marina in the southern reaches of the 30,000 Islands. That's a good thing, because the islands can take out a lot of lower units, according to Will Reichenbacher, who has owned **Killbear Marina** with his brother, Ed, and their father, Dieter, for many years. The marina can haul out boats up to 35 tons for repair and features a well-stocked chandlery and all the charts you'll need.

Necessity was the impetus for the marina, located 12 miles west of Parry Sound at the intersection of the north and south Small-Craft Route and the east and west shipping channel.

The marina has a restaurant featuring European cuisine and a million-dollar view of the islands. The chief attraction in the area is **Killbear Provincial Park**, just around the corner to the south. The park boasts the five best sandy beaches on the east coast of Georgian Bay. It has created a special marine area designated by yellow markers that enables boaters to anchor for the day, to swim in the warm, crystal-clear waters and to enjoy the view of the high granite cliffs of the distant points and islands. Boaters can go ashore and use the beach and shower facilities as well.

Where to Eat

Café Waldo 705-342-5505
A Canadian/German restaurant famous for its schnitzel and sauerbraten.

Ship 'n Shore Restaurant 705-342-5203
Located at Killbear Marina, this eatery is accessible by land or by water. Serving fish and schnitzel, chicken and linguine. It also offers a kids' menu. In late summer, it's only open on weekends until Canadian Thanksgiving.

More Info

Emergency 911
Parry Sound General Hospital 705-746-9321
Killbear Provincial Park 705-342-5492
Chamber of Commerce 800-461-4261
Parry Sound Taxi 705-746-1221
Detour Store 705-342-1611
Killbear Park Mall 705-342-5747

LAKELAND BOATING · LAKE HURON CRUISING GUIDE

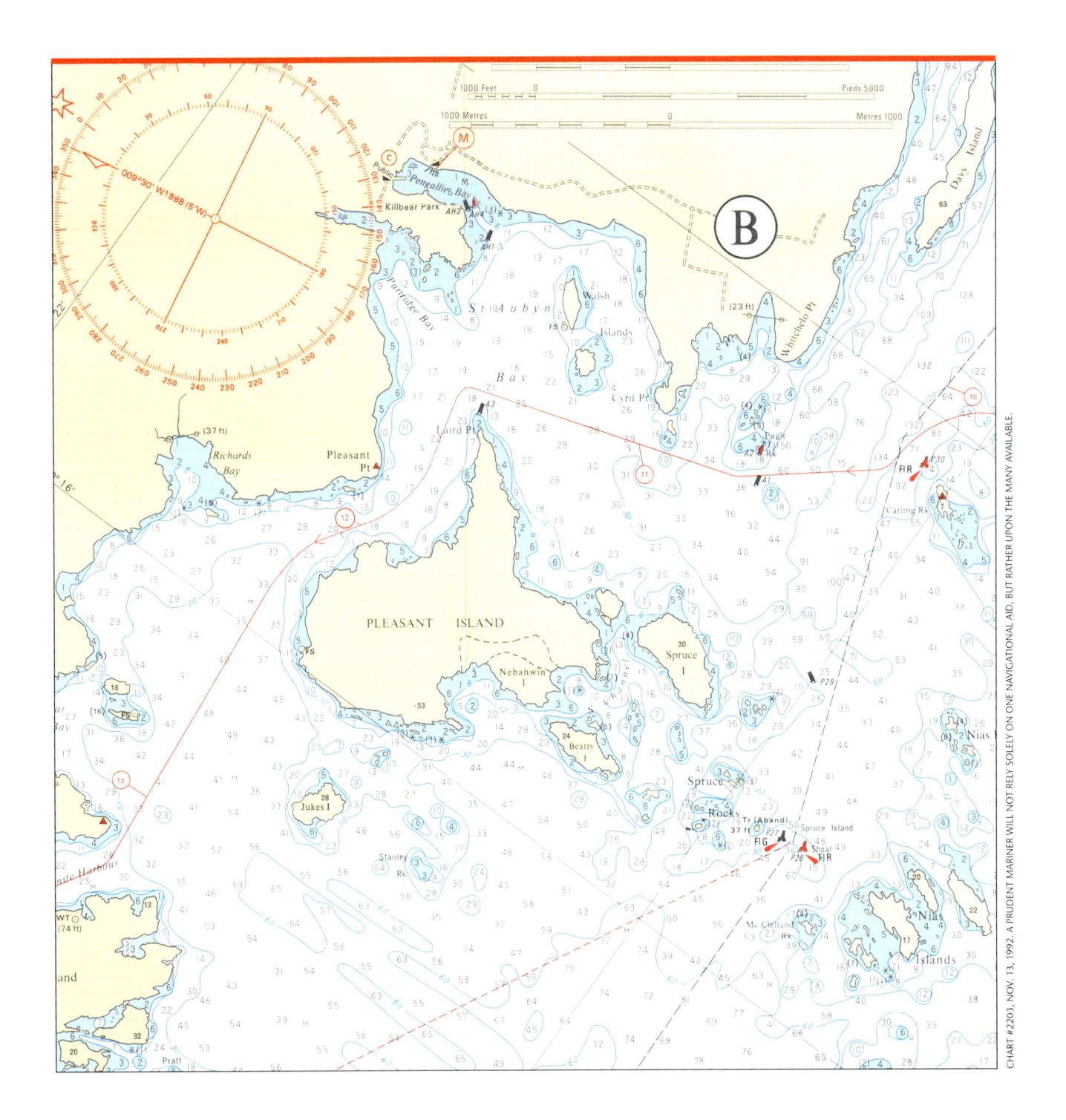

Facilities information subject to change. We suggest you call ahead.

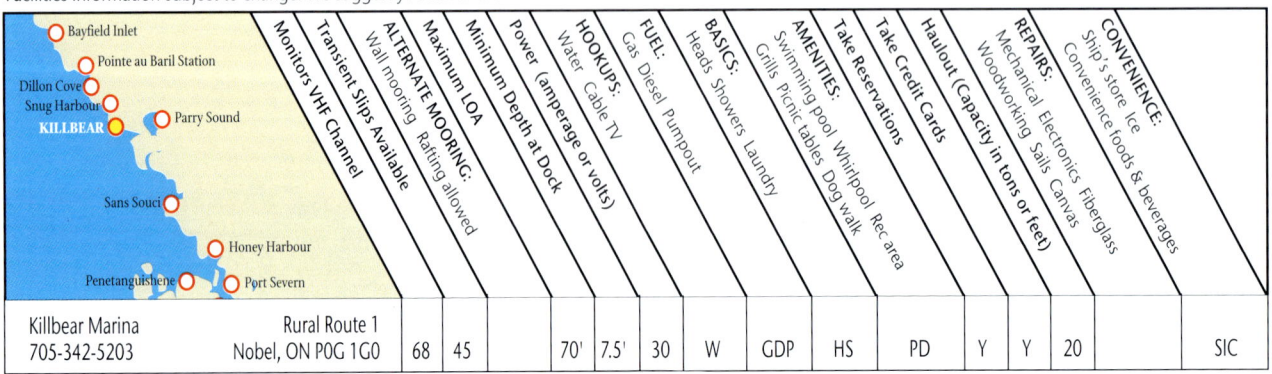

	Monitors VHF Channel	Transient Slips Available	ALTERNATE MOORING: Wall mooring Rafting allowed	Maximum LOA	Minimum Depth at Dock	Power (amperage or volts)	HOOKUPS: Water Cable TV	FUEL: Gas Diesel Pumpout	BASICS: Heads Showers Laundry	AMENITIES: Swimming pool Whirlpool Rec area Grills Picnic tables Dog walk	Take Reservations	Take Credit Cards	Haulout (Capacity in tons or feet)	REPAIRS: Mechanical Electronics Fiberglass Woodworking Sails Canvas	CONVENIENCE: Ship's store Ice Convenience foods & beverages
A Killbear Marina 705-342-5203 Rural Route 1 Nobel, ON P0G 1G0		68	45		70'	7.5'	30	W	GDP	HS	PD	Y	Y	20		SIC

314   LAKELAND BOATING · LAKE HURON CRUISING GUIDE

Snug Harbour, Ontario

Lat 45°22.40
Lon 80°18.68

Keep your eyes peeled for the fixed green Snug Harbour Light on the southern tip of Snug Island **A**. Pass between Snug Island and Middle Island **B** and continue east into the harbor. Snuggle up to Snug Harbour Marina **C** if you're a powerboater. Keel boats can tie up at the Government Dock **D** just 50 yards away.

Facilities information subject to change. We suggest you call ahead.

		Monitors VHF Channel	Transient Slips Available	ALTERNATE MOORING: Wall mooring, Rafting allowed	Maximum LOA	Minimum Depth at Dock	HOOKUPS: Water, Cable TV	Power (amperage or volts)	FUEL: Gas, Diesel, Pumpout	BASICS: Heads, Showers, Laundry	AMENITIES: Swimming pool, Whirlpool, Rec area, Grills, Picnic tables, Dog walk	Take Reservations	Take Credit Cards	Haulout (Capacity in tons of feet)	REPAIRS: Mechanical, Electronics, Fiberglass, Woodworking, Sails, Canvas	CONVENIENCE: Ship's store, Ice, Convenience foods & beverages
D	Government Dock 1 Noble No phone Snug Harbour, ON P0G 1G0				6		WR	100'+	10'				D	N	N	N
C	Gilly's Snug Harbour Marina 1 Noble 705-342-5552 Snug Harbour, ON P0G 1G0	68	10	WR	58'	5'	15 30	W	GP	H	PD	Y	Y	26'	MSC	SIC

LAKELAND BOATING · LAKE HURON CRUISING GUIDE

GEORGIAN BAY

Snug Harbour, Ontario
An angler's paradise with a smuggler's vibe.

Both classic wooden boats and commercial fishing vessels grace the port of Snug Harbour, 17 miles north of Parry Sound by water. After **Killbear Provincial Park**, boaters will see Snug Harbour Lighthouse beyond the Mile 15 spar buoy. Follow the channel straight east to the **government dock** or to **Gilly's Snug Harbour Marina**.

Though the whole place looks and feels like a smugglers' cove—you might expect to see someone like Humphrey Bogart's character in *Key Largo* chug by—the port is in reality an angler's paradise and a gourmet delight to boot.

While the government dock lacks a phone and offers no-frills dockage, Snug Harbour Marina is a full-service establishment with a friendly staff and most amenities, including a launch ramp. If you happen to have a question about either facility, you can call Terry Gilbert, owner of Snug Harbour Marina and neighbor to the government dock.

Gilbert is also the proprietor of **Gilly's Snug Harbour Restaurant,** a fresh fish restaurant located on the second floor of a 100-year-old log cabin. Of course, if you're the do-it-yourself-type, you can catch your own bass, pike, walleye and muskie in the area's bountiful waters.

Regatta Bay, an excellent though often crowded anchorage, lies just beyond the Snug Harbour light at the end of Snug Island. You'll see its buoyed entrance to port.

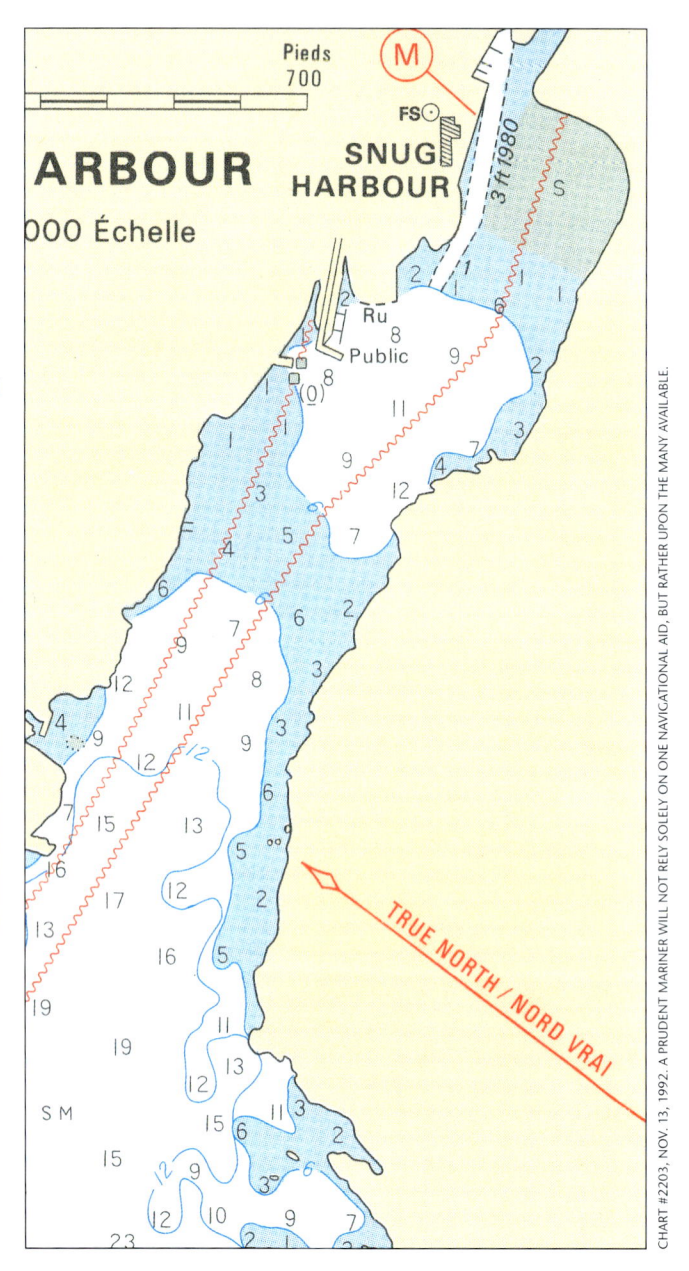

Distances from Snug Harbour (statute miles)
Parry Sound: 17 SE
Dillon Cove: 5 S
Pointe au Baril: 19 N
Byng Inlet/Britt: 47 N

Where to Eat

Gilly's Snug Harbour Restaurant . . . 705-342-5552
Known for its steak and smoked fish. Open for lunch and dinner.

More Info

Emergency 800-661-6777
Parry Sound General Hospital 705-746-9321
Sudbury & District
Chamber of Commerce 800-761-42610
Rainbow Country Travel 800-465-6655
Parry Sound Taxi 705-746-1221

316 LAKELAND BOATING · LAKE HURON CRUISING GUIDE

Dillon Cove, Ontario

Dillon Cove is a tiny village about five miles from Snug Harbour. Mooring is available at Dillon Cove Marina & Resort **A** for both power and sail. Gas is sold here, along with basic provisions. Transient boat traffic can also tie up at one of two Government Docks on either side of a launch ramp **B**. Water is about 3 to 5 feet deep at these docks.

Facilities information subject to change. We suggest you call ahead.

		Monitors VHF Channel	Transient Slips Available	ALTERNATE MOORING: Wall mooring / Rafting allowed	Maximum LOA	Minimum Depth at Dock	Power (amperage or volts)	HOOKUPS: Water / Cable TV	FUEL: Gas / Diesel / Pumpout	BASICS: Heads / Showers / Laundry	AMENITIES: Swimming pool / Whirlpool / Rec area / Grills / Picnic tables / Dog walk	Take Reservations	Take Credit Cards	Haulout (Capacity in tons or feet)	REPAIRS: Mechanical / Electronics / Woodworking / Sails / Canvas / Fiberglass	CONVENIENCE: Ship's store / Ice / Convenience foods & beverages
A	Dillon Cove Marina & Resort Dillon Road 705-342-5431 Dillon, ON P0G 1G0	68	4		35'	4'	30	W	G	HS	SPD	Y	Y	28'	MEFC	SIC
B	Government Docks Dillon Road No phone Dillon, ON P0G 1G0		5	WR	100'	3'				H	D	N	N			

GEORGIAN BAY

Dillon Cove, Ontario

Dillon Cove, Ontario
Enjoy the solitude in this island paradise.

A hidden treasure on the mainland northeast of Galna Island, Dillon Cove is a pristine spot with plenty of wildlife. Two government docks and **Dillon Cove Marina and Resort** offer transient docking; alternatively, find a night's mooring off of some lonely island amidst this hauntingly beautiful chain.

Fishermen come to this area for the good bass, pike, perch and walleye fishing, and the cove is visited by both power and sail cruisers. The full-service marina has a store, a pool, cottages and campsites, and a good launch ramp. There's another ramp between the two public docks.

Fixed-keel boats are advised to stay right in the buoyed channel and boaters are warned to follow their charts closely to avoid the numerous rocks and shoals. Channel depth varies from 10 to 20 feet. Dillon is on the small-craft channel network running north and south along the eastern shore of Georgian Bay. Anchorages abound in this area; try Sand Bay (enter on the east side of Sunnyside Island).

More Info

Emergency	911
Parry Sound General Hospital	705-746-9321
Chamber of Commerce	800-461-4261
Parry Sound Taxi	705-746-1221

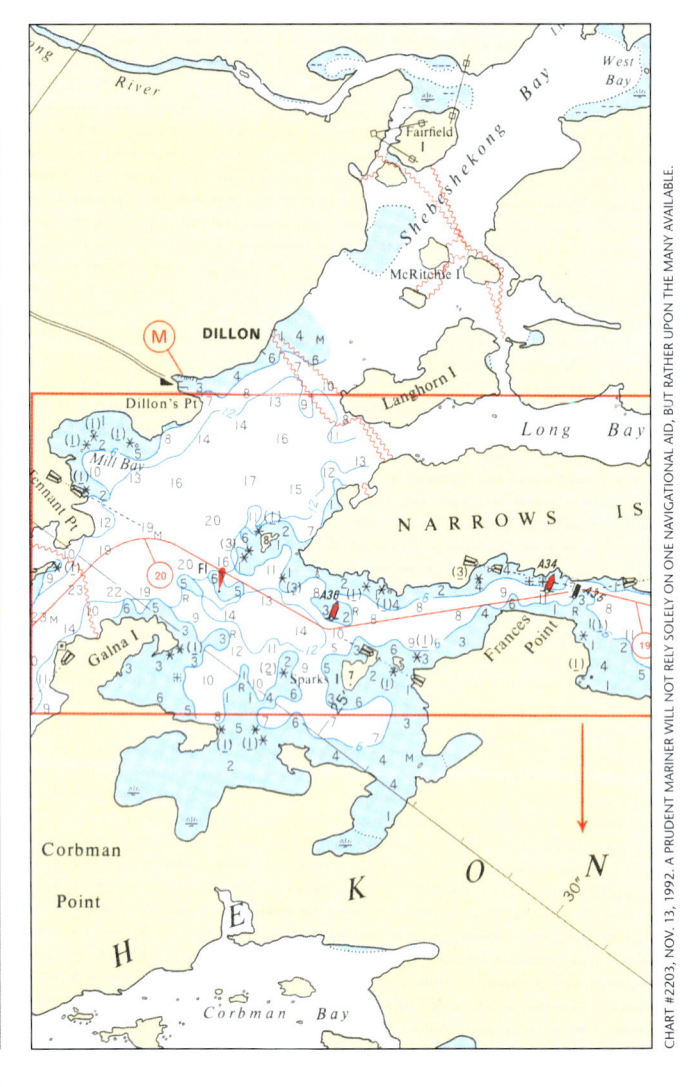

Distances from Dillon Cove (statute miles)
Snug Harbour: 5 N
Parry Sound: 21 SE
Pointe au Baril: 16 N
Byng Inlet/Britt: 42 N

318 LAKELAND BOATING · LAKE HURON CRUISING GUIDE

Pointe au Baril Station, Ontario

Lat 45°35.72
Lon 80°23.02

Once you've reached the end of the narrow inlet, you're right in the heart of Point Au Baril Station. The best bet for transient space and services is Desmasdon's Marine **A** on the north shore. If you pick the south shore as your point of entry, Adamanda Services **B** offers propane and a few slips if you want to make a quick run to the post office. Beacon Marina **C** does not cater to transients but will do emergency repairs. The Government Wharf **D** can accommodate up to 15 transients. Evoy's Marine **E**, located adjacent to the government docks, offers a couple of slips for overnighters. For groceries and hardware, visit C.C. Kennedy Home Hardware **F**, located between the government basin and the high C.P. Railway Bridge **G**. Highway 69 **H** runs parallel to the tracks.

Distances from Pointe au Baril Station (statute miles)
Pointe au Baril: 6 S
Parry Sound: 36 S
Byng Inlet/Britt: 26 N
Tobermory: 69 W

LAKELAND BOATING · LAKE HURON CRUISING GUIDE

GEORGIAN BAY

Pointe au Baril Station, Ontario
Roll out the 'baril' in this friendly fishing hamlet.

The French name Pointe au Baril, which translates as Barrel Point, dates back to the days when the first boats navigated these tricky island passages. In the mid-1800s, very few of the channels were marked with anything more than a pile of stones. But on an extreme northern point of land here, barrels filled with pitch were set on fire to guide ships. This point later became the site of the red-roofed lighthouse and keeper's cottage that boaters pass coming in from the north passage.

One of this area's main landmarks is the private 100-room **Ojibway Club**, a historic former hotel on Ojibway Island at Mile 33 on the Small-Craft Channel.

Pointe au Baril Station, the nucleus of the area's activity, is located about six miles north of the lighthouse at the end of a narrow inlet. The northern passage comes by Lookout Island and the lighthouse on the mainland, while the passage from the south slips between the mainland and picturesque Shawanaga Island. Follow the Pointe Au Baril Channel east, hugging the mainland coast to port.

When you pass **Payne Marine, Ltd.**, a friendly transient marina with fuel and pumpout, you're two miles from Pointe au Baril Station. You know you've reached it when you spot the C.P. Railway Bridge near Highway 69. The hamlet is right on the highway, which is the main Ontario roadway running along the eastern shore of Georgian Bay. Pointe au Baril Station, though rather out-of-the-way, has a variety of transient options. Highway 69 provides easy access for those meeting or being picked up by cruisers, and several marinas—notably **Desmasdon's Boat Works** and **Beacon Marine**—have launch ramps. Desdamon's is to port near the inlet's terminus. The rest of the marine facilities line the opposite bank. **Adamanda Services** is not a marina per se, but temporary dockage is available if you stop in to pick up some propane or run into the post

Where to Eat

Forest Glen **705-366-2841**
The clam chowder is delightful. Serving a full menu for lunch and dinner, this place will even drop off your meals at the dock.

The Harbour View . . . . **No phone**
Outdoor eating on swings and Adirondack chairs overlooking the water. The menu consists mostly of barbecue, with some soups and salads. There are also two chip wagons, which serve french fries, within walking distance.

The Haven **705-366-2525**
Don't miss the roast beef special on Thursday nights. Also try the famous wings, fried chicken and burgers. Only a quarter mile walk from the dock.

More Info

Emergency 888-310-1122
Coast Guard Search & Rescue 800-867-7270
Pointe au Baril Nursing Station 705-366-2376
Chamber of Commerce 705-366-2331

office. Beacon Marine can handle emergency repairs but doesn't offer dockage. The **Government Wharf** has ample transient dockage, and the nearby **Evoy's Marine** has a few slips as well.

If you need to do some provisioning, stop in at **C.C. Kennedy Co. Ltd. Home Hardware**, which is also a grocery store and a butcher shop, located between the Government Wharf and the C.P. Railway Bridge. The **Harbour View** restaurant is on the south shore, and a **nursing station** and a **liquor, wine and beer store** are on the highway, about a quarter of a mile from the wharf.

There's a **museum** in the lighthouse, which is open in July and August and is accessible by water only.

French explorer Samuel de Champlain, who called Georgian Bay *la mer douce* ("the sweet sea"), landed nearby. Travel to **Champlain Monument Island**, where a memorial has been erected to him and his place in Canadian history.

This is real cruising country, and the "Station" is the main provisioning center for a sprawling cottage area. Anchorages abound in Shawanaga Inlet and the surrounding area—notably the Shawanaga River at the northeast end of the inlet, Shawanaga Island at Frederick Inlet northwest of Anchor Island or Cairn Bay, further west.

POINTE AU BARIL STATION

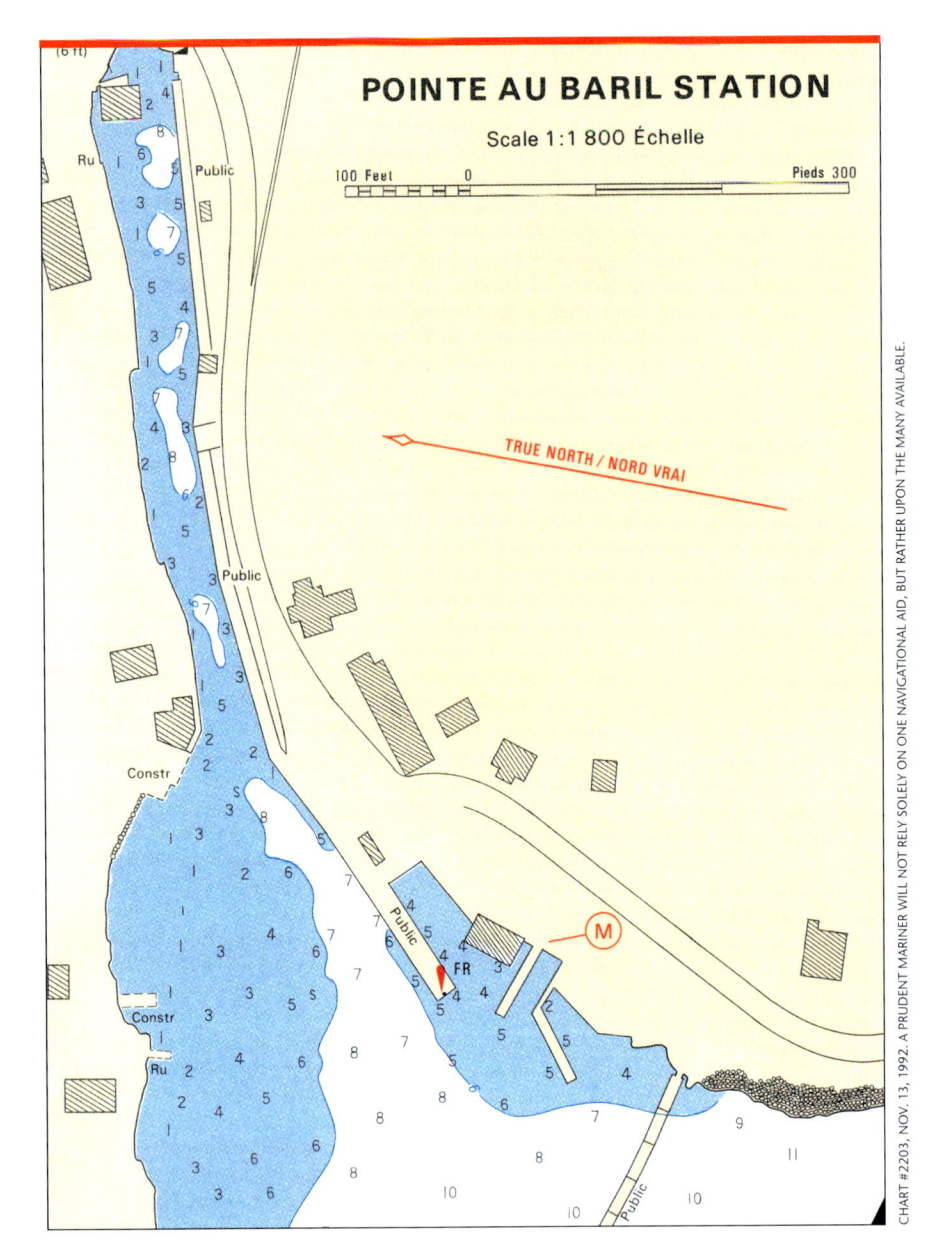

Scale 1:1 800 Échelle

Facilities information subject to change. We suggest you call ahead.

			Monitors VHF Channel	Transient Slips Available	ALTERNATE MOORING: Wall mooring / Rafting allowed	Maximum LOA	Minimum Depth at Dock	Power (amperage or volts)	HOOKUPS: Water / Cable TV	FUEL: Gas / Diesel / Pumpout	BASICS: Heads / Showers / Laundry	AMENITIES: Swimming pool / Whirlpool / Rec area / Grills / Picnic tables / Dog walk	Take Reservations	Take Credit Cards	Haulout (Capacity in tons or feet)	REPAIRS: Mechanical / Electronics / Fiberglass / Woodworking / Sails / Canvas	CONVENIENCE: Ship's store / Ice / Convenience foods & beverages
A	Desmasdon's Boat Works 705-366-2581	North Shore Road Pointe Au Baril, ON P0G 1K0				4			50'	12'		G	H	D	N	Y	M	SIC
E	Evoy's Marine 705-366-2361	Highway 644 No. 38 Pointe Au Baril, ON P0G 1K0			4	R	40'	5'		G	H	PD	Y	Y	30'	MESC	IC
D	Government Wharf 705-746-4243	Highway 644 Pointe Au Baril, ON P0G 1K0			15	WR	95'	4'		P	H	D	Y	N			IC
	Payne Marine Ltd. 705-366-2296	59 Payne's Road Pointe Au Baril, ON P0G 1K0	68	10		150'	10'	30		G	HS	GPD	Y	Y	8	ME	SIC

Bayfield Inlet, Ontario

There are a few transient options in the Bayfield Inlet area. Thompson Marine **A** offers slips, gas, diesel and full repairs. The Public Wharf **B** has only one slip and is primarily for launching. Another good bet for dockage is Hangdog Marina **C**, located west of here in nearby Georgian Inlet. Bayfield Lodge **D** offers accommodations.

Facilities information subject to change. We suggest you call ahead.

			Monitors VHF Channel	Transient Slips Available	ALTERNATE MOORING: Wall mooring, Rafting allowed	Maximum LOA	Minimum Depth at Dock	Power (amperage or volts)	HOOKUPS: Water, Cable TV	FUEL: Gas, Diesel, Pumpout	BASICS: Heads, Showers, Laundry	AMENITIES: Swimming pool, Whirlpool, Rec area Grills, Picnic tables, Dog walk	Take Reservations	Take Credit Cards	Haulout (Capacity in tons or feet)	REPAIRS: Mechanical, Electronics, Fiberglass Woodworking, Sails, Canvas	CONVENIENCE: Ship's store, Ice Convenience foods & beverages
C	Hangdog Marina 705-366-2000	P.O. Box 300 Pointe au Baril, ON P0G 1K0		15		65'	10'			GDP	HSL	D	Y	Y	20	ME	SIC
A	Thompson Marine 705-366-2361	P.O. Box 87, Bayfield Wharf Pointe au Baril, ON P0G 1K0	11	10	WR	50'	7'	15 30		GDP	HS	D	Y	Y	50	MEFW	SIC

LAKELAND BOATING · LAKE HURON CRUISING GUIDE **323**

GEORGIAN BAY

Bayfield Inlet, Ontario
Relax and explore this water wonderland.

In the heart of the 30,000 Islands—the largest concentration of freshwater isles in the world—Bayfield Inlet is a deep, narrow inlet that's great for exploring or fishing for bass, northern pike, walleye and musky. This water wonderland also offers countless adventures for punting around in small boats.

The thousands of channels and little bays and inlets must have offered a real challenge for the British Navy's Lt. Henry Wolsey Bayfield, the first cartographer to make accurate maps of Georgian Bay. There are many lovely spots to moor for the night or for several days, but consult the locals before you head out. There are numerous shoals and shallow areas, so boats of anything more than moderate draft should proceed cautiously.

Transient dockage can be found at **Hangdog Marina** and **Thompson Marine**, which also has a launch ramp. The **Public Wharf** is a bare-bones launch with just one slip.

Bayfield Lodge, which offers meals in a unique, casual setting, also has a fully stocked store with home-baked goods. For meals, call ahead for a reservation. The lodge also provides accommodations and can handle a few boats at its dock, which has a depth of about 8 feet.

Distances from Bayfield Inlet (statute miles)
Byng Inlet/Britt: 10 N
Pointe au Baril: 8 S

Where to Eat

Bayfield Lodge **705-366-2523**
A casual family dining experience. Breakfast and dinner by reservation. Call for a ride by boat or car.

More Info

Emergency 888-310-1122
Coast Guard Search & Rescue 800-267-7270
Pointe au Baril Nursing Station 705-366-2376
West Parry Sound Health Centre 705-746-9321
Rainbow Country
Travel Association 800-465-6655

BAYFIELD INLET

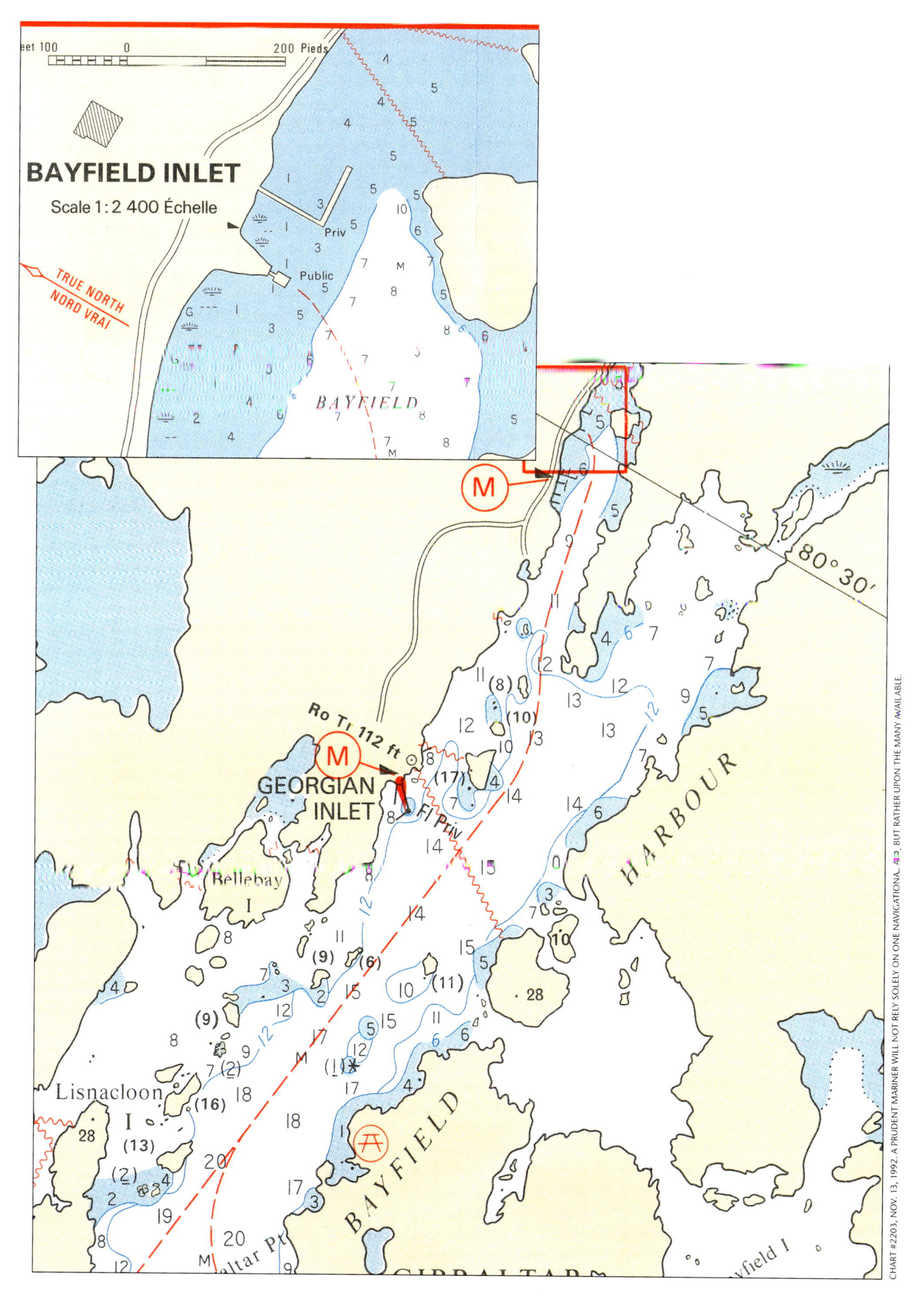

Scale 1:2 400 Échelle

TRUE NORTH / NORD VRAI

GEORGIAN INLET

LAKELAND BOATING · LAKE HURON CRUISING GUIDE

Byng Inlet & Britt, Ontario

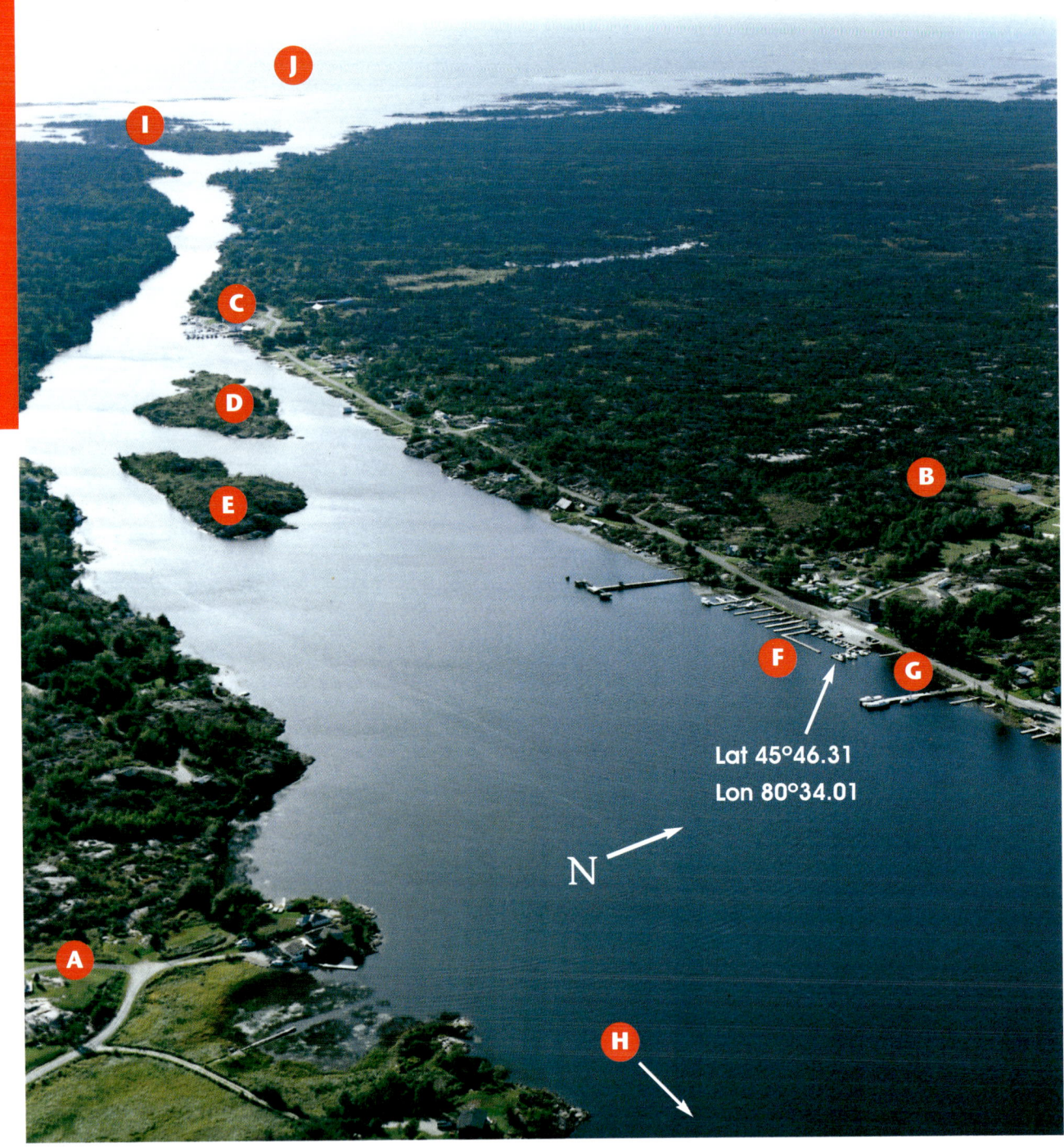

The five-mile-long Byng Inlet separates the towns of Byng Inlet **A** and Britt **B**. In Britt, transients can dock at Wright's Marina **C**, or continue past Old Mill Island **D** and Rabbit Island **E** to St. Amant's Marine **F**. The Britt Town Dock **G** offers no services—just mooring space in front of the Little Britt Inn. In Byng Inlet, there's dockage, lodging and dining at the Sawmill Lodge **H**. Clark Island **I** and Georgian Bay **J** are visible in the distance.

Distances from Byng Inlet (statute miles)
Killarney (inside route): 65 WNW
Killarney (open water): 43 WSW
Key Harbour: 17 N
Pointe au Baril: 29 S
Parry Sound: 63 S

Lat 45°46.31
Lon 80°34.01

326 LAKELAND BOATING · LAKE HURON CRUISING GUIDE

GEORGIAN BAY

Byng Inlet & Britt, Ontario
A tale of two hamlets.

The hamlets of Byng Inlet and Britt offer a taste of history and plenty of outdoor activities. Byng Inlet, the five-mile-long body of water that separates them, is on the northwest Small-Craft Route inside the 30,000 Islands. Clark Island splits the big open bay leading into the narrow inlet. The well-marked south channel is the main entrance to Byng Inlet. The north channel entrance has low water levels and is recommended only for those with local knowledge. When coming from or heading north, take the bypass from the south channel west of Bigwood Island and east of McNab Rocks; the channel is marked with green buoys on your port going north (and the reverse coming south). This is clearly marked on the charts.

The inlet and the many islands nearby provide mooring for boats with moderate drafts. There is plenty of water in the marked channels. The entry to the bay is dotted with thousands of islands. Named after Admiral Byng of the British Royal Navy, the inlet gained importance in the early days of Canada's industrial boom due to its natural harbor and its immense stands of timber. Old Mill Island, mid-channel between the hamlets of Britt and Byng Inlet, once provided cut timber from

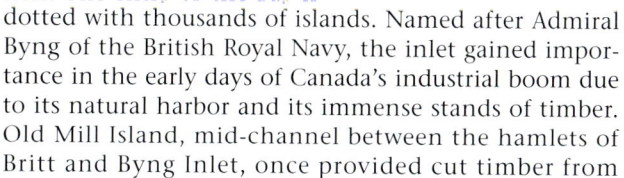

LAKELAND BOATING · LAKE HURON CRUISING GUIDE **327**

Where to Eat

Little Britt Inn 705-383-0028
Billed as "the hidden pearl of Georgian Bay," this historic establishment is open for lunch and dinner.

Sawyer's Loft 705-383-0068
Located in the Sawmill Lodge. Open for a hearty breakfast starting at 7 a.m. Also serving lunch and dinner. Specialties include steak, pickerel, whitefish and pork chops.

Waterfront Inn 705-383-2434
Patrons eagerly gobble down the fresh pickerel and whitefish when available. There are also drinks in the **Boathouse Sports Bar,** which has pool tables and a dartboard. Located at St. Amant's Marine.

Gibson's Sawmill.

On the north shore, in Britt, some of the best dockage is at **Wright's Marina**, a full-service facility. Visit the **Upper Deck** and enjoy the rec room with table shuffleboard, ping pong, darts and pool, or just relax and take in the view from the deck. Forget your worries in the sauna and take a dip in the inlet before turning in. Visit the craft shop in the **Net Shed**. Daily boat and motor rentals are available for those who want to explore or fish among the many islands and inlets of the bay. Transportation to local restaurants is complimentary.

Another great overnight spot is **St. Amant's Marine**, which offers a faxing and photocopying service, allowing boaters who are feeling isolated the opportunity to check up on the situation at the office. This self-dubbed "one-stop marina" has an inn, restaurant-bar and general store that sells fresh meat and produce. The water here comes from an artesian well. Both St. Amant's and Wright's have launch ramps.

Dockage can also be found at **The Sawmill Lodge** in the hamlet of Byng Inlet on the south shore. The Sawmill offers motel lodging, cottages and showers.

Sit with Jim Sorrenti and Teri McLean, owners of the historic **Little Britt Inn**, in the second-floor dining room to get a sense of local history. The old inn on the north shore has that perfect old-time atmosphere. Built in 1940, the building has a delightful dining room overlooking the inlet. Dock at the inn or walk the short distance from the neighboring marinas along the shore road. The inn serves full lunches and dinners and features fine European cuisine. There are four suites if you're looking for shore accommodations. The bar in the basement is open every noon for an hour or two, but will stay open longer if patrons are still thirsty.

Britt is the larger of the two hamlets, with a **post office** and a **nursing station** that boaters can walk to. Immediately opposite on the south shore, Byng Inlet is mainly a cottage and residential area. **Donna's Grocery & Post Office** is a short walk from the Sawmill Lodge.

You won't find a cinema or arcade in Byng Inlet, but it's great for more important pastimes like fishing, hiking, canoeing and picnicking.

More Info

Emergency 800-661-6777
Parry Sound General Hospital 705-746-9321
Pointe Au Baril
Chamber of Commerce 705-366-2331

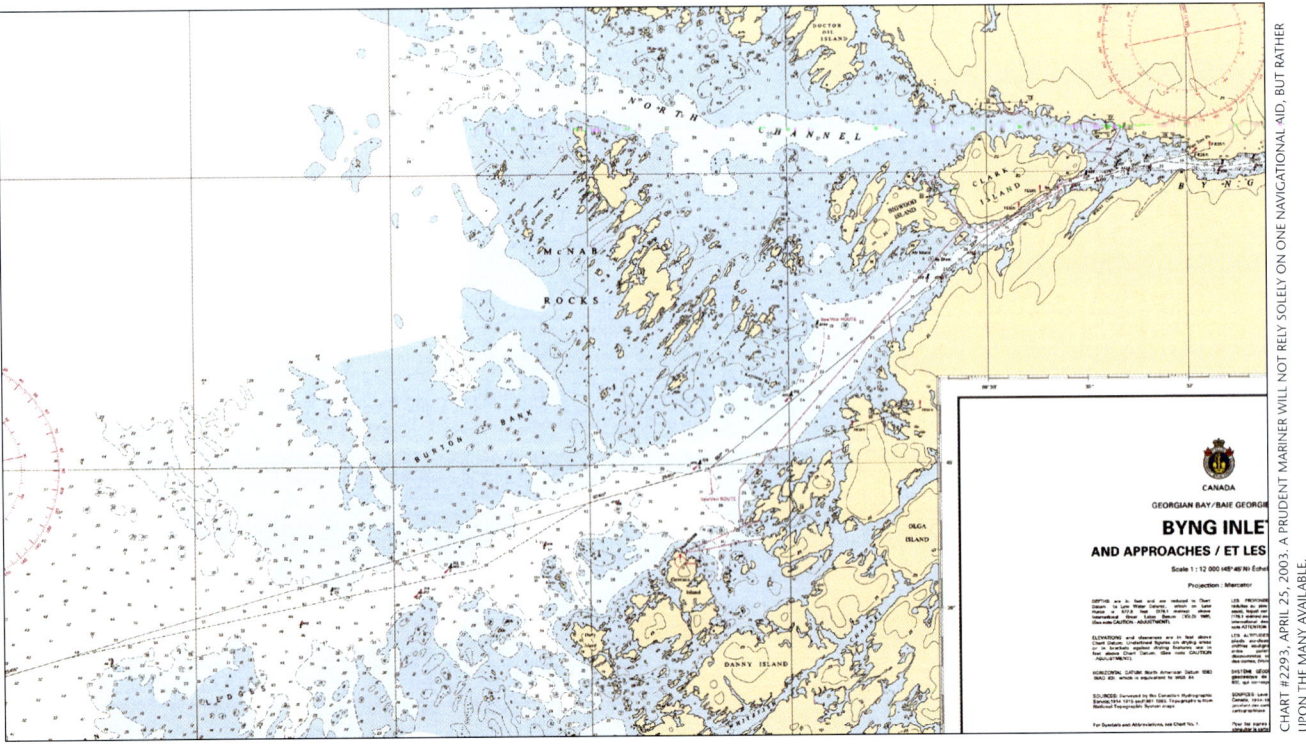

Facilities information subject to change. We suggest you call ahead.

			Monitors VHF Channel	Transient Slips Available	ALTERNATE MOORING: Wall mooring / Rafting allowed	Maximum LOA	Minimum Depth at Dock	Power (amperage or volts)	HOOKUPS: Water / Cable TV	FUEL: Gas Diesel Pumpout	BASICS: Heads Showers Laundry	AMENITIES: Swimming pool Whirlpool Rec area / Grills Picnic tables Dog walk	Take Reservations	Take Credit Cards	Haulout (Capacity in tons or feet)	REPAIRS: Mechanical Electronics Fiberglass / Woodworking Sails Canvas	CONVENIENCE: Ship's store Ice / Convenience foods & beverages
A	The Sawmill Lodge 705-383-0068	General Delivery Byng Inlet, ON P0G 1B0	68	10	R	45'	9'	15	W	G	H	GD	Y	Y			IC
B	St. Amant's Marine 705-383-2434	P.O. Box 10 Britt, ON P0G 1A0	68	4	R	64'	10'	30	W	GDP	HSL	GPD	Y	Y			SIC
C	Wright's Marina 705-383-2295	P.O. Box 160 Britt, ON P0G 1A0	68	25		70'	10'	30	W	GDP	HSL	RGPD	Y	Y	20	MF	SIC

Northeastern Shore
Find hidden treasures in this breathtaking cruising ground.

The cruising ground stretching from Byng Inlet and Britt to Key Harbour and from Key Harbour to the French River is probably the most beautiful and pristine in Georgian Bay and offers innumerable anchorages. Chart 2204, sheets 1 and 2, covers the area, and the Great Lakes Cruising Club's *Log Book* lists many of the anchorages.

Key Harbour is at the northeastern corner of Georgian Bay. A scenic trip of 8 miles (well buoyed but not recommended for anything but outboards and dinghies when water levels are low) takes you to Key River at Highway 69, where Key River Marina provides gasoline. There is also a large general store here.

The area between Key Harbour and the French River has many unexplored wilderness sites for anchorage on the inside track, or you can take the main route outside to the Bustard Islands. The two routes separate just west of Dead Island on chart 2204, sheet 2. The inside route is navigable by cruisers up to 40 feet.

As you round Dock Island on the inside, scenic route and head toward the French River, you will see a group of white buildings with red roofs to port. This is the Georgian Bay Fishing Camp (705-383-2810), which offers dockage, a dining room (by reservation), groceries and gasoline and can accommodate boats up to 40 feet. The camp monitors channel 68.

As you come out of the inside passage at Magee Island, the mainland is due west, with the primary outlet of the French River just past Sabine Island. It was here that explorer Samuel de Champlain—and earlier his servant Etienne Brulé—first saw the Great Lakes as they came by canoe up the Ottawa River, across Lake Nippising and down the French River in 1615. From here he went down the eastern shore of Georgian Bay as far as Owen Sound. For 200 years this was the main route to the "Upper Lakes."

Going north up the French River, you will find on your starboard side the remains of a lumbering community of 2,000 people that thrived here from 1850 to 1907. On the west bank is McIntosh Camp, a fishing lodge offering gas, some dockage and meals. You'll find an excellent anchorage north of the camp in MacDougal Bay.

Significant Waypoint
Georgian Bay Fishing Camp:
45°55.03N/80°51.90W

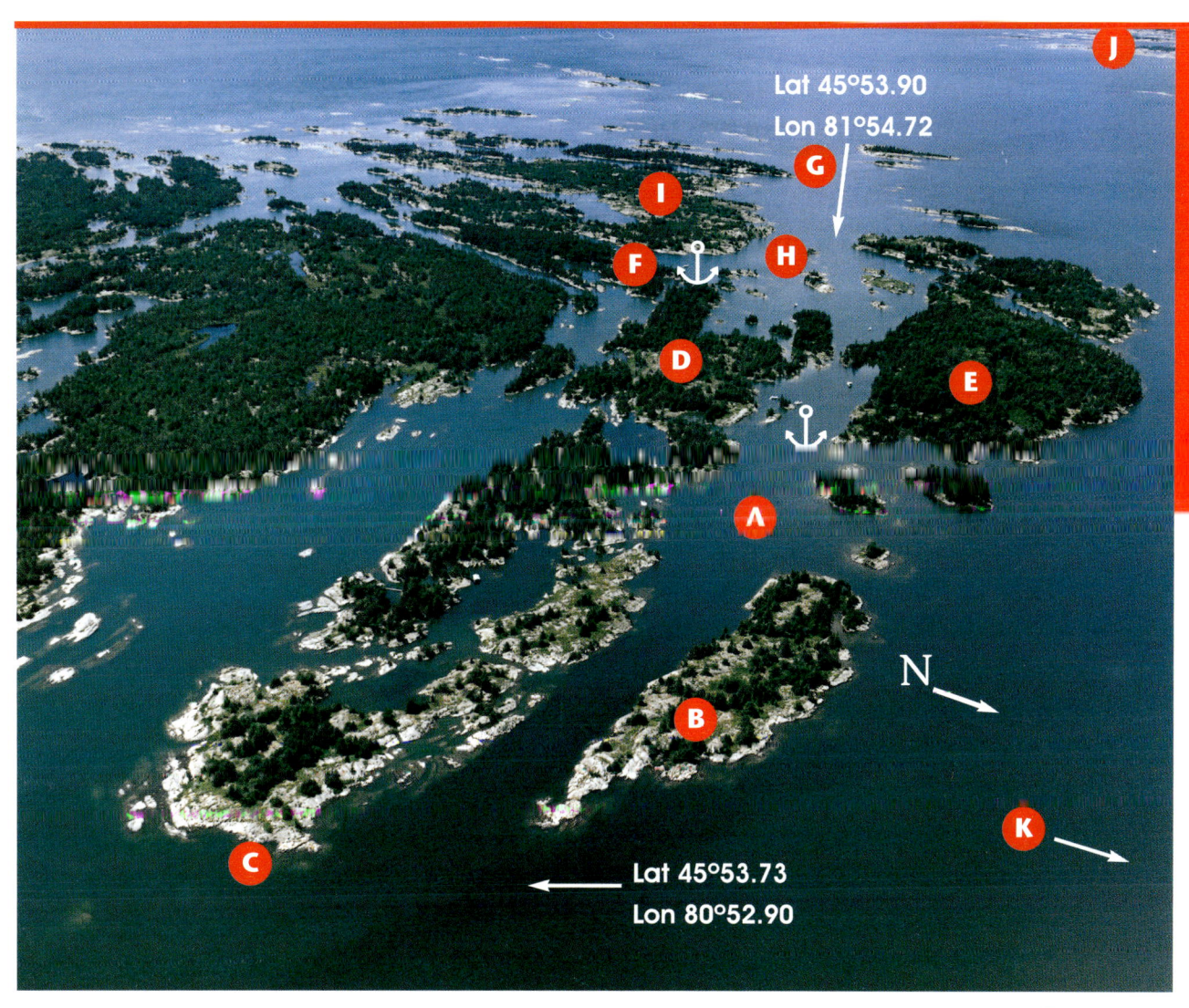

The Bustard Islands, Ontario

Lat 45°53.90
Lon 81°54.72

Lat 45°53.73
Lon 80°52.90

Generally the water is deeper to the north and west of the Bustards and quite foul to the south and east. Enter Northeast Harbour **A** from the main track out of Key Harbour. Favor the shore of Northeast Island **B** to avoid the shoals off Northeast Point **C** and anchor just north of Strawberry Island **D**. Dinghies can squeeze through the channel between Strawberry Island and Tie Island **E** to reach the main Bustard Islands Harbour **F**, but larger boats will have to enter from the west via the Gun Barrel **G**, curling around the south side of tiny Pearl Island **H** to find good anchoring ground just east of Burnt Island **I**. At the top of the picture, to the northwest, is the entrance to Bad River Channel **J**. Many cruisers head north up the mouth of the French River **K** to find terrific anchorages.

Significant waypoint
Northwest of the Bustard Rocks: 45°53.20N/81°58.80W

Facilities information subject to change. We suggest you call ahead.

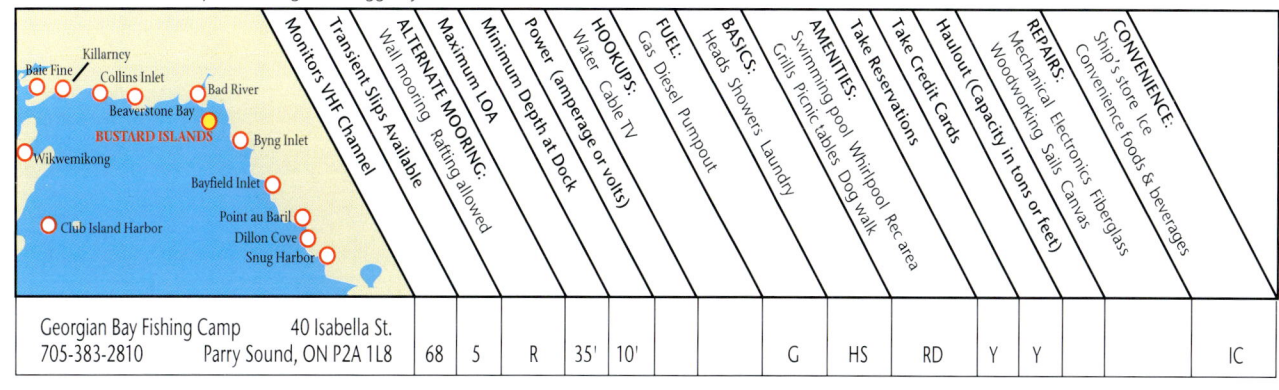

	Monitors VHF Channel	Transient Slips Available	ALTERNATE MOORING: Wall mooring Rafting allowed	Maximum LOA	Minimum Depth at Dock	Power (amperage or volts)	HOOKUPS: Water Cable TV	FUEL: Gas Diesel Pumpout	BASICS: Heads Showers Laundry	AMENITIES: Swimming pool Whirlpool Rec area Grills Picnic tables Dog walk	Take Reservations	Take Credit Cards	Haulout (Capacity in tons or feet)	REPAIRS: Mechanical Electronics Fiberglass Woodworking Sails Canvas	CONVENIENCE: Ship's store Ice Convenience foods & beverages
Georgian Bay Fishing Camp 40 Isabella St. 705-383-2810 Parry Sound, ON P2A 1L8	68	5	R	35'	10'			G	HS	RD	Y	Y			IC

GEORGIAN BAY

The Bustard Islands, Ontario
Kick back and enjoy the views in this stunning anchorage.

At the close of your tour of the 30,000 Islands, you'll be rewarded by the spectacular scenery of the Bustard Islands, a picturesque, tightly clustered archipelago. As you come out of the inside passage at Key River or French River, you'll reach the island group, which covers an area four miles long by two miles wide and offers two primary anchorages.

You pass Northeast Harbor, located two-thirds of the way up the east side, close to port if you take the outside route coming out of Key Harbour. Bustard Islands Harbour provides excellent protection and must be approached from the west by running up the "gun barrel" to Pearl Island (which features a prominent cottage), and curling around the south side of the island into the harbor. There are anchorages on both sides here, but these are foul areas, so be careful. A passage between Tie and Green islands that connects Northeast and Bustard Islands harbors is navigable by dinghy or, with care, by boats under 25 feet with 3-foot drafts. Think of Northeast Harbour as the sunrise anchorage and Bustard Islands Harbour as the sunset anchorage—take your pick!

There are also many channels for fishing and dinghy exploration, although the terrain is generally not suited for hiking. The nearest point for restocking is the **Georgian Bay Fishing Camp**, a lodge with limited provisions, gas, dockage and a dining room. It's a few miles northeast of the Bustards on the inland route.

If you approach via the inland route, you may want to swing northward up into the French River for another excellent anchorage before heading toward the Bustards. **McIntosh Fishing Camp** offers gas, dockage and meals.

On the outside route from the Bustards to Killarney, those worried about the narrows at the northeastern end of Beaverstone Bay (which are 5 feet deep but mud-bottomed) should plot a course between Green Island (44°54.93N, 81°19.60W) and Smooth Rock and then, keeping southwest of the rocks off the Fox Islands, continue to Killarney. Green Island is a great blue heron rookery that attracts hundreds of these magnificent birds early in the summer.

Where to Eat
Georgian Bay Fishing Camp. 705-383-2810
Sit down to some home cooking with the anglers at the camp. Specializing in pickerel and walleye.

More Info
Emergency 911
Parry Sound General Hospital 705-746-9321
Rainbow Country Travel 800-465-6655
Parry Sound
Chamber of Commerce 800-461-4261

Distances from the Bustard Islands (statute miles)
French River: 5 N
Bad River: 8 WNW
Key Harbour: 9 E
Beaverstone Bay: 18 W
Killarney: 41 W

332 LAKELAND BOATING · LAKE HURON CRUISING GUIDE

Bad River Channel, Ontario

Lat 45°56.12
Lon 80°58.37

Although this photo of the Bad River doesn't show the range lights and the initial approach past Bad River Point Island, it does roughly mark the narrow spot where several underwater rock pinnacles **A** demand your undivided attention. If you keep a third of the channel to starboard at this point, you should be fine. But post a bow lookout (the water can be murky here) and take it dead slow.

There is a tiny anchorage to the west of the last big island **B**, or you can head into the main part of the harbor. Avoid the rocks off the island's northeast side **C** and you're home free. The Devil's Door Rapids **D** separate the anchorage from Cross Channel **E**, which cuts (left to right in this picture) across the Bad River Channel. It makes for a great dinghy side trip.

Distances from Bad River (statute miles)
Bustard Islands: 8 ESE
Beaverstone Bay: 18 W
Killarney: 41 W

Significant waypoint
On track into the Bad River off the main track:
45°53.00N/81°00.00W

LAKELAND BOATING · LAKE HURON CRUISING GUIDE   **333**

GEORGIAN BAY

Bad River Channel, Ontario
A hidden paradise for modern-day explorers.

West of the Bustards, where the chopped-up French River makes one of its five exits into Georgian Bay, you'll find another beautiful anchorage amid a maze of pink granite isles and bluffs.

The Bad River Channel and the maze of cuts here are all outflows of the French River as it makes its exit into Georgian Bay. The Bad River has a fairly tricky approach, so you'll want to closely consult your charts, idle in and post a bow lookout. Only those who disregard the charts and travel to port have problems here. A set of range lights will bring you in bearing 18°T off the Small-Craft Route. The first caution spot is just past Bad River Point Island, where shoals extend off its northern end. Also keep a close lookout for a narrow pass between underwater pinnacles, just beyond the point where you can see the old Bad River Fishing Lodge (no longer in operation) through the islands to port. The narrow spot is roughly identified in the photo, but the pinnacles (3 to 6 feet underwater) can be hard to spot. Go dead slow. Also watch out for two rocks off the northeast side of the last large island defining the harbor.

Once inside, this is a glorious anchorage, offering many opportunities for rock-climbing, blueberry-picking and exploring by dinghy. There's even a small set of rapids emptying into the anchorage that you can run in a sufficiently powered dinghy. The Cross Channel and its cuts beyond the rapids are great for exploring.

This is where de Champlain and the priests who accompanied him first laid eyes on the Great Lakes in 1615. After leaving Ottawa via the Ottawa River, they crossed Lake Nipissing and traveled down the French River to its mouth on the bay.

A word to the wise: This whole area is the home of the Massasauga rattlesnake, so traverse with care!

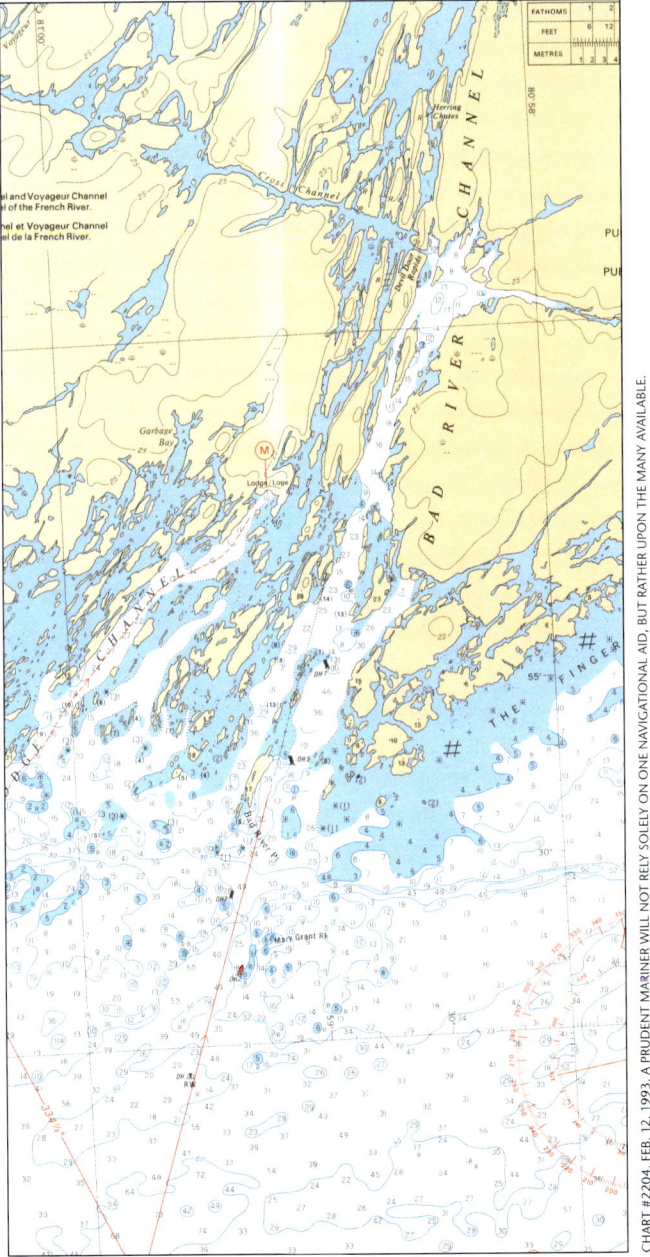

More Info

Emergency	911
Sudbury General Hospital	705-674-3181
Rainbow Country Travel	800-465-6655
Parry Sound Chamber of Commerce	800-461-4261

334 LAKELAND BOATING · LAKE HURON CRUISING GUIDE

Lat 45°58.26
Lon 81°11.69

C

Lat 45°57.62
Lon 81°11.34

B

N

A

Lat 45°57.40
Lon 81°10.69

Beaverstone Bay, Ontario

Dotted with many islands, Beaverstone Bay is an ideal holding and resting place for transient cruisers. The best anchorages are found northeast of Nobles Island **A**, northwest of Barto Island **B** and west of Burnt Island **C**. These are only a few—get out your dinghy and find some of your own.

Distances from Beaverstone Bay (statute miles)

Killarney: 22 W

Bustard Islands: 18 E

LAKELAND BOATING · LAKE HURON CRUISING GUIDE   **335**

GEORGIAN BAY

Beaverstone Bay, Ontario
Anchorages abound in this scenic stopover.

While Beaverstone Bay lacks the plentiful marine options of neighboring Killarney, this beautiful area offers lots of treats for the senses, so get in your dinghy and explore. Running north and south at the eastern outlet of Collins Inlet, it is a delightful holding and resting place for both motor and sailing cruisers. Except for the northwest corner entrance to Collins Inlet, the depth in most of the bay area is about 12 to 20 feet. The bay, which is about four miles long and two miles wide, features some 80 islands of varying size that boats can shelter behind. When heading toward Killarney, many cruisers will make the 20-plus-mile run fully "inside" through the protected waters of Beaverstone Bay, Mill Lake and Collins Inlet.

Once known as a lumbering center, the area hosted the thriving settlement of Beaverstone Bay near the Beaverstone River. It died out with the decline of the timber industry. Very few traces of the village can now be found, but there are scattered cottages along the shorelines. In 1920, after the forests had been stripped bare and the mill had closed, the Paterek family settled on the northeast corner of the bay, where members of the family still reside.

Many cruisers heading up the east side of Georgian Bay to Killarney, or the reverse, stop in Beaverstone to rest for a few days and maybe do some fishing—the area is great for bass, perch and pike. The channel is well-marked by spar buoys at both the Collins Inlet entrance and the entrance up from the south. To approach from the south, from the Bustard Islands area, start by locating the red bell marker about two miles south of Toad

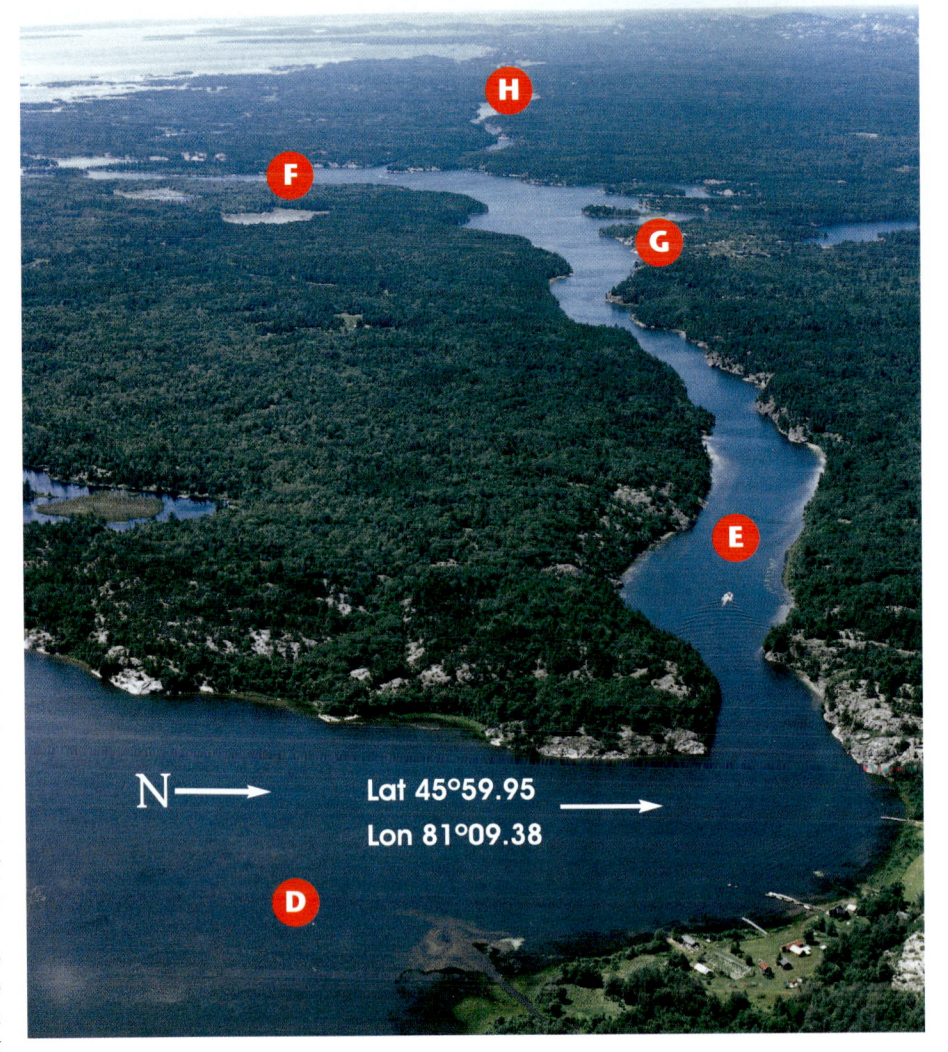

Lat 45°59.95
Lon 81°09.38

This shot from Beaverstone Bay shows the protected trek to Killarney. From the north end of the bay **D**, enter Collins Inlet **E** and proceed toward Mill Lake **F** with its many superb anchorages. Stop off at the Mahzenazing Lodge **G**, where dockage is available. Beyond Mill Lake, Collins Inlet continues westerly **H** toward Killarney.

336 LAKELAND BOATING · LAKE HURON CRUISING GUIDE

Island. It's regarded as the marker of the southern entrance into the bay. Head north from the bell marker and watch for range markers on Toad Island.

More Info

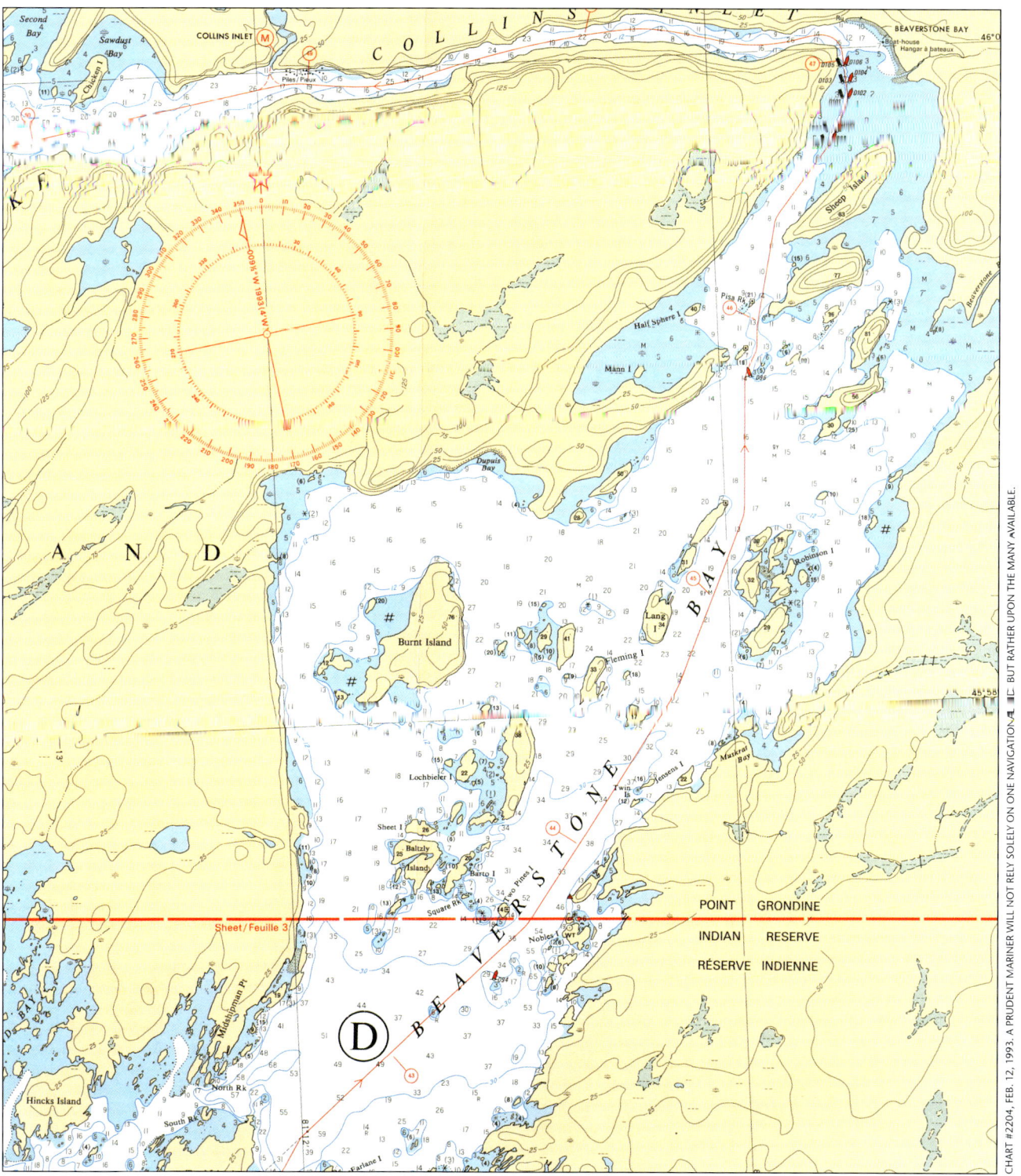

Emergency	800-661-6777
Killarney Health Centre	705-287-2300
Municipality of Killarney	705-287-2424

Collins Inlet, Ontario

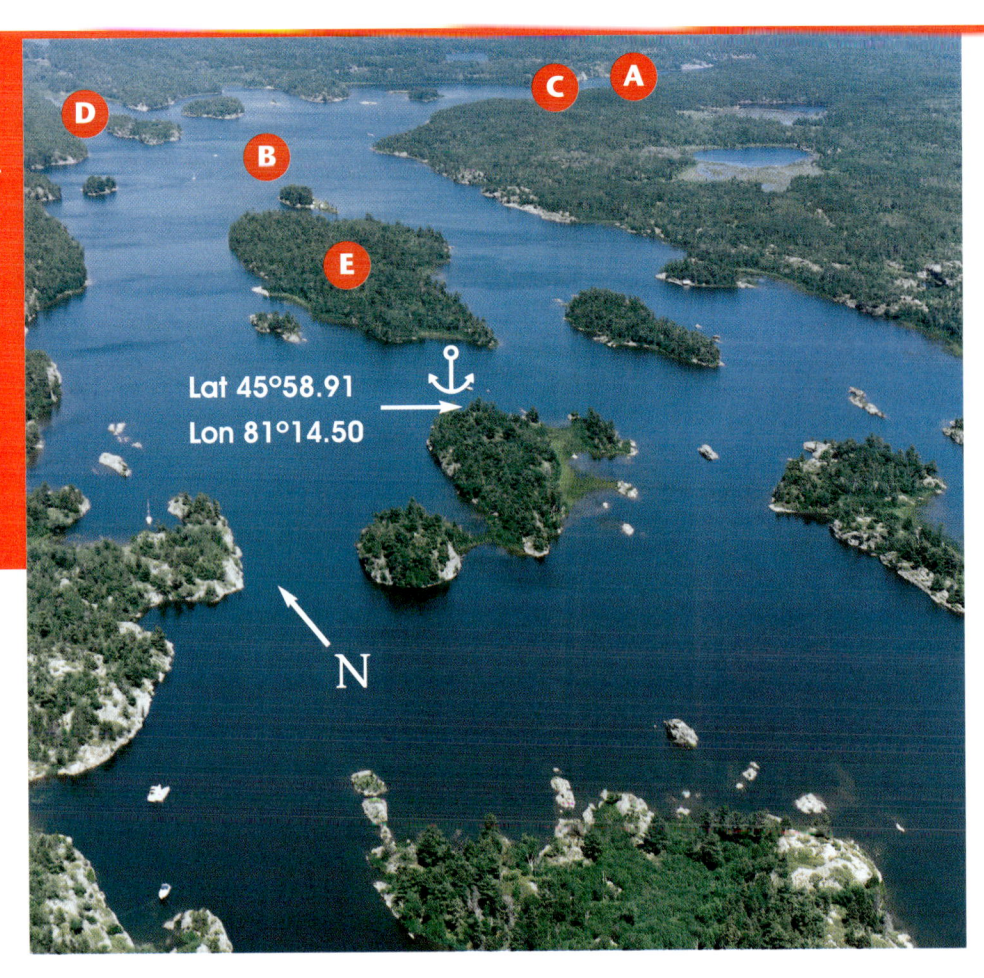

Looking northeast from the west end of Mill Lake, we see where Collins Inlet **A** joins the lake **B**. Mahzenazing River Lodge **C** stands on the north shore of the inlet just east of the lake. The inlet resumes **D** just southwest of Muskie Island. The most popular anchorage is just southwest of Green Island **E**, but anywhere in Mill Lake is a great place to drop the hook.

Here, on the northwest shore of Mill Lake **F**, Collins Inlet **G** resumes its westward course toward Killarney. Once between Muskie **H** and Salisbury **I** islands, bend to starboard to enter the inlet.

GEORGIAN BAY

Collins Inlet, Ontario
Enter a passage of breathtaking beauty.

Collins Inlet, Ontario

Stretching between Beaverstone Bay and Killarney, Collins Inlet is a 20-mile-long sliver of water running through the heart of the granite mainland. It's a protected passage when weather and waves kick up on Georgian Bay, but that's only part of its appeal. This idyllic north shore refuge also offers amazing natural scenery.

Many cruisers slip into the inlet, which runs east-west, from either end to make a safe passage. The western entrance is marked with two spar buoys, plus two red and green markers on two islands. The first island, on the port side, is Flat Rock. The other, ahead and to starboard, is One Tree Island. Early in the season and after heavy rains, you may be able to make out waterfalls shrouded in mist. Occasionally, you'll also see eye-rings driven by loggers and fishermen into the walls for anchoring.

From the east, enter Collins Inlet from Beaverstone Bay. Boaters cannot rely on anything more than 6-foot depths above chart datum, so stay within the narrow but well-marked channel. The entrance is easy to spot, thanks to the silt that accumulates on either side of the channel. Halfway along, between the western entrance to Collins Inlet and Beaverstone Bay, is a bulge of water called Mill Lake, where many cruisers stop and anchor to take in the scenic rock hills and cliffs of the

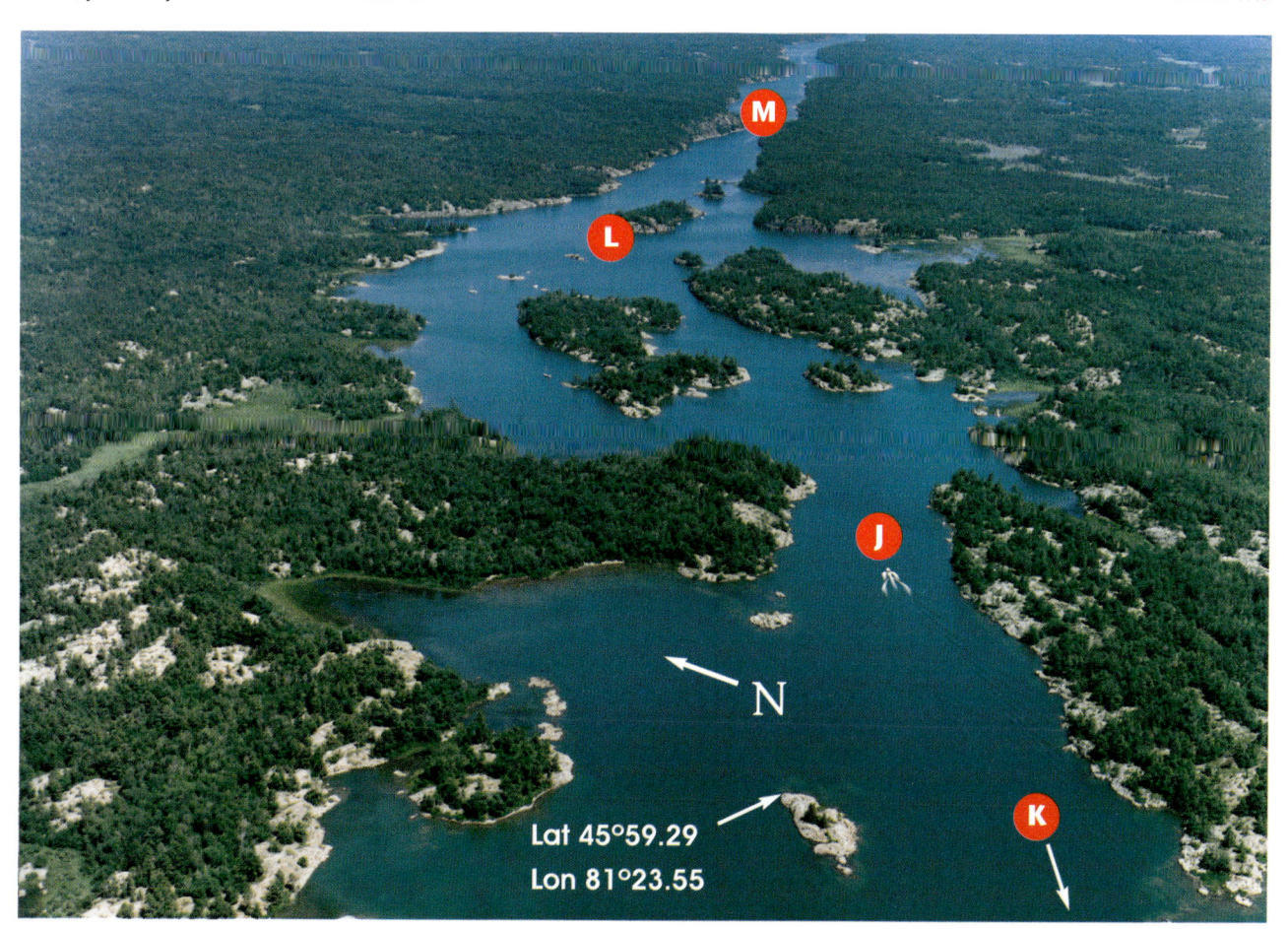

Lat 45°59.29
Lon 81°23.55

Looking northeast up Collins Inlet **J** from where it joins Georgian Bay **K**, you can see Keyhole Island **L** and Mill Lake **M**.

LAKELAND BOATING · LAKE HURON CRUISING GUIDE   **339**

Where to Eat

Mahzenazing River Lodge........705-287-2089
This fishing lodge on the inlet just east of Mill Lake runs its dining room like Mom did. There's a set menu for breakfast at 7 a.m., lunch at noon and dinner at 6 p.m. Lunch is usually soup and a sandwich; expect a roast, chops or chicken for dinner. If you're not staying at the lodge, make reservations at least a day ahead.

passage. Good anchorages abound in the southern half of the lake, an area known historically for lumbering and commercial fishing.

From the eastern approach, the ruins of the town of Collins Inlet are located at the north side near the entrance to Mill Lake. Although old pilings are all that remain of the town wharf, this was once a busy lumbering center of 2,000 people.

Mahzenazing River Lodge is a hunting and fishing lodge owned by Bill Pitfield. With depths of about 4 feet, there is dockage, canoe and boat rentals, a picnic area and a dining room, as well as gas in an emergency. You'll also find good fishing for musky, bass, perch and pike. You'll need to reserve a slip at least a day in advance. What Collins Inlet lacks in services, it makes up for as an ideal hidey-hole and rest area for cruisers. From here, you can continue on to Killarney.

Distances from Collins Inlet (statute miles)
Beaverstone Bay: 6 E
Killarney: 14 W

More Info

Emergency................................. 911
Killarney Health Centre 705-287-2300
Rainbow Country Travel 800-465-6655

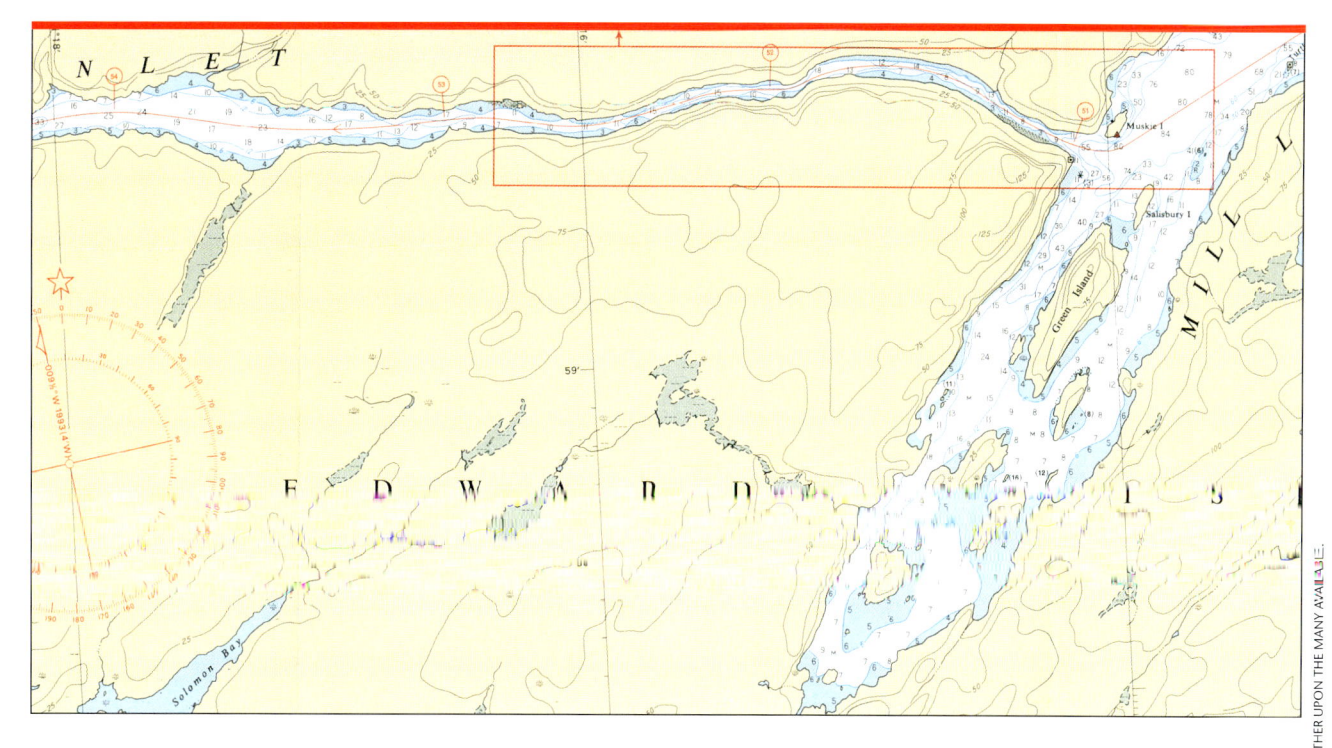

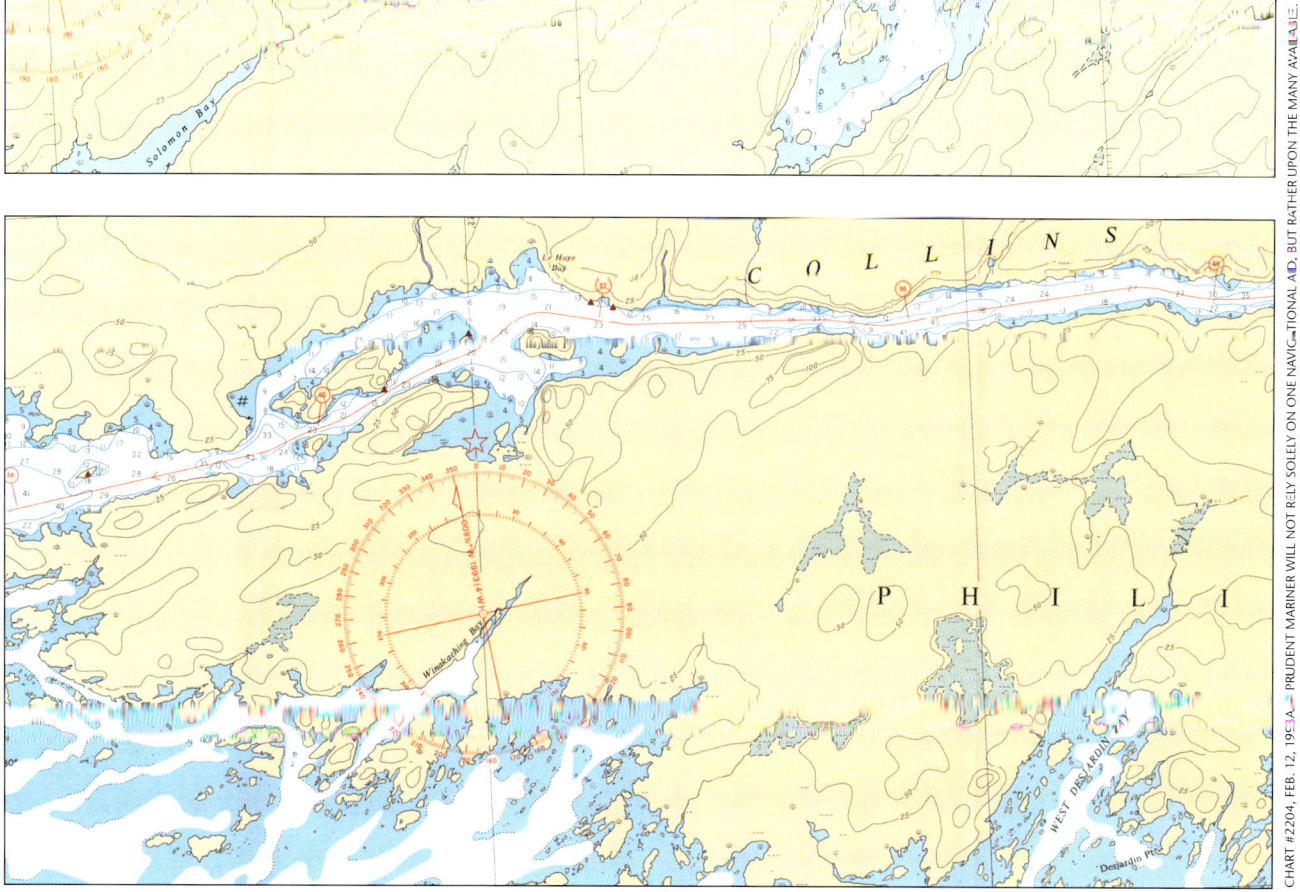

Facilities information subject to change. We suggest you call ahead.

	Monitors VHF Channel	Transient Slips Available	ALTERNATE MOORING: Wall mooring Rafting allowed	Maximum LOA	Minimum Depth at Dock	Power (amperage or volts)	HOOKUPS: Water Cable TV	FUEL: Gas Diesel Pumpout	BASICS: Heads Showers Laundry	AMENITIES: Swimming pool Whirlpool Rec area Grills Picnic tables Dog walk	Take Reservations	Take Credit Cards	Haulout (Capacity in tons or feet)	REPAIRS: Mechanical Electronics Fiberglass Woodworking Sails Canvas	CONVENIENCE: Ship's store Ice Convenience foods & beverages
C Mahzenazing River Lodge 705-287-2089 Collins Inlet Killarney, ON P0M 2A0		4		30'	4'		W	G	H	PD	Y	Y			IC

LAKELAND BOATING · LAKE HURON CRUISING GUIDE   **341**

Club Island, Ontario

E — Lat 45°32.03 Lon 81°42.73

F

Lat 45°33.90 Lon 81°35.30

N →

Club Island **A** is a good, unimproved harbor of refuge for cruisers heading to Killarney and Little Current. The harbor **B** is entered from the east. On approach, beware the North Reef **C** and South Spit **D**. The anchorage is better at Rattlesnake Harbour **E**, at the northeast tip of Fitzwilliam Island to the west (lat/lon given above is for the northwest entrance to the harbor) or Squaw Island to the north-northeast (the northeast end of the island just before the opening to the bay is at 45°49.93N, 81°26.37W). Manitoulin Island **F** is visible in the distance.

Distances from Club Island (statute miles)
Tobermory: 19 S
Killarney: 24 N

Rattlesnake Harbour: 7 W
Squaw Island: 10 NNE

342 LAKELAND BOATING · LAKE HURON CRUISING GUIDE

GEORGIAN BAY

Club Island, Ontario
Head for this port in a storm.

Club Island is uninhabited and has no facilities, but boaters headed to Killarney or Little Current may find it a handy harbor of refuge in case of bad weather. There's no telephone service on the Island, but thanks to the cells in Tobermory and Gore Bay, your wireless phone should work here.

There are cliffs on the northeast, north and east sides, while the west shore is stony. The entrance to the harbor is located on the east side, and there's a strong surge when winds are from the east. As you enter, be sure to avoid two rocky ledges known as North Reef and South Spit. Anchoring is possible in 9 to 15 feet of water. Holding ground is reported to be poor, with a rock, gravel and mud bottom. A better-protected anchorage is Rattlesnake Harbour on Fitzwilliam Island to the east or Squaw Island to the north.

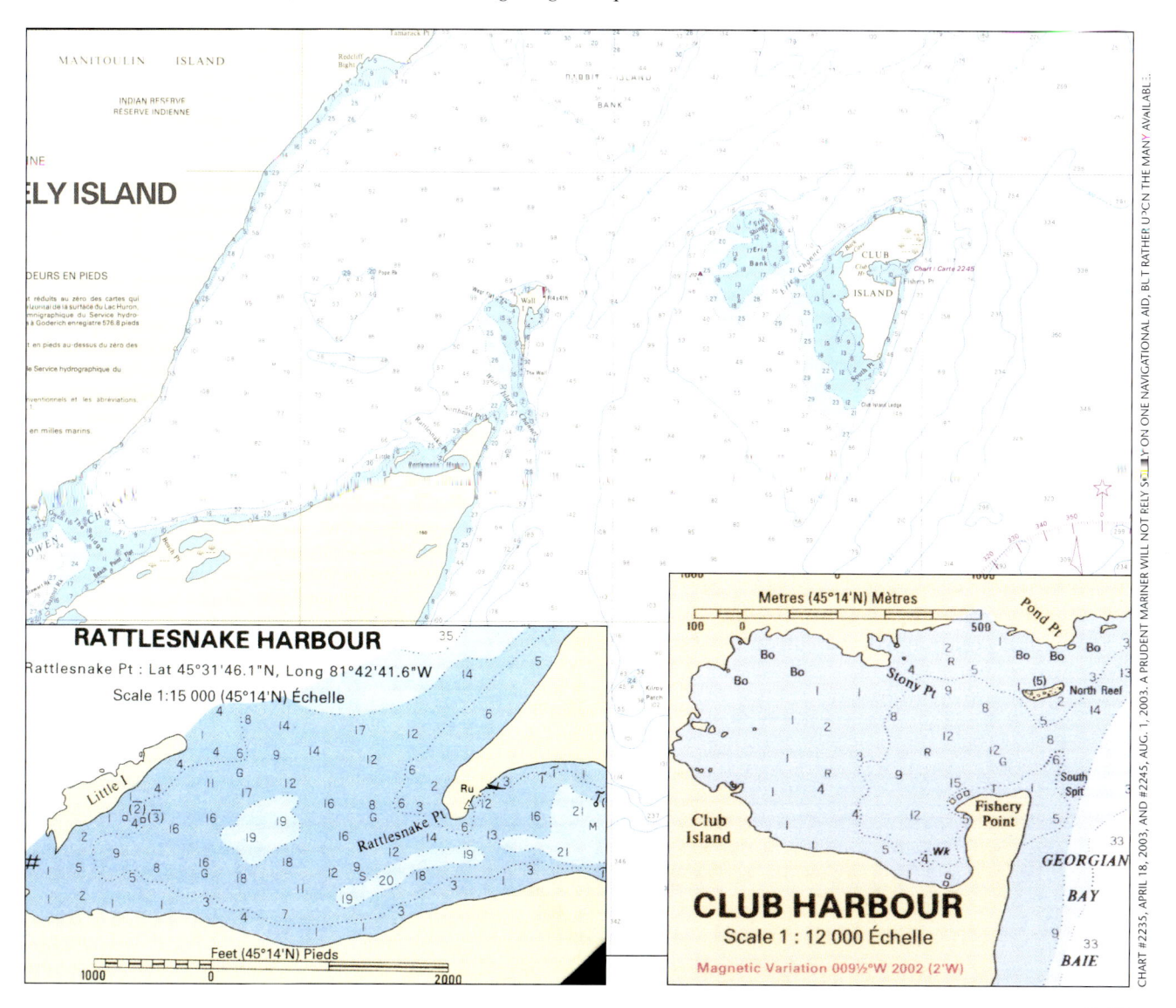

LAKELAND BOATING · LAKE HURON CRUISING GUIDE   **343**

Reference List

Government Publications

Navigation Rules, International and Inland
(Required on boats 12 meters or more)
Superintendent of Documents, U.S. Government Printing Office, Washington, D.C. 20402; 202-783-3238

Sailing Directions: Great Lakes, Vol. II
Canadian Hydrographic Service, Ottawa, ON; 613-998-4931

United States Coast Guard Light List
Superintendent of Documents, U.S. Government Printing Office, Washington, D.C. 20402

United States Coast Pilot No. 6, Great Lakes: Lakes Ontario, Erie, Huron, Michigan, and Superior and St. Lawrence River
U.S. Department of Commerce, National Oceanic and Atmospheric Administration, and National Ocean Service, Office of Coast Survey, 1315 East-West Highway, Room 6242, Silver Spring, MD 20910-3233; 301-713-1910 ext. 201; www.nauticalcharts.noaa.gov

Charts

MapTech, Inc. 888-839-5551
Region 22. CD-ROMs of charts; www.maptech.com

Richardsons' Chartbook & Cruising Guide: Lake Huron Edition, including North Channel and Georgian Bay 800-873-4057

Bluewater Books & Charts 800-942-2583
www.bluewaterweb.com

General

Chapman Piloting: Seamanship and Small Boat Handling
Elbert S. Maloney. Hearst Books, 959 Eight Ave., New York, NY 10019; 212-649-2000

Nautical Charts
You'd be lost without us!℠

When planning a trip on the Great Lakes, inland rivers or anywhere in the world, give us a call. We can provide you with up-to-date nautical charts as well as information about your selected route.

At Marine Navigation, we specialize in nautical charts and can make specific recommendations that will save you in the long-run.

As usual, you may shop at our store or order by phone. Please give us a call for chart catalogs and planning materials.

Store Hours: (Central Time)
Mon-Fri 9:00 AM to 5:00 PM
Sat 10:00 AM to 2:00 PM
Ask for Laura Cannell

- River Charts • NOAA Charts
- DMA Charts • Canadian Hydro Charts
- Richardson's Chartbooks • Imray-Iolaire
- BBA Chartkits • Waterway Guide Products
- Plotting Tools • Videos
- Houston Study Aids • Steiner Binoculars
- Log Books and lots more...

VISA®   Discover   MasterCard®

Lat.: 41° 48'10" Lon.: 87° 52' 08"

Marine Navigation
613 South La Grange Rd., La Grange, IL 60525

(708) 352-0606
Fax: (708) 352-2170

NOAA & CHS Chart Locator

Alpena, Michigan **14864**
Au Gres, Michigan **14863**
Au Sable, Michigan **14863**
Bad River Channel, Ontario **2204**
Baie Fine, Ontario 2205, **2206**
Bay City, Michigan 14863, **14867**
Bay of Islands, Ontario **2207**
Bayfield, Ontario 2228, 2260, **2261**, 2290, **14862**
Bayfield Inlet, Ontario **2203**
Beardrop Harbour, Ontario **2268**
Beaverstone Bay, Ontario **2204**
Benjamin Islands Group, Ontario **2257**, 2299
Blind River, Ontario 2259, **2268**
Britt, Ontario **2203**, 2204
Bruce Mines, Ontario 2250, **2251**
The Bustard Islands, Ontario **2204**
Byng Inlet, Ontario 2203, **2204**
Campbell Bay, Ontario **2258**, 2299
Caseville, Michigan **14885**
Cedarville, Michigan 14881, **14885**
Cheboygan, Michigan 14880, **14881**, 14886
Club Island, Ontario 2235, **2245**
Cockburn Island, Ontario **2251**
Collingwood, Ontario **2283**
Collins Inlet, Ontario **2204**
Covered Portage Cove, Ontario **2205**, 2245
Croker Island, Ontario **2257**, 2299
Desbarats, Ontario **2250**, 14883
DeTour Village, Michigan 14880, 14881, **14882**
Dillon Cove, Ontario **2203**
Drummond Is., Michigan 2251, 2297, 14880, **14882**
Dyer Bay, Ontario **2282**
East Tawas, Michigan **14863**
Fox Island, Ontario **2257**, 2299
Gidley Point, Ontario 2241, **2283**
Goderich, Ontario **2228**
Gore Bay, Ontario **2257**, 2299
Grand Bend, Ontario 2228, **2260**, 2290, **14862**
Grindstone City, Michigan **14863**
Hammond Bay Harbor, Michigan 14880, **14881**
Harbor Beach, Michigan **14862**
Harbor Island, Michigan 2295, 14880, **14882**
Harrisville, Michigan **14864**
Hessel, Michigan **14885**
Heywood Island, Ontario **2205**, 2245, 2286
Hilton Beach, Ontario **2250**
Honey Harbour, Ontario **2202**, 2241
John Island, Ontario **2259**
Kagawong, Ontario **2257**, 2299
Killarney, Ontario 2204, 2205, **2245**
Killbear, Ontario **2203**
Kincardine, Ontario 2261, **2291**
Lansdowne Channel, Ontario **2205**, 2245
Les Cheneaux Islands, Michigan ...14880, 14881, **14885**
Lexington, Michigan **14862**
Lion's Head, Ontario **2282**
Little Current, Ontario 2205, **2207**, 2286
Little Detroit, Ontario 2257, **2268**, 2286
Mackinac Island, Michigan 14880, **14881**
Mackinaw City, Michigan 14880, **14881**
MacGregor Bay, Ontario **2207**, 2282

Manitoulin Island, Ontario **14860**
Manitowaning, Ontario **2245**, 2286
McBean Channel, Ontario **2207**, 2286, 2287, 2299
McGregor Harbour, Ontario **2205**, 2245, 2286
Meaford, Ontario **2283**
Meldrum Bay, Ontario **2251**, 2299
Midland, Ontario 2202, **2221**, 2241
Mill Lake, Ontario **2204**
Oliphant, Ontario **2292**
Oscoda, Michigan **14863**
Owen Sound, Ontario 2201, **2202**, **2283**
Parry Sound, Ontario **2202**, 2203
Penetanguishene, Ontario **2202**, **2218**, 2241
Pointe au Baril, Ontario **2203**
The Pool, Ontario **2205**, 2245
Port Austin, Michigan **14863**
Port Elgin, Ontario **2291**
Port Franks, Ontario 2228, **2260**, 2290, **14862**
Port Hope, Michigan 2228, **14862**
Port Huron, Michigan 14852, 14853, 14860, **14865**
Port Sanilac, Michigan **14862**
Port Severn, Ontario **2202**, 2241
Presque Isle, Michigan 14864, **14869**, 14880
Providence Bay, Ontario **2266**, 2298
Rattlesnake Harbour, Ontario **2235**
Richards Landing, Ontario **2250**
Rogers City, Michigan **14864**, 14880
St. Ignace, Michigan 14880, **14881**
Sans Souci, Ontario **2202**
Sarnia, Ontario 2200, **2290**, 14852, 14853, 14865
Sauble Beach, Ontario **2292**
Sault Ste. Marie, Michigan & Ontario...14883, **14884**, 14962
Sebewaing, Michigan **14863**
Sheguiandah, Ontario **2205**
South Bay Cove, Ontario 2202, **2241**
South Baymouth, Ontario 2235, **2273**, 2286, 2298
Southampton, Ontario 2291, **2292**
Snug Harbour, Ontario **2203**
Snug Harbour, Ontario (anchorage) ...**2205**, 2245, 2286
Spanish, Ontario 2257, **2268**
Spragge, Ontario 2251, 2259, **2268**
Stokes Bay, Ontario **2292**
Tawas City, Michigan **14863**
Thessalon, Ontario **2251**
The 30,000 Islands, Ontario 2202, 2203, 2204, **14860**
Thornbury, Ontario **2283**
Thunder Bay, Michigan **14864**
Thunder Beach, Ontario **2218**, 2241
Tobermory, Ontario 2235, **2274**
Victoria Harbour, Ontario 2202, **2223**, 2241
Wasaga Beach, Ontario **2283**
Waubaushene, Ontario **2202**, 2241
West Bay, Ontario **2207**, 2286
West Coast of the Bruce Peninsula, Ontario **2292**
Whalesback Channel, Ontario...**2268**, 2257, 2259, 2299
Whitefish Falls, Ontario **2207**
Wiarton, Ontario 2282, **2283**
Wikwemikong, Ontario **2245**, 2286
Wingfield Basin, Ontario 2235, **2282**

Port Index

Port	page
Alpena, Michigan	56
Au Gres, Michigan	72
Au Sable, Michigan	64
Bad River Channel, Ontario	333
Baie Fine, Ontario	201
Bay City, Michigan	76
Bay of Islands, Ontario	205
Bayfield, Ontario	114
Bayfield Inlet, Ontario	323
Beardrop Harbour, Ontario	216
Beaverstone Bay, Ontario	335
Benjamin Islands Group, Ontario	206
Blind River, Ontario	220
Britt, Ontario	326
Bruce Mines, Ontario	228
The Bustard Islands, Ontario	331
Byng Inlet, Ontario	326
Campbell Bay, Ontario	170
Caseville, Michigan	84
Cedarville, Michigan	143
Cheboygan, Michigan	40
Club Island, Ontario	342
Cockburn Island, Ontario	163
Collingwood, Ontario	278
Collins Inlet, Ontario	338
Covered Portage Cove, Ontario	199
Croker Island, Ontario	209
Desbarats, Ontario	236
DeTour Village, Michigan	154
Dillon Cove, Ontario	317
Drummond Is., Michigan	158
Dyer Bay, Ontario	253
East Tawas, Michigan	68
Fox Island, Ontario	209
Gidley Point, Ontario	285
Goderich, Ontario	118
Gore Bay, Ontario	172
Grand Bend, Ontario	110
Grindstone City, Michigan	90
Hammond Bay Harbor, Michigan	46
Harbor Beach, Michigan	92
Harbor Island, Michigan	159
Harrisville, Michigan	60
Hessel, Michigan	146
Heywood Island, Ontario	204
Hilton Beach, Ontario	232
Honey Harbour, Ontario	300
John Island, Ontario	216
Kagawong, Ontario	176
Killarney, Ontario	194
Killbear, Ontario	312
Kincardine, Ontario	122
Lansdowne Channel, Ontario	200
Les Cheneaux Islands, Michigan	142
Lexington, Michigan	98
Lion's Head, Ontario	256
Little Current, Ontario	180
Little Detroit, Ontario	212
MacGregor Harbour, Ontario	256
Mackinac Island, Michigan	30
Mackinaw City, Michigan	36
McGregor Bay, Ontario	202
Manitoulin Island, Ontario	166
Manitowaning, Ontario	186
McBean Channel, Ontario	209
Meaford, Ontario	270
Meldrum Bay, Ontario	167
Midland, Ontario	288
Mill Lake, Ontario	338
Oliphant, Ontario	134
Oscoda, Michigan	64
Owen Sound, Ontario	266
Parry Sound, Ontario	308
Penetanguishene, Ontario	284
Pointe au Baril, Ontario	319
The Pool, Ontario	202
Port Austin, Michigan	88
Port Elgin, Ontario	126
Port Franks, Ontario	106
Port Hope, Michigan	91
Port Huron, Michigan	102
Port Sanilac, Michigan	94
Port Severn, Ontario	296
Presque Isle, Michigan	52
Providence Bay, Ontario	140
Rattlesnake Harbour, Ontario	343
Richards Landing, Ontario	238
Rogers City, Michigan	48
St. Ignace, Michigan	149
Sans Souci, Ontario	306
Sarnia, Ontario	102
Sauble Beach, Ontario	134
Sault Ste. Marie, Michigan & Ontario	241
Sebewaing, Michigan	80
Sheguiandah, Ontario	184
South Bay Cove, Ontario	300
South Baymouth, Ontario	138
Southampton, Ontario	130
Snug Harbour, Ontario	315
Snug Harbour, Ontario (anchorage)	200
Spanish, Ontario	212
Spragge, Ontario	218
Stokes Bay, Ontario	134
Tawas City, Michigan	68
Thessalon, Ontario	224
30,000 Islands, Ontario	299
Thornbury, Ontario	274
Thunder Bay, Michigan	58
Thunder Beach, Ontario	291
Tobermory, Ontario	248
Victoria Harbour, Ontario	293
Wasaga Beach, Ontario	282
Waubashene, Ontario	295
West Bay, Ontario	179
West Coast of the Bruce Peninsula, Ontario	134
Whalesback Channel, Ontario	215
Whitefish Falls, Ontario	205
Wiarton, Ontario	262
Wikwemikong, Ontario	190
Wingfield Basin, Ontario	253

Editorial Index

Abandoning Ship	125
Airports	189
Anchoring Tips	19
Chart Locator	345
Chi-Cheemaun Ferry	252
Crossing the Lake	109
Customs	22
Distance Chart	27
Fishing	
10 Commandments	231
Hotspots	260
Recipes	
Fathom Five National Marine Park	252
Flowerpot Island	250
Great Lakes Cruising Club	175
Helicopter Rescue	83
How to Use this Book	16
Hypothermia	113
Les Cheneaux Islands	142
Lightning	75
Manitoulin Island	166
Maps	
Lake Huron	28
Georgian Bay	246
North Channel	152
Museums	235
Paul Bunyan	66
Pollution Regulations	101
Powwows	199
Reference List	344
Resource Guide	20
Shipwrecks of Thunder Bay	59
30,000 Islands	299
Trailerboating	265
VHF Radio Frequencies	277
Weather	
Heavy Weather Preparation	113
Lake Huron Forecast	20
Yacht Clubs	101

Advertiser Index

Anchor In Marina, Cheboygan, Michigan	44
Bayport Marina, Midland, Ontario	291
BoatU.S.	8
Boyle, Little Current, Ontario	182
Crusader Engines	Inside Back Cover
Doral Marine Resort	290
Drummond Island Resort	161
Four Winns	3
George Kemp Marina	247
Grand Banks	17
Great Lakes Angler	348
Harbour Lights, Bayfield, Ontario	116
Harbour Vue, Little Current, Ontario	182
Hunter	Back Cover
Huron North	210
Intracoastal Waterway Made Easy	18
Lakeland Boating Cruising Guides	160, 349
Lakeland Boating Website	97
Maitland Valley Marina, Goderich, Ontario	120
Marine Navigation	344
Mertaugh Boat Works, Hessel, Michigan	147
Michigan Boating Indutries Association	25
Midland, Ontario	290
Navionics	45
Nordhavn	10
North Country Sports	156
Ocean Yachts	7
Ontario Marine Operators Association	240
Owen Sound Marina, Owen Sound, Ontario	268
Penetanguishene, Ontario	286
Pier 7, Bay City, Michigan	78
Pursuit	21
Queen's Cove Marina, Victoria Harbour, Ontario	295
Raymarine	2
Rinker	15
Rogers City	49
Sarnia Bay Marina, Sarnia, Ontario	104
Sault Ste. Marie, Canada	245
Sea Ray	Inside Front Cover, 1
South Bay Cove, Honey Harbour, Ontario	305
Sportsman's Inn, Killarney, Ontario	196
St. Ignace City Marina, St. Ignace, Michigan	151
Sunset Harbor, Harsen's Island, Michigan	104
Tiara Yachts	4
Toledo Beach Yacht Sales, LaSalle, Michigan	352
Volvo	13
Walstrom, Cheboygan, Michigan	44
Worldwide Marine	18

Extend your boating season AND bring home the bacon.

Well, it's not really bacon, but salmon, perch and walleye are tasty treats—and **GREAT LAKES ANGLER**, from the publishers of Lakeland Boating, will show YOU how to catch them!

More and more Great Lakes boaters are finding that by rigging their vessels for fishing, they can enjoy them nearly year-round! **GREAT LAKES ANGLER** shows you how!

Each information-packed issue covers the basics—and intricacies—of rigging up for our freshwater sportfish. Try an issue for **FREE**! Just call 800-214-5558 and use the code S502. You won't have to pay unless you like it and nearly 90 percent of our subscribers renew each year.

For a free, no-obligation issue of GREAT LAKES ANGLER
CALL 800-214-5558 NOW

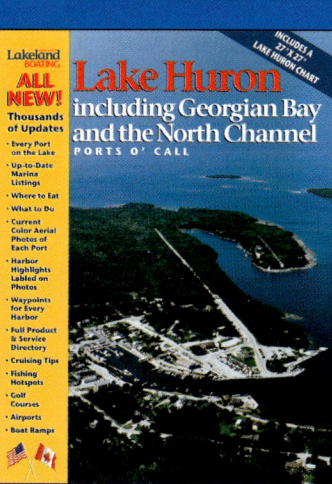

Lakeland BOATING PORTS O' CALL

The Complete Cruising Guides for the Great Lakes

FREE Poster-Size Nautical Chart With Each Guide

- **Every Port Featured**
- **Waypoints for Every Port**
- **Aerial Photos of Every Port**
- **Updated Nautical Charts**
- **Marina Listings and Contact Info**
- **Where to Eat and What to Do**

Be privy to the latest word on marinas, restaurants, attractions, events and important boater amenities in each port, all presented in an attractive, well-organized design. Nobody knows the Great Lakes like *Lakeland Boating*...and there is no one better to help you plan your next Great Lakes cruise.

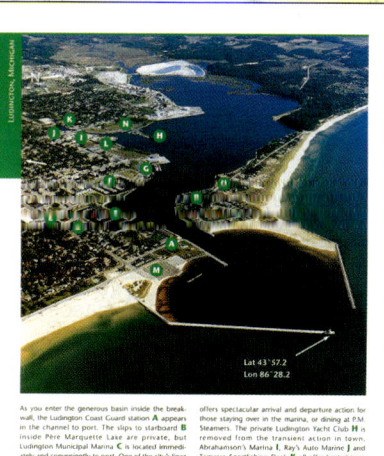

- **Order Online at: www.lakelandboating.com**
- **Order by Phone: 800-589-9491**
- **Mail Request to:**

Lakeland Boating, c/o Retail Services
P.O. Box 704 • Mt. Morris, IL 61054

Lake Erie & Ontario Updated!

Notes

Notes

Toledo Beach & Great Lakes yacht sales

Do The Math!

SILVERTON   CABO YACHTS   HUNTER SAILBOATS   ISLAND PACKET   PROLINE BOATS   OCEAN YACHTS

6 ### Great Manufacturers
Silverton • Hunter • Cabo • Island Packet • Proline • Ocean Yachts

3 ### Great Locations
• Monroe, Mi. on Lake Erie. • St. Clair Shores, Mi. on Lake St Clair.
• Grosse Ile, Mi on the Detroit River.

2 ### Impressive Awards
Best Silverton customer service world wide for 2 years running.

1 ### Cool Mascot!
Charlie, the boating wonder dog!

= ### One Great Boating Experience!

Toledo Beach Yacht Sales
11840 Toledo Beach Rd. LaSalle, Mi 48145   Ph 734 243-3830   Fax 734 243-3815

Great Lakes Yacht Sales
24770 E. Jefferson Ave. St. Clair Shores, Mi. 48080   Ph. 586 445-9191   Fax 586 445-1347

Grosse Ile Marine Sales
24201 Meridian, Grosse Ile, Mi 48138   Ph 734 675-7447   Fax 734 675-7444

Check out our extensive brokerage listings at:
www.ToledoBeachYachtSales.com